Springer Series in
MATERIALS SCIENCE 50

Springer
Berlin
Heidelberg
New York
Hong Kong
London
Milan
Paris
Tokyo

Physics and Astronomy ONLINE LIBRARY

http://www.springer.de/phys/

Springer Series in
MATERIALS SCIENCE

Editors: R. Hull R. M. Osgood, Jr. J. Parisi

The Springer Series in Materials Science covers the complete spectrum of materials physics, including fundamental principles, physical properties, materials theory and design. Recognizing the increasing importance of materials science in future device technologies, the book titles in this series reflect the state-of-the-art in understanding and controlling the structure and properties of all important classes of materials.

51 **Microscopic and Electronic Structure of Point Defects in Semiconductors and Insulators**
Determination and Interpretation of Paramagnetic Hyperfine Interactions
Editors: J. M. Spaeth and H. Overhof

52 **Polymer Films with Embedded Metal Nanoparticles**
By A. Heilmann

53 **Nanocrystalline Ceramics**
Synthesis and Structure
By M. Winterer

54 **Electronic Structure and Magnetism of Complex Materials**
Editors: D.J. Singh and A. Dimitrios

55 **Quasicrystals**
An Introduction to Structure, Physical Properties and Applications
Editors: J.-B. Suck, M. Schreiber, and P. Häussler

56 **SiO_2 in Si Microdevices**
By M. Itsumi

57 **Radiation Effects in Advanced Semiconductor Materials and Devices**
By C. Claeys and E. Simoen

58 **Functional Thin Films and Functional Materials**
New Concepts and Technologies
Editor: D. Shi

Series homepage – http://www.springer.de/phys/books/ssms/

Volumes 1–50 are listed at the end of the book.

F. Ernst M. Rühle (Eds.)

High-Resolution Imaging and Spectrometry of Materials

With 211 Figures
Including 32 Color Figures

Springer

Dr. Frank Ernst
Department of Materials Science and Engineering
Case Western Reserve University
414 White Building
10900 Euclid Avenue, Cleveland, OH 44106-7204, USA

Professor Manfred Rühle
MPI für Materialforschung
Heisenbergstrasse 3, 70569 Stuttgart, Germany

Series Editors:

Professor R. M. Osgood, Jr.
Microelectronics Science Laboratory, Department of Electrical Engineering
Columbia University, Seeley W. Mudd Building, New York, NY 10027, USA

Professor Robert Hull
University of Virginia, Dept. of Materials Science and Engineering, Thornton Hall
Charlottesville, VA 22903-2442, USA

Professor Jürgen Parisi
Universität Oldenburg, Fachbereich Physik, Abt. Energie- und Halbleiterforschung
Carl-von-Ossietzky-Strasse 9-11, 26129 Oldenburg, Germany

ISSN 0933-033X

ISBN 3-540-41818-0 Springer-Verlag Berlin Heidelberg New York

Library of Congress Cataloging-in-Publication Data: High-resolution imaging and spectrometry of materials/ F. Ernst and M. Rühle (eds.). p. cm. – (Springer series in materials science, ISSN 0933-033X; 50) Includes bibliographical references and index. ISBN 3540418180 (alk. paper) 1. Materials–Microscopy. 2. Transmission electron microscopy. I. Rühle, Manfred. II. Ernst, F. (Frank), 1938– III. Springer series in materials science ; v. 50. TA417.23.H534 2002 620.1'1299–dc21 2002030612

Springer-Verlag Berlin Heidelberg New York
a member of BertelsmannSpringer Science+Business Media GmbH

http://www.springer.de

Printed in Germany

Typesetting by the editors using a Springer TeX macro package
Data conversion: Frank Herweg, Leutershausen
Cover concept: eStudio Calamar Steinen
Cover production: *design & production* GmbH, Heidelberg

Printed on acid-free paper SPIN: 10717049 57/3141/ba 5 4 3 2 1 0

Preface

The characterisation of materials and material systems is an essential aspect of materials science. A few decades ago it became obvious that, because the properties of materials depend so critically on the microstructure of their components, this characterisation must be determined to the atomic level.

This means that the position – as well as the nature – of individual atoms has to be determined at "critical" regions close to defects such as dislocations, interfaces, and surfaces. The great impact of advanced transmission electron microscopy (TEM) techniques became apparent in the area of semiconducting materials, where the nature of internal interfaces between silicon and the corresponding silicides could be identified, and the results used to enhance the understanding of the properties of the compounds studied.

At that time, advanced TEM techniques existed predominantly in the US. However, advanced TEM instrumentation was not available in the materials science and solid-state science communities in Germany. This gap was bridged by the late Peter Haasen who, after a visit to the US, initiated a Priority Programme on Microstructural Characterisation at the Volkswagen Foundation (Hannover). The programme was in effect from 1985 to 1997 and supported a wide range of research projects – from fundamental, trendy, innovative projects to projects in applied materials science.

Within the Priority Programme, several innovative concepts and techniques for instrumentation, imaging, and analysis were developed. In instrumentation, a major breakthrough occurred through the development of electron-optical elements, which allowed the correction of fundamental aberrations of rotational symmetrical lenses. Quantitative techniques in high resolution transmission electron microscopy (HRTEM) were introduced. Through these HRTEM techniques, the positions of atom columns could be determined with high accuracy. Through the use of analytical electron microscopy (AEM), the field of electron energy-loss spectroscopy (EELS) was introduced. Techniques were established for obtaining information on the bonding of individual atoms by experimental and theoretical studies of the energy-loss near-edge structures (ELNES).

At the beginning of the Priority Programme on Microstructural Characterisation, the fields of scanning tunneling microscopy (STM) and atom force microscopy (AFM) were in their infancy. Supported activities included instrumentation and the application of surface-science techniques. By the end of the programme, because of the introduction of these advanced TEM techniques, scientists outside the US had obtained many interesting and important results at the atomic level in real materials research as well as through characterisation of model systems.

This book contains 10 chapters describing major contributions that were supported by the Volkswagen Foundation through this programme. The authors hope that this presents a comprehensive overview of the field of advanced microstructural characterisation.

The authors and the editors would like to thank Peter Haasen for initiating the programme, Herbert Steinhardt for his unbureaucratic help with individual projects, and the Volkswagen Foundation for its generous support of major proposals, and for supporting this book.

All authors wish to acknowledge the invaluable help of Prof. P.W. Hawkes for the revision of the English text and many helpful comments and Dipl.-Ing. (FH) U. Eigenthaler for the time-consuming and painstaking editing of the text.

The authors dedicate this book to the scientific community, and hope that it will serve as an inspiration for innovative research ideas in the area of microstructural characterisation of materials.

Stuttgart, *Frank Ernst*
July 2002 *Manfred Rühle*

Contents

List of Contributors

Dr. Taalat Al-Kassab
Universität Göttingen,
Institut für Metallphysik,
Hospitalstr. 3–7,
37073 Göttingen, Germany
alkassab@umpa07.gwdg.de

Prof. Dr. Ernst Bauer
Department of Physics and Astronomy,
Arizona State University,
PO Box 871504,
Tempe, AZ 85287-1504, USA
Ernst.Bauer@asu.edu

Prof. Dr. R.W. Cahn
Department of Materials Science and Metallurgy,
New Museums Site,
Pembroke Street,
Cambridge CB2 3QZ, UK
rwc12@cam.ac.uk

Prof. Dr. Frank Ernst
Department of Materials Science and Engineering,
414 White Building,
Case Western Reserve University,
10900 Euclid Avenue,
Cleveland, OH 44106-7204, USA
fxe5@po.cwru.edu

Prof. Dr. Gerhard Ertl
Fritz-Haber-Institut der
Max-Planck-Gesellschaft,
Faradyweg 4–6,
14195 Berlin, Germany
ertl@fhi-berlin.mpg.de

Margit Hanbücken
CRCM2-CNRS, Campus de Luminy,
Case 913,
13288 Marseille, Cedex 9, France
margrit@crmc2.univ-mrs.fr

Prof. Dr. Johannes Heydenreich
Max-Planck-Institut für Mikrostrukturphysik,
Weinberg 2,
06120 Halle/Saale, Germany
hdr@mpi-halle.de

Prof. Dr. Rainer Kirchheim
Universität Göttingen,
Institut für Metallphysik,
Hospitalstr. 3–7,
37073 Göttingen, Germany
rkirch@umpa01.gwdg.de

Dr. Philipp Kohler-Redlich
Infineon Technologies
Villach AG Abt. P1,
Siemenstr. 2,
Villach, Austria
Philipp.Kohler-Redlich@infineon.com

Prof. Dr. Joachim Mayer
Gemeinschaftslabor für Elektronenmikroskopie (GFE),
RWTH Aachen,
Ahornstr. 55,
52074 Aachen, Germany
mayer@gfe.rwth-aachen.de

Dr. Günter Möbus
Department of Materials,
Oxford University,
Parks Road,
Oxford OX13PH, UK
guenter.moebus@materials.ox.ac.uk

Heiko Müller
CEOS GmbH,
Englerstrasse 28,
69126 Heidelberg, Germany
mueller@ceos-gmbh.de

Prof. Dr. Henning Neddermeyer
Martin-Luther-Universität,
Halle-Wittenberg,
Fachbereich Physik,
Fachgruppe Oberflächen- und Grenzflächenphysik,
Friedemann-Bach-Platz 6,
06108 Halle/Saale, Germany
neddermeyer@ep3.uni-halle.de

Prof. Dr. Jürgen Rödel
FG Nichtmetallisch-Anorganische Werkstoffe,
Tu Darmstadt, FB 21,
Petersenstr. 23, Geb. 73 a,
64287 Darmstadt, Germany
roedel@ceramics.tu-darmstadt.de

Prof. Dr. Harald Rose
Institut für Angewandte Physik,
TU Darmstadt,
Hochschulstr. 6,
64289 Darmstadt, Germany

Prof. Dr.Dr.h.c. Manfred Rühle
Max-Planck-Institut für Metallforschung,
Heisenbergstr. 3,
70569 Stuttgart, Germany
ruehle@mf.mpg.de

Dr. Thomas Schmidt
Universität Würzburg,
Experimentelle Physik II,
Am Hubland,
97074 Würzburg, Germany
thomas.schmidt@physik.uni-wuerzburg.de

Dr. Guido Schmitz
Universität Göttingen,
Institut für Metallphysik,
Hospitalstr. 3–7,
37073 Göttingen, Germany
guido@umpa01.gwdg.de

Dr. Wilfried Sigle
Max-Planck-Institut für Metallforschung,
Heisenbergstr. 3,
70569 Stuttgart, Germany
sigle@mf.mpg.de

Prof. Dr. H. Wollenberger
Schlierbergstr. 84,
79100 Freiburg, Germany

1 Microcharacterisation of Materials

F. Ernst and W. Sigle

No material has a perfect structure. On the contrary: most of the materials we encounter in daily life are full of microscopic *defects* – regions where the actual arrangement of the atoms deviates from the ideal structure. The atomic configuration in the bulk of a crystalline material, for example, may contain defects such as vacancies, atoms on interstitial sites, impurity atoms, dislocations, stacking faults, grain boundaries, phase boundaries, voids, or cracks. Similar types of defects exist in non-crystalline materials. The entirety of these bulk defects, except for those that occur in thermodynamic equilibrium, such as a certain concentration of vacancies, constitutes the *microstructure* of the respective material.

Figure 1.1 illustrates the structure of materials at different length scales. The largest features of the microstructure typically have dimensions in the micrometre range, much larger than the typical distances between neighbouring atoms (less than a nanometre), and much smaller than the dimensions relevant for macroscopic properties of the material (more than a millimetre). In many materials, the fraction of the atoms that belong to a microscopic defect is small compared to the fraction of the atoms that reside in regions of perfect structure. For the macroscopic behaviour of the material, however, the atoms that belong to defects often play a key role. Materials research over the last century has provided overwhelming evidence that the microstructure

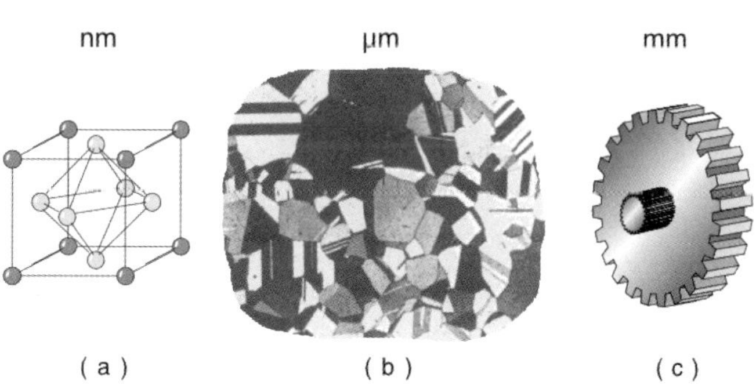

Fig. 1.1. Structure of materials on different length scales. (**a**) On the nanometre scale: atomic structure. (**b**) On the micrometre scale: microstructure – the entirety of *defects* in the atomic structure. (**c**) On the millimetre scale and above: macroscopic structure

of a material can substantially influence the properties of the material at the macroscopic level.

In fact, many materials become technically useful only if they have an appropriate microstructure. The mechanical strength of high-purity, defect-free single crystalline aluminium, for example, is much too small for the structural applications in which aluminium alloys are used today. Only the presence of alloying atoms, dislocations, and grain boundaries makes "technical" aluminium an important material for structural applications. Since the density and the mobility of defects depend on the thermal treatment, temperature can affect material properties drastically. For example, only by careful thermal treatment can a steel reach the high strength necessary for its manifold technical applications. In semiconducting materials, the *electrical conductivity* can be increased by orders of magnitude by small concentrations of substitutional foreign atoms. Similarly, microscopic defects influence the *optical properties* of semiconducting and insulating materials. Trace impurities of chromium in α-Al_2O_3, for instance, give the red colour to ruby. Furthermore, the microstructure affects superconductivity and magnetism, particularly the superconducting and the magnetic hardness.

Among the various components of the microstructure, internal interfaces (grain boundaries and heterointerfaces) have a particularly strong impact on materials properties. Such interfaces occur in every material, unless it happens to be either single crystalline or single-phase amorphous. In reality, nearly all materials that serve in structural applications or as functional materials are polycrystalline, or contain several phases, or both. In particular, many "advanced" materials owe their special properties to the very interfaces they contain. Examples of interface-controlled, advanced materials include functional ceramics, epitaxial multilayer structures, nanocrystalline metals, and artificially structured semiconductor materials for quantum-effect electronic devices. Furthermore, interfaces play an important role in ferroic materials, like ferroelectric, ferroelastic, or ferromagnetic materials, and in high-T_c superconductors. The presence of interfaces affects materials in many different ways. Interfaces interact with the transport of atoms (or ions) and electrons through the material, and thus influence the diffusivity and the electrical conductivity of "real" materials. Sintering of ceramics and superplasticity of metals are examples where diffusion reactions across or along interfaces determine the kinetics.

Apart from defects in the bulk, a solid material usually has a surface, at which it is in contact with a liquid or gas phase or with vacuum. Like internal interfaces, surfaces not only affect the atoms directly at the surface but also influence the properties of the material in a three-dimensional layer below the surface. The thickness one associates with this surface layer depends on the property under consideration and on the sensitivity by which one can measure the influence of the surface. Typical thicknesses quoted for the surface layer range between 1 μm and 1 mm. Naturally, the influence of the surface on the local properties decays with increasing distance from the surface. In analogy to the bulk, the surface of a material has a unique equilibrium structure, in which the atom positions usually differ from the bulk arrangement of atoms. Surfaces may contain such defects as vacancies, adsorbed atoms, steps, or facets. Just like the microstructure of bulk material, the entirety of the surface defects may have a strong impact on the macroscopic properties of the surface region, such as its hardness, its chemical reactivity, its electrical conductivity, and its optical, dielectric and magnetic properties. These surface properties play a key role in epitaxy, corrosion, friction, wear, tribology, catalysis, and sensors. The trend towards miniaturization in practically all fields of artificially structured materials has significantly increased the importance of materials research on surfaces in the recent years.

These considerations show that, in order to improve our fundamental understanding of materials and to optimize materials processing or to control materials behaviour, one needs to understand not only the ideal theoretical properties of the material but also its real consistency, including the nature of the microscopic defects in the bulk and the nature of the surface. Although some defects exert long-range or medium-range forces on the surrounding atoms (for example strain fields of dislocations or Coulomb fields of electrically charged interfaces), it is often the structure and chemistry on the atomic scale that determines the behaviour of materials. For example, the interatomic bonding at interfaces differs from the bonding in bulk material, but this difference is confined to only a few atomic layers (Fig. 1.2).

As a consequence, even a small concentration of impurity atoms located ("segregated") at an interface can lead to a dramatic change in the interface properties. This phenomenon is responsible, for example, for the embrit-

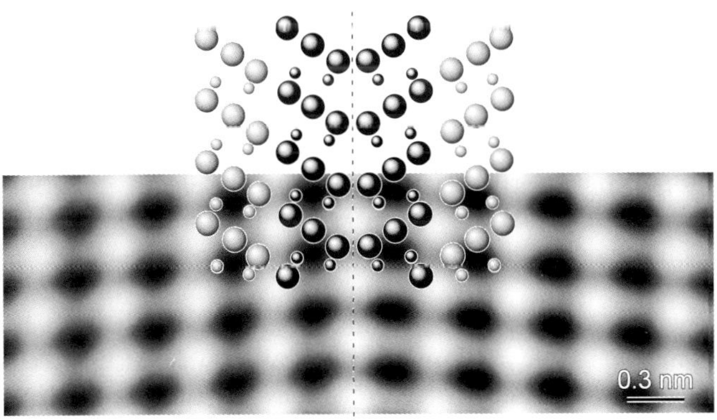

Fig. 1.2. High-resolution transmission electron micrograph of a rhombohedral twin boundary in α-Al_2O_3 [14]. The atom positions are shown in the overlayed structure model. Small spheres represent aluminium atoms, large spheres oxygen atoms. Atoms for which the interatomic bonding deviates from that in bulk α-Al_2O_3 are shown by black spheres

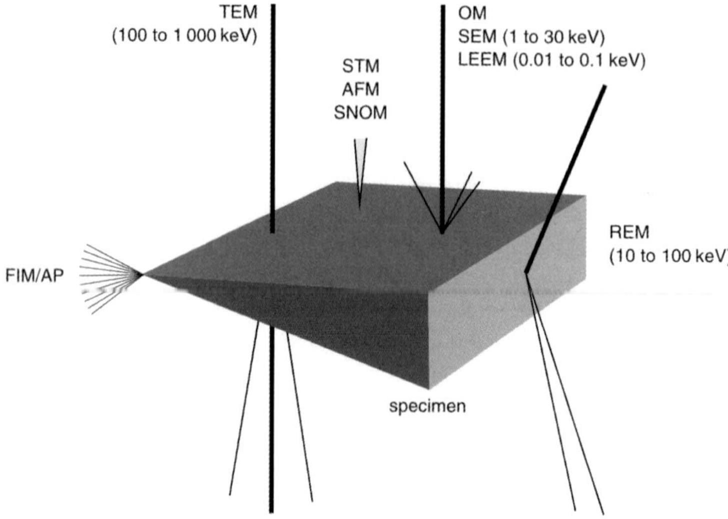

Fig. 1.3. Microscopy techniques and their application to the study of materials. The energies given in parentheses refer to electron kinetic energies employed in electron beam techniques

tlement of steels by phosphorous or sulphur atoms, or the increased creep resistance of α-Al_2O_3 obtained by adding small amounts of yttrium. Similarly, segregation of impurity atoms at crystal dislocations is well known to change the dislocation mobility and thus the capability of plastic deformation (ductility). Even in the absence of impurity atoms, however, the dislocation mobility is often determined by the particular atomic arrangement in the immediate vicinity of the dislocation line (the "dislocation core"). The brittle behaviour of most body-centred cubic metals (such as Fe, Ta, Nb, Mo, or W) at low temperatures, or the anomalous plastic behaviour of many intermetallic compounds, such as Ni_3Al, for example, can be attributed to the particular structure of the dislocation core.

Apart from the importance of the atomic structure of defects, materials scientists are nowadays also confronted with materials consisting of structural entities with dimensions of only a few nanometres (for example nanotubes, nanowires, or catalyst particles). Clearly, such materials require a characterisation on the atomic scale just because of the small size of their constituents. Experimentally, local information can be obtained by several microscopy techniques, which can be subdivided into surface-sensitive and volume sensitive methods (Fig. 1.3).

Historically, the first microstructural studies on the scale between a millimetre and a micrometer were carried out in 1864 by Sorby in Sheffield [1], using optical microscopy (OM). Imaging the surface of materials using reflected light, he discovered that steel is actually composed of microcrystalline grains. This remarkable observation was possible only as a result of very careful mechanical grinding, polishing, and chemical etching of the metallic specimens. Today, more than ever before, sophisticated specimen preparation techniques are the key step towards successful microcharacterisation. Up to the early 20th century, it was not generally accepted that materials are composed of atoms. The observation of sharp X-ray diffraction spots from a zincblende specimen by Friedrich, Knipping, and von Laue [2] in 1912 was the first proof of the regular atomic arrangement in crystals. This was the birth of modern solid state physics. It was several decades before it became possible to obtain actual images of the atomic structure of crystalline materials.

A first major step towards a lateral resolution substantially better than that provided by OM was the invention of the scanning electron microscope (SEM) by von Ardenne in 1938 [3]. In SEM, an electron beam is focused onto an area of only a few nanometres in diameter – significantly smaller than the wavelength of visible light, which ranges between 400 and 800 nm. Images obtained by this technique feature a corresponding resolution as well as an outstanding depth of focus. The first images of the *atomic* structure of solid surfaces were only obtained, however, after the development of field-ion microscopy (FIM) by Müller in 1955 [4]. In a field-ion microscope, the specimen needs to have the shape of a sharp tip, which is mounted opposite a fluorescent screen. The space between the tip and the screen is filled with an inert gas at low pressure. When a voltage is applied between the screen and the tip, a strong electric field is created at the tip, particularly where atoms protrude from the surface – notably, at edges of atomic terraces or facets. At these particular sites, the electric field is strong enough to ionize the gas atoms, which are then accelerated radially towards the fluorescent screen, where they cause light emission indicating the locations of the protruding atoms at very high magnification.

At about the same time, the preparation of thin electron-transparent specimens for transmission electron microscopy was developed in the group of Hirsch in Cambridge. Although the electron microscope had been invented as early as 1932 by Knoll and Ruska [5], it is this advance in specimen preparation that made the transmission electron microscope (TEM) an invaluable tool for the study of defects in the interior of materials. In 1956, Bollmann [6] and the Cambridge group [7] demonstrated that dislocations could be made visible in the TEM, and in the early 1970s, the first atomic resolution TEM images of inorganic materials were published [8]. Continuous improvement of the electron optical design then lead to the first commercial transmission electron microscopes capable of routinely achieving atomic resolution in the 1980s.

Another technique for atom-resolved imaging of the three-dimensional structure of materials evolved from FIM in 1974 [9]. In the atom-probe field-ion microscope (APFIM), it is not the imaging gas (as in the FIM), but the surface atoms of the specimen itself that are ionized and accelerated towards the fluorescent screen and form the image. In combination with a time-of-flight mass spectrometer, the elemental composition of the specimen can be probed atom by atom within a cylinder of atomic dimensions below the initial specimen surface.

During the 1980s, new surface microscopy techniques were developed that revolutionized surface science. Under the guidance of Bauer, the low-energy electron microscope (LEEM) was developed [10,11]. At kinetic energies of a few tens of eV, the interaction volume extends only a few atomic layers into the material, which makes this method extremely surface-sensitive. In 1984, Pohl et al. [12] built the first scanning near-field optical microscope (SNOM). This technique pushed the lateral resolution to between 20 and 50 nm – way below the wavelength of visible light. An alternative to FIM for surface imaging at atomic resolution was developed in 1985 by Binnig and Rohrer [13]. In scanning tunneling microscopy (STM), a sharp metallic tip is scanned across a flat surface with extremely fine control via piezo-electric transducers, and the surface topography is measured either by monitoring the tunneling current at constant height of the tip or by measuring the transducer voltage required to maintain a constant tunneling current. During the past decades, a bewildering variety of scanning probe microscopy techniques (SPM) has emerged. For instance, by making the tip out of a ferromagnetic material, the magnetic properties of the sample can be probed. Or by measuring the force as a function of distance, a measure of the sample hardness is obtained. Using the same concept, measurable properties now include roughness, stiffness, electrical effects, capacitance, adhesion, and friction.

The above historical outline shows that, by the mid 1980s, experimental techniques with lateral resolutions on the atomic length scale found their way into materials science, both for studying surfaces and for studying the bulk properties of materials. At this time, the Volkswagen Foundation established their priority programme dedicated to "Microcharacterisation of Materials and Devices", which strived for further instrumental and methodical developments in high-resolution imaging and spectrometry. Without striving for completeness, the present book presents some of the important achievements that were obtained in the framework of this programme in the field of high-resolution imaging and spectrometry of materials.

Chapters 2–5 focus on the present state of TEM techniques. Starting from first principles, the process of image formation will be shown considering the partial coherence of the electron beam, the dynamic scattering process, including elastic and inelastic scattering (Chap. 2). On this basis, is it possible to determine precise positions of atom columns in high-resolution transmission electron micrographs. The accuracy of such methods is discussed in Chap. 3. Complementary to the structural information obtained by high-resolution TEM, *analytical* transmission electron microscopy (ATEM) makes extensive use of inelastic electron-electron scattering. Inelastic scattering, thus an energy-loss of the primary electrons in the specimen, leads to the emission of Auger electrons as well as element-characteristic X-rays, which can be used for elemental analysis. The energy loss of the primary electrons can be analysed with high resolution by passing them through a magnetic spectrometer, or by passing them through an imaging energy filter, a sophisticated electron-optic component that enables, for example, elemental mapping via mapping of particular energy losses. Apart from elemental concentrations, the energy-loss spectrum also contains information about optical properties of the material as well as localized information about the oxidation state and atomic coordination of the atom involved in the inelastic scattering event. Chapter 4 explains how this wealth of chemical information can be obtained with sub-nanometre lateral resolution. Recent developments in ATEM actually aim at atom-column resolution. Owing to the spherical and the chromatic aberration of the electromagnetic lenses used in TEM, a resolution close to or even below 0.1 nm is possible only by incorporating aberration-correcting and/or monochromating components into the microscope column. The realization of these components has become possible only by careful theoretical design, which will be addressed in more detail in Chap. 5.

An inherent drawback of TEM is that only projections of the three-dimensional atomic and chemical structure are obtained. This is different in APFIM, where the position and nature of every single atom in the studied volume is detected. Although destructive, APFIM makes a three-dimensional atomic and element-specific reconstruction of materials possible, even if the microstructure is very complex. While the conventional atom probe only explores a small part of the specimen volume (an axial cylindrical region below the initial surface), the recent development of the *tomographic* atom probe enables the tip to be analysed completely, atom by atom. Chapter 6 presents the current status and applications of this exciting development. Scanning-probe microscopy techniques, which are discussed in Chap. 7, have now reached such a level of sophistication that they can be used in combination with other surface-sensitive techniques or even for in situ studies of atomic motion. The latter application is of particular importance in the study of surface reactions, crystal growth, or gas-adsorption kinetics. Moreover, scanning-probe techniques can provide spectrometric information by probing the electron states at the specimen surface. This aspect of SPM is also described in Chap. 7. Electron beam methods that are sensitive to only the near-surface atomic layers require very low kinetic energies of the incident electrons. Since it is difficult to form a fine electron probe at such low energies, high-resolution imaging in LEEM is achieved by employing parallel illumination with low-energy electrons. In Chap. 8, this exciting and powerful experimental technique will be discussed together with the closely related electron emission microscopy techniques. Ultimately, the goal of all microcharacterisation in materials science is to understand, explain, and predict the macroscopic properties of materials from their microscopic structure and chemistry. The feasibility of bridging the gap from atoms to macroscopic properties constitutes the subject of Chap. 9 and is demonstrated there for the important class of materials containing metal/ceramic interfaces. The book closes with a review and summary of the priority programme on "Microcharacterisation of Materials and Devices", which has been sponsored by the Volkswagen Foundation for a total of 12 years.

References

1. Sorby H.C. (1887) On the microscopical structure of iron and steel. J Iron Steel Inst **1**, 255–288
2. Friedrich W., Knipping P., von Laue M. (1912) Interferenz–Erscheinungen bei Röntgenstrahlen. Sitzungsbericht der Math-Phys Klasse der Königl Bayer Akad Wiss 303–322; (1952) Naturwiss **39**, 361–368
3. von Ardenne M. (1938) The scanning electron microscope: theoretical fundamentals (in German). Z Physik **109**, 553–572; The scanning electron microscope: practical construction (in German). Z Tech Phys **19**, 407–416
4. Müller E.W. (1960) Field ionization and field ion microscopy. Adv Electron Electron Phys **13**, 83–179
5. Knoll M., Ruska E. (1932) Das Elektronenmikroskop. Z Physik **78**, 318–339
6. Bollmann W. (1956) Interference effects in the electron microscopy of thin crystal foils. Phys Rev **103**, 1588–1589
7. Hirsch P.B., Horne R.W., Whelan M.J. (1956) Direct observations of the arrangement and motion of dislocations in aluminium. Phil Mag **1**, 677–684.
8. Iijima S. (1971) High resolution electron microscopy of crystal lattice of titanium–niobium oxide. J Appl Phys **42**, 5891–5893
9. Panitz J.A. (1973) The 10 cm atom probe. Rev. Sci. Instrum. **44**, 1034–1038
10. Telieps W., Bauer E. (1985) An analytical reflection and emission UHV surface electron microscope. Ultramicroscopy **17**, 57–66
11. Bauer E. (1994) Field ionization and field ion microscopy. Rep Prog Phys **57**, 895–938
12. Pohl D.W., Denk W., Lanz M. (1984) Optical stethoscopy – image recording with resolution lambda/20. Appl Phys Lett **44**, 651–653
13. Binnig G., Rohrer H. (1985) The scanning tunneling microscope. Scientific American **253/2**, 40–46
14. Nufer S., Marinopoulos A.G., Gemming T., Elsässer C., Kurtz W., Köstlmeier S., Rühle M. (2000) Quantitative atomic-scale analysis of interface structures: transmission electron microscopy and local density functional theory. Phys Rev Lett **86**, 5066–5069

2 Electron Scattering

H. Müller and H. Rose

2.1 Introduction

The propagation of electrons is governed by the Schrödinger wave equation. Owing to this wave property, the scattering of electrons differs appreciably from that of classical solid particles. In particular, the interference between different partial waves affects the intensity distribution of an electron micrograph and prevents straightforward interpretation in many cases. Without such interference, the formation of an image would not be possible. The description of image formation in an electron microscope must, therefore, account for the possibility of interference, which is determined by the degree of coherence of the electron wave-field. Interference effects must be considered in order to extract correctly the information about the spatial structure of an object from the image. The modulation of the image intensity depends on the partial coherence of the electron wave-field. Partial coherence is caused by the finite energy width and the lateral extent of the effective electron source, by parasitic incoherent perturbations, and by unavoidable inelastic scattering processes within the object. Inelastic scattering generally decreases the degree of coherence. Even for energy-filtered high-resolution imaging, inelastic scattering effects are important because electrons that have suffered a very small energy-loss cannot be separated from the unscattered or elastically scattered electrons by a conventional energy filter [1]. Thermal diffuse scattering, for example, produces this kind of very small energy-losses below 0.1 eV and contributes appreciably to the intensity at high scattering angles [2].

The interaction of the imaging electrons with the atoms of the object must be described in terms of quantum theory [3–5]. Owing to the quantum nature of the interaction process, we must distinguish carefully between elastic scattering and inelastic scattering. For elastic scattering, the quantum state of the object remains unchanged; it suffices to assume a static scattering potential without any internal degrees of freedom. The state of the object is altered in an inelastic scattering process. This change is accompanied by a transfer of energy between the scattered electron and the object.

The intensity recorded by an electron micrograph does not directly represent the object. The signal produced by the scattered electrons at some kind of detector does not yield any information about the final object state after the scattering process. This fact has grave consequences for the theory of image formation with elastically and inelastically scattered electrons. The different energy eigenstates of the object are mutually orthogonal and, hence, partial waves belonging to scattering processes resulting in different final object states cannot interfere with each other. They contribute to the image signal in an incoherent manner. This incoherence is due to the change of the object state and not to different energy losses of the scattered electrons. In the presence of degenerate final object states, inelastic scattering processes with the same energy losses may result in different object eigenstates. Since it is always possible to choose an orthogonal basis in the multi-dimensional subspace of the final eigenfunctions, the corresponding partial waves of the scattered electron must be considered as incoherent [6]. The situation becomes much more involved in the case of multiple scattering. The plane partial waves of an inelastically scattered electron, which has excited a distinct object transition, are mutually coherent and can interfere with each other after subsequent elastic scattering. This mechanism shows that the inelastically scattered wave contains high-resolution spatial information about the object [7–9].

The interaction between electrons and matter is rather strong, owing to the intense Coulomb interaction. This causes some well-established approximation methods, like the first-order Born approximation or even the phase-object approximation, to fail in most cases of electron imaging [10]. With a few exceptions (e.g. thin amorphous foils consisting of light elements such as biological specimens) electron imaging is dominated by plural scattering events. As a result of these dynamical effects, the theoretical treatment of electron scattering is more involved than that of x-ray or neutron scattering. On the other hand, the strong sensitivity of electrons to the electric fields of the atoms makes electron scattering a powerful high-resolution imaging method [11,12]. This advantage becomes even more important if we consider the possibilities of energy-resolved analytical electron microscopy [1].

2.2 The Schrödinger Equation

A quantum-mechanical system is associated with a complex-valued wavefunction Ψ, which depends on all internal degrees of freedom of the isolated system and on time. This wavefunction is called the probability amplitude of the system because the probability density of finding the system at a given time in a certain state is the square of the absolute value of the corresponding probability amplitude. The time evolution of this wavefunction is governed by the Schrödinger equation

$$i\hbar\, \partial_t\, \Psi = \hat{\mathbf{H}}\, \Psi\,. \tag{2.1}$$

The self-adjoint differential operator $\hat{\mathbf{H}}$ on the right-hand side of (2.1) is the Hamilton operator of the entire system. The Hamiltonian thus considers the scattered electron and all atoms of the specimen. The total wavefunction $\Psi = \Psi(\boldsymbol{r}, \boldsymbol{R}, t)$ depends on the position of the scattered electron $\boldsymbol{r}$

and on all internal degrees of freedom $\boldsymbol{R}$ of the object. Here the vector set $\boldsymbol{R} = (\boldsymbol{R}_0, \ldots, \boldsymbol{R}_l)^t$ comprises the positions $\boldsymbol{R}_i$, $i = 0, \ldots, l$, of all constituent particles of the object. Since the object state may change during the scattering process, we cannot rewrite the total wavefunction Ψ as a simple product of an object wavefunction and an electron wavefunction. This complication is an important property of interacting quantum systems. The Hamiltonian of the entire system adopts the form

$$\hat{\mathbf{H}} = \hat{\mathbf{H}}_{\mathrm{E}} + \hat{\mathbf{H}}_{\mathrm{O}} + W(\boldsymbol{r}, \boldsymbol{R}) \; . \tag{2.2}$$

In this equation the operators $\hat{\mathbf{H}}_{\mathrm{E}}$ and $\hat{\mathbf{H}}_{\mathrm{O}}$ denote the Hamiltonians of the scattered electron and of the object, respectively. The interaction between the incident electron and the object is governed by the interaction potential $W = W(\boldsymbol{r}, \boldsymbol{R})$. The Hamiltonian of the object acts only on the coordinates of the vector $\boldsymbol{R}$, whereas the Hamiltonian of the electron acts solely on the position vector $\boldsymbol{r}$. Nevertheless, equation (2.1) does not separate with respect to the coordinates of the object and the scattered electron because the interaction potential depends on both $\boldsymbol{r}$ and $\boldsymbol{R}$. Although inelastic interaction is taken into account and the object state may be altered during the scattering process, the total Hamiltonian (2.2) of the system does not depend on time. Accordingly, the Schrödinger equation can be separated with respect to space and time by means of the Bernoulli product ansatz.

Energy conservation including relativistic effects, connects the momentum $\boldsymbol{p}$ of the electron to the acceleration voltage U via the relations

$$(\gamma - 1)\; m_0 c^2 = eU \; , \quad \gamma^2 = 1 + \frac{\boldsymbol{p}^2}{m_0^2 c^2} \; , \tag{2.3}$$

where c is the velocity of light, m_0 the rest mass, and e the charge of an electron. This gauge of the electric potential assumes that the accelerated electron is initially at rest at the tip of the cathode. To account for the finite energy-width of a real electron gun, we must consider an ensemble of electrons accelerated by slightly different acceleration voltages.

By combining the expression (2.3) with the de Broglie relation, we obtain

$$p = \hbar k = \frac{h}{\lambda} = \sqrt{2em_0 U \left(1 + \frac{eU}{2m_0 c^2}\right)} \; . \tag{2.4}$$

The electron wavelength λ is thus determined by the acceleration voltage. In transmission electron microscopy, the acceleration voltage U typically ranges between 100 kV and 1.2 MV with the result that the wavelength of the electron is shorter than 3.70 pm. This relativistically corrected result differs from the non-relativistic value by about 4.8%. In the laboratory system the mass of the electron seems to be increased by about 19.6% for an acceleration voltage of $U = 100$ kV. In order to provide atomic resolution, uncorrected electron microscopes must operate at highvoltages of about 1.2 MV. For this

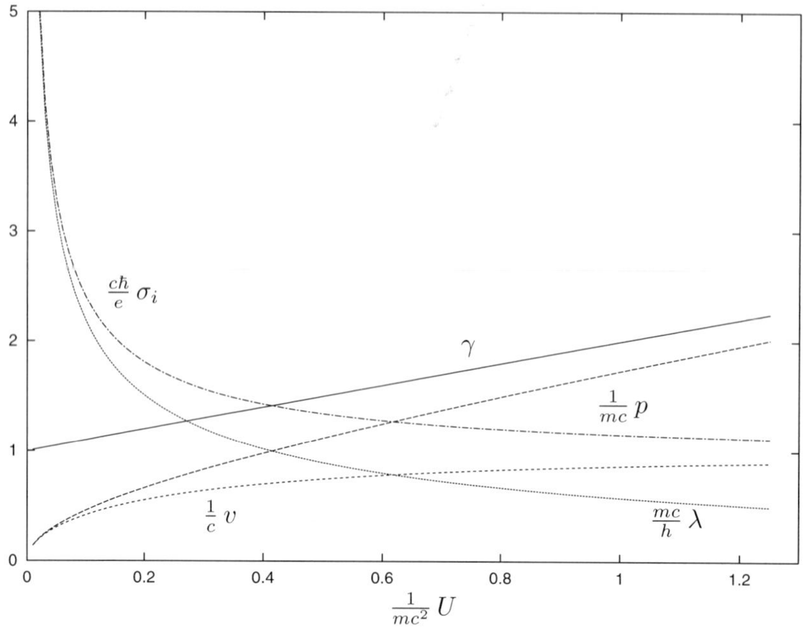

Fig. 2.1. Relativistic kinematics of an accelerated electron. The relativistic mass increase γ, the momentum p, the velocity v, and the wavelength λ are appropriately normalized and plotted versus the normalized acceleration voltage U/m_0c^2. The normalized interaction constant σ_i governing the interaction between the electron and the object potential is also shown

acceleration voltage, the relativistic correction of the wavelength is approximately 47% and the mass of the accelerated electron is about 3.3 times higher than its rest mass. This result demonstrates that relativistic kinematics becomes important in electron imaging theory for high acceleration voltages. In Fig. 2.1 the fundamental kinematical relations for relativistic electrons are sketched. The wavelength falls off rapidly for moderate acceleration voltages and approaches the Compton wavelength $\lambda_C = \hbar/m_0c \approx 2.43\pm$ at about $U_C = 212\,\text{kV}$. For energies $eU > m_0c^2$ the wavelength is approximately inversely proportional to the acceleration voltage.

In free space the propagation of the electron wave is rather simple. In the absence of an electromagnetic field, the Schrödinger equation (2.1) reduces to

$$i\hbar\,\partial_t\,\psi = -\frac{\hbar^2}{2m_0}\,\Delta\,\psi\,. \tag{2.5}$$

The simplest possible solution for a single quantum-mechanical particle satisfying this equation is a plane wave

$$\psi(\boldsymbol{r},t) = \exp i(\boldsymbol{k}\cdot\boldsymbol{r} - \omega t) \tag{2.6}$$

propagating in the direction $\boldsymbol{k}/k$. By substituting (2.6) into (2.5), we obtain the dispersion relation for a free electron

$$\frac{\hbar^2 \boldsymbol{k}^2}{2m_0} = \hbar\omega = eU\left(1 + \frac{eU}{2m_0c^2}\right) = eU^* \, . \tag{2.7}$$

This equation relates the relativistically corrected kinetic energy eU^* of the electron to the momentum or, alternatively, to the wave-vector $\boldsymbol{k}$ of the corresponding plane wave [13]. The combination of the non-relativistic Schrödinger equation with a relativistically correct energy–momentum relation is sufficiently accurate for most calculations in electron imaging [14].

For the general case, which includes the presence of an object, the form of the electron wave is affected by the microscopic electric fields of the atoms of the specimen

$$\Psi = \Psi(\boldsymbol{r}, \boldsymbol{R}) \exp\left(-i\frac{Et}{\hbar}\right) \, . \tag{2.8}$$

The time-dependent exponential factor remains in the general solution. In this case E denotes the total energy of the entire system. The factorization in equation (2.8) is a direct consequence of the fact that the Hamiltonian (2.2) of the total system does not depend on the time. The total energy of the system

$$E = E_0 + m_0c^2\sqrt{1 + \frac{\hbar^2 k_0^2}{m_0^2c^2}} = E_n + m_0c^2\sqrt{1 + \frac{\hbar^2 k_n^2}{m_0^2c^2}} \, , \tag{2.9}$$

for $n = 0, 1, \ldots$, is thus conserved. Here E_0 and E_n denote the energy of the object before and after the scattering process, respectively. Accordingly, k_0 and k_n are the corresponding wavenumbers of the interacting electron in front of and behind the object. For elastic scattering, the object remains in the initial state. Since in this case no energy is transferred, we find from (2.9) that the wavenumber does not change when we consider only the final and initial states of the scattering process.

Within the range of the object potential the kinetic energy of an elastically scattered electron is not conserved. Here, only the sum of the potential energy and of the kinetic energy is constant and in the presence of electromagnetic fields the wave-vector of the electron becomes a function of its position. Therefore, the electron wave suffers an additional phase shift when it travels a short distance through the object. For weak potentials, this phase shift is proportional to the strength of the object potential integrated along the trajectory of the electron. Under the assumption that the electric potential of the object $\varphi_O = \varphi_O(\boldsymbol{r})$ is small compared to the kinetic energy of the incident electron, we can expand the momentum–energy relation (2.4) as a Taylor series with respect to a small deviation of the kinetic energy. This yields the first-order relation

$$\frac{1}{\lambda(U + \varphi_O)} = \frac{1}{\lambda(U)} + \sigma_i\varphi_O + \ldots \, , \qquad \sigma_i = \frac{1}{U\lambda(U)}\frac{1 + eU/m_0c^2}{2 + eU/m_0c^2} \, . \tag{2.10}$$

The interaction constant σ_i is hence a useful measure of the strength of the elastic interaction between the imaging electrons and the object. Figure 2.1 shows the dependence of the interaction constant on the acceleration voltage. The interaction constant decreases rapidly for moderate acceleration voltages and it is nearly constant for wavelengths below the Compton wavelength. The divergence for very small acceleration voltages has no physical meaning because in this case the first-order approximation (2.10) is no longer valid.

To develop the quantum theory of elastic and inelastic scattering further, we assume that the dynamics of the object is completely known. The analysis of the possible quantum states of an object consisting of many atoms is a very demanding task of solid-state physics, especially if bulk effects have to be included in the calculations [15]. For our purpose it often suffices to employ simple quantum-mechanical models of the object dynamics, which account approximately for the possible object transitions. We employ the set $\phi_n = \phi_n(\boldsymbol{R})$, $n = 0, 1, \ldots$ of object eigenfunctions, which satisfy the Schrödinger equation for the isolated object,

$$\hat{\mathbf{H}}_{\mathrm{O}}\, \phi_n = E_n\, \phi_n \,. \tag{2.11}$$

These eigenfunctions are mutually orthogonal and, moreover, fulfil the completeness relation

$$\int \phi_m(\boldsymbol{R})^*\, \phi_n(\boldsymbol{R})\, d^{3l}\boldsymbol{R} = \delta_{mn}\,, \quad m, n = 0, 1, \ldots\,, \tag{2.12}$$

where the integration extents over the total configuration space of the object and δ_{mn} denotes the Kronecker symbol. This property allows us to expand the spatial part of the wavefunction $\Psi = \Psi(\boldsymbol{r}, \boldsymbol{R})$ as a series with respect to the eigenstates of the object

$$\Psi(\boldsymbol{r}, \boldsymbol{R}) = \sum_{n=0}^{\infty} \psi_n(\boldsymbol{r})\phi_n(\boldsymbol{R})\,. \tag{2.13}$$

The coefficients $\psi_n(\boldsymbol{r})$, $n = 0, 1, \ldots$, describe the projections of the total wavefunction $\Psi(\boldsymbol{r})$ onto the eigenstates of the object. The influence of the object on the scattered electron is completely described by the matrix elements

$$U_{nm}(\boldsymbol{r}) = \frac{2m_0\gamma}{\hbar^2} \int \phi_n^*(\boldsymbol{R})\, W(\boldsymbol{r}, \boldsymbol{R})\, \phi_m(\boldsymbol{R})\, d^{3l}\boldsymbol{R} \tag{2.14}$$

of the interaction potential. These matrix elements only depend on the position vector of the scattered electron. For simplicity we assume that the object is in the ground state ϕ_0 prior to the scattering process. Moreover, we approximate the incident wave by a plane wave. In this case the initial wavefunction of the entire system has the form

$$\Psi_i(\boldsymbol{r}, \boldsymbol{R}, t) = \exp\left(i\boldsymbol{k}_0\boldsymbol{r}\right)\, \phi_0(\boldsymbol{R})\, \exp\left(-i\frac{Et}{\hbar}\right)\,. \tag{2.15}$$

The scattering process may either excite the object from its ground state ϕ_0 to a certain excited state ϕ_n or, as in the case of elastic scattering, leave the object state unchanged. Each final state satisfies the Schrödinger equation (2.1) with the Hamiltonian (2.2). In general the final wavefunction of the scattered electron will be a superposition of partial waves $\psi_m = \psi_m(\boldsymbol{r})$, $m = 0, 1, \ldots$, belonging to different final object states. To find the differential equation for each final object state we substitute the expansion (2.13) into the Schrödinger equation (2.1), multiply by all different eigenfunctions ϕ_m, and integrate over the configuration space of the object. By employing the relations (2.2), (2.11) and (2.12), we eventually obtain the set of equations

$$\left(\Delta + \boldsymbol{k}_n^2\right) \psi_n = \sum_{m=0}^{\infty} U_{nm}\, \psi_m\ , \quad n = 0, 1, \ldots\ . \tag{2.16}$$

The indices m and n enumerate the eigenstates of the object. We may interpret the partial wavefunctions ψ_m as the projections of the total wavefunction onto the final object states. The matrix elements U_{mn} measure the probability amplitude for a transition of the object from state ϕ_m to state ϕ_n. This probability depends on the internal structure of the object.

The choice of the initial object state is arbitrary. It may be different from the ground state. Equation (2.16) also describes scattering at any excited state ϕ_n, $n \neq 0$, because its derivation does not depend on the special choice made in (2.15) for the initial object state. For a system in thermal equilibrium, we do not know the initial object state precisely but we know the probability P_m that the object initially is in state ϕ_m with the energy eigenvalue E_m. Since the set of object states is complete, the relation

$$\sum_{m=0}^{\infty} P_m = 1 \tag{2.17}$$

holds true. In order to determine the probabilities P_m, we assume that the object is in thermal contact with a heat reservoir at temperature T_{H}. The specimen is thus a subsystem of a much bigger system. Since the energy of the object fluctuates, the probability of finding the system in the m-th state is proportional to the Boltzmann factor

$$P_m = \frac{1}{Z} \exp\left(-\frac{E_m}{k_{\mathrm{B}} T_{\mathrm{H}}}\right)\ , \quad Z = \sum_{n=0}^{\infty} \exp\left(-\frac{E_n}{k_{\mathrm{B}} T_{\mathrm{H}}}\right)\ . \tag{2.18}$$

Here k_{B} denotes Boltzmann's constant and Z is the canonical partition sum. This probability distribution can be employed to describe image formation in electron microscopy with a sufficient degree of accuracy as long as object heating and object damage are negligibly small. A system that can be described appropriately by time-independent probabilities P_m is called a stationary system. Since each scattering process is associated with a real or virtual fluctuation of the object state, the state of the object is not static. However, the object can be considered as stationary during the time of observation.

The system of coupled partial differential equations (2.16) describes elastic and inelastic scattering. The case of purely elastic scattering is obtained if the interaction potential $W(\boldsymbol{r}, \boldsymbol{R})$ depends only on $\boldsymbol{r}$. In this case, the interaction matrix (2.14) is diagonal: $U_{mn} = 0$ for $m \neq n$. In the case of a distinct initial state (2.15), the system (2.16) reduces to the single equation

$$\left(\Delta + \boldsymbol{k}_0^2\right) \psi_0 = \psi_0 \, U_{00}(\boldsymbol{r}) \, . \tag{2.19}$$

This is the Schrödinger equation for scattering at a static potential $U = U_{00}$. It can be transformed into an integral equation if we employ the free-space Green function

$$G(\boldsymbol{r}, \boldsymbol{r}') = -\frac{1}{4\pi} \frac{\exp\left(ik_0|\boldsymbol{r} - \boldsymbol{r}'|\right)}{|\boldsymbol{r} - \boldsymbol{r}'|} \, , \tag{2.20}$$

which represents the solution of equation (2.19) for a point scatterer located at $\boldsymbol{r}' \neq \boldsymbol{r}$. The resulting integral equation for the scattered wave,

$$\psi(\boldsymbol{r}) = \exp\left(i\boldsymbol{k}_0 \cdot \boldsymbol{r}\right) - \frac{1}{4\pi} \int U_{00}(\boldsymbol{r}') \, \psi(\boldsymbol{r}') \, \frac{\exp\left(ik|\boldsymbol{r}' - \boldsymbol{r}|\right)}{|\boldsymbol{r}' - \boldsymbol{r}|} d^3\boldsymbol{r}' \, , \tag{2.21}$$

is known as the Lippmann–Schwinger integral equation [16].

This equation can be generalized in order to consider inelastic scattering processes. By applying the Green function approach (2.20) to each of the differential equations (2.16), we find

$$\begin{aligned} \psi_n(\boldsymbol{r}) &= \exp\left(i\boldsymbol{k}_0 \cdot \boldsymbol{r}\right) \, \delta_{n0} \\ &\quad - \frac{1}{4\pi} \sum_{m=0}^{\infty} \int \psi_m(\boldsymbol{r}') \, U_{nm}(\boldsymbol{r}') \, \frac{\exp\left(ik_n|\boldsymbol{r}' - \boldsymbol{r}|\right)}{|\boldsymbol{r}' - \boldsymbol{r}|} d^3\boldsymbol{r}' \, , \end{aligned} \tag{2.22}$$

where we have assumed that the object is initially in the state ϕ_0. The set of coupled integral equations (2.23) together with the expansion of the total wavefunction (2.13) completely describe both elastic and inelastic scattering of an electron by an arbitrary object.

2.3 The Scattering Amplitude

Description of the image formation in electron microscopy requires a knowledge of the scattered electron wave at a great distance from the object. If this distance is large compared with the size of the imaged object, the electron wave can be written as a superposition of spherical waves modulated by complex, direction-dependent scattering amplitudes

$$\begin{aligned} \Psi(\boldsymbol{r}, \boldsymbol{R}) &= \exp(i\boldsymbol{k}_i \cdot \boldsymbol{r}) \, \phi_0(\boldsymbol{R}) \\ &\quad + \sum_{n=0}^{\infty} f_{n0}(\boldsymbol{k}_n, \boldsymbol{k}_i) \frac{\exp(ik_n r)}{r} \, \phi_n(\boldsymbol{R}) \, , \quad |\boldsymbol{k}_i| = k_0 \, , \end{aligned} \tag{2.23}$$

where a distinct scattering amplitude $f_{n0}(\boldsymbol{k}_n, \boldsymbol{k}_i)$ is associated with each final object state ϕ_n. The scattering amplitude depends on the initial and the final object state and on the directions $\boldsymbol{k}_i/k_0$ and $\boldsymbol{k}_n/k_n$ of the incident and the scattered electron, respectively. Hence the scattering characteristic depends on the energy loss and on the nature of the corresponding object excitation.

The scattering amplitudes are connected with the matrix elements of the corresponding scattering processes via the relation

$$f_{n0}(\boldsymbol{k}_n, \boldsymbol{k}_i) = -\frac{1}{4\pi} \sum_{m=0}^{\infty} \int U_{nm}(\boldsymbol{r}') \exp(-i\boldsymbol{k}_n \cdot \boldsymbol{r}') \psi_m(\boldsymbol{r}') d^3\boldsymbol{r}' \,. \tag{2.24}$$

We have derived this representation from the Lippmann–Schwinger integral equation (2.23) by employing the asymptotic approximation for the spherical wave

$$\frac{\exp(ik_n |\boldsymbol{r} - \boldsymbol{r}'|)}{|\boldsymbol{r} - \boldsymbol{r}'|} \approx \frac{\exp(ik_n r)}{r} \exp\left(-ik_n \frac{\boldsymbol{r} \cdot \boldsymbol{r}'}{r}\right) \tag{2.25}$$

far from the scatterer located at $\boldsymbol{r}'$. Substituting this approximation into (2.23) and comparing the result with the definition of the scattering amplitudes (2.24) yields the relation (2.24) above. If the scattering potential vanishes outside an area with diameter a around the origin of the coordinate system, the approximation (2.25) is valid for $r \gg a^2/\lambda$. This region is called the Fraunhofer domain [17]. The set of equations (2.24) yields an implicit representation of the the elastic and the inelastic scattering amplitudes because the scattered partial waves reappear on the right hand side of each equation. The scattering problem must therefore be solved before the integral expressions for the scattering amplitudes can be evaluated. Fortunately, (2.24) shows that each scattering amplitude depends only on the wavefunction in the close vicinity of the object, where the matrix elements of interaction $U_{mn} = U_{mn}(\boldsymbol{r})$ are non-vanishing. Owing to this particular circumstance, the scattering amplitudes can be calculated efficiently by iteration methods.

In standard scattering experiments, only the modulus of the scattering amplitude is measured and the phase information of the scattered wave is completely lost. Electron microscopy, however, utilizes this phase information because the image is an interference pattern formed by the unscattered and the scattered partial waves. It is therefore possible to gain high-resolution spatial information about the internal structure of the object by electron microscopical techniques. Unfortunately, the relation between the object potential and the scattering amplitude is highly nonlinear for most objects. This complication poses a serious obstacle for direct structure retrieval by means of electron microscopy.

In mathematical terms a scattering experiment corresponds to a map of a given object potential $V = V(\boldsymbol{r})$ onto a scattering amplitude $f(\boldsymbol{k}_f, \boldsymbol{k}_i)$ when only elastic scattering is concerned. As long as we know the scattering amplitude for only one k value, the mapping is not unambiguously invertible. The scattering amplitude is complex-valued and depends on two directions in physical space, whereas the scattering potential is a real function of a three-dimensional vector. The functions f admissible as scattering amplitudes constitute an extremely restricted subset of all possible complex-valued functions of the form $f = f(\boldsymbol{k}_f, \boldsymbol{k}_i)$. The data contained in the scattering amplitude must hence be highly redundant. It is clear that there can be no simple approach to the inverse scattering problem. In the short-wavelength limit $k \to \infty$, the object potential can be determined unambiguously from the scattering amplitude. Unfortunately, this result has no practical relevance because we cannot measure the scattering amplitude in the short-wavelength limit, where all information is contained in the forward scattering direction, even if the object remained undamaged. Although the potential is uniquely determined by the short-wavelength limit of the scattering amplitude, the result obtained by solving the inverse scattering problem is very sensitive to small errors in the data very close to the forward-scattering direction, where all information is contained. This difficulty illustrates the intricate fundamental problems one encounters when attempting to retrieve the structure of the imaged object by applying inverse scattering procedures [18].

The scattering amplitude satisfies distinct symmetry relations even if the object potential possesses no internal symmetry. In light optics, the reciprocity relation for the scattering amplitude states that the position of the light source and the detector are interchangeable in a diffraction experiment. Assume a light amplitude $u(S)$ at the source point S and a light amplitude $u(P)$ at the image point P. If we put a source at position P and measure the resulting amplitude at position S then according to the reciprocity theorem of light optics the light amplitude will be $u(P)$ [19]. This reciprocity also holds for elastic electron scattering, which implies that the roles of the incident and the scattered electron can be interchanged without altering the scattering amplitude of the static scattering potential.

$$f_{mm}(\boldsymbol{k}_f, \boldsymbol{k}_i) = f_{mm}(-\boldsymbol{k}_i, -\boldsymbol{k}_f) , \quad m = 0, 1, \dots . \tag{2.26}$$

To apply the reciprocity in the case of inelastic scattering, we must consider that the interchange of the initial and final scattering state not only refers to the direction and wavenumber of the scattered electron but also to the initial and final state of the object. During an inelastic scattering process, the object state is changed from Φ_m to Φ_n with transfer of an energy $E_{nm} = E_n - E_m$ from the electron to the object. The reciprocal scattering event changes the state of the object back from Φ_n to Φ_m and the energy difference E_{mn} is transferred in the opposite direction from the object to the electron. Hence, for electrostatic interactions the generalized reciprocity relation

$$f_{nm}(\boldsymbol{k}_n, \boldsymbol{k}_m) = f_{mn}(-\boldsymbol{k}_m, -\boldsymbol{k}_n) \tag{2.27}$$

is valid for elastic and inelastic scattering [3,6]. This relation reveals that the meaning of reciprocity becomes only slightly more involved if inelastic scattering is taken into account. For the reversed scattering process, the object must be initially raised into a specific excited state. This state must be identical with the final object state of the initial scattering process. This situation can hardly be achieved in a real experiment.

In an ideal scattering experiment, the total probability current must be conserved. This requirement is a consequence of the conservation of the number of particles during the interaction process if we neglect absorption effects and relativistic pair generation for very high energies above $2\,m_0c^2$. For transmission electron microscopy the influence of these effects is often negligibly small. The probability current density is given by the symmetric expectation value of the momentum operator

$$\boldsymbol{j}(\boldsymbol{r}) = \frac{\hbar}{2im_0} \int \left(\Psi \frac{\partial}{\partial \boldsymbol{r}} \Psi^* - \Psi^* \frac{\partial}{\partial \boldsymbol{r}} \Psi \right) d\boldsymbol{R}^{3l} \tag{2.28}$$

with respect to the total wavefunction $\Psi = \Psi(\boldsymbol{r}, \boldsymbol{R}, t)$. The conservation of the probability current implies that $\mathrm{div}\boldsymbol{j} = 0$ everywhere in space. We transform this relation into an integral equation by applying Gauss's theorem. The resulting integral

$$\frac{\hbar}{2im_0} \int_S r^2 \int \left(\Psi \frac{\partial}{\partial r} \Psi^* - \Psi^* \frac{\partial}{\partial r} \Psi \right) d\boldsymbol{R}^{3l}\, d^2\Omega = 0 \tag{2.29}$$

over the unit sphere S vanishes. Assuming that our object is initially illuminated by a superposition of two plane waves with wave vectors $|\boldsymbol{k}| = |\boldsymbol{k}'| = k_0$

$$\Psi_i(\boldsymbol{r}, \boldsymbol{R}) = [c_1 \exp{(i\boldsymbol{k} \cdot \boldsymbol{r})} + c_2 \exp{(i\boldsymbol{k}' \cdot \boldsymbol{r})}]\, \phi_0(\boldsymbol{R}) \;, \qquad c_1, c_2 \in \mathbb{C} \;, \tag{2.30}$$

and rearranging we obtain from (2.28), (2.24) and (2.12) for large $r \to \infty$ the asymptotic expression

$$\lim_{r \to \infty} \frac{\hbar}{2im_0} \; \left[|c_1|^2 J(\boldsymbol{k}, \boldsymbol{k}) + c_1 c_2^* J(\boldsymbol{k}', \boldsymbol{k}) \right. \\ \left. + c_1^* c_2 J(\boldsymbol{k}, \boldsymbol{k}') + |c_2|^2 J(\boldsymbol{k}', \boldsymbol{k}') \right] = 0 \;, \tag{2.31}$$

in the Fraunhofer domain (2.25), where the function J is defined as the sum of four integrals:

$$\begin{aligned} J(\boldsymbol{k}, \boldsymbol{k}') = {} & ir^2 \int_{S_2} \boldsymbol{\theta} \cdot (\boldsymbol{k} + \boldsymbol{k}') \exp{[-i(\boldsymbol{k} - \boldsymbol{k}') \cdot \boldsymbol{\theta}]}\, d^2\Omega \\ & - \int f_{00}(\boldsymbol{\theta}, \boldsymbol{k}) \exp{(ik_0 r - i\boldsymbol{k}' \cdot \boldsymbol{r})}\, (1 - i\boldsymbol{k}' \cdot \boldsymbol{r} - ik_0 r)\, d^2\Omega \\ & + \int f_{00}^*(\boldsymbol{\theta}, \boldsymbol{k}') \exp{(-ik_0 r + i\boldsymbol{k} \cdot \boldsymbol{r})}\, (1 + i\boldsymbol{k}'\boldsymbol{r} + ik_0 r)\, d^2\Omega \\ & + \sum_{n=0}^{\infty} \int f_{n0}(\boldsymbol{\theta}, \boldsymbol{k}) f_{n0}^*(\boldsymbol{\theta}, \boldsymbol{k}') \left(ik_n - \frac{1}{r} + ik_n + \frac{1}{r} \right) d^2\Omega \;. \end{aligned} \tag{2.32}$$

The integration must be performed with respect to the direction vector $\boldsymbol{\theta}$. The first integral in this expression vanishes owing to symmetry. The second and third integral can be evaluated by partial integration. We eventually find that

$$\lim_{r\to\infty} J(\boldsymbol{k},\boldsymbol{k}') = -4\pi\,\left(f_{00}(\boldsymbol{k},\boldsymbol{k}') - f_{00}^*(\boldsymbol{k}',\boldsymbol{k})\right) - \sum_{n=0}^{\infty} 2ik_n \int f_{n0}(\boldsymbol{k},\boldsymbol{\theta}) f_{n0}^*(\boldsymbol{k}',\boldsymbol{\theta})\, d^2\Omega\ . \tag{2.33}$$

The integration over the solid angle comprises all directions $\boldsymbol{\theta}$ of the final scattering vectors $\boldsymbol{k}_n$ for $n = 0, 1, \ldots$.

The condition (2.32) must be fulfilled for arbitrary pairs of complex constants c_1 and c_2. This is only the case if the function J vanishes identically in the asymptotic limit, thus

$$\frac{1}{2i}\left[f_{00}(\boldsymbol{k},\boldsymbol{k}') - f_{00}^*(\boldsymbol{k}',\boldsymbol{k})\right] = \sum_{n=0}^{\infty} \frac{k_n}{4\pi} \int f_{n0}(\boldsymbol{k},\boldsymbol{\theta}) f_{n0}^*(\boldsymbol{k}',\boldsymbol{\theta})\, d^2\Omega\ . \tag{2.34}$$

This fundamental property of the scattering amplitude is known as the generalized optical theorem [5]. It is a consequence of the unitarity of the mapping from the initial states onto the final states of the scattering process. The generalized optical theorem is valid for elastic and inelastic scattering. In the special case $\boldsymbol{k} = \boldsymbol{k}'$ the relation (2.34) reduces to the well-known optical theorem of scattering theory

$$\begin{aligned}\frac{1}{2i}\left[f_{00}(\boldsymbol{k},\boldsymbol{k}) - f_{00}^*(\boldsymbol{k},\boldsymbol{k})\right] &= \sum_{n=0}^{\infty} \frac{k_n}{4\pi} \int f_{n0}(\boldsymbol{k},\boldsymbol{\theta}) f_{n0}^*(\boldsymbol{k},\boldsymbol{\theta})\, d^2\Omega \\ &= \frac{k_0}{4\pi}\left(\sigma_{\mathrm{el}} + \sigma_{\mathrm{in}}\right)\ .\end{aligned} \tag{2.35}$$

The left-hand side is the imaginary part of the elastic scattering amplitude in the forward direction, while the right-hand side is proportional to the total elastic and inelastic scattering cross-section. Hence the elastic scattering amplitude contains information about all possible excitations of the object. The second relation on the right-hand side of (2.35) is obtained by considering the definition of the differential scattering cross sections

$$\frac{d\sigma_n}{d\Omega} = \frac{k_n}{k_0}|f_{n0}(\boldsymbol{k},\boldsymbol{k}')|^2\ , \quad n = 0, 1, \ldots \tag{2.36}$$

for elastic ($n = 0$) and inelastic ($n \neq 0$) scattering. It should be noted that the differential scattering cross-section generally depends on the initial $\boldsymbol{k}_i$ and the final $\boldsymbol{k}_f$ scattering direction. Only in a few special cases does it become a function of the transferred momentum $\hbar\,(\boldsymbol{k}_f - \boldsymbol{k}_i)$. The elastic scattering amplitude can be written as a sum of a symmetric and an anti-symmetric contribution

$$f_{00}(\boldsymbol{k},\boldsymbol{k}') = F_{\mathrm{s}}(\boldsymbol{k},\boldsymbol{k}') + iF_{\mathrm{a}}(\boldsymbol{k},\boldsymbol{k}') \tag{2.37}$$

with $F_s(\boldsymbol{k},\boldsymbol{k}') = F_s^*(\boldsymbol{k}',\boldsymbol{k})$ and $F_a(\boldsymbol{k},\boldsymbol{k}') = -F_a^*(\boldsymbol{k}',\boldsymbol{k})$. In particular, this decomposition shows that the generalized optical theorem (2.34) relates the anti-symmetric part of the elastic scattering amplitude to the quadratic terms of the elastic and inelastic scattering amplitudes. Since the generalized optical theorem is a direct consequence of the conservation of the number of particles, any reliable approximation method employed for the calculation of electron micrographs must consider the contributions arising from the second-order Born approximation of the scattering amplitude in order to describe the nonlinear contributions to the image intensity correctly.

2.4 The Born Approximation

The system of coupled integral equations (2.23) can be solved in principle by employing the method of successive approximations. In the case of a weak scattering potential, it suffices to perform the first step of the iteration. We replace the wavefunction ψ on the right-hand side of (2.23) by the zeroth-order approximation $\psi_m^{(0)} = \delta_{m0} \exp(i\boldsymbol{k}_0 \cdot \boldsymbol{r})$ which represents the undisturbed incident wave. From the resulting first-order approximation of the scattered electron wave

$$\begin{aligned}\psi_m^{(1)}(\boldsymbol{r}) &= \exp(i\boldsymbol{k}_0 \cdot \boldsymbol{r})\ \delta_{m0} \\ &\quad - \frac{1}{4\pi}\int \exp(i\boldsymbol{k}_0 \cdot \boldsymbol{r}')\ U_{m0}(\boldsymbol{r}')\frac{\exp(ik_m|\boldsymbol{r}-\boldsymbol{r}'|)}{|\boldsymbol{r}-\boldsymbol{r}'|}\, d^3\boldsymbol{r}' \qquad (2.38)\end{aligned}$$

we can find the first-order approximation of the scattering amplitude by evaluating the integral in the Fraunhofer domain. Inserting the asymptotic expression (2.25) into (2.38) and comparing the result with (2.21), we obtain the first-order Born approximation for the elastic ($m = 0$) and inelastic ($m \neq 0$) scattering amplitudes

$$\begin{aligned}f_m^{(1)}(\boldsymbol{k}_m, \boldsymbol{k}_0) &= f_{m0}^B(\boldsymbol{k}_m - \boldsymbol{k}_0) \\ &= -\frac{1}{4\pi}\int U_{m0}(\boldsymbol{r}')\exp[-i(\boldsymbol{k}_m - \boldsymbol{k}_0)\cdot\boldsymbol{r}']\, d^3\boldsymbol{r}' . \qquad (2.39)\end{aligned}$$

Thus, in the first-order approximation, the scattering amplitudes are proportional to the three-dimensional Fourier transformation of the corresponding matrix elements of the scattering potential U_{m0}. In this approximation the scattering amplitudes depend linearly on the object potential. The kinematic theory of scattering, which is based on the first-order Born approximation, is therefore a straightforward and – compared to the more general dynamic approach – a much simpler theory. Unfortunately, its range of validity in electron microscopy is limited to thin amorphous objects. The Born approximation of the scattering amplitude (2.39) is the first-order term of the expansion of the scattering amplitude into a von Neumann series. The first-order Born approximation yields reasonable results for the scattering of fast electrons at single atoms with low atomic numbers. However, this approximation fails if we consider the scattering at an assembly of atoms because it does not account for *shadowing effects* causing multiple scattering. In the case of a spherically symmetric atomic potential, the first-order Born approximation of the scattering amplitude is real-valued for all scattering angles, whereas the exact scattering amplitude is always complex-valued. Accordingly, the first-order Born approximation can never account for the phase shift, which depends on the scattering angle. As a result, the interference between the partial waves originating from different atoms of the assembly is not correctly described. This explains why the kinematical theory of electron diffraction may fail even for relatively thin specimens. In high-resolution imaging especially, which makes use of high scattering angles, this inaccuracy is particularly noticeable because the high-angle contributions suffer the largest phase shifts since the corresponding electrons penetrate deep into the atomic potential.

To calculate higher-order Born approximations of the scattering amplitude, it is advantageous to employ the Fourier representation of the Green function

$$\lim_{\varepsilon\to 0}\frac{1}{(2\pi)^3}\int\frac{1}{k^2-K^2+i\varepsilon}\exp\left(i\boldsymbol{K}\cdot\boldsymbol{r}\right)d^3\boldsymbol{K}=-\frac{1}{4\pi}\frac{\exp\left(ikr\right)}{r}\,. \tag{2.40}$$

The integration on the left hand side must be performed with care because the small parameter ϵ must approach zero from positive values to account for the fact that the Green function corresponds to an outgoing spherical wave. Using the preceding form of the Green function, we can formally calculate the higher-order Born approximation of the scattering amplitude by successive iteration. Starting from (2.23) and comparing it with (2.24), we eventually find the $(n+1)$th-order expansion term to be

$$f_m^{(n+1)}(\boldsymbol{k}_m,\boldsymbol{k}_0)=\left(-\frac{1}{2\pi^2}\right)^n\sum_{l_1=0}^{\infty}\cdots\sum_{l_n=0}^{\infty}\int\cdots\int\frac{f^{\mathrm{B}}_{l_1 0}(\boldsymbol{k}'_1-\boldsymbol{k}_0)}{k^2_{l_1}-k'^2_1}\cdots \tag{2.41}$$
$$\times\frac{f^{\mathrm{B}}_{l_n l_{n-1}}(\boldsymbol{k}'_n-\boldsymbol{k}'_{n-1})}{k^2_{l_n}-k'^2_n}f^{\mathrm{B}}_{ml_n}(\boldsymbol{k}_m-\boldsymbol{k}'_n)\,d\boldsymbol{k}'_1\ldots d\boldsymbol{k}'_n\,.$$

Although this expression is rather involved, its structure is rather simple. The higher-order terms in (2.42) can be understood as the contributions of multiple scattering. Each integral in (2.42) must be evaluated in the same way as the integral (2.40) in order to fulfil the boundary conditions of the scattering problem correctly.

The first-order Born approximation of the elastic scattering amplitude satisfies the symmetry relation $f^{\mathrm{B}}_{00}(\boldsymbol{K})=f^{\mathrm{B}*}_{00}(-\boldsymbol{K})$. This function is thus real-valued for scattering in the forward direction $\boldsymbol{K}=0$. Accordingly, the first-order Born approximation violates the optical theorem (2.35). However, the generalized optical theorem (2.34) can be utilized to determine the antisymmetric part of the elastic scattering amplitude in the second-order Born approximation. This part depends quadratically on the scattering potential. To be consistent in the approximation, we expand the scattering amplitudes on both sides of (2.34) with respect to the strength of the scattering potential, drop all terms of third and higher order [7], and find

$$\frac{1}{2i}\left[f_0^{(2)}(\boldsymbol{k}_0,\boldsymbol{k}_0') - f_0^{(2)*}(\boldsymbol{k}_0',\boldsymbol{k}_0)\right]$$
$$= \sum_{n=0}^{\infty} \frac{k_n}{4\pi} \int f_{n0}^{\mathrm{B}}(\boldsymbol{k}_0 - \boldsymbol{k}) f_{n0}^{\mathrm{B}*}(\boldsymbol{k}_0' - \boldsymbol{k})\, d^2\Omega \,. \tag{2.42}$$

This result can also be obtained from the general representation (2.42) of the Born approximation. In second order we find

$$f_0^{(2)}(\boldsymbol{k}_0,\boldsymbol{k}_0')$$
$$= -\frac{1}{2\pi^2} \lim_{\varepsilon\to 0} \sum_n \int \frac{1}{k_n^2 - \boldsymbol{k}^2 + i\varepsilon} f_{n0}^{\mathrm{B}}(\boldsymbol{k}_0 - \boldsymbol{k})\, f_{n0}^{\mathrm{B}*}(\boldsymbol{k}_0' - \boldsymbol{k})\, d^3\boldsymbol{k} \,. \tag{2.43}$$

Using the representation of Dirac's delta function as the limit of a series of Lorentz functions

$$\lim_{\varepsilon\to 0}\left[\frac{1}{k_n^2 - \boldsymbol{k}^2 - i\varepsilon} - \frac{1}{k_n^2 - \boldsymbol{k}^2 + i\varepsilon}\right] = \lim_{\varepsilon\to 0} \frac{2i\varepsilon}{(k_n^2 - \boldsymbol{k}^2)^2 + \varepsilon^2}$$
$$= 2\pi i \delta(k_n^2 - \boldsymbol{k}^2) \,, \tag{2.44}$$

the $\boldsymbol{k}$-integration is confined to the surface of a sphere with the radius $k = k_n$. The result then coincides with the right-hand side of (2.42). The remaining, symmetric part of the second-order Born approximation (2.43) has the form

$$\frac{1}{2}\left\{f_0^{(2)}(\boldsymbol{k}_0,\boldsymbol{k}_0') + f_0^{(2)*}(\boldsymbol{k}_0',\boldsymbol{k}_0)\right\}$$
$$= \frac{1}{2\pi^2} \sum_{n=0}^{\infty} P \int \frac{f_{n0}^{\mathrm{B}}(\boldsymbol{k} - \boldsymbol{k}_0) f_{n0}^{\mathrm{B}*}(\boldsymbol{k} - \boldsymbol{k}_0')}{k_n^2 \quad \boldsymbol{k}^2}\, d^3\boldsymbol{k} \,. \tag{2.45}$$

For the evaluation of the volume integral in the last expression the Cauchy principal value must be taken with respect to the integration over the variable $\boldsymbol{k}$, as indicated by the letter P in front of the integral sign [7]. The different structures of the expressions (2.42) and (2.45) furnish an instructive interpretation of the contribution of the second-order Born approximation to the scattering amplitude. Both expressions describe the contribution of the double-scattering processes to the elastic scattering amplitude. For elastic scattering the initial and final momentum states have equal modulus $|\boldsymbol{k}_i| = |\boldsymbol{k}_f|$. Since the intermediate states are not observable, energy conservation is not required for these *virtual* states. The anti-symmetric part (2.42) of the second-order Born approximation accounts only for the *real* intermediate states on the energy shell $\boldsymbol{\kappa}_n^2 = \boldsymbol{k}_n^2$, while the symmetric part (2.45) also includes the scattering events off the energy shell. In the case of purely elastic scattering all internal degrees of freedom are frozen-in and do not contribute. In this case the contribution of the second-order Born approximation reduces to elastic double-scattering. The effect of inelastic scattering on the elastic scattering amplitudes appears in the second-order Born approximation. The optical theorem ideally illustrates the interrelation between the elastic and inelastic scattering processes.

The first-order Born approximation yields analytical expressions for the scattering amplitudes and the scattering cross-sections for elastic and inelastic scattering for simple potentials. We should therefore not expect exact agreement between the analytical results and the experimental observations. Nevertheless, the resulting equations can serve as a first approximation since they describe the overall properties of the scattering amplitudes rather well.

The Wentzel potential

$$U(\boldsymbol{r}) = -\frac{2\gamma Z}{a_{\mathrm{H}}} \frac{\exp(-|\boldsymbol{r}|/R)}{|\boldsymbol{r}|} \tag{2.46}$$

is a good approximation for the potential of a single atom if we disregard effects resulting from the shell structure of the electron density. The shielding radius R is a measure of the shielding of the nucleus by the electron cloud; $a_{\mathrm{H}} = \hbar^2/m_0 e^2$ denotes the Bohr radius and Z the atomic number. The shielding radius is a free parameter, which can be used to match the Wentzel potential to more accurate models [20]. The statistical Thomas–Fermi atomic model [3] yields $R = a_{\mathrm{H}} Z^{-1/3}$. Another reasonable choice is $R^2 = \langle r^2 \rangle$, where $\langle r^2 \rangle$ denotes the mean-square radius of the electron cloud of the scattering atom. The scattering amplitudes calculated from the Wentzel potential are quite realistic for small scattering angles because the exponential decrease of the Wentzel potential is fairly accurate. Employing this model, we obtain from (2.39) the elastic scattering amplitude

$$f_{00}^{\mathrm{B}}(\boldsymbol{k}_0 - \boldsymbol{k}_0') = \frac{2\gamma Z R^2}{a_{\mathrm{H}}} \frac{1}{R^2|\boldsymbol{k}_0 - \boldsymbol{k}_0'|^2 + 1} \tag{2.47}$$

in the first Born approximation. In high-resolution electron imaging, when using an acceleration voltage U above 60 kV, the shielding radius satisfies the relation $R \gg \lambda$. We can thus employ the small-angle approximation for the scattering vector

$$|\boldsymbol{k}_0 - \boldsymbol{k}_0'| = 2k_0 \sin(\theta/2) \approx k_0\theta\ , \tag{2.48}$$

where $\theta \ll 1$ denotes the angle enclosed by the initial $\boldsymbol{k}_0$ and final $\boldsymbol{k}_0'$ scattering direction. Within the frame of validity of the small-angle approximation, the elastic differential cross-section has the form

$$\frac{d\sigma_{\mathrm{el}}^{\mathrm{B}}}{d\Omega} = |f_{00}^{\mathrm{B}}(\theta)|^2 = \left(\frac{2\gamma Z R^2}{a_{\mathrm{H}}}\right)^2 \frac{\theta_0^4}{\left(\theta^2 + \theta_0^2\right)^2}\ , \quad \theta_0 = \frac{1}{k_0 R}\ . \tag{2.49}$$

By integrating over the full angular range, we obtain the result

$$\sigma_{\mathrm{el}} = \frac{4\pi\,\gamma^2}{k_0^2}\,\frac{Z^2 R^2}{a_{\mathrm{H}}^2} \tag{2.50}$$

for the total elastic scattering cross section. This approximation agrees quite well with the result of more accurate calculations and represents a very useful rule-of-thumb [1].

According to the results we have derived by employing the first-order Born approximation and the Wentzel potential, elastic scattering is confined in a cone with half-angle θ_0, which is called the characteristic angle of elastic scattering. The larger the extension of the atomic potential the sharper will be the peak of elastic scattering in forward direction. For large scattering angles, $\theta \gg \theta_0$, the scattering amplitude decreases as $1/\theta^2$. As an example we consider 60 keV electrons scattered at silicon ($Z = 14$). The characteristic scattering angle is then $\theta_0 = 18$ mrad and about 50% of the incident electrons are scattered into angles smaller than θ_0.

For a more realistic treatment of electron scattering by single atoms, we must take into account the exact interaction potential

$$U(\boldsymbol{r}, \boldsymbol{R}_0, \ldots, \boldsymbol{R}_Z) = -\frac{2\gamma}{a_{\mathrm{H}}}\left(\frac{Z}{|\boldsymbol{r}-\boldsymbol{R}_0|} - \sum_{\nu=1}^{Z}\frac{1}{|\boldsymbol{r}-\boldsymbol{R}_\nu|}\right) \tag{2.51}$$

between a scattered electron and a single atom with atomic number Z. The first term in this expression results from the charge of the atomic nucleus located at the position $\boldsymbol{R}_0$ and the other terms describe the interaction of the incident electron with the electrons of the atom at the positions $\boldsymbol{R}_\nu$, for $\nu = 1, \ldots, Z$. The distribution of the atomic electrons is determined by the Z-particle electron wavefunctions $\phi_n = \phi_n(\boldsymbol{R}) = \phi_n(\boldsymbol{R}_1, \ldots, \boldsymbol{R}_Z)$ of the free atom. The index n denotes the energy eigenstate of the atom.

Inelastic scattering mainly results from an interaction of the incident electron with the electrons of the object, except for thermally diffuse or phonon scattering. If we neglect such non-electronic excitations, the position of the nucleus $\boldsymbol{R}_0 = 0$ remains fixed. Inelastic scattering caused by electronic excitations is confined to small scattering angles. The energy loss and the scattering angle depend on each other. The nearer the electron passes to the nucleus the higher is the excitation energy that must be transferred to the atomic electron in order to make a transition, because the probability to hit a tightly bound electron increases with decreasing distance from the atomic nucleus. In the first-order Born approximation, we can verify this intuitive argument more quantitatively.

By inserting the interaction potential (2.51) into the expression (2.39) for the scattering amplitude in the first-order Born approximation, we can perform the integration with respect to $\boldsymbol{r}$ analytically. We find the approxi-

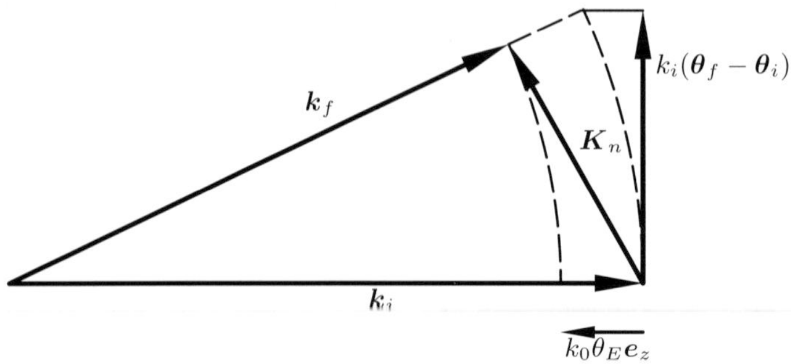

Fig. 2.2. The scattering vector $\boldsymbol{K}_n$ is the difference vector between the final $\boldsymbol{k}_f$ and initial $\boldsymbol{k}_i$ wave-vectors of the scattered electron wave. The scattering vector decomposes into a lateral component $\boldsymbol{q}$ perpendicular to the optic axis and a longitudinal component parallel to the optic axis. For small energy losses and small scattering angles, the longitudinal component is given by $-k_0\theta_{\mathrm{E}}\boldsymbol{e}_z$. Within the frame of validity of this small-angle approximation, the lateral component is not affected by the energy loss

mation

$$f^{\mathrm{B}}_{n0}(\boldsymbol{K}_n) = -\frac{1}{4\pi}\int \exp(-i\boldsymbol{K}_n\cdot\boldsymbol{r})\,U_{n0}(\boldsymbol{r})d^3\boldsymbol{r} \tag{2.52}$$
$$= \frac{2\gamma}{a_{\mathrm{H}}}\frac{1}{|\boldsymbol{K}_n|^2}\left[Z\,\delta_{n0} - \sum_{\nu=1}^{Z}\int \phi_n^*(\boldsymbol{R})\phi_0(\boldsymbol{R})\,\exp(-i\boldsymbol{K}_n\cdot\boldsymbol{R}_\nu)\,d^{3Z}\boldsymbol{R}\right]$$

for the scattering amplitudes for elastic and inelastic scattering.

The scattering vector $\boldsymbol{K}_n$ is a function of the energy loss $\varepsilon_n = E_n - E_0$ of the scattered electron. For sufficiently small energy losses $|\varepsilon_n| \ll eU$ the scattering vector is approximately given by [21,22]

$$\boldsymbol{K}_n = \boldsymbol{q} - k_0\theta_{\mathrm{E}}\,\boldsymbol{e}_z\,, \quad \theta_{\mathrm{E}} = \frac{\varepsilon_n}{2eU}\,\frac{eU+mc^2}{eU/2+mc^2}\,. \tag{2.53}$$

The relations between the scattering vector $\boldsymbol{K}_n$, the initial and final wave-vector and the characteristic angle of inelastic scattering θ_{E} are depicted in Fig. 2.2. The relation (2.53) between the energy loss ϵ_n and the angle θ_{E} follows from the energy relation (2.9) if we assume that $k_n \approx k_0$ for small energy losses. Within the frame of validity of the small-angle approximation (2.48), the vector $\boldsymbol{q}$ in (2.53) denotes the lateral component of the scattering vector. Apart from a relativistic factor, the characteristic scattering angle θ_{E} for inelastic scattering equals one half of the relative energy loss. For inelastic scattering, therefore, the momentum transfer between the scattered electron and the object is always non-zero. The minimum momentum transfer $\hbar k_0\theta_{\mathrm{E}}$ occurs for forward scattering ($\theta = 0$).

The differential cross-section for inelastic scattering is a function of the energy loss. Using (2.53) and definition (2.36), we obtain the following expression for the double-differential scattering cross-section:

$$\frac{d^2\,\sigma}{d\Omega d\varepsilon} = \left(\frac{2\gamma}{a_{\mathrm{H}}}\right)^2 \sum_{n=0}^{\infty} \frac{1}{|\boldsymbol{K}_n|^4} \frac{k_n}{k_0} \tag{2.54}$$

$$\times \left| Z\delta_{n0} - \sum_{\nu=1}^{Z} \int \phi_n^*(\boldsymbol{R})\phi_0(\boldsymbol{R}) \exp\left(-i\boldsymbol{K}_n \cdot \boldsymbol{R}_\nu\right) d^{3Z}\boldsymbol{R} \right|^2 \delta(\varepsilon - E_n + E_0)\,.$$

Owing to the presence of the delta function on the right-hand side, the double-differential scattering cross-section is non-zero only if the energy loss ε matches one of the possible excitation energies $\varepsilon_n = E_n - E_0$ with $n = 0, 1, \ldots$. The first term ($n = 0$) of the sum accounts for elastic scattering. Accordingly, the elastic scattering amplitude in the first-order Born approximation is described by the Bethe–Mott [23] formula

$$f_{00}^{\mathrm{B}}(\boldsymbol{K}_0) = \frac{2\gamma}{a_{\mathrm{H}}} \frac{Z - F(\boldsymbol{K}_0)}{\boldsymbol{K}_0^2}\,. \tag{2.55}$$

The electronic form-factor

$$\begin{aligned} F(\boldsymbol{K}) &= \int \phi_0^*(\boldsymbol{R})\phi_0(\boldsymbol{R}) \left(\sum_{\nu=1}^{Z} \exp\left(-i\boldsymbol{K}\cdot\boldsymbol{R}_\nu\right)\right) d^{3Z}\boldsymbol{R} \\ &= \int \varrho(\boldsymbol{r}')\,\exp\left(-i\boldsymbol{K}\cdot\boldsymbol{r}\right)\,d^3\boldsymbol{r} \end{aligned} \tag{2.56}$$

is the Fourier transform of the electron density

$$\varrho(\boldsymbol{r}) = \int \phi_0^*(\boldsymbol{R}) \left[\sum_{\nu=1}^{Z} \delta(\boldsymbol{r} - \boldsymbol{R}_\nu)\right] \phi_0(\boldsymbol{R})\, d^{3Z}\boldsymbol{R} \tag{2.57}$$

of the atom in the ground state. For x-ray scattering, this form-factor represents the scattering amplitude. The electron scattering amplitude and the atomic form-factor in the first-order Born approximation are intimately related by the Bethe–Mott formula (2.55), which is a reformulation of the Poisson equation of electrodynamics in reciprocal space.

If (2.55) is evaluated numerically, care must be taken for small scattering angles because of the vanishing denominator. For very small scattering vectors $|\boldsymbol{K}_0| \ll k_0$ and a spherically symmetric charge density, the expansion of (2.55) as a power series with respect to $|\boldsymbol{K}_0|$ yields

$$f_{00}^{\mathrm{B}}(\boldsymbol{K}_0) = \frac{2\gamma Z}{a_{\mathrm{H}}} \left(\frac{\langle r^2\rangle}{6} - \frac{\langle r^4\rangle}{120}\boldsymbol{K}_0^2 + \mathcal{O}(\boldsymbol{K}_0^4)\right)\,. \tag{2.58}$$

Here $\langle r^2\rangle$ is the mean-square radius of the atomic electron density, $\langle r^4\rangle$ its next higher moment. Comparison of the small-angle behaviour of the electron scattering amplitude with the corresponding result for the less accurate

Wentzel potential suggests that the choice $R^2 = \langle r^2 \rangle$ should improve the Wentzel approximation. With hindsight, this justifies the ad hoc assumption for the shielding radius of the Wentzel potential in (2.46).

A much better approximation for the elastic scattering amplitude in the first-order Born approximation can be obtained by Hartree–Dirac–Fock–Slater methods. These numerical computations yield the charge density of a single atom in its ground state. The scattering amplitudes obtained by means of these data are tabulated in the literature [24]. It is convenient to parametrize the numerical data by introducing appropriate fit-functions. For analytical calculations, it is crucial to find a representation in terms of a small number of simple basis functions that provide a sufficient accuracy and have the correct asymptotical behaviour for large scattering angles. To describe the atomic form-factor, a Gaussian fit as introduced by Doyle and Turner is very useful. The corresponding parametrization has the form [24]

$$F_{\mathrm{DT}}(\boldsymbol{K}) = \sum_{n=1}^{N_{\mathrm{DT}}} a_n \exp\left(-b_n \boldsymbol{K}^2\right) . \tag{2.59}$$

The real and positive parameters $a_n, b_n > 0$, $n = 1, \ldots, N_{\mathrm{DT}}$, must be determined for each element by a least-squares fit to the numerically obtained form-factors. The Bethe–Mott formula (2.55) relates the atomic form-factor to the elastic scattering amplitude in the first-order Born approximation

$$F(\boldsymbol{k}) = Z - \frac{a_{\mathrm{H}}}{2\gamma} \boldsymbol{K}^2 f_{00}^{\mathrm{B}}(\boldsymbol{K}) . \tag{2.60}$$

This relations suggests that the elastic scattering amplitudes in first-order Born approximation should be parametrized in terms of a modified set of basis functions

$$f_{00}^{\mathrm{B}}(\boldsymbol{K}) \approx \sum_{n=1}^{N_{\mathrm{KW}}} a_n \frac{1 - \exp\left(-b_n \boldsymbol{K}^2\right)}{\boldsymbol{K}^2} \tag{2.61}$$

with real and positive parameters $a_n, b_n > 0$, $n = 1, \ldots, N_{\mathrm{KW}}$. This fit has been introduced by Weickenmeier and Kohl [25] and provides the correct asymptotic behaviour for small and large scattering vectors. With a view to improving the accuracy of the calculation, a number of alternative fitting procedures have been proposed.

The differential inelastic scattering cross-section is given by the terms $n > 0$ in (2.55). These terms can also be calculated numerically in a realistic approximation. Unfortunately, these calculations are very laborious and moreover, the inelastic scattering factors obtained by this method depend in a non-trivial way on the element [26].

Fortunately, it is possible to find a simple approximation for the inelastic contribution to the total scattering cross section in the case of high-energy scattering by making use of the completeness relation of the atomic wave functions. For this purpose we neglect the details of the electronic excitation and approximate the energy transfer between the scattered electron and the atom by a mean energy loss $\varepsilon \approx \overline{\varepsilon}$. A reasonable choice [21] for the mean energy loss is $\overline{\varepsilon} = ZE_H/2$, where $E_H = 13.6\,\mathrm{eV}$ denotes the Rydberg energy. This value is proportional to the atomic number Z and equals about half of the excitation energy of the atomic resonance line. Within the frame of validity of this approximation, the inelastic scattering vectors $\boldsymbol{K}_n$, $n = 1, 2, \ldots$ are replaced by an average scattering vector $\overline{\boldsymbol{K}}$. The characteristic scattering angle θ_{E} defined by the mean energy loss has the effect of a shielding radius, limiting the range of interaction between the scattered electron and the atomic electrons. This spatial cut-off of the Coulomb interaction prevents the total inelastic cross-section [6] from diverging.

With these assumptions the summation over the final states ϕ_n in expression (2.55) for the double differential inelastic-scattering cross-section can be performed. The additional integration over the energy loss yields

$$\begin{aligned} \frac{d\sigma_{\mathrm{in}}}{d\Omega} &= \left(\frac{2\gamma}{a_{\mathrm{H}}}\right)^2 \frac{1}{|\overline{\boldsymbol{K}}|^4} \sum_{n=1}^{\infty} \left| \sum_{\nu=1}^{Z} \int \phi_n^*(\boldsymbol{R})\phi_0(\boldsymbol{R}) \exp\left(-i\overline{\boldsymbol{K}} \cdot \boldsymbol{R}_\nu\right) d\boldsymbol{R}^{3Z} \right|^2 \\ &= \left(\frac{2\gamma}{a_{\mathrm{H}}}\right)^2 \frac{1}{|\overline{\boldsymbol{K}}|^4} \left\{ \int \sum_{\mu,\nu=1}^{Z} \exp\left[i\overline{\boldsymbol{K}} \cdot (\boldsymbol{R}_\mu - \boldsymbol{R}_\nu)\right] |\phi_0(\boldsymbol{R})|^2 d^{3Z}\boldsymbol{R} \right. \\ &\quad \left. \left| \sum_{\nu=1}^{Z} \int |\phi_0(\boldsymbol{R})|^2 \exp\left(i\overline{\boldsymbol{K}} \cdot \boldsymbol{R}_\nu\right) \right|^2 d^{3Z}\boldsymbol{R} \right\} . \end{aligned} \tag{2.62}$$

The expression in braces is the Fourier transformation of the variance of the mixed electron density of the atom averaged over its ground-state. Assuming that the electrostatic correlations between the electrons of the atom are negligibly small, we may approximate the mixed term in expression (2.62) by the square of the ground state electron density or, equivalently, by the electron form-factor (2.56). This yields an approximation for the differential inelastic scattering cross-section

$$\frac{d\sigma_{\mathrm{in}}}{d\Omega} = \left(\frac{2\gamma}{a_{\mathrm{H}}}\right)^2 \frac{Z - \frac{1}{Z}F^2(\overline{\boldsymbol{K}})}{|\overline{\boldsymbol{K}}|^4} , \tag{2.63}$$

where $F(\boldsymbol{K})$ denotes the electronic form-factor defined in (2.56). This last expression is equivalent to that derived classically by Raman and Compton [27] for x-ray scattering. It has been successfully used to describe inelastic electron scattering by Lenz [20] and Rose [22]. Equation (2.63) is therefore called the modified Raman–Compton approximation with mean energy loss.

In order to derive a simple analytic expression for the inelastic scattering cross-section it is advantageous to employ the Wentzel potential (2.46). This potential yields the expression

$$\frac{d\sigma_{\mathrm{in}}}{d\Omega} = 2\pi \left(\frac{2\gamma}{k_0^2 a_{\mathrm{H}}}\right)^2 \frac{Z}{(\theta^2 + \theta_{\mathrm{E}}^2)^2} \left(1 - \frac{\theta_0^4}{(\theta^2 + \theta_0^2 + \theta_{\mathrm{E}}^2)^2}\right) \tag{2.64}$$

for the small-angle approximation of the differential inelastic scattering cross-section. The integration of the differential scattering cross-section over the full solid angle is performed by the same method as in (2.50) with the result

$$\begin{aligned}\sigma_{\mathrm{in}} &= 2\pi \left(\frac{2\gamma}{a_{\mathrm{H}} k_0^2}\right)^2 Z \left[\frac{\ln\left(1+\theta_0/\theta_{\mathrm{E}}\right)}{\theta_0^2} - \frac{1}{2\left(\theta_0^2+\theta_{\mathrm{E}}^2\right)}\right] \\ &\approx 2\pi \left(\frac{2\gamma}{a_{\mathrm{H}} k_0^2}\right)^2 \frac{Z}{\theta_0^2}\left[\ln\left(\frac{\theta_0}{\theta_{\mathrm{E}}}\right) - \frac{1}{2}\right] ,\end{aligned} \tag{2.65}$$

where in the second expression it has been assumed that $\theta_0 \gg \theta_{\mathrm{E}}$. Comparison of the last result with the relation (2.50) for the elastic scattering cross-section shows that their ratio is proportional to $1/Z$ and depends only logarithmically on the acceleration voltage. More elaborate calculations based on the modified Raman–Compton approximation (2.63) using the Doyle–Turner fit (2.59) for tabulated x-ray scattering amplitudes confirm this behaviour. The relation

$$\sigma_{\mathrm{in}}/\sigma_{\mathrm{el}} \approx 20/Z \tag{2.66}$$

may serve as a useful rule-of-thumb in many cases [1]. It agrees surprisingly well with experimental results and shows that inelastic scattering is predominant for light elements. The contrast in electron micrographs of biological specimens is hence strongly affected by inelastically scattered electrons if no energy filter is used.

The factor $1/(\theta^2+\theta_{\mathrm{E}}^2)^2$ in (2.64) shows that, the inelastic scattering cross-section is significantly more peaked in the forward direction than the elastic scattering cross-section. The effect of the spatial delocalization d_i in an electron micrograph due to inelastic scattering is often discussed in terms of the characteristic angle θ_{E} of inelastic scattering. However, it should be emphasized that the simple conclusion $d_i = 0.6\,\lambda/\theta_{\mathrm{E}}$ is erroneous. From (2.64) we find that only about 50 percent of the inelastically scattered electrons are actually contained in a forward cone with half-angle θ_{E}. The other half contains spatial information with much higher resolution and, therefore, diminishes the delocalization considerably [22].

2.5 The Mutual Dynamic Object Spectrum

In a real electron microscope, the illumination is never perfectly coherent. Perfectly coherent illumination would require a monochromatic point source. Even a modern field-emission gun equipped with a monochromator, which will become available in the near future, fulfils these requirements only imperfectly. The finite energy-width and the lateral extension of any real source and the incoherent perturbations resulting from electromagnetic and mechanical instabilities during the time of exposure reduce the degree of coherence of the electron wave and thus the information limit of the instrument.

The inelastic interaction between the imaging electrons and the object introduces another source of incoherence, but this must not be considered as a factor limiting the retrievable information. On the contrary, the inelastically scattered electrons provide very sensitive analytical information about the chemical composition and even the local electronic structure of the object. Moreover, the inelastically scattered electrons also carry some spatial information because they form low-resolution images. Electrons that are scattered elastically and inelastically even convey high-resolution spatial information. This fundamental fact is often termed the conservation of elastic contrast in inelastic electron imaging.

In Sect. 2.2 we have shown that the quantum-mechanical state of the scattered electron is not completely described by the wavefunction of a pure quantum state. Accordingly, we have to consider the total state of the imaging electron and of the object as a mixed quantum state. This situation is similar to that encountered for partially coherent imaging in light optics. Therefore, the concept of the light-optical mutual coherence function is also very suitable for expressing the influence of inelastic scattering and partially coherent illumination in electron microscopy [28,22,29].

The physical quantity recorded by the detector in an electron microscope is neither the quantum-mechanical wavefunction nor the scattering amplitude of the object but the z-component of the probability current density at the plane of detection perpendicular to the optic axis. The detector integrates the measured signal over the time of detection T: the recorded image intensity is a time average. The probability current is related to a time-independent wavefunction only in the ideal case of elastic scattering and fully coherent illumination. In reality microscopic fluctuations within the object, the source, and the optical instrument during the time of exposure affect the detected image signal. The frequency of these fluctuations is much higher than the reciprocal time of detection $1/T$. If we neglect subsidiary effects like object damage and object heating, the time variation of the current density can be regarded as a stationary stochastic process. The probability amplitudes at two points $\boldsymbol{r}_1$ and $\boldsymbol{r}_2$ in free space are given by the time-dependent functions $\psi_1 = \psi(\boldsymbol{r}_1, t)$ and $\psi_2 = \psi(\boldsymbol{r}_2, t)$. The time-averaged signal at another point $\boldsymbol{r}_D$ is given by the superposition of the partial waves emanating from the points $\boldsymbol{r}_1$ and $\boldsymbol{r}_2$ as illustrated in Fig. 2.3. To account for the propagation of the partial waves between their points of origin $\boldsymbol{r}_1$ and $\boldsymbol{r}_1$ and the point of detection $\boldsymbol{r}_D$, we introduce the complex constants C_1 and C_2. The intensity recorded at the point of detection $\boldsymbol{r}_D$ is then given by

$$I_D = |C_1|^2 \langle \psi_1^* \psi_1 \rangle_T + |C_2|^2 \langle \psi_2^* \psi_2 \rangle_T + 2\mathrm{Re}\left[C_1^* C_2 \langle \psi_1^* \psi_2 \rangle_T \right] . \qquad (2.67)$$

The symbol $\langle \ldots \rangle_T$ indicates the time average taken over the time of detection T. The third term in expression (2.67) is proportional to the correlation between the temporal variations of the probability amplitude at the positions $\boldsymbol{r}_1$ and $\boldsymbol{r}_2$. For completely uncorrelated sources at $\boldsymbol{r}_1$ and $\boldsymbol{r}_2$ and sufficiently

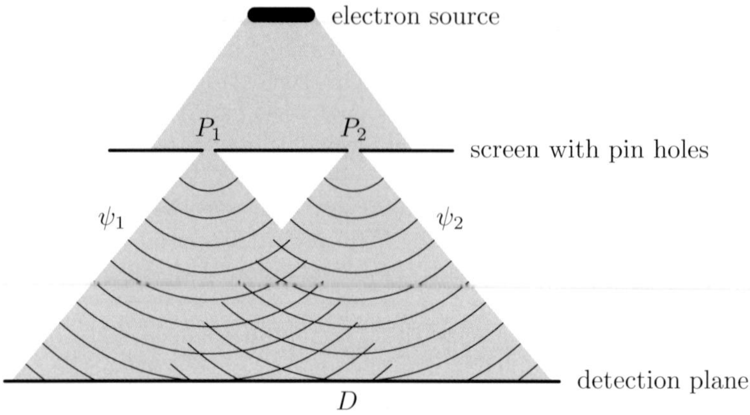

Fig. 2.3. Schematic arrangement illustrating the formation of partial coherence. The pinholes P_1 and P_2 are illuminated by an extended, incoherent source. The partial waves emanating from the two pinholes form a partially coherent wavefield. The interference patterns resulting from different points of the source superimpose incoherently. The lateral extension of the source can be used to adjust the degree of coherence of the wavefield behind the screen

long detection times, it disappears. In contrast, this term reaches its maximum value for full correlation between the partial waves emanating from the two pinholes [19].

The setup discussed above represents a largely simplified version of an electron imaging experiment. It demonstrates that the concept of partial coherence accounts for the correlation between distinct points in the wavefield. This correlation is described most conveniently by the mutual coherence function

$$\Gamma = \Gamma(\boldsymbol{r}, \boldsymbol{r}', \tau) = \langle \psi^*(\boldsymbol{r}, t)\psi(\boldsymbol{r}', t - \tau)\rangle_T \ . \tag{2.68}$$

This function allows one to calculate the time-averaged image intensity in the plane of detection. Owing to the stationarity of the imaging process, Γ depends only on the time difference $\tau = t - t'$ and not on the absolute time. Quantum-mechanically the time average corresponds to an average over the internal degrees of freedom of the total system and an ensemble average over differently prepared initial states.

The complex-valued mutual coherence function (2.68) is a bilinear and hermitian time average of the wavefunction evaluated at two different points in space and time. Hence, the physical laws governing the propagation and the transmission of the mutual coherence function are closely related to that for ordinary wavefunctions. Although the mutual coherence function does not contain complete information about the wavefunction $\psi = \psi(\boldsymbol{r}, t)$, it carries all the information necessary to describe the image formation. Accordingly, the time averages of both the electron intensity and the z-component of the probability current can be expressed in terms of the mutual coherence function

$$
\begin{aligned}
j_z(\boldsymbol{r}) &= \frac{\hbar}{m} \operatorname{Im}\left[\frac{\partial}{\partial z'} \Gamma(\boldsymbol{r}, \boldsymbol{r}', \tau = 0)\right]\Bigg|_{\boldsymbol{r}=\boldsymbol{r}'} , \\
I(\boldsymbol{r}) &= \langle \psi(\boldsymbol{r}, t)\psi^*(\boldsymbol{r}', t)\rangle_T = \Gamma(\boldsymbol{r}, \boldsymbol{r}' = \boldsymbol{r}, \tau = 0) .
\end{aligned} \tag{2.69}
$$

The free-space propagation of the mutual coherence function is governed by the stationary Schrödinger equation. The relative energy-width of the electron beam in a high-resolution electron microscope is very small, even if inelastic scattering is taken into account. First we consider a quasi-monochromatic wave-field emanating from an extended incoherent source. The corresponding mutual coherence function $\Gamma(\boldsymbol{r}_{\mathrm{A}}, \boldsymbol{r}'_{\mathrm{A}}, \tau)$ is assumed to be known for all pairs of points $(\boldsymbol{r}_{\mathrm{A}}, \boldsymbol{r}'_{\mathrm{A}})$ in a plane $z = z_{\mathrm{A}}$ perpendicular to the optic axis. In this case, the mutual coherence function at any subsequent plane $z = z_{\mathrm{B}}$ in the field-free region behind the plane z_{A} can be determined by employing Sommerfeld's diffraction formula [19]

$$
\begin{aligned}
\Gamma(\boldsymbol{r}_{\mathrm{B}}, \boldsymbol{r}'_{\mathrm{B}}, \tau) = &\left(\frac{k}{2\pi}\right)^2 \int\!\!\int \Gamma(\boldsymbol{r}_{\mathrm{A}}, \boldsymbol{r}'_{\mathrm{A}}, \tau) \\
&\times \frac{\exp\left(ik|\boldsymbol{r}_{\mathrm{B}} - \boldsymbol{r}_{\mathrm{A}}|\right)}{|\boldsymbol{r}_{\mathrm{B}} - \boldsymbol{r}_{\mathrm{A}}|} \frac{\exp\left(-ik|\boldsymbol{r}'_{\mathrm{B}} - \boldsymbol{r}'_{\mathrm{A}}|\right)}{|\boldsymbol{r}'_{\mathrm{B}} - \boldsymbol{r}'_{\mathrm{A}}|} \, d^2\boldsymbol{r}_{\mathrm{A}} d^2\boldsymbol{r}'_{\mathrm{A}} .
\end{aligned} \tag{2.70}
$$

The wavenumber k corresponds to the mean energy of the imaging electrons. This integral equation neglects retardation effects caused by different geometrical distances and the consequences of large inclination angles. Nevertheless this approximation describes the propagation of the mutual coherence function sufficiently accurately within the field-free region. In an electron microscope the electron beam is confined to a narrow region about the optic axis. The distance

$$
\begin{aligned}
|\boldsymbol{r}_{\mathrm{B}} - \boldsymbol{r}_{\mathrm{A}}| &= \sqrt{(\boldsymbol{\rho}_{\mathrm{B}} - \boldsymbol{\rho}_{\mathrm{A}})^2 + (z_{\mathrm{B}} - z_{\mathrm{A}})^2} \\
&\approx |z_{\mathrm{B}} - z_{\mathrm{A}}| \left(1 + \frac{1}{2}\frac{(\boldsymbol{\rho}_{\mathrm{B}} - \boldsymbol{\rho}_{\mathrm{A}})^2}{(z_{\mathrm{B}} - z_{\mathrm{A}})^2} + \dots\right)
\end{aligned} \tag{2.71}
$$

can therefore be expanded as a power series in the lateral distance of $|\boldsymbol{\rho}_{\mathrm{B}} - \boldsymbol{\rho}_{\mathrm{A}}|$. Since the planes z_{A} and z_{B} are parallel, we only need information about the correlation between different lateral positions. Therefore, Γ reduces to a four-dimensional correlation function $\Gamma(\boldsymbol{\rho}, \boldsymbol{\rho}', \tau; z) = \Gamma(\boldsymbol{r}, \boldsymbol{r}', \tau)|_{z=z'}$ with the z coordinate as a parameter. The dependence of $\Gamma(\boldsymbol{\rho}, \boldsymbol{\rho}', \tau; z)$ on $\boldsymbol{\rho}$ and $\boldsymbol{\rho}'$ accounts for the lateral coherence and the dependence on the time difference τ describes the longitudinal or temporal coherence of the wave-field. Retaining only terms up to second order in the expansion (2.71), we obtain for the free-space propagation between two parallel planes, separated from each other by the distance $d = z_B - z_A$, the approximation

$$\Gamma(\boldsymbol{\rho}_B, \boldsymbol{\rho}'_B, \tau; z_B) = \left(\frac{k}{2\pi d}\right)^2 \int\int \Gamma(\boldsymbol{\rho}_A, \boldsymbol{\rho}'_A, \tau; z_A) \times \exp\left\{\frac{ik}{2d}\left[(\boldsymbol{\rho}_B - \boldsymbol{\rho}_A)^2 - (\boldsymbol{\rho}'_B - \boldsymbol{\rho}'_A)^2\right]\right\} d^2\boldsymbol{\rho}_A d^2\boldsymbol{\rho}'_A \tag{2.72}$$

This behaviour resembles the Fresnel approximation of light optics. However, owing to the quantum nature of the electrons, we cannot directly apply the light-optical theory of image formation under partially coherent illumination conditions to imaging with electrons.

The right-hand side of (2.73) describes a linear mapping of the mutual coherence function from the plane $z = z_A$ onto the plane $z = z_B$; it can be read as a convolution of a propagator function P_F with the mutual coherence function at the initial plane A. The four-dimensional free-space propagator

$$P_F(\boldsymbol{\rho}, \boldsymbol{\rho}'; d) = \left(\frac{k}{2\pi d}\right)^2 \exp\left[i\frac{k}{2d}\left(\boldsymbol{\rho}^2 - \boldsymbol{\rho}'^2\right)\right] \tag{2.73}$$

decomposes into a product of two two-dimensional Fresnel propagators.

The Fresnel approximation describes the real free-space propagation in electron imaging surprisingly well, even for small distances d. To understand this behaviour, we consider a high-energy electron moving in the direction $\boldsymbol{k}$. Its spatial wavefunction satisfies the three-dimensional Helmholtz equation

$$\Delta\psi + \boldsymbol{k}^2\,\psi = 0\,. \tag{2.74}$$

The z dependence of ψ can be expressed as a slightly distorted plane wave

$$\psi(\boldsymbol{\rho}, z) = \exp(ikz)\;\tilde{\psi}(\boldsymbol{\rho}, z)\,. \tag{2.75}$$

Inserting this ansatz into the differential equation (2.74) and recalling that $\tilde{\psi}(\boldsymbol{\rho}, z)$ changes only slowly over a distance of several wavelengths, we obtain the approximation

$$\Delta_\rho\,\tilde{\psi} + 2ik\,\partial_z\,\tilde{\psi} = 0\,. \tag{2.76}$$

This differential equation is the high-energy approximation of the Schrödinger equation. It neglects backscattering and has the form of the two-dimensional diffusion equation with a complex diffusion coefficient. By employing the Green-function technique, we find the solution

$$\psi(\boldsymbol{\rho}_B, z_B) = -\frac{ik}{2\pi d}\exp\left[ik(z_B - z_A)\right] \times \int \psi(\boldsymbol{\rho}_A, z_A)\exp\left[i\frac{k}{2d}(\boldsymbol{\rho}_B - \boldsymbol{\rho}_A)^2\right] d^2\boldsymbol{\rho}_A\,. \tag{2.77}$$

Comparison of this result with (2.73) shows that the Fresnel approximation is an exact solution of the Schrödinger equation in the high-energy approximation.

If $\Gamma = \Gamma(\boldsymbol{\rho}_{\mathrm{A}}, \boldsymbol{\rho}'_{\mathrm{A}}, \tau; z_{\mathrm{A}})$ vanishes in the plane z_{A} outside the central region with maximum diameter D_{A}, one may replace the Fresnel approximation of the mutual coherence function at distances $d > D_{\mathrm{A}}^2/\lambda$ from the initial plane z_{A} by the Fraunhofer approximation [19,17]

$$\Gamma(\boldsymbol{\rho}_{\mathrm{B}}, \boldsymbol{\rho}'_{\mathrm{B}}, \tau; z_{\mathrm{B}}) = \left(\frac{k}{2\pi d}\right)^2 \exp\left[i\frac{k}{2d}\left(\boldsymbol{\rho}_{\mathrm{B}}^2 - \boldsymbol{\rho}'^2_{\mathrm{B}}\right)\right] \tag{2.78}$$
$$\times \int\int \Gamma(\boldsymbol{\rho}_{\mathrm{A}}, \boldsymbol{\rho}'_{\mathrm{A}}, \tau; z) \exp\left[-i\frac{k}{d}\left(\boldsymbol{\rho}_{\mathrm{B}} \cdot \boldsymbol{\rho}_{\mathrm{A}} - \boldsymbol{\rho}'_{\mathrm{B}} \cdot \boldsymbol{\rho}'_{\mathrm{A}}\right)\right] d^2\boldsymbol{\rho}_{\mathrm{A}}\, d^2\boldsymbol{\rho}_{\mathrm{A}} \,.$$

The four-dimensional Fresnel propagator cannot be used to describe the propagation of the electron wave through the lens system or through the object since the electromagnetic field acts as an inhomogeneous and anisotropic refracting medium for the imaging electrons. In this case we must employ a different approach, as outlined in Sect. 2.7.

Nevertheless, the mutual coherence function behind the object is linearly related to the mutual coherence function in front of the object. The mapping between the entrance and the exit plane can be described quite generally in terms of the mutual dynamic object spectrum $T = T(\boldsymbol{q}_i, \boldsymbol{q}'_i, \boldsymbol{q}_f, \boldsymbol{q}'_f, z_i, z_f, \tau)$. This function depends on the primed and unprimed coordinates and connects the mutual coherence function in front of the object with that behind the object via the relation

$$\Gamma(\boldsymbol{q}_f, \boldsymbol{q}'_f, \tau; z_{\mathrm{B}}) = P_{\mathrm{F}}(\boldsymbol{q}_f, \boldsymbol{q}'_f; D/2) \int T(\boldsymbol{q}_i, \boldsymbol{q}'_i, \boldsymbol{q}_f, \boldsymbol{q}'_f, \tau)$$
$$\times P_{\mathrm{F}}(\boldsymbol{q}_i, \boldsymbol{q}'_i; D/2)\, \Gamma(\boldsymbol{q}_i, \boldsymbol{q}'_i, \tau; z_{\mathrm{A}})\, d^2\boldsymbol{q}_i\, d^2\boldsymbol{q}'_i \,. \tag{2.79}$$

The functions

$$\tilde{\Gamma}(\boldsymbol{q}, \boldsymbol{q}', \tau) = \frac{1}{(2\pi)^2} \int \Gamma(\boldsymbol{\rho}, \boldsymbol{\rho}', \tau) \exp\left(i\boldsymbol{q} \cdot \boldsymbol{\rho} - i\boldsymbol{q}' \cdot \boldsymbol{\rho}'\right) d^2\boldsymbol{\rho}\, d^2\boldsymbol{\rho}' \,, \tag{2.80}$$

and

$$\tilde{P}_{\mathrm{F}}(\boldsymbol{q}, \boldsymbol{q}'; d) = \exp\left(-\frac{id}{2k}\left(\boldsymbol{q}^2 - \boldsymbol{q}'^2\right)\right) \tag{2.81}$$

denote the four-dimensional Fourier transforms of the mutual coherence function and of the Fresnel propagator (2.73), respectively.

The relation (2.79) is based on the assumption that the object is situated between the planes $z = z_{\mathrm{A}}$ and $z = z_{\mathrm{B}}$, as depicted in Fig. 2.4. For a thick object it is crucial to differentiate between the object plane $z = z_{\mathrm{O}}$, the entrance plane $z = z_{\mathrm{A}}$, and the exit plane $z = z_{\mathrm{B}}$. For simplicity we choose the central plane of the object as the object plane. For a thin object the

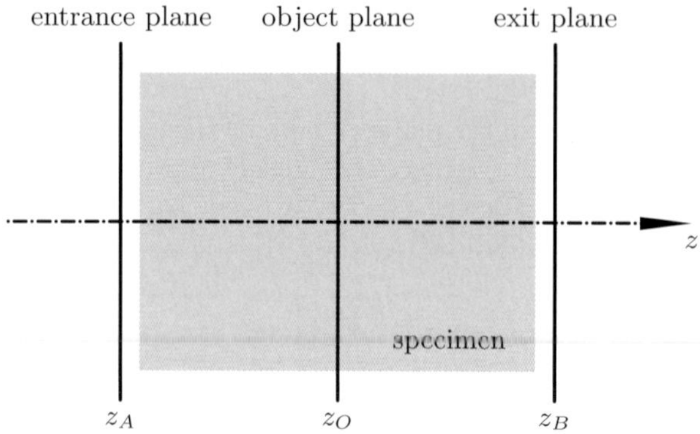

Fig. 2.4. Definition of the object plane $z = z_{\mathrm{O}}$, the entrance plane $z = z_{\mathrm{A}}$, and the exit plane $z = z_{\mathrm{B}}$ for an object thickness $D = z_{\mathrm{B}} - z_{\mathrm{A}}$. The object plane is situated in the centre of the object at $z_{\mathrm{O}} = (z_{\mathrm{A}} + z_{\mathrm{B}})/2$

thickness $D = z_{\mathrm{B}} - z_{\mathrm{A}}$ is small and the three planes become approximately equivalent.

Equation (2.79) only holds within the frame of validity of the small-angle approximation. For larger scattering angles the mutual dynamic object spectrum depends on the three-dimensional incident and final scattering vectors $\boldsymbol{k}_i$, $\boldsymbol{k}'_i$, $\boldsymbol{k}_f$, and $\boldsymbol{k}'_f$, and not just on their lateral projections. In this more general case the mutual object spectrum $T = T(\boldsymbol{k}_i, \boldsymbol{k}'_i, \boldsymbol{k}_f, \boldsymbol{k}'_f, \tau)$ also accounts for back-scattering and connects the full six-fold spatial Fourier transform of the mutual coherence function behind the object with that in front of the object. Accordingly, the theory of image formation described in terms of the mutual dynamic object transparency is not restricted to small scattering angles, at least in principle [7]. The transfer function (2.79) decomposes into a product of two functions, one dependent only on the primed and the other only on the unprimed coordinates, if the influence of partial coherence and inelastic scattering is neglected. It should be noted that (2.79) is applicable only if the stochastic fluctuations of the wavefunction in front of the object and those of the object state are uncorrelated. Fortunately, this requirement is fulfilled in high-resolution electron imaging, where backscattering can be neglected.

To derive an explicit expression for the mutual dynamic object spectrum in terms of the scattering amplitudes of the object, we assume for the present that the object is illuminated by a coherent superposition of slightly tilted plane waves. Accordingly, the initial mutual coherence function in front of the object adopts the form

$$\Gamma(\boldsymbol{\rho}, \boldsymbol{\rho}', \tau; z_{\mathrm{A}}) = \exp(i\omega\tau)\, \exp[-i\boldsymbol{q}_i \cdot (\boldsymbol{\rho} - \boldsymbol{\rho}')]\,, \quad \tau = t - t'\,, \tag{2.82}$$

where $\omega = eU/\hbar$ is determined by the mean acceleration voltage U. Using the expression (2.69) and the definition (2.24) of the scattering amplitude and assuming parallel illumination we find

$$j_z(\boldsymbol{r}) = \frac{\hbar k_0}{m} \left\{ 1 + 2\,\mathrm{Re}\left[f_{00}(\boldsymbol{k}_i, \boldsymbol{r}/r)\, \frac{\exp i(k_0 r - \boldsymbol{k}_i \cdot \boldsymbol{r})}{r} \right] + \sum_{n=0}^{\infty} \frac{k_n}{k_0} \frac{|f_{0n}(\boldsymbol{k}_i, \boldsymbol{r}/r)|^2}{r^2} \right\} . \tag{2.83}$$

for the z-component of the probability current density far from the object. The direction of incidence is given by the wave vector $\boldsymbol{k}_i = \boldsymbol{q}_i + k_0 \boldsymbol{e}_z$ with $|\boldsymbol{k}_i| \approx k_0$. The origin of the coordinate system is situated at the object plane on the optic axis. Within the paraxial domain the spherical wave in the second and third terms of (2.83) may be approximated by the Fresnel propagator (2.73). Moreover, the wavenumber k_0 can be substituted for k_n, $n = 1, 2, \ldots$, because most energy losses are small compared with the mean beam energy eU. Within the frame of validity of this approximation the z-component of the current density (2.69) is proportional to the diagonal elements of the mutual coherence function

$$j_z(\boldsymbol{\rho}, \tau; z) \approx k_0 \Gamma(\boldsymbol{\rho}, \boldsymbol{\rho}' = \boldsymbol{\rho}, \tau; z) . \tag{2.84}$$

This result and the Fraunhofer approximation (2.79) for the wavefunction far from the object allow one to calculate the mutual coherence function at the object plane $z = z_O$. Using the expression (2.82) for the mutual coherence function in front of the object and the assumption that the object is in its ground-state before the scattering occurs, we eventually find for the Fourier transform of the mutual coherence function at the object plane [7]

$$\begin{aligned} \tilde{\Gamma}_f(\boldsymbol{q}, \boldsymbol{q}', \tau; z_O) = \exp(i\omega\tau) \big[& \delta(\boldsymbol{q} - \boldsymbol{q}_i)\delta(\boldsymbol{q}' - \boldsymbol{q}_i) \\ & - \frac{1}{2\pi i k_0} \delta(\boldsymbol{q}' - \boldsymbol{q}_i) f_{00}^*(\boldsymbol{q}_i, \boldsymbol{q}) + \frac{1}{2\pi i k_0} \delta(\boldsymbol{q} - \boldsymbol{q}_i) f_{00}(\boldsymbol{q}_i, \boldsymbol{q}') \\ & + \frac{1}{4\pi^2 k_0^2} \sum_{n=0}^{\infty} f_{n0}^*(\boldsymbol{q}_i, \boldsymbol{q}) f_{n0}(\boldsymbol{q}_i, \boldsymbol{q}') \exp(-i\omega_{n0}\tau) \big] . \end{aligned} \tag{2.85}$$

In the case of partially coherent illumination, we must replace the initial wave function (2.82) by a more general initial mutual coherence function.

The small-angle approximation (2.85) neglects backscattering. The stochastic fluctuations of the wavefunction in front of the object and within the object are hence uncorrelated and the thermal average over different initial object states can be performed independently. The resulting expression for the mutual dynamic object spectrum can be written as

$$T(\boldsymbol{q}_{\mathrm{A}}, \boldsymbol{q}'_{\mathrm{A}}, \boldsymbol{q}_{\mathrm{B}}, \boldsymbol{q}'_{\mathrm{B}}, \omega) = \sum_{m=0}^{\infty} P_m \big[\delta(\boldsymbol{q}_{\mathrm{B}} - \boldsymbol{q}_{\mathrm{A}})\delta(\boldsymbol{q}'_{\mathrm{B}} - \boldsymbol{q}'_{\mathrm{A}})\delta(\omega)$$
$$-\frac{1}{2\pi i k_0}\delta(\boldsymbol{q}'_{\mathrm{A}} - \boldsymbol{q}'_{\mathrm{B}}) f^*_{mm}(\boldsymbol{q}_{\mathrm{A}}, \boldsymbol{q}_{\mathrm{B}})\delta(\omega)$$
$$+\frac{1}{2\pi i k_0}\delta(\boldsymbol{q}_{\mathrm{A}} - \boldsymbol{q}_{\mathrm{B}}) f_{mm}(\boldsymbol{q}'_{\mathrm{A}}, \boldsymbol{q}'_{\mathrm{B}})\delta(\omega)$$
$$+\frac{1}{4\pi^2 k_0^2}\sum_{n=0}^{\infty} f^*_{nm}(\boldsymbol{q}_{\mathrm{A}}, \boldsymbol{q}_{\mathrm{B}}) f_{nm}(\boldsymbol{q}'_{\mathrm{A}}, \boldsymbol{q}'_{\mathrm{B}})\delta(\omega - \omega_{nm})\big] \,. \tag{2.86}$$

This result is obtained by comparing (2.85) with the more general equation (2.79) and then taking the thermal average with respect to the initial state of the object. We have then performed the Fourier transformation with respect to the time lag τ. The argument of the resulting delta functions accounts for the discrete energy losses of the scattered electron caused by specific excitations of the object.

The representation (2.79) of the mutual dynamic object transparency elucidates the different nature of the two averaging processes. The thermal average over the possible initial states with probability P_m is an incoherent average, while the summation over all final states of the object is a coherent average. The mutual dynamic object spectrum completely describes the influence of a thick object on the coherence function. It depends on the spatial coordinates at the object plane and on the direction of incidence. The latter dependence is caused by the strong dynamic effects in electron imaging due to multiple scattering. Equation (2.86) clearly demonstrates that a factorization of the mutual dynamic object transparency is impossible because partial coherence causes temporal correlations between different off-axial points. Primarily, this correlation carries the information about the inelastic interaction between the imaging electrons and the object.

The mutual dynamic object spectrum (2.86) can be written as a sum over four different contributions

$$T = T_0 + T_1 + T_{2,\mathrm{el}} + T_{2,\mathrm{in}} \,. \tag{2.87}$$

The first term, which accounts for the unscattered part of the wavefunction forms the central spot in the diffraction pattern. In the absence of an object this is the only non-vanishing contribution

$$T = T_0 = \delta(\boldsymbol{q}_f - \boldsymbol{q}_i)\, \delta(\boldsymbol{q}'_f - \boldsymbol{q}'_i)\, \delta(\omega) \,. \tag{2.88}$$

This condition guarantees that the wave-vector of an outgoing plane wave coincides with that of the corresponding incident plane wave. The second term

$$T_1 = -\frac{1}{2\pi i k_0} \sum_{m=0}^{\infty} P_m \left[\delta(\boldsymbol{q}'_A - \boldsymbol{q}'_B) f^*_{mm}(\boldsymbol{q}_A, \boldsymbol{q}_B) \right.$$
$$\left. -\delta(\boldsymbol{q}_A - \boldsymbol{q}_B) f_{mm}(\boldsymbol{q}'_A, \boldsymbol{q}'_B)\right] \delta(\omega) \tag{2.89}$$

is linear in the scattering amplitudes and results from elastic scattering at a thermally averaged object potential. Inelastic scattering does not contribute linearly to the mutual coherence function. Hence the linear theory of image formation must neglect inelastic scattering in order to be consistent in its approximations. The thermal average accounts for thermal diffuse scattering and modifies the elastic scattering amplitude by a Debye–Waller factor, even if the quadratic terms are neglected. The last two terms

$$T_{2,\mathrm{el}} = \frac{1}{4\pi^2 k_0^2} \sum_{m=0}^{\infty} P_m f^*_{mm}(\boldsymbol{q}_A, \boldsymbol{q}_B) f_{mm}(\boldsymbol{q}'_A, \boldsymbol{q}'_B)\delta(\omega) \, ,$$
$$T_{2,\mathrm{in}} = \frac{1}{4\pi^2 k_0^2} \sum_{m=0}^{\infty} \sum_{\substack{n=0 \\ n\neq m}}^{\infty} P_m f^*_{nm}(\boldsymbol{q}_A, \boldsymbol{q}_B) f_{nm}(\boldsymbol{q}'_A, \boldsymbol{q}'_B)\delta(\omega - \omega_{nm}) \tag{2.90}$$

on the right-hand side of (2.79) consider the contributions that are quadratic in the scattering amplitudes. Unlike the linear part, these contributions are affected by partially coherent imaging conditions and inelastic processes.

The mutual dynamic object transparency is not invariant under an exchange of the initial and final scattering vectors. This behaviour is caused by plural scattering and known from experiments as the top-bottom effect. In particular, details at the exit surface of thick specimens facing the objective lens are imaged with better resolution than those located on the entrance surface. By inserting expression (2.88) for the first term T_0 into (2.79) and recalling that two consecutive Fresnel propagators are equivalent to a single propagation, we can show that in the absence of an object the entrance and the exit planes are connected by a simple Fresnel propagator.

If inelastic scattering is considered, it is impossible to rewrite the mutual dynamic object transparency as a product of two factors, one depending only on the primed and the other one depending only on the unprimed lateral position vector. Unfortunately, analytical expressions for the mutual dynamic object spectrum do not exist for realistic objects. Nevertheless, general and quite efficient numerical procedures have been developed to evaluate (2.86) for realistic objects.

2.6 Mixed Dynamic Form-Factor

Owing to the extremely large number of internal degrees of freedom, it is not possible to calculate the exact mutual dynamic object spectrum explicitly for any realistic object. Unfortunately, the representation (2.87) of the mutual dynamic object spectrum in terms of the scattering amplitudes only demonstrates its over all structure. The numerical evaluation of this expression necessitates further approximations. For weakly scattering specimens it is possible to derive a second-order approximation with respect to the strength of the interaction potential. The linear part in the expression (2.89) is exclusively related to elastic scattering. We must therefore replace the elastic scattering amplitude by its second-order Born approximation. Inelastic scattering does not contribute to the linear terms because the inelastically scattered partial waves cannot interfere with the unscattered wave owing to the orthogonality of the final object states. The first-order Born approximation may therefore be used for the inelastic scattering amplitudes in the quadratic terms of (2.90). The resulting approximation to the mutual dynamic object transparency for weakly scattering objects

$$T^{(2)} = T^{(0)} + T^{(1)} + T^{(2)}_{\mathrm{el},1} + T^{(2)}_{\mathrm{el},2} + T^{(2)}_{\mathrm{in}} \tag{2.91}$$

is consistent in second order with respect to the strength of the interaction potential [22,7,6].

The so-called weak phase-object approximation only contributes the terms $T^{(0)}$ and $T^{(1)}$. This approximation completely neglects the quadratic terms, which contain the information about the electronic properties of the object. Hence the linear approximation can only describe the phase contrast of thin amorphous objects with any degree of accuracy. Inelastic scattering solely affects the quadratic terms $T^{(2)}_{\mathrm{el},1}$ and $T^{(2)}_{\mathrm{in}}$. The second-order approximation describes the interaction of the incident electron wave with a thin object much better than the linear approximation because it considers the nonlinear dependence of the current density on the scattering amplitude. If we express the scattering amplitudes in (2.87) by the Born series and drop all terms of third and higher order in the interaction potential, we eventually obtain

$$\begin{aligned}
T^{(0)} &= \delta(\boldsymbol{K}_\perp)\,\delta(\boldsymbol{K}'_\perp)\delta(\omega)\,,\\
T^{(1)} &= \frac{1}{2\pi i k_0}\sum_{m=0}^{\infty} P_m \left[\delta(\boldsymbol{K}_\perp)\, f^{(1)*}_{mm}(\boldsymbol{K}') - \delta(\boldsymbol{K}'_\perp)\, f^{(1)}_{mm}(\boldsymbol{K})\right]\,\delta(\omega)\,,\\
T^{(2)}_{\mathrm{el},1} &= \frac{1}{2\pi i k_0}\sum_{m=0}^{\infty} P_m \left[\delta(\boldsymbol{K}_\perp)\, f^{(2)*}_{mm}(\boldsymbol{k}_f,\boldsymbol{k}_i) - \delta(\boldsymbol{K}'_\perp)\, f^{(2)}_{mm}(\boldsymbol{k}_f,\boldsymbol{k}_i)\right]\,\delta(\omega)\,,\\
T^{(2)}_{\mathrm{el},2} &= \frac{1}{4\pi^2 k_0^2}\sum_{m=0}^{\infty} P_m f^{(1)*}_{mm}(\boldsymbol{K})\, f^{(1)}_{mm}(\boldsymbol{K}')\,\delta(\omega)\,,\\
T^{(2)}_{\mathrm{in}} &= \frac{1}{4\pi^2 k_0^2}\sum_{m=0}^{\infty}\sum_{\substack{n=0\\ n\neq m}}^{\infty} P_m f^{(1)*}_{nm}(\boldsymbol{K})\, f^{(1)}_{nm}(\boldsymbol{K}')\,\delta(\omega-\omega_{nm})\,,
\end{aligned} \tag{2.92}$$

where $\boldsymbol{K}_\perp$ denotes the lateral part of the scattering vector $\boldsymbol{K} = \boldsymbol{k}_f - \boldsymbol{k}_i$. It is important to note that the first-order Born approximation of the inelastic scattering amplitude is sufficient to calculate the mutual dynamic object spectrum up to the second order [7,6]. This is not true of the elastic scattering amplitude, which must be known up to the second-order Born approximation. Fortunately, the second-order contribution can readily be obtained from the integral relation (2.43).

Inelastic scattering results from electronic excitations and from thermal diffuse scattering. An electronic excitation is initiated by the scattering of an incident electron at a bound electron of the object. Since the masses of the two particles are equal, the scattering process is accompanied by an appreciable energy transfer. Thermal diffuse scattering results from scattering at the potentials of the nuclei, which oscillate about their positions of equilibrium. These vibrations always affect the intensity distribution in electron micrographs and diffraction patterns. The energy loss of the scattered electrons due to thermal diffuse scattering is very low. It is for this reason that it is not possible to separate the thermal diffusely scattered electrons from the unscattered and the elastically scattered electrons by means of a conventional energy filter [1].

Single inelastic scattering at an assembly of fixed atoms can be described conveniently by means of the mixed dynamic form-factor $S = S(\boldsymbol{K}, \boldsymbol{K}', \omega)$. This quantity considers scattering at the electronic charge distribution of the specimen. The mixed dynamic form-factor is defined as the thermal average of the Fourier transform of the electron density of the object [22]:

$$S(\boldsymbol{K}, \boldsymbol{K}', \omega) = \frac{1}{2\pi} \int_{-\infty}^{\infty} \langle \varrho(\boldsymbol{K}, t)^* \varrho(\boldsymbol{K}', t-\tau) \rangle_T \exp(i\omega\tau)\, d\tau \; . \tag{2.93}$$

In this equation $\boldsymbol{K} = \boldsymbol{q}_f - \boldsymbol{q}_i - k_0 \theta_{\mathrm{E}} \boldsymbol{e}_z$ denotes the scattering vector, which is related to the energy loss ΔE via the characteristic angle θ_{E} for inelastic scattering (2.53). The Fourier transform of the electrostatic density operator is given by

$$\varrho(\boldsymbol{K}, t) = \sum_{j=1}^{Z} \exp\left[i\boldsymbol{K} \cdot \boldsymbol{R}_j(t)\right] \; , \tag{2.94}$$

where $\boldsymbol{R}_j(t)$, $j = 1, \ldots, Z$ denotes the time-dependent position operator in the Heisenberg representation of the j-th object electron. The density operator (2.94) implicitly depends on the internal degrees of freedom of the object [3,4].

Since the interaction between the incident particle and all constituents of the object is governed by the Coulomb potential, the quadratic contribution to the mutual dynamic object transparency can also be expressed by means of the the mixed dynamic form-factor. Considering that the electric potential of a point charge satisfies the Poisson equation, we find

$$\int \frac{1}{|\boldsymbol{r} - \boldsymbol{R}|} \exp(-i\boldsymbol{K} \cdot \boldsymbol{r})\, d^3\boldsymbol{r} = \frac{4\pi}{\boldsymbol{K}^2} \exp(-i\boldsymbol{K} \cdot \boldsymbol{R}) \tag{2.95}$$

for the Fourier transform of the Coulomb potential. The matrix elements of the time-dependent density operator are related via

$$\int \phi_n^*(\boldsymbol{R})\, \varrho(\boldsymbol{K}, t-\tau)\, \phi_m(\boldsymbol{R})\, d^{3Z}\boldsymbol{R}$$
$$= \exp\left(-i\omega_{nm}\tau\right) \int \phi_n^*(\boldsymbol{R})\, \varrho(\boldsymbol{K}, t)\, \phi_m(\boldsymbol{R})\, d^{3Z}\boldsymbol{R} \tag{2.96}$$

to the eigenfunctions of the object. This expression, together with the Fourier representation (2.95) of the Coulomb potential, proves the relation [7]

$$\begin{aligned}\frac{S(\boldsymbol{K}, \boldsymbol{K}', \omega)}{\boldsymbol{K}^2 \boldsymbol{K}'^2} &= \frac{1}{2\pi} \int_{-\infty}^{\infty} \frac{\langle \varrho^*(\boldsymbol{K}, t)\varrho(\boldsymbol{K}', t-\tau)\rangle_T}{\boldsymbol{K}^2 \boldsymbol{K}'^2} \exp\left(i\omega\tau\right)\, d\tau \\ &= \frac{1}{2\pi} \frac{1}{\boldsymbol{K}^2 \boldsymbol{K}'^2} \int_{-\infty}^{\infty} \exp\left(i\omega\tau\right) \sum_{m=0}^{\infty} \sum_{n=0}^{\infty} \int \phi_n^*(\boldsymbol{R}) \varrho^*(\boldsymbol{K}, t) \phi_m(\boldsymbol{R})\, d\boldsymbol{R} \\ &\quad \times \int \phi_m^*(\boldsymbol{R}') \varrho(\boldsymbol{K}', t+\tau) \phi_n(\boldsymbol{R}')\, d\boldsymbol{R}'\, d\tau \\ &= \sum_{m=0}^{\infty} \sum_{n=0}^{\infty} P_m\, \delta(\omega - \omega_{nm}) f_{nm}^{(1)*}(\boldsymbol{K}) f_{nm}^{(1)}(\boldsymbol{K}') \\ &= \frac{k_0^2}{2\pi} T_{\text{in}}^{(2)}(\boldsymbol{K}, \boldsymbol{K}', \omega) + \frac{k_0^2}{2\pi} T_{\text{el},2}^{(2)}(\boldsymbol{K}, \boldsymbol{K}', \omega)\, .\end{aligned} \tag{2.97}$$

With this result we define the inelastic part of the mixed dynamic form-factor by

$$\frac{2\pi}{k_0^2}\, S_{\text{in}}(\boldsymbol{K}, \boldsymbol{K}', \omega) = \boldsymbol{K}^2 \boldsymbol{K}'^2\, T_{\text{in}}^{(2)}(\boldsymbol{K}, \boldsymbol{K}', \omega)\, . \tag{2.98}$$

The mixed dynamic form factor is a generalization of the conventional dynamic form factor $S = S(\boldsymbol{K}, \omega) = S(\boldsymbol{K}, \boldsymbol{K}, \omega)$. In contrast to the latter, the mixed dynamic form-factor is, in general, a complex-valued quantity because it also contains information about the spatial structure of the object. The mixed dynamic form-factor is therefore well suited to describe the image formation by the elastically and inelastically scattered electrons in the case of thin objects.

Equation (2.97) shows that the mixed dynamic form factor is related to a density–density correlation function. The fact that the scattering properties of an object are determined by such a correlation function was first discovered by van Hove [30]. The introduction of the mixed dynamic form-factor generalizes his result to include the phases of the partial waves, which carry the information about the atomic structure. Within the frame of validity of the second-order approximation (2.91), the result (2.97) shows that inelastic electron scattering caused by electronic excitations is proportional to the statistical fluctuation of the electron density within the specimen.

The double-differential scattering cross-section in the first-order Born approximation can be expressed in terms of the dynamic form-factor:

$$\frac{d^2\sigma}{d\Omega\, dE} = \frac{4\gamma^2}{a_{\mathrm{H}}^2} \frac{1}{\boldsymbol{K}^4} S(\boldsymbol{K}, E/\hbar) \ . \tag{2.99}$$

Thus the inelastic part of the dynamic form factor describes inelastic scattering. Starting from this relation, we may derive a more descriptive interpretation of the mixed dynamic form-factor. The scattering cross-section only accounts for the current density scattered into a specific direction. However, the information about the relative phases and the coherence of the partial waves emanating in other directions is lost. Nevertheless, a correct description of the image intensity distribution relies entirely on this information. For simplicity we consider an object illuminated by a plane wave with wave-vector $\boldsymbol{k}_i$. In addition, we decompose the scattered wavefront emerging from the object into a sum of plane waves propagating in different directions. In the case of inelastic scattering these plane waves are partially coherent. The mixed dynamic form-factor $S = S(\boldsymbol{k}_f - \boldsymbol{k}_i, \boldsymbol{k}'_f - \boldsymbol{k}_i, \omega)$ characterizes the degree of coherence between plane waves that propagate in the directions $\boldsymbol{k}_f$ and $\boldsymbol{k}'_f$. For a single atom we can approximately calculate the mixed dynamic form-factor for inner shell excitations. To perform this task we need the matrix elements of the interaction potential (2.51) with respect to the electron eigenfunctions of the atom. It is possible to obtain these data by employing the central-field approximation for electronic eigenfunctions. A less accurate but more efficient approach is to calculate the mixed dynamic form-factor for a mean energy loss $\overline{E} = \hbar\overline{\omega}$. Although this approach cannot distinguish between different energy losses, it still describes the overall effect of inelastic scattering on the image intensity surprisingly well. This method is a generalization of the Raman–Compton approximation introduced in Sect. 2.4. Within the frame of validity of this approximation, the inelastic part of the mixed dynamic form-factor of a single atom can be expressed by the atomic form-factor $F = F(\boldsymbol{K})$ of the ground state electron density [22] as

$$S_{\mathrm{in}}(\boldsymbol{K}, \boldsymbol{K}', \omega) = \left[F(\boldsymbol{K} - \boldsymbol{K}') - \frac{F(\boldsymbol{K})F(\boldsymbol{K}')}{Z} \right] \delta(\omega - \overline{\omega}) \ . \tag{2.100}$$

This approximation is reasonable as long as the resolution limit of the electron microscope exceeds the diameter of an atom. In the special case $\boldsymbol{K} = \boldsymbol{K}'$, the relations $F(0) = Z$ and (2.99) show that the expression (2.100) is consistent with the differential inelastic scattering cross-section (2.63) for the Raman–Compton model. Since this approximation treats the atoms of the object as statistically uncorrelated, it does not correctly account for collective excitations. Nevertheless, the generalized Raman–Compton approximation is very suitable for the simulation of image formation by inelastically scattered electrons.

For the Wentzel potential (2.46) we can derive an explicit formula for the inelastic part of the mixed dynamic form-factor for a mean energy loss $\hbar\overline{\omega}$

in small angle approximation. By substituting the atomic form-factor of the Wentzel potential into (2.100) we find

$$S_{\rm in}(\boldsymbol{q},\boldsymbol{q}',\omega) = \left(\frac{2\gamma}{a_{\rm H}}\right)^2 Z \frac{1}{\boldsymbol{q}^2 + k_0^2\theta_{\rm E}^2} \frac{1}{\boldsymbol{q}'^2 + k_0^2\theta_{\rm E}^2}\,\delta(\omega - \overline{\omega})$$
$$\times \left[\frac{k_0^2\theta_0^2}{(\boldsymbol{q}-\boldsymbol{q}')^2 + \theta_0^2 k_0^2} - \frac{k_0^2\theta_0^2}{\boldsymbol{q}^2 + k_0^2\theta_0^2 + k_0^2\theta_{\rm E}^2}\frac{k_0^2\theta_0^2}{\boldsymbol{q}'^2 + k_0^2\theta_0^2 + k_0^2\theta_{\rm E}^2}\right], \qquad (2.101)$$

where $\boldsymbol{q}$ denotes the two-dimensional lateral scattering vector. For a small energy loss the relation $\theta_{\rm E} \ll \theta_0$ holds and (2.101) adopts the form

$$S_{\rm in}(\boldsymbol{q},\boldsymbol{q}',\omega) = \left(\frac{2\gamma}{a_{\rm H}}\right)^2 Z \frac{1}{\boldsymbol{q}^2 + k_0^2\theta_{\rm E}^2} \frac{1}{\boldsymbol{q}'^2 + k_0^2\theta_{\rm E}^2}\,\delta(\omega - \overline{\omega})$$
$$\times \left\{\frac{\theta_0^2}{(\boldsymbol{\theta}-\boldsymbol{\theta}')^2 + \theta_0^2} - \frac{\theta_0^4}{(\boldsymbol{\theta}^2 + \theta_0^2)(\boldsymbol{\theta}'^2 + \theta_0^2)}\right.$$
$$\left.\times \left[1 - \theta_{\rm E}^2 \frac{\boldsymbol{\theta}^2 + \boldsymbol{\theta}'^2 + 2\theta_0^2}{(\boldsymbol{\theta}^2 + \theta_0^2)(\boldsymbol{\theta}'^2 + \theta_0^2)}\right]\right\}. \qquad (2.102)$$

This result demonstrates that a small finite value of $\theta_{\rm E}$ prevents the differential scattering cross-section

$$\left.\frac{d\sigma}{d\Omega}\right|_{\boldsymbol{q}=0} = \int_{-\infty}^{\infty} S(0,0,\omega)\,d\omega = \left(\frac{2\gamma}{a_{\rm H}k_0^2}\right)^2 \frac{2Z}{\theta_{\rm E}^2\theta_0^2} \qquad (2.103)$$

from diverging in the forward direction.

2.7 Coherence Function Approach

The first-order Born approximation assumes that each layer of the object is illuminated by the undistorted initial wave. However, in the case of electron scattering this assumption proves invalid for most real specimens. The initial energy of the imaging electrons in transmission electron microscopy is well above 60 keV. The average energy loss resulting from single electronic excitations is below 1 keV. Moreover, backscattering effects can be neglected because the electrons are predominantly scattered in the forward direction. The propagation of the electron wave through a very thin object can be described with a sufficient degree of accuracy by means of the Glauber high-energy approximation [5]. The range of validity of this approximation is larger than that of the first-order Born approximation because the high-energy approximation satisfies the optical theorem. In order to include inelastic scattering, we must generalize the conventional Glauber formalism by considering objects with internal degrees of freedom [7]. By combining this approach with the multislice formalism, we obtain an explicit representation of the mutual dynamic object spectrum for a thick object in terms of the object potential. Since this formulation accounts for multiple scattering, it correctly describes the influence of elastic and inelastic scattering on the image formation.

In order to determine the coherence function, we employ the quantum--mechanical interaction representation and assume that the dynamics of the object are completely known. The interaction representation is well-suited for handling interacting systems if the properties of only one of the systems are of interest. In our case the desired information is the probability current of the transmitted electrons behind the object. This quantity is completely determined by the mutual coherence function of the scattered electrons, as we have shown in Sect. 2.5. To describe inelastic scattering and partial coherence correctly, we have to account for all internal degrees of freedom of the object. Within the frame of the interaction representation, the object state is described by a time-dependent position operator $\hat{\mathbf{R}} = \hat{\mathbf{R}}(t) = (\hat{\mathbf{R}}_1(t), \ldots, \hat{\mathbf{R}}_l(t))$. The total wavefunction in the interaction representation is given by

$$\Psi_I = \Psi_I(\boldsymbol{r}, \boldsymbol{R}, t) = \exp\frac{i}{\hbar}\,\hat{\mathbf{H}}_{\mathrm{O}} t \Psi(\boldsymbol{r}, \boldsymbol{R}, t)\ , \tag{2.104}$$

where $\hat{\mathbf{H}}_{\mathrm{O}}$ denotes the Hamiltonian (2.11) of the undisturbed object. Then the time-evolution of the position operator of the object has the form

$$\hat{\mathbf{R}}_i(t) = \exp\frac{i}{\hbar}\hat{\mathbf{H}}_{\mathrm{O}} t\, \boldsymbol{R}_i\ \exp -\frac{i}{\hbar}\hat{\mathbf{H}}_{\mathrm{O}} t\ , \qquad i = 1, \ldots, l\ . \tag{2.105}$$

By substituting the transformations (2.104) and (2.105) into the Schrödinger equation (2.1), we eventually obtain the interaction representation

$$\left(\frac{\hbar^2}{2m\gamma}\nabla^2 + i\hbar\,\partial_t\right)\Psi_I = \hat{V}\Psi_I\ , \tag{2.106}$$

where

$$\hat{V} = V(\boldsymbol{r}, \hat{\mathbf{R}}(t)) = \sum_{i=0}^{l} V_i(\boldsymbol{\rho} + z\mathbf{e}_z - \hat{\mathbf{R}}_i(t)) \tag{2.107}$$

denotes the interaction operator. The relation (2.107) is based on the assumption of a two-particle interaction between the incident electron and each constituent particle $\boldsymbol{R}_i$, $i = 0 \ldots l$, of the object.

The interaction operator is obtained by substituting the time-dependent Heisenberg operator $\hat{\mathbf{R}} = \hat{\mathbf{R}}(t)$ for the object coordinates $\boldsymbol{R}$ in the interaction potential. The resulting expression is an operator-valued function of the position $\boldsymbol{r}$ of the scattered electron and of the time t. In (2.106), the object coordinates no longer occur explicitly. The electron wavefunction interacts with the object by means of the interaction operator. To determine this interaction we must solve the differential equation (2.106) with an operator-valued right-hand side [8,9].

Since the energy of the incident particle is much higher than the interaction energy and the wavelength of the incident electron is much smaller than the range of the interaction potential V, we can solve the differential equation (2.106) with the ansatz

$$\Psi_m(\boldsymbol{r}, \boldsymbol{R}, t) = \psi_0(\boldsymbol{r}, t)\,\hat{\varphi}(\boldsymbol{r}, t)\phi_m(\boldsymbol{R}) \tag{2.108}$$

for any initial energy eigenstate ϕ_m of the object. The operator-valued function $\hat{\varphi}(\boldsymbol{r}, t)$ describes the interaction between the incident electron and the object. It acts only on the $\boldsymbol{R}$ coordinates of the object and depends on the position of the scattered electron $\boldsymbol{r}$ and on the time t. This function varies slowly within the range of the potential V and depends implicitly on the internal degrees of freedom of the object. With these assumptions the differential equation (2.106) can be replaced by its high-energy approximation

$$\left(\frac{1}{v}\partial_t + \partial_z\right)\hat{\varphi}(\boldsymbol{r}, t) = -\frac{i}{\hbar v}\,\hat{V}\,\hat{\varphi}\,, \qquad v = \frac{\hbar k_i}{\gamma m_0}\,. \tag{2.109}$$

Since this equation neglects all second-order partial derivatives of $\hat{\varphi}$ with respect to x, y, and z, it can only be used for very thin objects. Moreover, this approximation ignores the interference between partial waves originating from different lateral points within the object. The first-order equation (2.109) can be solved analytically by integration along its characteristics. Employing the method of successive approximation, the solution is expressed by the Dyson series [4]

$$\begin{aligned}\hat{\varphi}(\boldsymbol{\rho}, z, t) = 1 &- \frac{i}{\hbar v}\int_{-\infty}^{z}\hat{V}\left(z_0, t - \frac{z - z_0}{v}\right)dz_0 \\ &+ \left(\frac{i}{\hbar v}\right)^2\int_{-\infty}^{z}\hat{V}\left(z_0, t - \frac{z - z_0}{v}\right) \\ &\int_{-\infty}^{z_0}\hat{V}\left(z_1, t - \frac{z_0 - z_1}{v}\right)dz_1\,dz_0 - \ldots\,.\end{aligned} \tag{2.110}$$

The function $\hat{V}$ acts as an operator. Its values, taken at different positions z, do not commute. Nevertheless, by using the time-ordering brackets $[\ldots]_+$ with the time replaced by the z coordinate, we can formally sum this power series:

$$\begin{aligned}\hat{\varphi}(\boldsymbol{r}, t) = \Bigg[\exp\Bigg\{&-\frac{i}{\hbar v}\int_{-\infty}^{z}\sum_{i=1}^{l} \\ &\times V_i\left(\boldsymbol{\rho} + z'\boldsymbol{e}_z - \hat{\mathbf{R}}_i\left(t - \frac{z - z'}{v}\right)\right)dz'\Bigg\}\Bigg]_+\,,\end{aligned} \tag{2.111}$$

in which we have already inserted the definition of the operator-valued potential (2.107). The use of the time-ordering brackets accounts for the fact that the potential of the object changes during the passage of the scattered electron. The time retardation has the same origin because the potential at a distinct point z_0 of the trajectory must be evaluated at the time of passage $t-(z-z_0)/v$. The time dependence can be neglected for thermal diffuse scattering because the transition time of the electron is much shorter than the time of vibration of a displaced nucleus. In this case the use of the time-ordering brackets is superfluous. The time retardation can always be neglected if the extension of $\hat{V}(t,z)$ in the z-direction is sufficiently small. This important fact will help us to find an approximation for the mutual dynamic object transparency of a thin object-slice.

The expression (2.111) yields the amplitude of the scattered electron at an arbitrary point $\boldsymbol{r}$ behind the object at a time t after the scattering event. The probability $|\psi_{mn}|^2$ that the object state has been changed from the initial state ϕ_m to the final state ϕ_n is determined by the function

$$\begin{aligned}\psi_{mn}(\boldsymbol{r},t) &= \int \phi_n^*(\boldsymbol{R})\hat{\varphi}(\boldsymbol{r},t)\phi_m(\boldsymbol{R})\, d\boldsymbol{R} \\ &= \psi_0(\boldsymbol{r},t) \int \phi_n^*(\boldsymbol{R}) \left[\exp\{i\,\hat{\chi}(\boldsymbol{\rho})\}\right]_+ \\ &\quad \times \phi_m(\boldsymbol{R}) \exp\left(i\frac{E_m-E_n}{\hbar v}(z-vt)\right) d\boldsymbol{R}\,, \end{aligned} \tag{2.112}$$

where the observation point $\boldsymbol{r}$ is located outside the range of the potential V. The function $\hat{\chi}(\boldsymbol{\rho})$ denotes the operator-valued projected potential

$$\hat{\chi}(\boldsymbol{\rho}) = -\frac{1}{\hbar v}\int_{-\infty}^{\infty} \sum_{i=1}^{l} V_i\left(\boldsymbol{\rho}+z'\,\boldsymbol{e}_z-\hat{\boldsymbol{R}}_i(z'/v)\right) dz'\,. \tag{2.113}$$

In the time-dependent case this expression is only meaningful in the context of an exponential function enclosed in time-ordering brackets, as in equation (2.111).

The mutual product of the wavefunction at two distinct points $\boldsymbol{\rho}$ and $\boldsymbol{\rho}'$ at an arbitrary plane behind the object can be expressed as a superposition of all final eigenstates of the object:

$$\begin{aligned}&\int \Psi_m(\boldsymbol{\rho},z,t,\boldsymbol{R})\, \Psi_m^*(\boldsymbol{\rho}',z,t',\boldsymbol{R})\, d\boldsymbol{R} \\ &= \int \left[\sum_{n=0}^{\infty} \psi_{mn}\, \phi_n(\boldsymbol{R})\right]^* \left[\sum_{n'=0}^{\infty} \psi_{mn'}\phi_{n'}(\boldsymbol{R})\right] d\boldsymbol{R} \\ &= \psi_0^*\psi_0 \sum_{n=0}^{\infty} \exp -i\omega_{mn}\tau \iint \phi_n^*(\boldsymbol{R}) \left[\exp\{i\hat{\chi}(\boldsymbol{\rho})\}\right]_+ \phi_m(\boldsymbol{R}) \\ &\quad \times \phi_m^*(\boldsymbol{R}') \left[\exp\{-i\hat{\chi}(\boldsymbol{\rho}')\}\right]_+ \phi_n(\boldsymbol{R}')\, d\boldsymbol{R}\, d\boldsymbol{R}'\,. \end{aligned} \tag{2.114}$$

During the time of exposure, an incident electron impinges on the object in the initial states ϕ_m with the probability P_m. Since the product (2.114) depends only on the difference $\tau = t - t'$, the time average in the definition of the mutual coherence function (2.68) can be replaced by an average taken over all initial states ϕ_m of the object. Accordingly, the mutual coherence function behind a very thin object adopts the stationary form

$$\begin{aligned} \Gamma(\boldsymbol{\rho}, \boldsymbol{\rho}', \tau) = \psi_0(\boldsymbol{\rho}, z, t)\, \psi_0^*(\boldsymbol{\rho}', z, t') \sum_{m,n=0}^{\infty} P_m \exp -i\omega\tau \\ \times \int\!\!\int \phi_n^*(\boldsymbol{R}) \left[\exp\{i\hat{\chi}(\boldsymbol{\rho})\}\right]_+ \phi_m(\boldsymbol{R})\, \phi_m^*(\boldsymbol{R}') \\ \times \left[\exp\{-i\hat{\chi}(\boldsymbol{\rho}')\}\right]_+ \phi_n(\boldsymbol{R}')\, d\boldsymbol{R}\, d\boldsymbol{R}' \,. \end{aligned} \tag{2.115}$$

This relation does not depend on any specific form of the incident wave. We can therefore replace the incident plane wave ψ_0 by a somewhat distorted plane wave. Since we have neglected backscattering, we can consistently assume that the fluctuations of both the initial wave and the object potential are stochastically uncorrelated when performing the time average.

The mutual dynamic object transparency of a thin object-slice is defined to be

$$\begin{aligned} M(\boldsymbol{\rho}, \boldsymbol{\rho}', \tau) = \sum_{m,n} P_m \exp -i\omega\tau \int\!\!\int \phi_n^*(\boldsymbol{R}) \left[\exp\{i\hat{\chi}(\boldsymbol{\rho})\}\right]_+ \\ \times \phi_m(\boldsymbol{R})\, \phi_m^*(\boldsymbol{R}') \left[\exp\{-i\hat{\chi}(\boldsymbol{\rho}')\}\right]_+ \phi_n(\boldsymbol{R}')\, d\boldsymbol{R}\, d\boldsymbol{R}' \,. \end{aligned} \tag{2.116}$$

This function relates the mutual coherence function directly behind the thin slice $\Gamma^{(f)}$ to $\Gamma^{(i)}$ at the plane directly in front of the slice via the relation

$$\Gamma^{(f)}(\boldsymbol{\rho}, \boldsymbol{\rho}', \tau) = M(\boldsymbol{\rho}, \boldsymbol{\rho}', \tau)\, \Gamma^{(i)}(\boldsymbol{\rho}, \boldsymbol{\rho}', \tau) \,. \tag{2.117}$$

The sum in (2.117) can be rewritten as the time average of the product of the time-dependent operator-valued phase factors:

$$M(\boldsymbol{\rho}, \boldsymbol{\rho}', \tau) = \left\langle \left[\exp\{i\hat{\chi}(\boldsymbol{\rho}, t)\}\right]_+ \left[\exp\{-i\hat{\chi}(\boldsymbol{\rho}', t')\}\right]_+ \right\rangle_T \,. \tag{2.118}$$

The time-dependent operator-valued projected potential $\hat{\chi}$ is defined by

$$\hat{\chi}(\boldsymbol{\rho}, t) = -\frac{1}{\hbar v} \int_{-\infty}^{\infty} V\left(\boldsymbol{\rho}, z, \hat{\boldsymbol{R}}(t + z/v)\right) dz \,, \tag{2.119}$$

in accordance with (2.113). The propagation of the electron wave through a thin slice is governed by (2.118). This function depends only on the two coordinates $\boldsymbol{\rho}$ and $\boldsymbol{\rho}'$. The integration over the initial plane is unnecessary in the case of a thin slice. Nevertheless, the transmission function still depends on both the primed and unprimed lateral coordinates. This property guarantees the validity of the generalized optical theorem.

The successive application of formula (2.117) allows one to calculate the mutual coherence function for a thick object by iteration [9]. The mutual dynamic object transparency for a sufficiently thin slice of the object represents

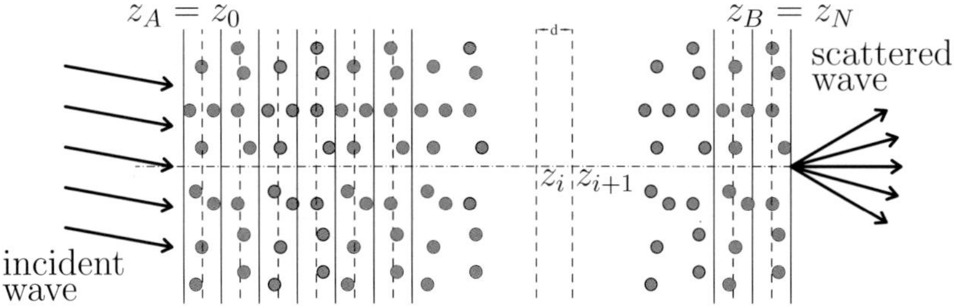

Fig. 2.5. Schematic illustration of the conventional multislice procedure. Each dot represents an atom of the object. The plane of projection is sketched for each slice by a dashed line

a generalization of the well-known transmission function used in the theory of purely elastic scattering. In this case the double sum in (2.117) reduces to a single sum, since the transition matrix is diagonal. If the object is in its ground-state, the mutual dynamic object transparency $M(\boldsymbol{\rho}, \boldsymbol{\rho}', \tau)$ reduces to the product $M_{\mathrm{el}}(\boldsymbol{\rho}, \boldsymbol{\rho}', \tau) = T(\boldsymbol{\rho})\, T^*(\boldsymbol{\rho}') \exp{(i\omega\tau)}$, where $T(\boldsymbol{\rho}) = \exp\{i\chi(\boldsymbol{\rho})\}$ is the complex transmission function for a static potential. We obtain this special case from the definition (2.117) by choosing $P_m = \delta_{m0}$, where δ_{mn} denotes the Kronecker symbol.

For thick objects the mutual dynamic object transparency behind the last object slice is a highly complicated function of the object potential owing to the multiple scattering of the electrons within the object. The conventional multislice algorithm [31–33,10] neglects inelastic scattering; the electron wave within a thin object slice is described by the elastic transmission function and the propagation of the electron wave through a thick object is solved iteratively by subdividing the object potential into a sequence of N thin slices. Each step in the iteration consists of a multiplication with the transmission function and a convolution with the free-space propagator, as illustrated in Fig. 2.5.

In order to incorporate inelastic scattering, we assume that (a) each object state ϕ_n is sufficiently well localized within a distinct slice of the object, (b) the ranges of the interaction potentials are not larger than the slice thickness and (c) the object is in thermal equilibrium [9]. In this case the wavefunction in front of each slice must be replaced by the mutual coherence function Γ, and the interaction with the object is described by a multiplication with the mixed dynamic object transparency (2.118) for each slice.

The propagation of the mutual coherence function between any two successive intermediate planes $z = z_i$ and $z = z_{i+1}$ is obtained by multiplying the mutual coherence function at the plane $z = z_i$ with the mutual dynamic object transparency M_i of the i-th slice and performing a four-dimensional

Fresnel propagation to the plane $z_{i+1} = z_i + d$. By iteration we find

$$\Gamma(\boldsymbol{\rho}_N, \boldsymbol{\rho}'_N, \tau) = \int \cdots \int \Gamma(\boldsymbol{\rho}_0, \boldsymbol{\rho}'_0, \tau) \times \prod_{i=0}^{N-1} M_i(\boldsymbol{\rho}_i, \boldsymbol{\rho}'_i, \tau)\, P_{\mathrm{F}}(\boldsymbol{\rho}_{i+1} - \boldsymbol{\rho}_i, \boldsymbol{\rho}'_{i+1} - \boldsymbol{\rho}'_i)\, d^2\boldsymbol{\rho}_i\, d^2\boldsymbol{\rho}'_i \,. \tag{2.120}$$

For the first and for the last step the free-space propagation must be performed over one half of the thickness of the first or last slice, respectively. This requirement arises from the fact that the object potential is projected onto the midplane of each slice. Hence we can omit these two propagations if we define the midplane of the first slice as the entrance plane and the corresponding plane of the last slice as the exit plane.

The proposed multislice formalism for the coherence function is a five-dimensional representation of the propagation of an electron through the object. If we insert the expression (2.120) into (2.86) and recall the definition (2.126) of the mutual dynamic object transparency, we eventually find the generalized multislice equation

$$T(\boldsymbol{\rho}_0, \boldsymbol{\rho}'_0, \boldsymbol{\rho}_N, \boldsymbol{\rho}'_N, \tau) = \int \cdots \int M_0(\boldsymbol{\rho}_0, \boldsymbol{\rho}'_0, \tau)\, P_{\mathrm{F}}(\boldsymbol{\rho}_1 - \boldsymbol{\rho}_0, \boldsymbol{\rho}'_1 - \boldsymbol{\rho}'_0) \times \prod_{i=1}^{N-1} M_i(\boldsymbol{\rho}_i, \boldsymbol{\rho}'_i, \tau)\, P_{\mathrm{F}}(\boldsymbol{\rho}_{i+1} - \boldsymbol{\rho}_i, \boldsymbol{\rho}'_{i+1} - \boldsymbol{\rho}'_i)\, d^2\boldsymbol{\rho}_i\, d^2\boldsymbol{\rho}'_i \,. \tag{2.121}$$

This relation is valid as long as backscattering can be neglected. The coordinates $\boldsymbol{\rho}_0$, $\boldsymbol{\rho}'_0$ refer to the entrance plane $z = z_{\mathrm{A}}$ and the coordinates $\boldsymbol{\rho}_N$, $\boldsymbol{\rho}'_N$ to the exit plane $z = z_{\mathrm{B}}$ of the thick object as depicted in Fig. 2.4.

The iterated integral representation of the mutual dynamic object transparency in (2.121) is formally equivalent to a Feynman path-integral representation of the propagator of the time-dependent Schrödinger equation. This correspondence becomes obvious if we recall that the high-energy approximation of the stationary Schrödinger equation has the same structure as the time-dependent Schrödinger equation, apart from its lower dimension. We can therefore use the same mathematical formalism to solve these equations. Nevertheless, there is an important difference. The Feynman path-integral approach to time-dependent quantum mechanics provides a formally exact solution of the theory, whereas the multislice method only holds true within the frame of validity of the high-energy approximation.

In the next step we have to evaluate the mutual dynamic object transparency approximately for a thin object-slice. This will provide us with analytical expressions for the absorption potential and for the terms describing inelastic and elastic scattering [7–9]. Using the fact that the projected potential can be considered as a small quantity if the object slices are sufficiently thin, we can approximate the mutual dynamic object transparency (2.118) by an expression that closely resembles the transmission function with an absorption potential $T_{\rm abs}(\boldsymbol{\rho}) = \exp(i[\chi(\rho) + i\mu_2(\boldsymbol{\rho})/2])$ [10,25]. In conventional image simulation the complex exponent in this expression is known as the optical potential. The term μ_2 accounts unrealistically for an absorption of the inelastically scattered electrons within the object. Unfortunately, this approximation violates the optical theorem of quantum mechanical scattering theory. To avoid this discrepancy we must employ the mutual dynamic object transparency. To obtain a more realistic approximation, we expand the logarithm of $M(\boldsymbol{\rho}, \boldsymbol{\rho}', \tau)$ as a power series. Using the expression (2.126) and considering only terms up to second order in $\hat{\chi}$, we find [7]

$$\begin{aligned}
\ln M(\boldsymbol{\rho}, \boldsymbol{\rho}', \tau) &\approx \ln \langle 1 + i\hat{\chi}(\boldsymbol{\rho}, t) - i\hat{\chi}(\boldsymbol{\rho}', t') + \hat{\chi}(\boldsymbol{\rho}, t)\hat{\chi}(\boldsymbol{\rho}', t') \\
&\quad - \left(\frac{1}{\hbar v}\right)^2 \int_{-\infty}^{\infty} \hat{V}(\boldsymbol{\rho}, z, t) \int_{-\infty}^{z} \hat{V}(\boldsymbol{\rho}, z', t)\, dz'\, dz \\
&\quad - \left(\frac{1}{\hbar v}\right)^2 \int_{-\infty}^{\infty} \hat{V}(\boldsymbol{\rho}', z, t') \int_{-\infty}^{z} \hat{V}(\boldsymbol{\rho}', z', t')\, dz'\, dz \Big\rangle \\
&\approx i\left[\langle\hat{\chi}(\boldsymbol{\rho}, t)\rangle - \langle\hat{\chi}(\boldsymbol{\rho}', t')\rangle\right] + \langle\hat{\chi}(\boldsymbol{\rho}, t)\hat{\chi}(\boldsymbol{\rho}', t')\rangle \\
&\quad + \frac{1}{2}\left[\langle\hat{\chi}(\boldsymbol{\rho}, t)\rangle - \langle\hat{\chi}(\boldsymbol{\rho}', t')\rangle\right]^2 \\
&\quad - \left(\frac{1}{\hbar v}\right)^2 \left\langle \int_{-\infty}^{\infty} \hat{V}(\boldsymbol{\rho}, z, t) \int_{-\infty}^{z} \hat{V}(\boldsymbol{\rho}, z', t)\, dz'\, dz \right\rangle \\
&\quad - \left(\frac{1}{\hbar v}\right)^2 \left\langle \int_{-\infty}^{\infty} \hat{V}(\boldsymbol{\rho}', z, t') \int_{-\infty}^{z} \hat{V}(\boldsymbol{\rho}', z', t')\, dz'\, dz \right\rangle \\
&= i\left[\mu_1(\boldsymbol{\rho}) - \mu_1(\boldsymbol{\rho}')\right] - \frac{1}{2}\left[\mu_2(\boldsymbol{\rho}) + \mu_2(\boldsymbol{\rho}')\right] + \mu_{11}(\boldsymbol{\rho}, \boldsymbol{\rho}', \tau) \,.
\end{aligned} \tag{2.122}$$

Here we have used the definitions

$$\mu_1(\boldsymbol{\rho}) = \langle\hat{\chi}(\boldsymbol{\rho}, t)\rangle \,, \tag{2.123}$$

$$\begin{aligned}
\mu_2(\boldsymbol{\rho}) &= \frac{2}{(\hbar v)^2} \left\langle \int_{-\infty}^{\infty} \hat{V}(\boldsymbol{\rho}, z, t) \int_{-\infty}^{z} \hat{V}(\boldsymbol{\rho}, z', t)\, dz'\, dz \right\rangle - \langle\hat{\chi}(\boldsymbol{\rho}, t)\rangle^2 \\
&\approx \langle\hat{\chi}^2(\boldsymbol{\rho}, t)\rangle - \langle\chi\rangle^2(\boldsymbol{\rho}) \,,
\end{aligned} \tag{2.124}$$

$$\mu_{11}(\boldsymbol{\rho}, \boldsymbol{\rho}', \tau) = \langle\hat{\chi}(\boldsymbol{\rho}, t)\hat{\chi}(\boldsymbol{\rho}', t')\rangle - \mu_1(\boldsymbol{\rho})\mu_1(\boldsymbol{\rho}') \,. \tag{2.125}$$

The second relation in (2.124) is obtained by considering that the slices are very thin. In this case the double integral approximately represents $\frac{1}{2}(\hbar v)^2\langle\chi^2\rangle$. With this result we find for the mutual dynamic object transparency of a thin slice the approximation

$$\begin{aligned}
M(\boldsymbol{\rho}, \boldsymbol{\rho}', \tau) \approx \exp\Big(& i\left[\mu_1(\boldsymbol{\rho}) - \mu_1(\boldsymbol{\rho}')\right] \\
& - \frac{1}{2}\left[\mu_2(\boldsymbol{\rho}) + \mu_2(\boldsymbol{\rho}')\right] + \mu_{11}(\boldsymbol{\rho}, \boldsymbol{\rho}', \tau)\Big) \,.
\end{aligned} \tag{2.126}$$

Elastic scattering is described by the first two terms of the exponent. Each of these terms represents a phase shift of the electron wave, which is proportional to the projected potential of the slice. The terms $\mu_2(\boldsymbol{\rho})$ and $\mu_2(\boldsymbol{\rho}')$ represent the so called absorption potential. The term $\mu_{11}(\boldsymbol{\rho}, \boldsymbol{\rho}', \tau)$ describes the contribution of the inelastically scattered electrons to the image intensity. This term is a direct consequence of the optical theorem which ensures the conservation of the number of particles:

$$M(\boldsymbol{\rho}, \boldsymbol{\rho}' = \boldsymbol{\rho}, \tau = 0) = 1 \ , \quad \mu_2(\boldsymbol{\rho}) = \mu_{11}(\boldsymbol{\rho}, \boldsymbol{\rho}' = \boldsymbol{\rho}, \tau = 0) \ . \tag{2.127}$$

We can restate the last relation in Fourier space. Following this line we obtain

$$\tilde{\mu}_2(\boldsymbol{q}) = \frac{1}{2\pi} \int \int \tilde{\mu}_{11}(\boldsymbol{q}+\boldsymbol{q}', \boldsymbol{q}', \omega) \, d^2\boldsymbol{q}' \, d\omega \ . \tag{2.128}$$

This expression is very useful for calculating the absorption potential for an interaction process described by $\tilde{\mu}_{11} = \tilde{\mu}_{11}(\boldsymbol{q}, \boldsymbol{q}', \omega)$. Conversely, for a pair of functions $\tilde{\mu}_2(\boldsymbol{q})$ and $\tilde{\mu}_{11}(\boldsymbol{q}, \boldsymbol{q}', \omega)$ that satisfy the relation (2.128), the validity of the optical theorem (2.127) is guaranteed.

The mutual dynamic object transparency (2.126) factorizes if $\mu_{11} = 0$. Each of the two factors represents the transmission function with an absorption potential for the standard multislice algorithm. The coherence function approach, therefore, is a true generalization of the conventional multislice theory.

For purely elastic scattering, the transmission function for each slice of the object depends only on the static projected object potential of the particular slice. A different situation arises when inelastic scattering is considered. In this case the object potential cannot be assumed to be static, owing to the excitation of its internal degrees of freedom. The information contained in the time-averaged generalized dynamic object potential (2.123) does not suffice to calculate the spatial distribution of the inelastically scattered electrons in the image plane. To account for the complete inelastic intensity distribution, we have introduced the time-dependent projected object potential through the interaction representation. This function provides the required dynamical information. Inelastic scattering contributes to the quadratic terms of the mutual dynamic object spectrum for thin specimens, as shown in Sect. 2.6. In this case we can describe the influence of inelastic scattering by the variance $\mu_2 = \mu_2(\boldsymbol{\rho})$ and the covariance $\mu_{11}(\boldsymbol{\rho}, \boldsymbol{\rho}', \tau)$ of the dynamic object potential. This result allows a somewhat more intuitive interpretation of the terms in (2.126). We may consider the dynamic object potential as a stochastic process in time. The effect of the inelastic scattering processes appears as the result of fast microscopic fluctuations of the object state. The magnitude of the contribution of inelastic scattering to the mutual coherence function for a very thin slice is, therefore, proportional to the covariance of the fluctuations of the dynamic object potential. We can verify this conclusion by expanding the exponential function in (2.126) as a Taylor series and considering the fact that both μ_1^2 and μ_2 are of the same order of magnitude as μ_{11}. The equations (2.120) and (2.126) reduce the problem of image formation to the calculation of the first and second stochastic moments of the time-dependent scattering potential of each slice of the object. With this information about a given object, we can employ the coherence function multislice procedure (2.120) to determine the mutual coherence function at the exit plane behind the object, if this function is known at the entrance plane in front of the object [9].

Unfortunately, the numerical evaluation of the five-dimensional coherence function multislice equation is not possible with present computers without further approximations. We show in the following section that the labour of calculating the coherence function can be drastically reduced by decomposing the mutual coherence function into a sum of symmetric products.

2.8 Generalized Multislice Method

The generalized multislice formula (2.120) describes the propagation of the mutual coherence function $\Gamma(\boldsymbol{\rho}, \boldsymbol{\rho}', \tau; z)$ through the object. It correctly accounts for elastic and inelastic plural scattering processes within the object. In most situations of practical importance, it suffices to consider plural elastic scattering together with single inelastic scattering. This approximation holds even for moderately thick specimens because partial waves originating from different inelastic scattering events are incoherent with each other and, consequently, do not interfere. Hence only elastic multiple scattering generates coherent dynamical scattering effects. Neglecting inelastic multiple scattering, the image intensity resulting from single inelastic scattering

$$I_{\text{in}}^{(1)}(\boldsymbol{\rho}) = \sum_{j=1}^{N} I_j(\boldsymbol{\rho}) \tag{2.129}$$

is composed of the partial intensities I_j, contributed by the individual slices $j = 1, \ldots, N$. If the object thickness is comparable with the inelastic mean free path-length, we must consider the contributions of multiple inelastic scattering to the sum on the right-hand side of (2.129). The second-order term

$$I_{\text{in}}^{(2)}(\boldsymbol{\rho}) = \sum_{j=1}^{N} \sum_{i=1}^{j-1} I_{ij}(\boldsymbol{\rho}) \tag{2.130}$$

accounts for inelastic double-scattering. In this expression the term I_{ij} denotes the part of the image intensity that is produced by inelastic scattering in the ith slice and subsequent inelastic scattering in slice $j > i$. Unfortunately, numerical evaluation of the higher-order terms requires a heavy computational expenditure. Since the partial waves resulting from different inelastic scattering processes cannot interfere with each other, we can express the higher-order contributions approximately by means of the single scattering intensities. This simple approximation describes the inelastic scattering cross-section correctly, even in the case of thick objects. Because the generalized multislice formalism is compatible with the optical theorem, no intensity is lost during the propagation of the mutual coherence function through the entire object. Within the frame of validity of the single inelastic scattering approximation, the sum of the total inelastically scattered intensity $\overline{I}_{\rm in}$ and the elastically scattered intensity $\overline{I}_{\rm el}$ is equal to the initial intensity $\overline{I}_0 = \overline{I}_{\rm el} + \overline{I}_{\rm in}$. The absorption term μ_2 in the exponent of the mutual dynamic object transparency for a single slice in the generalized multislice formalism generates an unphysical loss of intensity if we apply the single inelastic scattering approximation to objects with a thickness exceeding one half of the inelastic mean free path-length. If we do not need to distinguish between single and multiple inelastic scattering, we can use the relation

$$I \approx I_{\rm el} + \frac{\overline{I}_0 - \overline{I}_{\rm el}}{\overline{I}_{\rm in}^{(1)}} I_{\rm in}^{(1)} \tag{2.131}$$

to correct for this effect. For most crystalline specimens studied by high-resolution imaging this correction is not required since such objects are sufficiently thin.

The mutual coherence function depends on the spatial coordinates $\boldsymbol{\rho}$, $\boldsymbol{\rho}'$, and the time increment τ. This parameter distinguishes between the different energy losses within the object. Owing to the chromatic aberration of the objective lens, partial waves belonging to different energy losses are transferred differently by the microscope. It is therefore advantageous to discuss the generalized multislice formalism in terms of the spectral density of the mutual coherence function:

$$\Gamma(\boldsymbol{\rho}, \boldsymbol{\rho}', \omega; z) = \frac{1}{2\pi} \int \Gamma(\boldsymbol{\rho}, \boldsymbol{\rho}', \tau; z) \exp(i\omega\tau) \, d\tau \, . \tag{2.132}$$

If we assume a discrete set of possible excitations of the object with excitation energies ω_j, $j = 0, 1, \ldots$, the Fourier transform of the mutual dynamic object transparency with respect to τ is given by

$$\int M(\boldsymbol{\rho}, \boldsymbol{\rho}', \tau) \, \mathrm{e}^{i\omega\tau} \, d\tau = 2\pi \sum_j M(\boldsymbol{\rho}, \boldsymbol{\rho}', \omega_j) \, \delta(\omega + \omega_j) \, . \tag{2.133}$$

In this case the transmission of the mutual coherence function through a single slice adopts the form

$$\Gamma_f(\boldsymbol{\rho}, \boldsymbol{\rho}', \omega) = \sum_j M(\boldsymbol{\rho}, \boldsymbol{\rho}', \omega_j) \, \Gamma_i(\boldsymbol{\rho}, \boldsymbol{\rho}', \omega + \omega_j) \, . \tag{2.134}$$

This result has been obtained by inserting the approximation (2.133) into the Fourier transform of (2.117). Equation (2.134) shows that the image intensities belonging to different energy losses can be calculated separately and added up subsequently. Nevertheless, partial waves belonging to different energy losses must be propagated separately through the optical system up to the recording plane, where the incoherent superposition is performed [22,9]. This requirement is imposed by the chromatic aberration, which causes an energy-dependent transfer of the spatial frequencies. Considering the relation (2.134) and assuming single inelastic scattering, we derive from the Fourier transform of formula (2.120) the following expression for the mutual spectral density at the exit plane $z = z_{\mathrm{B}}$:

$$\Gamma(\boldsymbol{\rho}, \boldsymbol{\rho}', \omega; z_{\mathrm{B}}) = \sum_j \sum_{k=0}^{N-1} \int \cdots \int \Gamma(\boldsymbol{\rho}_0, \boldsymbol{\rho}'_0, \omega + \omega_j; z_{\mathrm{A}}) \tag{2.135}$$

$$\times \prod_{i=0}^{N-1} M_i(\boldsymbol{\rho}_i, \boldsymbol{\rho}'_i, \omega = \delta_{ik}\omega_j)\, P_{\mathrm{F}}(\boldsymbol{\rho}_{i+1} - \boldsymbol{\rho}_i, \boldsymbol{\rho}'_{i+1} - \boldsymbol{\rho}'_i)\, d^2\boldsymbol{\rho}_i\, d^2\boldsymbol{\rho}'_i\,.$$

In this representation we have assumed that the spectral density of the mutual coherence function at the entrance plane $z = z_{\mathrm{A}}$ of the object is known. If we neglect the chromatic aberration of the illumination system, the mutual coherence function in front of the object factorizes with respect to the spatial coordinates and the energy:

$$\Gamma(\boldsymbol{\rho}, \boldsymbol{\rho}', \omega; z_{\mathrm{A}}) = \Gamma(\boldsymbol{\rho}, \boldsymbol{\rho}'; z_{\mathrm{A}})\, p(\omega)\,. \tag{2.136}$$

The energy distribution $p(\omega)$ is determined by the electron source. In the case of a point-like effective source the spatial term $\Gamma(\boldsymbol{\rho}, \boldsymbol{\rho}'; z_{\mathrm{A}})$ also factorizes into two factors, one factor depending exclusively on the primed and the other solely on the unprimed coordinates. This situation is approximately realized for highly coherent illumination. In the general case of partially coherent illumination, the initial mutual coherence function does not factorize with respect to any of the variables. We encounter this situation in the scanning transmission microscope, where the chromatic and geometric aberrations of the probe forming lens and the finite size and energy width of the electron source cannot be neglected.

The sum over the index j in (2.136) comprises all possible energy losses. For purely elastic scattering the mutual dynamic object transparency M_i of slice i collapses to $M_i(\boldsymbol{\rho}, \boldsymbol{\rho}', \omega = 0) = T(\boldsymbol{\rho})\, T^*(\boldsymbol{\rho}')$. To include inelastic scattering we first consider a single excitation with energy loss $\hbar\omega \approx \hbar\omega_{\mathrm{ex}}$. This assumption does not restrict the general validity of our approach.

In order to compute efficiently the convolution of the four-dimensional free-space propagator with the mutual dynamic object transparency in (2.136) we appeal to the Fourier convolution theorem. For this purpose, we introduce the Fourier transform of the four-dimensional Fresnel propagator

$$\begin{aligned} \mathcal{F}\mathcal{F}'\ [P_{\mathrm{F}}(\boldsymbol{\rho}, \boldsymbol{\rho}')] &= \int\!\!\int P_{\mathrm{F}}(\boldsymbol{\rho})\, P_{\mathrm{F}}^*(\boldsymbol{\rho}')\, \exp(i\boldsymbol{\rho}\cdot\boldsymbol{q})\, \exp(-i\boldsymbol{\rho}'\cdot\boldsymbol{q}')\, d^2\boldsymbol{\rho}\, d^2\boldsymbol{\rho}' \\ &= \tilde{P}_{\mathrm{F}}(\boldsymbol{q})\, \tilde{P}_{\mathrm{F}}^*(\boldsymbol{q}')\,. \end{aligned} \tag{2.137}$$

Here $\mathcal{F}$ and $\mathcal{F}'$ denote the two-dimensional Fourier transformation with respect to the primed and the unprimed coordinates. To allow for the hermitian structure of the coherence function formalism, the primed Fourier transform is defined as the complex-conjugate of the unprimed transform. In the case of elastic scattering and coherent illumination, the mutual intensity $\Gamma(\boldsymbol{\rho}, \boldsymbol{\rho}', z_0) = \psi(\boldsymbol{\rho}, z_0)\psi^*(\boldsymbol{\rho}', z_0)$ factorizes into a bilinear, hermitian product of the stationary wavefunction ψ evaluated at the lateral positions $\boldsymbol{\rho}$ and $\boldsymbol{\rho}'$, respectively. In the general case the expression for the mutual intensity in front of the object can be expanded into a series of products

$$\Gamma(\boldsymbol{\rho}, \boldsymbol{\rho}'; z_{\rm A}) = \sum_l \Phi_l(\boldsymbol{\rho}; z_{\rm A})\Phi_l^*(\boldsymbol{\rho}'; z_{\rm A}) , \tag{2.138}$$

where each factor depends only on the primed or on the unprimed coordinates. The expansion functions Φ_l do not necessarily have a direct physical meaning. Only in the special case of an incident axial plane wave $\psi_0(z, t)$ does the sum in (2.138) reduce to a single term $l = 0$ with $\Phi_0(\boldsymbol{\rho}; z_0) = \psi_0(z_0, t = 0)$. For partially coherent Köhler illumination [19] it is convenient to decompose the initial mutual coherence function (2.138) into a product of two sums. The sum weighted over a discrete set of energies represents the energy spread of the source. The other sum over the directions of the incident plane waves accounts for the incoherently radiating points of the effective source. In the case of Köhler illumination each point of the effective source corresponds to a distinctly tilted plane wave in the front plane of the object.

Without loss of generality we restrict our further investigations to a single term of the sum (2.138). In this case the mutual spectral density at the plane $z = z_{\rm A} = z_0$ in front of the object has the form $\Gamma(\boldsymbol{\rho}, \boldsymbol{\rho}', \omega; z_{\rm A}) = \Phi(\boldsymbol{\rho}, \omega; z_{\rm A})\Phi^*(\boldsymbol{\rho}', \omega; z_{\rm A})$ and the mutual spectral density at the exit plane $z_{\rm B} = z_N = z_0 + N\,d$ is thus given by

$$\begin{aligned}\Gamma(\boldsymbol{\rho}_N, \boldsymbol{\rho}'_N, \omega; z_{\rm B}) &= \sum_{j=0}^{N-1} \int \cdots \int \Phi(\boldsymbol{\rho}_0, \omega + \omega_{\rm ex}; z_{\rm A})\, \Phi^*(\boldsymbol{\rho}'_0, \omega + \omega_{\rm ex}; z_{\rm A}) \\ &\times \prod_{i=0}^{N-1} P_{\rm F}(\boldsymbol{\rho}_{i+1} - \boldsymbol{\rho}_i, \boldsymbol{\rho}'_{i+1} - \boldsymbol{\rho}'_i; d)\; M_i(\boldsymbol{\rho}_i, \boldsymbol{\rho}'_i, \omega = \delta_{ji}\omega_{\rm ex})\; d^2\boldsymbol{\rho}_i\, d^2\boldsymbol{\rho}'_i \, .\end{aligned} \tag{2.139}$$

Although we start from the entrance plane with a product function, the coherence function at the exit plane does not factorize because the transparency $M_j(\boldsymbol{\rho}, \boldsymbol{\rho}', \omega)$, which describes the inelastic interaction in the $j^{\rm th}$ slice, depends on both the primed and unprimed position vectors.

Since the mutual dynamic object transparency of the j-th slice is a hermitian function $M_j(\boldsymbol{\rho}, \boldsymbol{\rho}', \omega) = M_j^*(\boldsymbol{\rho}', \boldsymbol{\rho}, \omega)$, it can be decomposed into a series of hermitian products

$$M_j(\boldsymbol{\rho}, \boldsymbol{\rho}', \omega_{\rm ex}) = \sum_{m=0}^{\infty} T_j^{(m)}(\boldsymbol{\rho}, \omega)\; T_j^{(m)\,*}(\boldsymbol{\rho}', \omega) , \tag{2.140}$$

as can the mutual coherence function. The individual modified transmission functions $T_j^{(m)}$, $m = 0, 1, \ldots$ have a physical meaning only if we neglect the mixed term $\mu_{11}(\boldsymbol{\rho}, \boldsymbol{\rho}', \tau)$ in the definition of the mutual dynamic object transparency (2.126). In this case the sum in (2.140) reduces to the first term $m = 0$. The remaining factor $T_i^{(0)}$ represents the conventional elastic transmission function with an absorption potential:

$$T^{(\mathrm{abs})}(\boldsymbol{\rho}) = \exp\left(i\mu_1(\boldsymbol{\rho}) - \frac{1}{2}\mu_2(\boldsymbol{\rho})\right) . \tag{2.141}$$

For practical computations we have developed a rapidly converging decomposition (2.140), which requires only a few terms of this sum.

If we insert the expansion (2.140) into the generalized multislice formula (2.139), use the factorization of the incident mutual intensity, and exchange the order of summation and integration, we find that the propagation of each term in the expansion of the final mutual intensity can be computed by means of a conventional two-dimensional multislice formalism. Since the four-dimensional Fresnel propagator $P_{\mathrm{F}}(\boldsymbol{\rho}, \boldsymbol{\rho}') = P_{\mathrm{F}}(\boldsymbol{\rho})\, P_{\mathrm{F}}^*(\boldsymbol{\rho}')$ decomposes into a product of two two-dimensional Fresnel propagators, the propagation of the function $\Phi(\boldsymbol{\rho}; z)$ through the object is governed by the modified transmission functions $T_i^{(m)}$. At the plane $z = z_{\mathrm{B}} = z_N$ behind the last slice we find

$$\Phi^{(m,j)}(\boldsymbol{\rho}_N; z_N) = \int \ldots \int \Phi(\boldsymbol{\rho}_0; z_0) \prod_{i=0}^{N-1} P_{\mathrm{F}}(\boldsymbol{\rho}_{i+1} - \boldsymbol{\rho}_i) \tag{2.142}$$
$$\times\, T_i^{(m)}(\boldsymbol{\rho}_i, \omega = \delta_{ji}\omega_{\mathrm{ex}})\, d^2\boldsymbol{\rho}_i .$$

This formulation closely resembles that of the conventional multislice formula [31,32]. By substituting the stationary wavefunction for $\Phi^{(m)}$ and the conventional transmission function for $T_i^{(m)}$, we readily obtain the well-known multislice formula of elastic scattering. In the case of a single energy loss ω_{ex} the mutual intensity at the plane $z = z_{\mathrm{B}} = z_N$ is given by

$$\Gamma(\boldsymbol{\rho}, \boldsymbol{\rho}'; z_N) = \sum_{j=0}^{N-1} \sum_{m=0}^{\infty} \Phi^{(m,j)}(\boldsymbol{\rho}; z_N)\Phi^{(m,j)\,*}(\boldsymbol{\rho}'; z_N) . \tag{2.143}$$

With this result it is possible to reduce the five-dimensional generalized multislice formalism to an iterative three-dimensional procedure. This crucial reduction renders practical the numerical evaluation of the generalized multislice formula for a realistic object with present-day computers.

To find a sufficiently rapidly converging representation of the type (2.140) in terms of partial transmission functions, we start from the five-dimensional mutual dynamic object transparency of a single slice:

$$M(\boldsymbol{\rho}, \boldsymbol{\rho}', \tau) = \exp\left(\sum_{j=1}^{M} \left\{ i \left[\mu_1^{(j)}(\boldsymbol{\rho} - \boldsymbol{\rho}_j) - \mu_1^{(j)}(\boldsymbol{\rho}' - \boldsymbol{\rho}'_j) \right] \right.\right.$$
$$- \frac{1}{2} \left[\mu_2^{(j)}(\boldsymbol{\rho} - \boldsymbol{\rho}_j) + \mu_2^{(j)}(\boldsymbol{\rho}' - \boldsymbol{\rho}_j) \right]$$
$$\left.\left. + \mu_{11}^{(j)}(\boldsymbol{\rho} - \boldsymbol{\rho}_j, \boldsymbol{\rho}' - \boldsymbol{\rho}_j, \tau) \right\} \right) . \tag{2.144}$$

The atoms are situated at the lateral positions $\boldsymbol{\rho}_j$, $j = 1, \ldots, M$. This formula follows from equation (2.126) with the assumption that the temporal fluctuations of the projected potentials of different atoms are stochastically uncorrelated $\langle \chi_i \, \chi'_j \rangle = \langle \chi_i \rangle \, \langle \chi'_j \rangle$ for $i \neq j$. The functions $\mu_2^{(j)}$ and $\mu_{11}^{(j)}$ depend on the atomic number of the atom located at the position $\boldsymbol{\rho}_j$. This assumption considerably simplifies the single-slice transparency function. In principle, the formalism can be extended to account for the most dominant nearest-neighbour correlations. However, the evaluation requires a simple model of these correlations.

The terms $\mu_2^{(j)}$ and $\mu_{11}^{(j)}$ in the exponent of expression (2.144) are small compared to unity. In order to obtain the decomposition (2.140), we expand the exponential function with respect to this quantity μ_{11}. Retaining only the first-order terms, we derive

$$M(\boldsymbol{\rho}, \boldsymbol{\rho}', \tau) = \exp\left(\sum_{j} i \left[\mu_1^{(j)}(\boldsymbol{\rho} - \boldsymbol{\rho}_j) - \mu_1^{(j)}(\boldsymbol{\rho}' - \boldsymbol{\rho}'_j) \right] \right.$$
$$\left. - \frac{1}{2} \left[\mu_2^{(j)}(\boldsymbol{\rho} - \boldsymbol{\rho}_j) + \mu_2^{(j)}(\boldsymbol{\rho}' - \boldsymbol{\rho}_j) \right] \right)$$
$$+ \sum_j \mu_{11}^{(j)}(\boldsymbol{\rho} - \boldsymbol{\rho}_j, \boldsymbol{\rho}' - \boldsymbol{\rho}_j, \tau) \, \exp\left(\sum_j i\{\langle \chi_j \rangle - \langle \chi'_j \rangle\} \right)$$
$$= T^{(\mathrm{abs})}(\boldsymbol{\rho}) \, T^{(\mathrm{abs})\,*}(\boldsymbol{\rho}')$$
$$+ \sum_j \mu_{11}^{(j)}(\boldsymbol{\rho} - \boldsymbol{\rho}_j, \boldsymbol{\rho}' - \boldsymbol{\rho}_j, \tau) \, T^{(\mathrm{abs})}(\boldsymbol{\rho}) \, T^{(\mathrm{abs})\,*}(\boldsymbol{\rho}') \, . \tag{2.145}$$

To simplify the previous expression we have used the definition of the conventional transmission function (2.141). This choice of the transmission function is somewhat arbitrary because we have performed the expansion of the exponential only with respect to μ_{11} but not with respect to μ_2. Since both quantities have the same order of magnitude, the expansion is not consistent with respect to the order of the expansion parameters. However, the approximation (2.145) simplifies the decomposition of the mutual transparency, as we will show later. The approximation (2.145) violates the optical theorem with an error of fourth-order in the projected potential. This fourth-order error vanishes for the improved approximation

$$M(\boldsymbol{\rho},\boldsymbol{\rho}',\tau) = \exp\left(i\left[\mu_1(\boldsymbol{\rho}) - \mu_1(\boldsymbol{\rho}')\right]\right)\left[\exp\left(-\frac{1}{2}\left[\mu_2(\boldsymbol{\rho}) + \mu_2(\boldsymbol{\rho}')\right]\right)\right.$$
$$\left. + \mu_{11}(\boldsymbol{\rho},\boldsymbol{\rho}')\exp\left(-\frac{1}{4}\left[\mu_2(\boldsymbol{\rho}) + \mu_2(\boldsymbol{\rho}')\right]\right)\right] . \qquad (2.146)$$

The first term in each of the decompositions (2.145) and (2.146) represents a product of two transmission functions with an absorption potential, each factor depending exclusively on the primed or on the unprimed coordinates. The second term on the right-hand side of (2.145) does not decompose in this simple manner because $\mu_{11}(\boldsymbol{\rho},\boldsymbol{\rho}',\tau)$ depends on both the primed and the unprimed coordinates. By employing the Fourier convolution theorem, we can conveniently calculate the spatially shifted functions for each atom

$$\mu_{11}^{(j)}(\boldsymbol{\rho} - \boldsymbol{\rho}_j, \boldsymbol{\rho}' - \boldsymbol{\rho}_j, \tau)$$
$$= \mathcal{F}^{-1}\mathcal{F}'^{-1}\left[\tilde{\mu}_{11}^{(j)}(\boldsymbol{q},\boldsymbol{q}',\tau)\,\mathcal{F}\mathcal{F}'\left[\delta(\boldsymbol{\rho} - \boldsymbol{\rho}_j)\,\delta(\boldsymbol{\rho}' - \boldsymbol{\rho}_j)\right]\right] , \qquad (2.147)$$

where $\tilde{\mu}_{11}^{(j)}(\boldsymbol{q},\boldsymbol{q}',\tau)$ denotes the Fourier transform of $\mu_{11}^{(j)}(\boldsymbol{\rho},\boldsymbol{\rho}',\tau)$ with respect to $\boldsymbol{\rho}$ and $\boldsymbol{\rho}'$. It is then only necessary to find a decomposition of $\tilde{\mu}_{11}^{(j)}$ in Fourier space.

In the following we derive such representations for the special cases of thermal diffuse scattering and inelastic scattering caused by electronic excitations. For simplicity we employ the Einstein model for thermal diffuse scattering [15] and the Raman–Compton approximation [27,20,22] for inelastic scattering. These simple models are well-suited to demonstrate the feasibility of the generalized multislice method because the numerical computations are not unduly laborious.

Thermal diffuse scattering is described with a sufficient degree of accuracy by the Einstein model because the resulting image is only affected by the time average of the lattice dynamics over the time of exposure. For our purpose it suffices to consider the individual atoms of the object as independent two-dimensional harmonic oscillators. The mean square elongation $u_j^2/2$ depends on the atomic number of the j-th atom. The time average of the oscillating projected object potential has the form

$$\langle\chi\rangle(\boldsymbol{\rho}) = \frac{2\pi}{k}\,\mathcal{F}^{-1}\left[f_{\mathrm{el}}(\boldsymbol{q})\,\exp\left(-\frac{u^2}{4}\boldsymbol{q}^2\right)\right] , \qquad (2.148)$$

$$\langle\chi\chi'\rangle(\boldsymbol{\rho},\boldsymbol{\rho}') = \left(\frac{2\pi}{k}\right)^2\mathcal{F}^{-1}\mathcal{F}'^{-1}\left[f_{\mathrm{el}}(\boldsymbol{q})\,\exp\left(-\frac{u^2}{4}(\boldsymbol{q} - \boldsymbol{q}')^2\right)\,f_{\mathrm{el}}(\boldsymbol{q}')\right] .$$

Here $f_{\mathrm{el}}(\boldsymbol{q})$ denotes the elastic scattering amplitude of a single atom in the first-order Born approximation. To retain the analytical character of the equations (2.149), it is advantageous to use the Kohl–Weickenmeier approximation (2.61) for the elastic scattering amplitudes.

The three-dimensional Fourier transform of the object potential is proportional to the first-order Born approximation of the elastic scattering amplitude. The Fourier projection theorem connects the two-dimensional Fourier transform of the projected object potential with the scattering amplitude:

$$\chi(\boldsymbol{q}) = \int \exp(-i\boldsymbol{q}\cdot\boldsymbol{\rho}) \int \phi(\boldsymbol{\rho}, z)\, dz\, d^2\boldsymbol{\rho} = f^{\mathrm{B}}(\boldsymbol{q})\,. \tag{2.149}$$

The relations (2.149) should therefore not be regarded as first-order Born approximations.

The mixed exponential term in the second expression of (2.149) can be written as a sum of products with the aid of the generating function of the modified Bessel function I_n [34]:

$$\begin{aligned} \exp\left(2(\frac{u}{2})^2\mathbf{q}\cdot\mathbf{q}'\right) &= \sum_{|n|} I_n\left(2\left(\frac{u}{2}\right)^2 qq'\right)\exp\left(in(\phi-\phi')\right) \\ &= \sum_{|n|,k} \frac{[(u/2)^2qq']^{n+2k}}{k!(k+|n|)!}\exp\left(in(\phi-\phi')\right)\,. \end{aligned} \tag{2.150}$$

Inserting this expansion into (2.149) yields the decomposition

$$\begin{aligned} \langle\chi\chi'\rangle(\boldsymbol{q},\boldsymbol{q}') &= \left(\frac{2\pi}{k}\right)^2 \\ &\times \sum_{|n|,k} f_{\mathrm{el}}(\boldsymbol{q})\frac{\exp\left(-(u/2)^2q^2\right)(uq/2)^{|n|+2k}}{\sqrt{k!(k+|n|)!}}\exp\left(in\phi\right)\{q',\phi'\}^*\,, \end{aligned} \tag{2.151}$$

where $\{q',\phi'\}^*$ denotes the complex-conjugate of the first factor with the unprimed coordinates replaced by the primed ones. In most cases only a few terms of the sum in (2.152) need to be taken into account because the product expansion converges sufficiently fast [9].

Inelastic scattering results in electronic excitations of the atoms. To describe this effect, we use the modified Raman–Compton approximation for the inelastic part $S_j^{(\mathrm{in})}(\boldsymbol{K},\boldsymbol{K}',\omega)$ of the mixed dynamic form-factor $S_j(\boldsymbol{K},\boldsymbol{K}',\omega)$ of the j-th atom with atomic number Z_j. We have already discussed this approximation in the Sects. 2.4 and 2.6. Assuming an average excitation energy $\hbar\overline{\omega}$, we obtain

$$S_j^{(\mathrm{in})}(\boldsymbol{K},\boldsymbol{K}',\omega) = \delta(\omega-\overline{\omega})\left\{F_j(\boldsymbol{K}-\boldsymbol{K}') - \frac{F_j(\boldsymbol{K})F_j(\boldsymbol{K}')}{Z_j}\right\}\,, \tag{2.152}$$

where $\boldsymbol{K}$, $\boldsymbol{K}'$ denote the three-dimensional scattering vectors.

The x-ray scattering amplitude F_j in (2.152) depends on the electronic charge distribution of the j-th atom. In order to perform several integrations analytically, it is advantageous to use the Doyle-Turner Gaussian fit 5

$$F(\boldsymbol{K}) = \sum_{i=0}^{4} a_i\, \exp\left(-b_i\,\boldsymbol{K}^2\right) \tag{2.153}$$

for the atomic form factor F. For most elements the constants a_i, b_i, $i = 0, 1, 2, 3, 4$ are tabulated [24,10].

The factorization of the analytic expression for $\mu_{11}^{(\mathrm{in},j)}$ follows from the representation

$$\mu_{11}^{(\mathrm{in},j)}(\boldsymbol{\rho}, \boldsymbol{\rho}', \tau) = \left(\frac{4\pi\alpha}{\beta}\right)^2 \times \mathcal{F}^{-1}\mathcal{F}'^{-1}\left[\int \frac{S_j^{(\mathrm{in})}(\boldsymbol{q} - k\theta_{\mathrm{E}}\,\boldsymbol{e}_z, \boldsymbol{q}' - k\theta_{\mathrm{E}}\,\boldsymbol{e}_z, \omega)}{(\boldsymbol{q}^2 + k^2\theta_{\mathrm{E}}^2)(\boldsymbol{q}'^2 + k^2\theta_{\mathrm{E}}^2)} \exp(-i\omega\tau)\,d\omega\right], \tag{2.154}$$

where $\alpha \simeq 1/137$ and $\beta = v/c$ are the fine-structure constant and the relative velocity of the incident electron, respectively. Inserting the approximation (2.152) into (2.155), we eventually obtain

$$\tilde{\mu}_{11}^{(j)}(\boldsymbol{q}, \boldsymbol{q}', \tau) = \left(\frac{4\pi\,\alpha}{\beta}\right)^2 \times \frac{1}{\boldsymbol{K}^2\boldsymbol{K}'^2}\left[F_j(\boldsymbol{K} - \boldsymbol{K}') - \frac{1}{Z_j}F_j(\boldsymbol{K})F_j(\boldsymbol{K}')\right] \exp(-i\omega_{\mathrm{ex}}\tau). \tag{2.155}$$

To factorize this expression, we only need to decompose the first term in the brackets on the right-hand side. Employing the formula (2.150), we find

$$\begin{aligned}\frac{F(\boldsymbol{K} - \boldsymbol{K}')}{\boldsymbol{K}^2\boldsymbol{K}'^2} &= \frac{F(\boldsymbol{q} - \boldsymbol{q}')}{(\boldsymbol{q}^2 + k\theta_{\mathrm{E}}^2)(\boldsymbol{q}'^2 + k^2\theta_{\mathrm{E}}^2)} \\ &= \sum_{i,|n|,k}\left[\frac{\sqrt{a_i}\exp\left(-b_i q^2\right)}{(\boldsymbol{q}^2 + k^2\theta_{\mathrm{E}}^2)}\frac{(\sqrt{b_i}q)^{|n|+2k}}{\sqrt{k!(k+|n|)!}}\exp(in\phi)\right]\{\boldsymbol{q}', \phi'\}^*,\end{aligned} \tag{2.156}$$

in close analogy with the phonon case [9].

The result (2.157) has been obtained with the assumption of an average energy loss $\hbar\overline{\omega}$ for all atomic excitations. This excitation energy depends on the atomic number of the atom. To derive a more accurate approximation, we must consider that the electron energy-loss spectrum (EELS) of each atom is a continuous function of the excitation energy. In a first approximation we describe this spectrum by a delta function $\delta(\omega - \omega_{\mathrm{ex}})$, where the specific excitation value $\omega_{\mathrm{ex}} = \overline{\omega}$ is chosen to correspond to the average energy loss of the scatterer. The parameter ω_{ex} enters the expression (2.155) through the definition of the characteristic scattering angle θ_{E}. The smaller the mean energy-loss, the more delocalized is the scattering process. Therefore, we can abandon the identifaction of ω_{ex} with $\overline{\omega}$ and reinterpret ω_{ex} as a formal parameter measuring the degree of localization of the scattering process.

This interpretation allows us to obtain an improved approximation for the term $\tilde{\mu}_{11}$ of each atom. For this purpose we use the energy-loss spectrum $f_j(\omega_{\mathrm{ex}})$ of the j-th atom, which one can, for example, look up in the EELS Atlas [35]. Moreover, we replace the average energy-loss $\hbar\overline{\omega}$ by the energy-loss $\hbar\omega_{\mathrm{ex}}$. In addition we substitute $f_j(\omega_{\mathrm{ex}})$ for $\delta(\omega - \overline{\omega})$ in relation (2.152).

By taking the Fourier transform of the resulting expression, we obtain the improved approximation

$$\tilde{\mu}_{11}^{(\mathrm{in},j)}(\boldsymbol{q},\boldsymbol{q}',\tau) \approx \left(\frac{4\pi\,\alpha}{\beta}\right) \int f_j(\omega_{\mathrm{ex}})\, \frac{\exp\left(-i\omega_{\mathrm{ex}}\tau\right)}{\boldsymbol{K}^2\boldsymbol{K}'^2} \times \left[F_j(\boldsymbol{K}-\boldsymbol{K}') - \frac{1}{Z_j}\, F_j(\boldsymbol{K})F_j(\boldsymbol{K}')\right] d\omega_{\mathrm{ex}}\,. \tag{2.157}$$

This formula is especially useful for simulating energy-filtered images obtained with rather large energy windows.

Inelastic partial waves emanating from different atoms are incoherent with each other. To account for this incoherence we must sum up the inelastic image intensities of all atoms. The incoherent superposition of the partial waves emanating from atoms situated in different slices is guaranteed by the generalized multislice formalism. In the case of a crystalline specimen the summation over the atoms situated in the same slice reduces to a sum over all atoms of the unit cell owing to the translational symmetry of the crystal lattice. For crystalline structures with a small unit cell we can perform this summation directly. For amorphous object details, such as interfaces, defect structures, or biological specimens, we must employ an alternative method.

To implement the generalized multislice procedure efficiently, the random phase method [8,9] can be used to compute sums of the form

$$I(\boldsymbol{q}) = \sum_{j=1}^{M} \Phi_j(\boldsymbol{q})\Phi_j^*(\boldsymbol{q})\,. \tag{2.158}$$

Here Φ_j, $j = 1,\ldots,M$ are arbitrary complex-valued functions. We consider a set α_{jk}, $j = 1,\ldots,M$ and $k = 1,\ldots,N$ of equally distributed random phases within the range $0 \le \alpha_{jk} < 2\pi$. For each k we define the expression

$$I_k(\boldsymbol{q}) = \left(\sum_{i=1}^{M} \Phi_i(\boldsymbol{q})\,\exp\left(i\alpha_{ik}\right)\right)\left(\sum_{j=1}^{M} \Phi_j(\boldsymbol{q})\,\exp\left(i\alpha_{jk}\right)\right)^{*}\,. \tag{2.159}$$

The complex exponential factors $\exp\left(i\alpha_{ik}\right)$ are called random phase factors. The value of $I_k(\boldsymbol{q})$ differs from $I(\boldsymbol{q})$ only by the contribution of the mixed factors originating from the product in (2.159):

$$I_k(\boldsymbol{q}) - I(\boldsymbol{q}) = \sum_{i=1}^{M}\sum_{j=1}^{i-1} 2\,\mathrm{Re}\left[\Phi_i(\boldsymbol{q})\Phi_j^*(\boldsymbol{q})\,\exp\left(i(\alpha_{ik}-\alpha_{jk})\right)\right]\,. \tag{2.160}$$

The terms on the right-hand side cancel out if we average I_k over a series of different sets of random phases α_{jk}, $k = 1,\ldots,N$. As a result we find

$$I(\boldsymbol{q}) \approx \frac{1}{N}\sum_{k=1}^{N} I_k(\boldsymbol{q}). \tag{2.161}$$

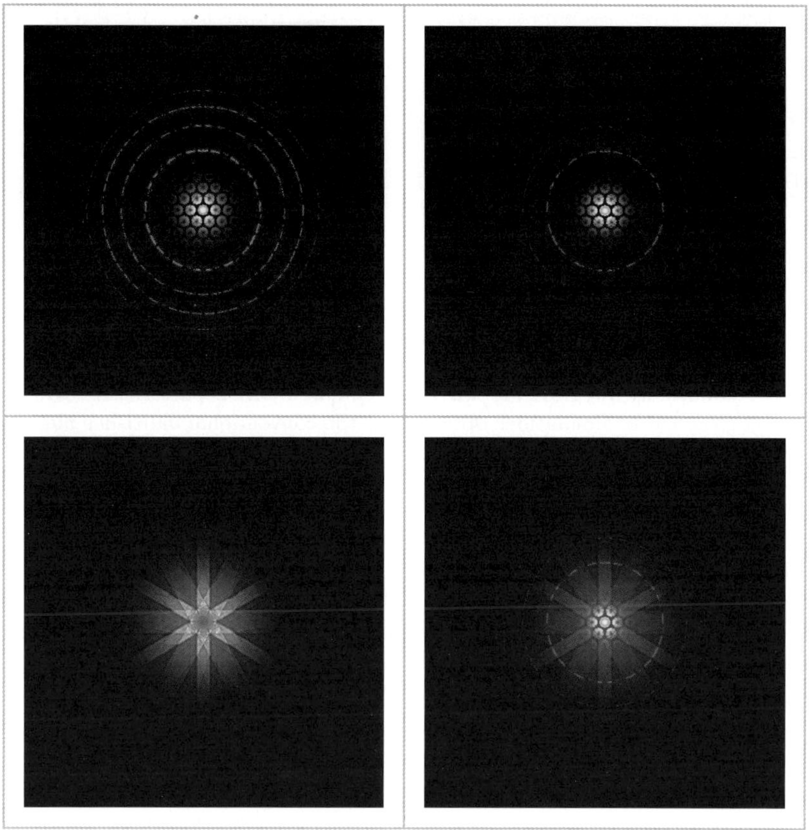

Fig. 2.6. Simulation of zero-loss filtered convergent beam electron diffraction (CBED) patterns for a 49 nm thick silicon crystal in $\langle 111 \rangle$ orientation. The imaging parameters are $U = 100$ kV, $C_s = 3.3$ mm and an objective aperture of $\vartheta_{\mathrm{A}} = 8$ mrad. The top images show the result of a conventional multislice simulation. The left image is calculated without considering the absorption potential for thermal diffuse scattering, whereas it is included in the image on the right side. At the bottom the results of the generalized multislice method are shown. The right image shows the complete result of the simulation considering thermal diffuse scattering. The left-hand image shows only the diffuse background resulting from this scattering

In the case of the generalized multislice formalism the function Φ_j represents a modified partial wave of the inelastically scattered wave packet emanating from the j-th atom. Since we employ a linear algorithm to calculate these functions from the input data, we can calculate any linear combination of these functions by applying the generalized multislice formalism to an appropriate linear combination of the input data. Therefore, we only need a single multislice calculation to evaluate (2.159) for a fixed k. In this case

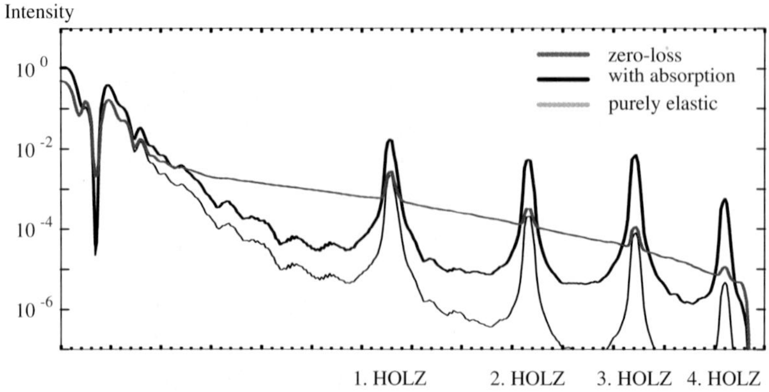

Fig. 2.7. Azimuthally integrated intensity of simulated CBED patterns. The different plots correspond to simulations obtained by the conventional multislice method with and without an absorption potential and the generalized multislice method

the numerical effort for determining the intensity $I(\boldsymbol{q})$ is proportional to the number of sets of random phases N, which is much smaller than the number of different inelastic scattering events occurring in a single slice. Therefore, the random phase method is especially useful if M is large and the calculation of the functions Φ_j in (2.158) is very time-consuming. This method is very advantageous for non-periodic objects where the multislice algorithm requires the implementation of large supercells. The number N of sets of random phases that is necessary to obtain a sufficient degree of accuracy must be determined by numerical experiments. For crystalline specimens the result of the exact calculation agrees very well with that obtained by means of the random-phase approximation.

The feasibility of the generalized multislice method is demonstrated in the numerical examples presented in Figs. 2.6, 2.7 and 2.8. Figure 2.6 shows simulations of convergent-beam diffraction patterns of silicon in the $\langle 111 \rangle$ direction. The top left image shows the simulated diffraction pattern using the standard multislice approach. The images on the right are calculated with an absorption potential for thermal diffuse scattering. The bottom left image shows the contribution of thermal diffuse scattering as given by the generalized multislice formalism. The bottom right image is the sum of the second and third images. It corresponds to a zero-loss filtered image containing the elastically and the thermal diffusely scattered intensity. It is the result of a full coherence-function calculation. Comparison of the first, second and fourth images shows that the best fit with the experimental diffraction pattern is obtained for the zero-loss image calculated with the coherence function method. The azimuthally integrated intensities in the convergent-beam diffraction patterns calculated by the conventional multislice method with and without an absorption potential and by the coherence function method

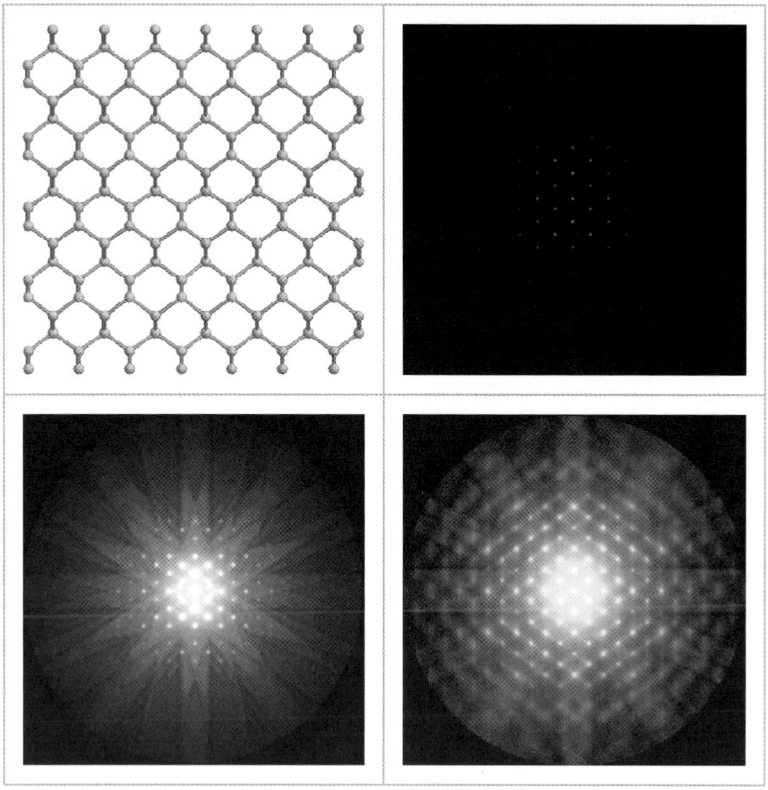

Fig. 2.8. Comparison of an experimental diffraction image (*bottom right*) of a 110 nm thick silicon crystal in the ⟨110⟩ direction with the results of the conventional simulation (*top*) and the coherence function method (*bottom left*). The symmetrically structured background signal caused by thermal diffuse and inelastic scattering is clearly visible in the simulation [Experimental image courtesy of W.D. Rau and P. Schwander, IHP Frankfurt/Oder, Germany]

are plotted in Fig. 2.7. The curves clearly demonstrate that a considerable amount of intensity is located between the diffraction disks. The conventional multislice method does not yield this intensity. Our result agrees with calculations based on the frozen-phonon method [36].

Figure 2.8 presents a comparison between an experimental diffraction pattern and calculated patterns using (a) the conventional multislice method and (b) the coherence-function method for a silicon specimenin the ⟨110⟩ orientation. The comparison clearly demonstrates that the characteristic background intensity visible in the experimental image produced by thermal diffuse and inelastic scattering is well reproduced by the simulation shown in the bottom left image. The symmetric structuring of the diffuse background is clearly visible in the simulated images obtained by the coherence-function method.

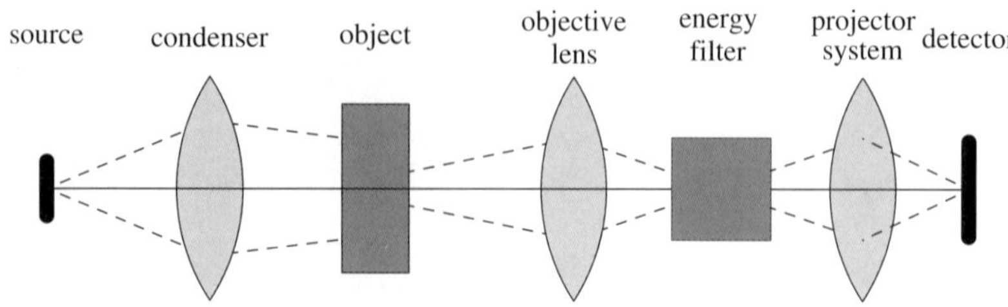

propagation processes	lateral and temporal coherence	illumination mode	elastic and inelastic scattering	imaging, phase shift	energy selection	intensity recording
parameters affecting the image	source size, energy width	illumination aperture, phase shift	atomic structure, mixed dynamic form factor	lens aberrations, defocus	width of energy window	saturation, crosstalk between detector elements
present state of simulation	monochromatic point source, incoherent superposition of images	small tilt angle, entrance wave function	transmission functions for elastic, quasi elastic and inelastic scattering	aperture function, transmission function of the optic system	restriction to specific scattering processes	nonlinear response function, point spread function

Fig. 2.9. Schematic representation of the propagation processes in an EFTEM and of the parameters affecting the image formation. The bottom row of the table lists the capabilities of the generalized multislice method [From [9]. Copyright 1998, The Royal Microscopical Society, London]

This result shows that the combination of elastic and inelastic scattering is correctly described.

The generalized multislice method outlined in this section is a straightforward generalization of the conventional multislice formalism. Within the context of the coherence-function approach, it can be used to account for the influence of partially coherent illumination and of inelastic scattering with a sufficient degree of accuracy. We conclude that this method provides a realistic description of the actual image formation in the electron microscope.

The present implementation of the generalized multislice software developed during the last years is illustrated in Fig. 2.9. Unfortunately, the computational demands method are still rather high for all realistic objects. Nevertheless, the ongoing improvement of high-speed computers continuously expands the range of applicability of our procedure.

References

1. Reimer L. (1995) Energy-filtering Transmission Electron Microscopy. Springer, New York
2. Hartel P., Rose H., Dinges C. (1996) Conditions and reasons for incoherent imaging in STEM. Ultramicroscopy 63: 93 - 114
3. Landau L.D., Lifschitz E.M. (1979) Lehrbuch der theoretischen Physik, Vol. 3. Akademie Verlag, Berlin
4. Sakurai J.J. (1985) Modern Quantum Mechanics. Addison-Wesley, New York, 325-326

5. Glauber R.J. (1959) High-energy collision theory. In: Lectures in Theoretical Physics, Boulder Vol.1. Interscience, New York, 315-414
6. Kohl H., Rose H. (1985) Theory of image formation by inelastically scattered electrons in the electron microscope. In: Advances in Electronics and Electron Physics, Vol.65. Academic Press, London, 173-227
7. Rose H. (1984) Information transfer in transmission electron microscopy. Ultramicroscopy 15: 173-192
8. Dinges C., Berger A., Rose H. (1995) Simulation of TEM and STEM images considering phonon and electronic excitations. Ultramicroscopy 6: 49-70
9. Müller H., Rose H., Schorsch P. (1998) A coherence function approach to image simulation. J Microsc 190: 73-88
10. Kirkland E.J. (1998) Advanced Computing in Electron Microscopy. Plenum Press, New York
11. Reimer L. (1992) Transmission Electron Microscopy. Springer, Berlin
12. Spence J.C.H., Zuo J.M. (1982) Electron Microdiffraction. Plenum Press, New York
13. Glaser W. (1952), Grundlagen der Elektronenoptik. Springer, Wien
14. Ferwerda H.A., Hoenders B.J., Slump C.H. (1986) Fully relativistic foundation of linear transfer theory in electron optics based on the Dirac equation. Optica Acta 33: 159-183
15. Kittel C. (1996) Introduction to Solid State Physics. Wiley, New York
16. Lippmann B.A., Schwinger J. (1940) Variational principles for scattering processes. Phys Rev 79: 469-480
17. Goodman J.W. (1996) Introduction to Fourier Optics. McGraw-Hill, New York, 63-95
18. Newton R.G. (1989) Inverse Schrödinger Scattering in Three Dimensions. Springer, New York
19. Born M., Wolf E. (1999) Principles of Optics, 7th edn.. Cambridge University Press, Cambridge
20. Lenz F. (1957) Zur Streuung mittelschneller Elektronen in kleinste Winkel. Z Naturforsch 9a: 185-204
21. Koppe H. (1948) Der Streuquerschnitt von Atomen für unelastische Streuung von schnellen Elektronen. Z Physik 124: 658-664
22. Rose H. (1976) Image formation by inelastically scattered electrons in electron microscopy. Optik 45: 139-158
23. Mott N.F. (1930) The scattering of electrons by atoms. Proc Roy Soc London A 127: 658-665
24. Doyle P.A., Turner P.S. (1968), Relativistic Hartree–Fock x-ray and electron scattering factors. Acta Cryst A 24: 390-397
25. Weickenmeier A., Kohl H. (1991) Computation of absorptive form factors for high-energy electron diffraction. Acta Cryst A 47: 590-603
26. Eusemann R., Rose H., Dubochet J. (1982) Electron scattering in ice and organic materials. J Microsc 128: 239-249
27. Compton A.H. (1930) The determination of electron distributions from measurements of x-rays. Phys Rev 35: 925-938
28. Hawkes P.W. (1978) Coherence in electron optics. In: Advances in Optical and Electron Microscopy, Vol.7. Academic Press, London, 101-184
29. Fertig J., Rose H. (1977) A reflection on partial coherence in electron microscopy. Ultramicroscopy 2: 269-279

30. Van Hove L. (1954) Correlations in space and time and Born approximation scattering in systems of interacting particles. Phys Rev 95: 249-262
31. Cowley J.M., Moodie A.F. (1957) The scattering of electrons by atoms and crystals. Acta Cryst 10: 609-619
32. Self P.G., O'Keefe M.A., Buseck P.R., Spargo A.E.C. (1983) Practical computation of amplitudes and phases in electron diffraction. Ultramicroscopy 11: 35-52
33. Stadelmann P.A. (1987) EMS – A software package for electron diffraction analysis and HREM image simulation in materials science. Ultramicroscopy 21: 131-146
34. Abramowitz M., Stegun I.A. (1970) Handbook of Mathematical Functions. Dover Publications, New York
35. Ahn C.C., Krivanek O.L. (1983) EELS Atlas. Center for Solid State Science, Arizona State University
36. Loane R.F., Xu P., Silcox J. (1991) Thermal vibrations in convergent-beam electron diffraction. Acta Cryst A 47: 267-278

3 Structure Determination by Quantitative High-Resolution Transmission Electron Microscopy

G. Möbus

3.1 Introduction

High-resolution transmission electron microscopy (HRTEM) comprises techniques of image formation by bright-field phase contrast with the aim of resolving the lattice fringes of a crystal lattice. There is no specific resolution threshold value separating "HR"-TEM from conventional TEM. Instead, the distinction is based on the fact that several diffracted beams are necessary to form the image of crystal planes. HRTEM is most important and powerful for studies of crystal defect structures in real space. For this, the aperture of the objective lens must be large enough to allow the diffracted beams corresponding to the projected crystal planes to pass and the passband of the contrast-transfer function (aberrations and incoherence envelopes) must extend sufficiently far. Many non-equivalent diffracted electron waves then build up an interference pattern in the image plane according to Abbe's optical microscope theory. Conventional TEM, on the other hand, uses scattering around a single diffracted beam only. An HRTEM image pattern normally mirrors well the atomic geometry and symmetry of the material examined, but not necessarily the atom positions. Visual interpretation is therefore routinely assisted by computer simulations of the experimental image formation process. The image is correctly interpreted in terms of the atomic structure once the simulated and experimental image match sufficiently. For many years, experimental and simulated HRTEM images were compared visually, but recently interest has shifted towards digital and automated interpretation of the micrographs. This trend towards computer-controlled atomic-resolution structure retrieval (as opposed to just verifying or falsifying a few structure models) has three main motivations:

1. *Present trends in materials science.* The need for accurate knowledge of the structure of crystal defects at atomic resolution has increased rapidly along with the engineering and designing of materials down to the nanometre s[illegible] For example, the semiconductor device industry and fundamental research on quantum confinement rely on the knowledge and control of defect structure, such as interfaces, dislocations and composition fluctuations at the nanometre level. Nanostructured and nanocrystalline metals, alloys, and ceramics, and especially compound materials,

thin films, and multilayered coatings likewise depend on atomic scale characterisation and control. Furthermore, the history of the development of "novel materials" such as carbon nanotubes, high-temperature superconductors, magnetic nanostructures, or metallic quantum wires, was closely linked to their observation in the high-resolution electron microscope.

2. *Progress in instrumentation.* Several important milestones have been passed in the last ten years. New ultra-resolution lens designs combining low spherical aberration with acceptable tilting ranges have been introduced for 200 kV–300 kV instruments. Field-emission guns now combine very high brightness with high spatial and temporal coherence. The development of very stable and comfortable high-voltage/high-resolution instruments has resulted in a point-resolution of about 1 Å thanks to the reduction in wavelength at 1250 kV. Finally, the most recent milestone was the demonstration of sub-1.4 Å Scherzer-resolution by correcting the spherical aberration at 200 kV. See Chap. 6 for more details. All these modern microscope technologies share the common advantage of pushing the information resolution limit (the ultimate detectable spatial frequency, independent of associated aberrations) to regions near or even below 1 Å. The information resolution has become a major figure of merit now that image processing allows image evaluation techniques that are less dependent on a Rayleigh-type point resolution limit to be employed. The development of slow-scan CCD cameras and, more recently, the image plate system was another milestone; these complement the conventional photographic film, which can, however, still be reliably used for quantitative work combined with digitisation in a high-quality scanner.
   The revolutionary increase of CPU power over the last decade can be appreciated from the following numbers: The computing time to convert one small defect structure model into one HRTEM-image on a state-of-the-art laboratory workstation decreased from 2–3 hours in 1990 (e.g. DEC μVAX II) to 10–20 seconds in 1999 (e.g. DEC Alpha AXP).
3. *Progress in modelling of materials.* The development of large-scale atomic modelling of defect structures now allows thousands of atoms to be included in a calculation cell, covering whole misfit dislocation networks at heterophase boundaries, for example. Methods of quantum chemistry and solid-state physics thus generate another challenge but also provide an important boost for HRTEM quantification: getting theory and experiment to agree on the Å-scale with accuracy of atom location on the pm-scale.
   Shifts of an atom position of a tenth of an Angström already cause significant changes in the energy value of a structure model. On the other hand, equilibrium structure models calculated from competing techniques of molecular modelling (mostly differing only slightly in the results), also need to be judged against the experimental structure.

Organization of this Chapter: After a general introduction and classification of HRTEM of crystal defects and strategies for quantitative interpretation of their images, two main techniques for the determination of atom coordinates from HRTEM images are selected for detailed presentation: *iterative refinement* and *strain mapping.* In according with the objective of this book, we recall that both of these techniques have been developed as a central activity within the key-research area on microcharacterization of materials and devices of the Volkswagen-Foundation, Hannover. A survey of the literature on alternative approaches is given at the end of the chapter. The entire research field comprising the extraction of numerical structure data (with confidence levels) from digitized HRTEM images has recently become assimilated under the name "Quantitative high-resolution electron microscopy (Q-HRTEM)".

3.2 Strategies of Structure Retrieval

To understand the nature of the problem of structure determination from a HRTEM-image, the inverse process needs first to be understood. From the description of this *image simulation* process, a classification scheme for objects and problem classes can be established.

3.2.1 Simulation of Image Formation in HRTEM

The flow of operations within an electron microscope in high-resolution imaging mode, as is followed to a good approximation by typical simulation programmes, is sketched in Fig. 3.1. The image formation is a three-step process. It starts with a "thin slice" wavefunction, formed by passing a plane illumination wave through a complex (almost pure imaginary) transparency function, describing phase shifts proportional to the projected potential of that slice (e.g. a monolayer of crystalline material). Subsequently, the "thick object" exit wavefunction is calculated by alternate application of free-space Fresnel propagation of the wave and the further thin-slice transparencies to describe the full structure of the object. The third step is the imaging of the exit wavefunction by the objective lens under conditions of partial temporal and spatial coherence and its final recording as an intensity. Common image simulation software practice, as used in Q-HRTEM applications so far, deviates from the image formation theory outlined in Chap. 2 of this book in several aspects:

1. The interaction of electrons with materials is modelled by a standard multislice technique [1], which assumes that the illumination is a perfectly coherent plane wave. Crystal structures are entered as supercells for periodic continuation with the defect often in the centre. Atoms are placed by means of their two lateral coordinates (x, y-plane) in slices of

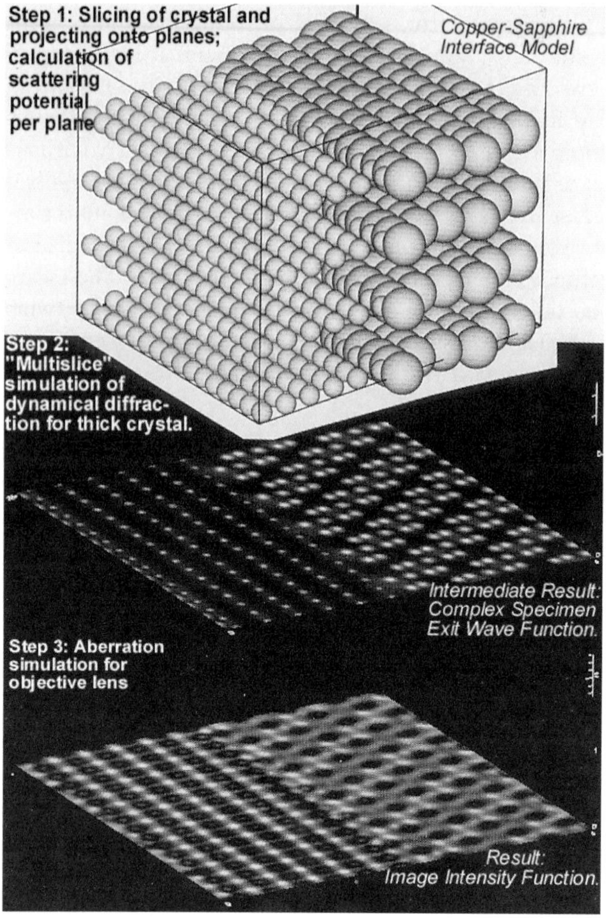

Fig. 3.1. Image formation in the high-resolution electron microscope as approximated by simulation software

height δz along the beam direction (z-axis). The thickness of the slices has to be adapted to the accelerating voltage (curvature of the Ewald sphere) and the complexity of the structure and may vary from 0.1 nm to 1.0 nm. Whether a particular slicing is sufficiently thin or whether contributions from higher-order Laue zones (HOLZ) become important, can be checked by comparison with Bloch wave calculations of equivalent defect-free crystal structures. The scattering amplitudes are then integrated over all atoms in every single slice using atom form factors taken from the tables of Doyle and Turner [2] or equivalent databases. Deviations from neutral-atom form factors have been found to be negligable [3,4] for Q-HRTEM purposes, even though the assumption of neutral atoms in a crystal seems inappropriate from a physical point of view (especially for covalent or ionic materials). The influence of atom vibrations, leading (strictly speaking) to thermal diffuse scattering and imaging, are normally, for reasons of simplification and speed of calculations, modelled only by Debye–Waller factors and the addition of an imaginary part (absorption index) to the real form factors. This treatment, though formally inaccurate, has been found to be adequate, at least for typical HRTEM specimen thicknesses of 5–15 nm [5].

2. For the imaging part of the simulations, which describes the aberrations of the objective lens, it is now standard practice to represent partial spatial and temporal coherence by envelope functions [1]; these account for beam convergence, energy spread of the gun and fluctuations in voltage and focusing current. Owing to the fact that it is intensities, rather than amplitudes, that are recorded, these envelopes are applied in the context of the transmission cross-coefficient [6,7]. Simple multiplicative damping envelopes in the sense of linear systems theory are, however, suitable to describe effects caused by mechanical damping and the resolution of the recording-medium (e.g. specimen drift, specimen vibration, modulation transfer function of CCD-cameras).
3. Another group of potential inaccuracies comprises numerical artefacts, such as limited sampling density and anti-aliasing cutoffs during repeated fast Fourier transforms (multislice). Periodic continuation of crystal supercells can lead to discontinuities at the frame (wrap-around artefacts). If the central region of interest within such a supercell is small, this may be acceptable; otherwise, artificial structure modification is required to restore periodicity, such as the generation of mirror dislocations, for example.

3.2.2 Object Classes and Object–Image Relations in HRTEM

The interesting question is, under what condition does the visual appearance of an image reflect the object structure? We restrict our classification of objects and imaging conditions to the case of objects that are periodic in the beam direction. This allows us to reduce the determination of three-dimensional atom coordinates to projected atom-column positions, with the columns assumed straight. Fig. 3.2 shows a typical example of a complex structure, consisting of a fcc lattice (copper) and a trigonal ceramic (sapphire). The metal is recorded under imaging conditions in which the black dots correspond to the atom positions. However, slight misalignments of sample axis and beam axis can introduce lateral shifts of these dots from the positions of atoms at the entrance plane of the specimen (as indicated in the figure). This matters when measuring the lattice displacement across the interface. The situation on the sapphire side, where neither the oxygen nor the aluminium positions are situated at contrast minima or maxima, is more complicated. The reason for this observation is derived in Fig. 3.2 by step-wise

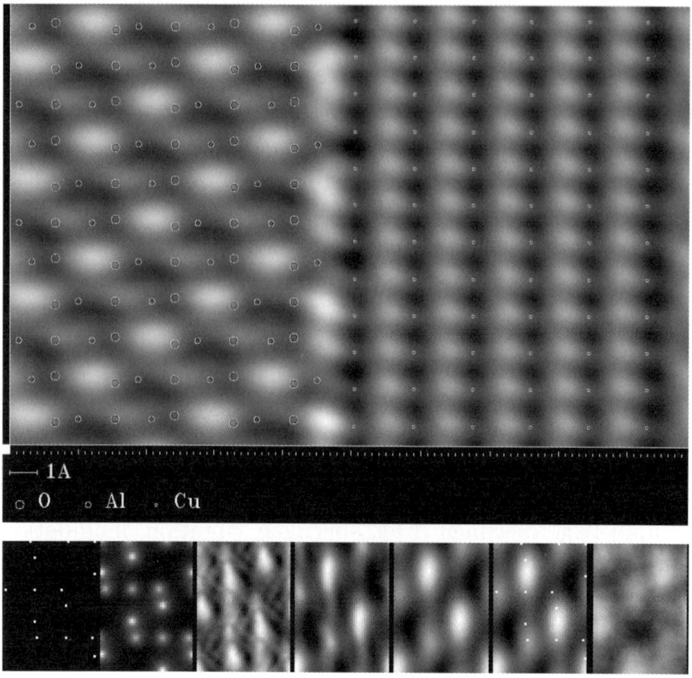

Fig. 3.2. Image formation for a copper/sapphire interface at 1250 kV (under tilted specimen condition). Main figure: interface image with superimposed atom column positions (*left*: sapphire, *right*: Cu) for the specimen entrance plane. *Bottom row*: stepwise image formation for the sapphire side (*from the left*): atom coordinates, single-slice scattering potential, specimen exit wave, image wave before recording, final simulated result (image wave with detector point–spread), coordinates superimposed on result, experimental image (all rotated by 90° with respect to the main image). All images are calculated for the microscope JEM-ARM-1250, a specimen thickness of 4.5 nm, a tilt of (0.8, 0.5°), viewing directions are $[2\bar{1}\bar{1}0]$ in sapphire and [211] in copper

simulation. It reveals that the track of the atomic structure in this (slightly) misaligned situation has already been lost when the specimen exit wave is calculated. The imaging by the lens only removes further contrast detail and resolution, thereby complicating the situation, since the positions of several atoms now need to be extracted from one elongated white spot.

With this description of the problem in mind, we shall speak of four classes of objects.

Weak-Phase Object. An object transparency representing a pure phase shift that can be approximated by

$$T(x,y) = \exp(-i\sigma\phi(x,y)) \simeq 1-\sigma\phi(x,y) \qquad (3.1)$$

is called a weak-phase object. Here, ϕ is the slice potential [eV] and σ an interaction constant [1/eV]. Its image can be interpreted straight-forwardly. After deconvolution by the contrast transfer function, it guarantees not only perfect coincidence of atom-column positions and the positions of image contrast extrema, it also provides a direct proportionality between reconstructed intensity and object potential. Such an object is sketched in Fig. 3.3a. The concentric iso-lines are meant to symbolize the interpretability of the whole intensity distribution as a local potential map. Unfortunately the restrictions on specimen thickness (roughly $< 1\,\text{nm}$), depending on atom number are hardly ever met in practice.

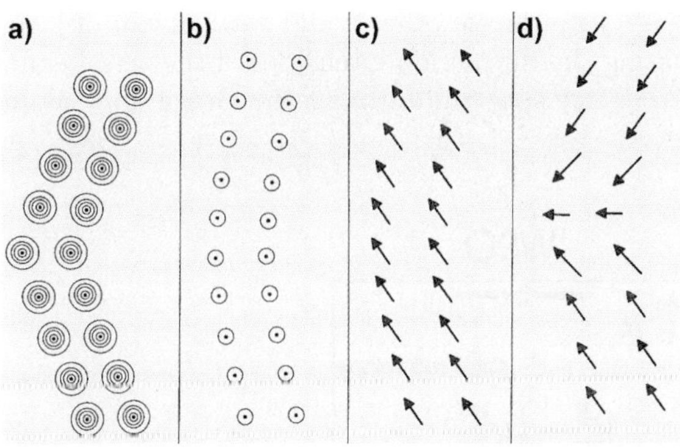

Fig. 3.3. Special objects in HRTEM image interpretation. (**a**) Weak-phase object, (**b**) structure image, (**c**) weak strain object, (**d**) general crystal defect

Objects for Structure Imaging. The expression "structure image" is used, e.g. in Appendix 4 of [1] for an HRTEM image in which every projected atom column of the crystal structure is seen as a separate spot, without lateral shifts between spots and atoms. Images are as directly interpretable in terms of the atomic structure as with weak phase objects. However, the intensity distribution around the spot maximum no longer has any direct meaning (symbolized in Fig. 3.3b). Structure images are often found in practice if the density of the projected structure is low compared to the point resolution of the microscope. There is no strict limitation on the specimen thickness, as long as the object–image relation remains local (no strong delocalization caused by diffraction).

Weak-Strain Objects. This very important next generalisation is essential for understanding the partitioning of this chapter into structure retrieval strategies. We now abandon the requirement of any coincidence between image detail (white or black spots) and atom columns. The lateral distance between column and spot is expressed as a vector $\boldsymbol{R} = \boldsymbol{r}_{\mathrm{atom}} - \boldsymbol{r}_{\mathrm{spot}}$ and referred to as the displacement error. An object is called a "weak-strain object" if $\boldsymbol{R}(x, y) = \mathrm{const}(x, y)$ across the whole field of view including crystal defects. Figure 3.3c symbolizes $\boldsymbol{R}(x, y)$ as a field of arrows of the same direction and length across an interface. In reality, this kind of object can be found only for crystal defects without sharp bends, like some domain boundaries, or the far-field of displacements around dislocation cores. The projected strain has to be small, hence the name of this object class. Interpretation of these images is straight-forward once $\boldsymbol{R}$ has been determined in strain-free regions near the defect of interest by the methods of Sect. 3.4. The specimen thickness is again basically unlimited while we still have the property that information remains local (object distortion = image distortion). An optical analogy can be drawn between weak-phase objects and weak-strain objects in reciprocal space by comparing the diffraction intensities of the zero beam, a major Bragg spot and the defect spectrum around the Bragg spot as in Fig. 3.4.

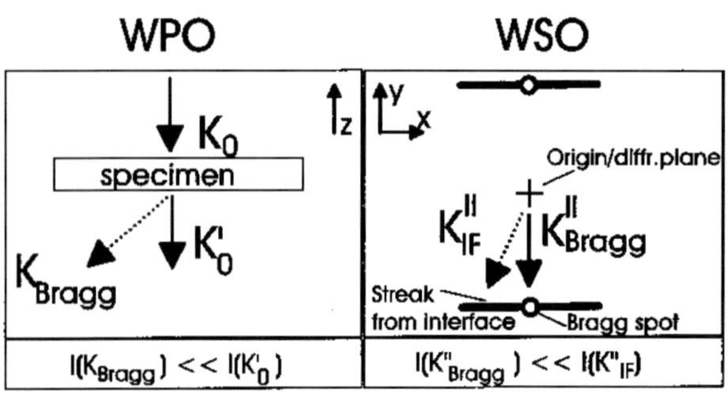

Fig. 3.4. Comparison of a **W**eak-**P**hase **O**bject and a **W**eak-**S**train **O**bject in reciprocal space

General Objects and Delocalization. For crystal defects exhibiting strong local strain fields the displacement between atom columns and image spots becomes spatially dependent: $\boldsymbol{R} = \boldsymbol{R}(x, y)$ (see Fig. 3.3d for a sharp grain boundary). For strong structural relaxations at crystal defects (e.g. dislocation cores), the number and resolution of image spots per atom column might change with position. Then "strain" is no longer expressible. In both cases image interpretation needs the assistance of HRTEM simulations (see Sect. 4 of Chap. 3). $\boldsymbol{R}(x, y)$ can change rapidly, from atom to atom, especially when caused by an atomically sharp (hetero-) interface as shown in Fig. 3.5 for an

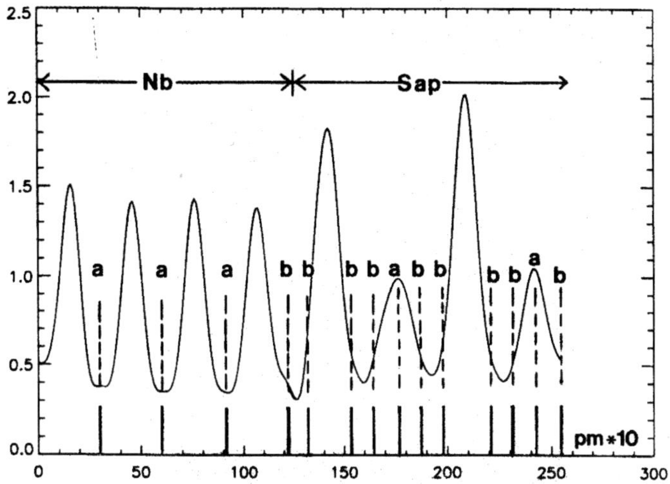

Fig. 3.5. Profile through a simulated HRTEM image of a metal-ceramic interface (8 nm thick specimen). Discrimination of "direct" atom location (at maxima or minima) and "indirect" positions (to be located by pattern matching)

ideal niobium-on-sapphire interface, where only unrelaxed bulk coordinates have been used to construct the interface.

We shall call the positions for which $\boldsymbol{R} = 0$ "direct" positions ("a" in Fig. 3.5) and those with $\boldsymbol{R} \neq 0$ "indirect" positions ("b") [8]. As can be seen, the displacement error builds up on the niobium side even in the absence of local relaxations. The neighbourhood of the sharply limited sapphire lattice with its differing properties (orientation, spatial frequencies, mean inner potential) leads to artefacts in the undisturbed niobium lattice. This phenomenon is called "delocalization" and is discussed in detail in [9–11]. The information about the position of an atom column will be spread out owing to contributions from dynamical diffraction in the crystal as well as from aberrations and the diffraction limit of the objective lens. The *optical delocalization* is defined with respect to the classical circle of least confusion in an aberrated lens system [11],

$$\rho = 0.25 C_s \lambda^3 k^3 \, , \tag{3.2}$$

in which C_s is the spherical aberration coefficient and k the spatial frequency. This quantity is most suitable for comparing TEMs independent of the particular specimen.

The *statistical definition* of delocalization on the other hand uses the second moment of a density function (spread function) and includes the spread within the object itself in accordance with information theory. The necessary spread function can be evaluated, (i) from the dynamically simulated intensity distribution of a single column image, (ii) from the mean square

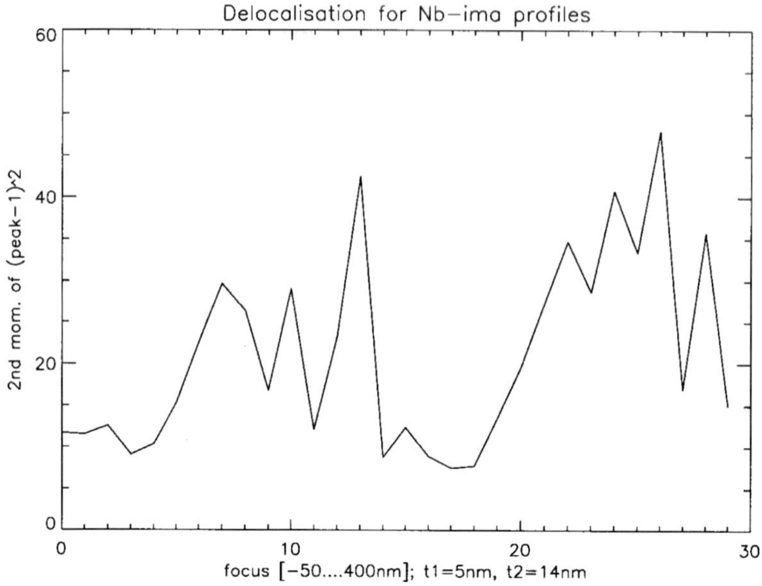

Fig. 3.6. Integral delocalization (second moment of point spread) as a function of defocus (data for the Oxford 300 kV JEM-3000F FEGTEM)

image difference between two structure models differing in one atom detail, or (iii), for thin light objects, from the squared lens point-spread function ($psf(\boldsymbol{x}; z)$). The latter is the Fourier transform of the contrast transfer function "$CTF(\boldsymbol{k}, z)$" and introduces a strong focal dependence (Fig. 3.6):

$$M_2(z) = \int (\boldsymbol{x} - \langle x \rangle)^2 p(\boldsymbol{x}; z) d\boldsymbol{x}\,, \tag{3.3}$$

with $\boldsymbol{x}$ the spatial coordinate, z the defocus and p one of the mentioned densities, e.g. $p(\boldsymbol{x}; z) = (I(\boldsymbol{x}; z) - 1)^2$ or $p = (\mathrm{FT}(CTF(\boldsymbol{k}, z))^2 * A^2_{\mathrm{exit}}(\boldsymbol{x})$. The latter spread function, obtained by convolution with the specimen exit wave amplitude A of a single atom column, is a good approximative formula to estimate the combined lens and specimen contributions. The spread functions for objective lens delocalisation described here are equivalent to the beam profiles obtained on the same TEM in STEM-mode.

3.3 Strain and Pattern Mapping

Strain is a tensor property of materials describing the deformation of crystal unit cells under internal or external stress. As in continuums mechanics, the property *strain* is defined on a scale that does not deal with positions of individual atoms within a unit cell. For strain to be defined, atomically abrupt defects, such as interfaces or dislocation cores, have to be excluded, because here (i) atoms in unit cells might be non-uniformly displaced, (ii) new prototypes of unit cells might form, and (iii) systematic or random occupancy changes might occur, such as half filled columns, interstitial atoms and substitutional elements. We recall that three-dimensional defects, where strain would be defined (e.g. foil bending), are excluded in addition for HRTEM-specific reasons [12]. The only accessible strain components are those of the two-dimensional projected strain field. The latter might again deviate from the distortion field ("apparent strain") in the image, as discussed in Sect. 3.2.

The concept of *pattern* is somewhat more artefact-tolerant. A pattern is a sub-area of an image and usually contains a typical local arrangement of image intensity. By definition, a pattern is normalised to unit average brightness and unit average contrast. On the object plane, the corresponding feature is a projected structural unit. This may be a crystal unit cell (with its atomic basis), which defines the space group and the crystal structure. It may also be a segment of a grain boundary/heterointerface or a dislocation core. While strain is the obvious quantity to be "measured" by quantitative HRTEM and image processing, pattern analysis tries to answer such questions as: Where are identical patterns distributed over the entire image (self-similarity) and how dissimilar are two (more or less) comparable patterns? Before discussing image-processing *techniques* for strain and pattern analysis, the general range of validity of the strain and pattern concepts in HRTEM is estimated.

3.3.1 Displacement Errors

In Sect. 3.2 we considered the object–image relations for a small area of a micrograph, near the optic axis. For strain mapping on substantial areas of the micrograph, projector lens distortion normally contributes by an additive strain field; this artefact can be measured on test samples and corrected for. More difficult to predict are the artefacts induced by changes in the imaging conditions, characterized by a set of parameters a_i, or expressible in terms of specimen parameters (e.g. thickness H) and lens parameters (non-isoplanacy, beam misalignment θ) that depend on image position: $H = H(x, y)$, $\Theta = \Theta(x, y)$.

We introduce the true displacement field $\boldsymbol{u} = \boldsymbol{x} - \boldsymbol{x}_{\text{ref}}$ measured from object coordinates in the object plane and the apparent displacement field $\boldsymbol{U} = \boldsymbol{X} - \boldsymbol{X}_{\text{ref}}$ measured from image spot positions (both relative to a reference area).

1. Strain mapping is based on the normally excellent approximation

$$\varepsilon_{xx} = \nabla_x \boldsymbol{u}_x \simeq \varepsilon_{XX} = \nabla_X \boldsymbol{U}_X \tag{3.4}$$

$$\boldsymbol{U}_{\boldsymbol{X}} = \boldsymbol{X} - \boldsymbol{X}_{\text{ref}} = (\boldsymbol{x} - \boldsymbol{R}) - (\boldsymbol{x}_{\text{ref}} - \boldsymbol{R}_{\text{ref}})\,. \tag{3.5}$$

We again used a coincidence error-vector $\boldsymbol{R}$ pointing locally from a spot position on the image to the correct underlying atom column coordinate. The approximation is that, for small ε, we assume $\boldsymbol{R} = \boldsymbol{R}_{\mathrm{ref}} \neq \boldsymbol{R}(x, y)$; while $\boldsymbol{u} = \boldsymbol{u}(x, y)$.

2. The image condition is expressed as a vector $\boldsymbol{a}$ collecting all simulation parameters (focus, thickness, misalignment, astigmatism,...). A spatially constant change in the image condition $\boldsymbol{a} = \boldsymbol{a}_0 + \delta\boldsymbol{a}$ could be due to insufficient knowledge about $\boldsymbol{a}$:

$$\boldsymbol{R} = \boldsymbol{R}_0 + (\nabla_a \boldsymbol{R})\delta\boldsymbol{a}\,, \tag{3.6}$$

$$\boldsymbol{U} = \mathrm{const}(\boldsymbol{a})\,. \tag{3.7}$$

   It hence does not influence the strain results.

3. An imaging condition that is continuously spatially varying, $\boldsymbol{a} = \boldsymbol{a}(x, y)$, will, however, change the measured strain by the artefact $\varepsilon_{\mathrm{af}}$:

$$\boldsymbol{R} = \boldsymbol{R}_0 + (\nabla_a \boldsymbol{R})\,\delta\boldsymbol{a}(x, y) = \boldsymbol{R}(x, y) \tag{3.8}$$

$$\varepsilon_{XX} = \nabla_X \boldsymbol{U}_X = \nabla_x \boldsymbol{u}_x - \nabla_x \boldsymbol{R} \tag{3.9}$$

$$= \nabla_x \boldsymbol{u}_x - (\nabla_a \boldsymbol{R})\nabla_x \boldsymbol{a} = \varepsilon_{xx} + \varepsilon_{\mathrm{af}}\,. \tag{3.10}$$

   The magnitude of the absolute changes in the imaging condition is less important than the local rate of change of $\nabla_x \boldsymbol{a}$ and the sensitivity of the displacement error $\nabla_a \boldsymbol{R}$ to the imaging condition.

4. Any sudden change in imaging condition such as a local thickness (surface) step or a heterointerface (change of crystallography) has a different effect. While local displacement fields plus strain remain free of artefacts according to

$$\varepsilon_{xx} = \varepsilon_{XX} - (\nabla_a \boldsymbol{R})\nabla_x \boldsymbol{a} = \varepsilon_{XX}\,, \tag{3.11}$$

   any long-range displacement measurements, which are important for estimating the translation vector (volume expansion) at heterointerfaces, are affected by:

$$\boldsymbol{T} = \boldsymbol{T}_0 + (\nabla_a \boldsymbol{R}_1)\delta\boldsymbol{a}_1 - (\nabla_a \boldsymbol{R}_2)\delta\boldsymbol{a}_2 = \boldsymbol{T}_0 + \delta\boldsymbol{T}_{\mathrm{af}}\,. \tag{3.12}$$

5. In this nomenclature we recall that for high strains artefacts will be introduced even without changes in the image conditions (delocalization):

$$\varepsilon_{xx} > \varepsilon^{\mathrm{threshold}} \Longrightarrow \boldsymbol{R} = \boldsymbol{R}(u_x, u_y; x, y) \tag{3.13}$$

$$\varepsilon_{xx} = \nabla_x \boldsymbol{u}_x = \nabla_X \boldsymbol{U}_X - \nabla_x \boldsymbol{R} \neq \nabla_X \boldsymbol{U}_X \tag{3.14}$$

Hitherto, we described certain limits on maximum object strain for a *given microscope* for which $\boldsymbol{R}$ relates to its contrast transfer function, CTF. Conversely, for a *given object* the limits on the maximum gradient of the CTF at the main lattice frequency can be derived as shown in [13].

Apart from the optical displacement error (objective lens and detector point-spread), a *numerical displacement error* is deliberately introduced during strain measurement on a noisy image. This is the real space equivalent of band restrictions in Fourier space (uncertainty principle). One has to decide during strain mapping whether one prefers to have a well-defined and noise-reduced spot position, measured from an average over several lattice fringes, or rather noisy geometric phase information, which remains as localized as possible (given by the microscope constraints). This delocalization must be left as a free parameter in the image processing code. It is easy to control and to document using for example the wrap-around effect for non-periodic interfaces occuring at the frame of the image.

3.3.2 Classification of Strain Mapping Techniques

The task to be accomplished in HRTEM displacement measurement techniques is simple: conversion of the locations of intensity maxima (or minima depending on focus) in the image into an array of atom coordinates, which can then be converted mathematically into the desired displacement or strain maps. However, numerical sampling problems, the presence of noise, and artefact minimization make things more complicated. The effort of image processing is justified by requiring "sub-pixel" accuracy for the peak locations. We are thus concerned with the following set of sub-problems, illustrated in Fig. 3.7:

1. noise suppression
2. image segmentation
3. interpolation for sub-pixel accuracy
4. selection and subtraction of a reference lattice (displacement maps)
5. gradient calculation (strain maps)

Several authors have solved these five steps in various combinations. Even more techniques could be derived by mixing their procedures. Some techniques combine two sub-problems in one numerical step. One technique, called geometric phase mapping, combines items 1 to 4 in a single step (Sect. 3.3.5 below).

3.3.3 Local Peak Detection

An early strain-mapping algorithm has been applied to quantum wells by Bierwolf et al., see Fig. 3.8 [15]. The micrograph of a sphalerite semiconductor in the [110] direction shows one bright spot per pair of atom columns. A noise filter in the form of a generally applicable bandpass is the first step. The filter function is a rectangle with unit transmission between an upper frequency u and a lower frequency l: $I'(k_x, k_y) = I(k_x, k_y) \cdot \mathrm{rect}_{l,u}(k_x, k_y)$. The bandwidth must include the peaks forming the lattice plus a surrounding

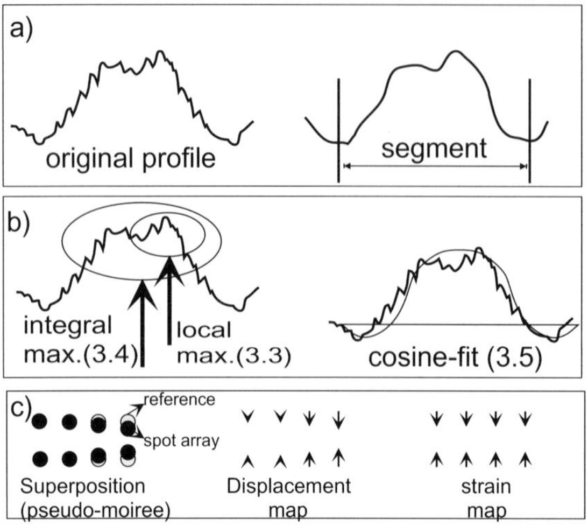

Fig. 3.7. Numerical steps during strain mapping: (**a**) Noise reduction and (optional) segmentation, (**b**) subpixel precise peak search (numbers refer to sections of the text), (**c**) superposition of reference lattice and postprocessing into maps

area large enough to pass the reciprocal shape function of the real space area to be averaged without falsifying the displacements. Here the term *local technique* denotes the fact that the maximum is sought from a limited region of intensity around the maximum not incorporating all pixel information within the peak or the rest of the unit cell. This region is determined by the band width. No elaborate image segmentation technique is necessary, therefore. Interpolation of higher order (parabolic, polynomial or spline – bilinear would be useless) is the preferred method of obtaining sub-pixel precision for coordinates (on noise-reduced data only). Most image-processing packages or numerical libraries contain a set of powerful routines, the description of which is beyond the scope of this book.

Once all the peak positions have been measured and tabulated, the displacement map is determined by calculation of pairwise differences of positions within the strained (defective) region and a selected (unstrained) perfect crystal region (this is similar to building an artificial moiré pattern). An arrow-vector map is the best way of displaying the results of this step. The final step is the numerical differentiation of the difference vectors to obtain the strain field from the displacement field.

3.3.4 Integral Peak Detection

Many, or even all of the experimentally available intensity-pixels can be evaluated to estimate a peak maximum within a neighbourhood. Such integral techniques can be combined with a *segmentation* of the image into subpatterns or unitcells. Rectangles for cubic projections, Wigner–Seitz cells [16]

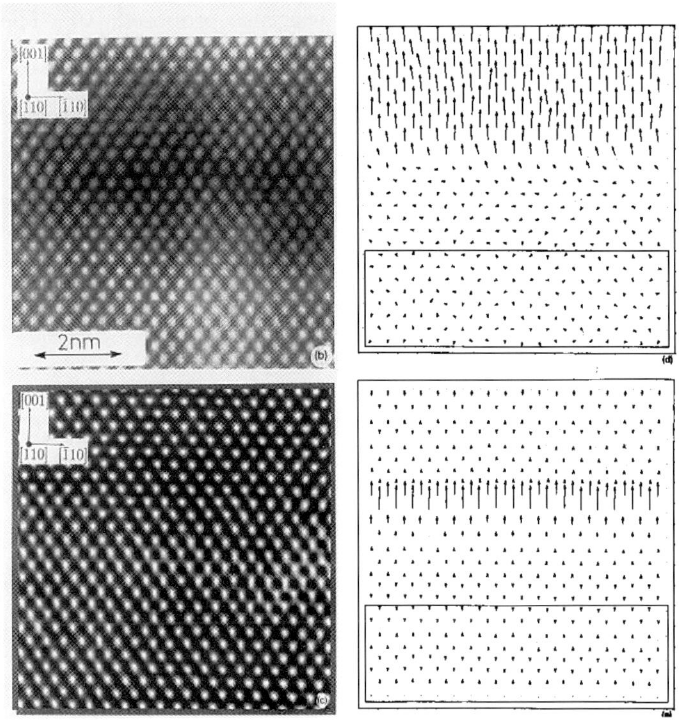

Fig. 3.8. Left column: Experimental image of quantum well in GaAlAs [110]; original and noise reduced versions (JEOL JEM4000EX). Right column: displacement and derivative maps (from [15])

for general space groups, or watershed-techniques [17] have in common that all pixels of the original image are used but none is used more than once.

1. Cross-correlation with a simulated or averaged experimental peak template $P(x, y)$ (correlation filtering) has been applied to grain boundaries in Cu [18].

$$I'(x,y) = I(x,y) * P(x,y) \tag{3.15}$$

$$I'(k_x,k_y) = I(k_x,k_y) \cdot \mathrm{FT}(P)(k_x,k_y). \tag{3.16}$$

Here the width, shape and symmetry of the template are the crucial parameters.

2. Interpolation and noise suppression in one step is performed by mass centre calculations [16,19,20], a powerful *integral* measure of peak positions. Deviations of mass centres from local peak maxima are expected for asymmetric (drop-like) peaks as often seen on misaligned images. Which of the two criteria is closer to the atom column position can only be assessed by full image simulation. For public-domain software, see [19].

3. Position fitting in the least-squares sense was also proposed [16]. The same templates are used as for correlation filtering, either simulated peak shapes or averaged experimental peaks.
4. Finally, one-dimensional line profiles have been used to derive the peak maximum iteratively [16].

For lack of space we simply mention further techniques of strain mapping such as CUSUM [21] or DALI [22] working on filtered one-dimensional lattice fringes.

3.3.5 Geometric Phase Mapping

An image pattern can be decomposed into its Fourier spectrum in order to establish what spatial frequencies are present. In addition, a distorted lattice (to first order, a cosine-square pattern) can be sorted by its local phases following Hÿtch [23,14]. Every pixel of the original image is replaced by the phase value of the locally best fitting cosine-square function:

$$I(x,y)_{ij} = \cos^2\{k_x x_{ij} + \phi_x(x,y)_{ij} + k_y y_{ij} + \phi_y(x,y)_{ij}\} \Longrightarrow \phi_{ij}\,. \qquad (3.17)$$

The term *geometric* phase is used here to distinguish it from the phase of the electron wave.

This transformation is obtained most easily (Fig. 3.9) by a procedure inspired by optical holography: after Fourier-transforming the image, a mask selects a Bragg peak and a circular neighbourhood (or Wigner–Seitz cell) in the spectrum. This peak-neighbourhood is then exclusively passed through the filter (one-sided!). The Bragg spot should be one of the fundamental lattice frequencies, and preferably one of the strongest, to improve the signal-to-noise ratio. The choice is, however, mainly determined by the direction desired for displacement measurements. Along with the back-transform, the spectrum may be shifted to the origin, or equivalently the cosine-lattice corresponding to the centre of the selected Bragg peak may be subtracted from the resulting filtered image. This filtered image is complex-valued and can be further decomposed into amplitude and phase images. The latter will be called the geometrical phase map. Several Bragg spots can be processed consecutively and the resulting images combined arithmetically.

The phase map contains the information about local lattice displacements in terms of the phase shifts $\phi(x,y) = \phi_0(x,y) + \delta\phi_g(x,y)$. Here ϕ_0 defines the chosen reference. The displacement phase map (Fig. 3.9b) is the equivalent to the popular arrow-diagrams commonly used for the display of strain, e.g. Fig. 3.8. Again, the displacement map can be further converted to a strain map by numerical differentiation [23].

This technique can be applied to the same problems as those discussed earlier. However, it has the unique property of consisting of pure image processing, relying at no intermediate step on segmentation of the image, of

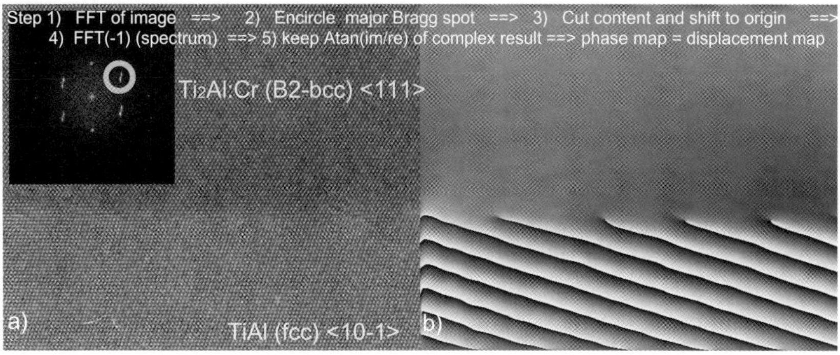

Fig. 3.9. (**a**) Interface $Ti_2Al(Cr)$–TiAl, recorded on the Stuttgart JEM-ARM1250. (**b**) Phase map derived using the peak encircled in the inset-spectrum of (**a**)

identification of specific peaks, or on a discrete coordinate field of local lattice parameters. Instead the image is converted "pixel by pixel" to a continuous displacement map or strain map. The high speed makes it possible to optimize all free parameters (e.g. the filter diameter) in rapid iteration loops.

For misfitting hetero-interfaces, two Bragg spots, one from each lattice, are to be included in the filter-window to obtain a moiré pattern. The sine of the phase map is then a mathematically exact representation of the optical moiré technique, previously proposed in [24].

The $Ti_2Al(Cr)$-TiAl hetero-interface in Fig. 3.9. forms a distinct semi-coherent region, consisting of segments of coherently matching lattices separated by well localised (misfit) dislocations [25]. Each dislocation in Fig. 3.9b is highlighted by a singularity. The top lattice in Fig. 3.9a has been selected as the reference lattice. The bottom lattice then shows the moiré type lines as 2π phase jumps. The continuing phase modulations within the next two atomic layers of the top lattice are particularly interesting, since they indicate semi-coherence. This example clearly demonstrates the display and feature-extraction capabilities of the technique (in addition to the quantitative displacement measurements), which prove to be more robust and widely applicable, even beyond some of the limits enumerated in Sect. 3.3.1.

As in some other real space image decomposition techniques (e.g. Hotelling transforms), several images can be derived from one original (phase/ amplitude, use of several peaks). According to information theory, we cannot, however, increase the original information content. A loss in reliability of pixel values by noise (e.g. serpent lines of phase jumps) and data redundancy within a map series will result.

As a transform, phase mapping is *always* applicable, "allowed" and more or less useful. Phase mapping is not peculiar to electron microscopy, it works for all kinds of photographic records. On the other hand, the interpretation of phase maps in terms of the underlying object structure is not part of image processing, but is subject to cystallography and microscopy conditions and requires care.

3.3.6 Comparison of Techniques and Postprocessing Steps

When judging the above-listed approaches by local/integral peak search and geometric phase mapping, it should be noted that none of them is mathematically distinct. Fourier filters (bandpass) and correlation filters are linked by the "Matched Filter" principle (see Sect. 3.3.7). Mass-centre methods and correlation filters are linked by their weighted sum of intensities within a working circle around the peak maximum. Phase mapping could be classified under integral peak search as well, since it is, seen in real space, nothing but a convolution with a point-spread function. The methods can yield different results when many high-frequency (especially non-linear) details are present in the HRTEM image per projected unit cell. Local techniques seek the brightest sub-spot, while integral search or phase mapping concentrate on a more averaged position of the peak intensity, Fig. 3.7. Expressed in reciprocal space, local techniques use all "higher harmonics", while integral techniques do not. Which one reveals the object strain better is, however, not clear. Sometimes, local methods have fewer systematic artefacts, but higher errors caused by poor signal-to-noise ratio (SNR). However, in materials with poorly resolved large unit cells containing many atoms, excessive image detail within the unit cell seems to confuse local techniques, making integral techniques preferable. For nearly structure-image conditions or for strain mapping on reconstructed exit waves, local peak detection is again preferable since it can reveal the atom positions directly [26]. To minimize artefacts and optimize precision, it is more important to adjust the free parameters (Fourier filter diameter, mass center circle) within a technique correctly, than to decide between the various techniques.

As well as displaying of displacement maps around crystal defects, the foregoing methods are useful for examining statistical properties of the sample: Paciornik et al. [27] used a mass-centre peak search on a perfect lattice (no distortions) to assess the amount of specimen noise. Standard criteria for SNR-measurements are based on intensity fluctuations in real space or segmental averaged Fourier-space intensities. The standard deviation of random peak displacements in a perfect crystal lattice is used. This method also defines a sensitivity limit for strain mapping in a noisy environment [27].

Seitz et al. [16] extended this procedure to derive comparative sensitivity criteria for various peak-finding methods, various sampling densities and various specimen preparation techniques.

Figure 3.10 shows the definition of 2D distance measurements on a micrograph digitized into a pixel array of N_{g} samples per unit cell length (g is a direction, e.g. [110]). After evalution of many peaks of numerous crystal units, a "finite-width" probability distribution of distances in a perfect lattice will

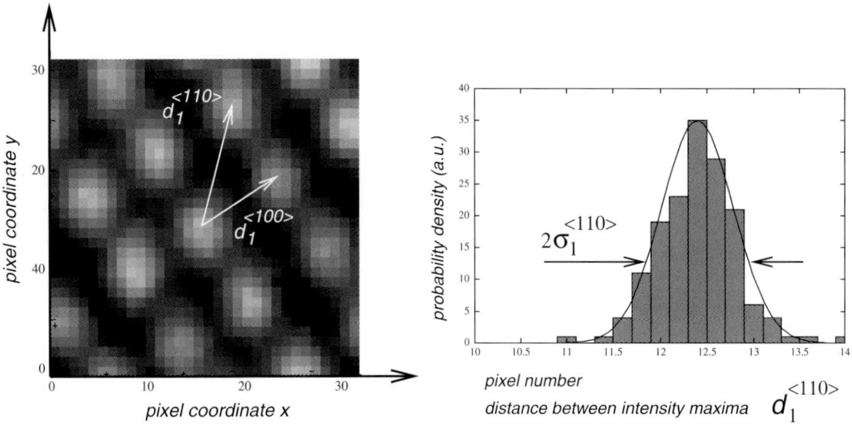

Fig. 3.10. (**a**) HRTEM micrograph of silicon [100] at 200 kV. The two basis vectors for the displacement-field analysis are indicated. (**b**) Resulting histogram of peak distances (from [16])

Table 3.1. Strain mapping sensitivity for various conditions, Si <100>

Peak find technique	Sample preparation	Sampling points	γ [%]
line profile	cleavage	>60	2.4
mass-centre	cleavage	>60	1.68
simulated peak	cleavage	>60	1.76
experimental peak	cleavage	>60	1.66
line search	cleavage	20	3.0
line search	cleavage	10	4.6
line search	ion mill	>60	6.6
line search	dual-angle ion mill	>60	5.0
line search	chemical etch	>60	3.8

result (owing to the noise). The sensitivity γ is here defined to be twice the ratio of the standard deviation and the lattice constant: $\gamma = 2 \cdot \sigma_g / d_g$. Table 3.1 contains a selection of the experimental γ-results. The lines in the table refer to the above-listed techniques, such as mass-centre, iterative line profile technique, fitting to a simulated or averaged experimental peak pattern. The other conditions in the table refer to specimen preparation (ion-beam milling, chemical etching, cleavage), and sampling density.

3.3.7 Pattern Mapping

The most basic pattern recognition technique is based on cross-correlating an image with a certain sub-sized pattern ("template"). Each occurence of the pattern in the image is then high-lighted by a bright spot. In reciprocal space this is equivalent to try to compensate the phases of the complex image transform by a filter which is complex conjugate. The more similar the pattern the better this phase match will succeed and produce a "delta-spot" upon back-transform ("Matched filter").

This procedure is quite common in light optics and in digital image processing in general, and some convincing applications in HRTEM are now available as well. It is a generalization of strain mapping, where a single spot is mostly used as template. Interface classification has been achieved for: hetero-interfaces and grain boundaries: For the former, a periodicity check is useful for both the quasi-coherent segments between misfit dislocations and also for the analysis of the distribution (and mutual similarity) of the dislocations themselves [28].

For grain boundaries, we attempt to detect the sequence of special structural units. Paciornik et al. examined a $\Sigma 5$ grain boundary in rutile [29]. Here two basis segments were cut from the micrograph and moved over the entire image, while recording the correlation values. Coincidence with the template results in a bright peak on a lower background level (caused by random subpattern coincidences).

3.3.8 Noise Filters with Structure-Determination Strategies

Noise suppression is described here under the subject strain mapping, since it is almost unavoidable as a first step, in particular for peak search. Noise suppression remains, however, a useful option (with appropriate adaptation in motivation and details) for structure refinement of the kind described in Sect. 3.4 [8].

While shot noise and film grain merely hinder the detection of a maximum within a peak, amorphous noise was shown to shift peaks, at least if the thickness of an amorphous film equals or exceeds that of the crystalline material [30].

Noise suppression algorithms can be encoded (via the convolution theorem) either in real space (*spatial averaging*) or in reciprocal space (*Fourier filter*). This theoretical equivalence may break down in practice owing to numerical limitations for non-infinite two-dimensional images or non-perfect periodic symmetry. As a rule-of-thumb, noise filters should preferably be applied in the space in which the information is most compact. For extended periodic objects or at least 1-D periodic objects, Fourier space (and thus Fourier filtering) is appropriate. For sharp local defects separating areas of translational symmetry (e.g. interfaces with steps, dislocations, precipitates), real space averaging has the advantage of being applicable right up to the defect, without including it.

The optimum shape of a Fourier filter mask (equivalently: the matrix of weighting factors in a spatial average) has been the subject of much debate. A systematic approach to optimum filters starts from the well defined

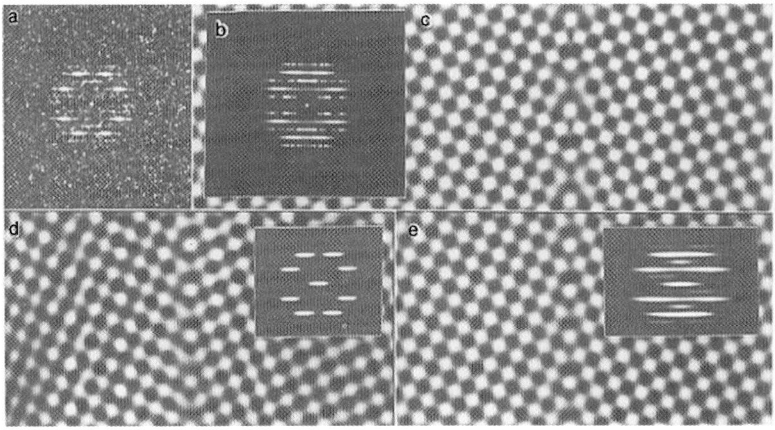

Fig. 3.11. Adaptive Fourier filter for interfaces (simulated $\Sigma 5$ grain boundary). (**a,b**) Original power spectrum with, without noise, (**c**) original image, (**d**) incomplete non-adaptive restoration with lost boundary superstructure, (**e**) optimum filter threshold, restoring correct boundary period (insets show Fourier filters)

"Matched Filter" and "Wiener filter" concepts. The former optimizes the signal-noise ratio in correlation-based pattern searches, while the latter restores the signal with the least possible deviation (in a least-square sense) from the estimated true signal, given the power-spectrum of both signal and noise. Neither is suitable for typical crystal defect problems in Q-HRTEM. Their filter functions depend heavily on the *region of interest* and *texture-anisotropy*. For an interfacial HRTEM-image, the pixels in the defect core area need to be filtered in a different way than the pixels in the neighbouring crystal cells. The ultimate solution to guarantee a patch-free filtered image of complex texture is the calculation of nearest-neighbourhood Wiener filters for every pixel of the original image. For a matrix of 1024×1024 pixels, one million optimum Wiener filters would need to be calculated and then applied to the whole image after which only one pixel in the centre of their original neighbourhood would be retained. As a compromise and a shortcut through this cumbersome pixelwise filtering scenario, an *adaptive filter* has been proposed for interfaces [28], which filters at least the most interesting area (the interfacial core) yielding a better SNR than the traditional Wiener filter. Figure 3.11 shows the concept of adaptive filtering on a calculated grain boundary image. With the noise-free original known in this case, the size and aspect ratio of the elliptical filter windows can be optimized specifically for the grain-boundary region (Fig. 3.11e). Figure 3.12 presents an experimental application of such a filter to a coherent segment of a hetero-interface.

Unlike strain mapping, where the noise disturbs locally the search for maxima, the main purpose of noise suppression in many other areas of Q-HRTEM is the *confinement of information*. An experimental image area of

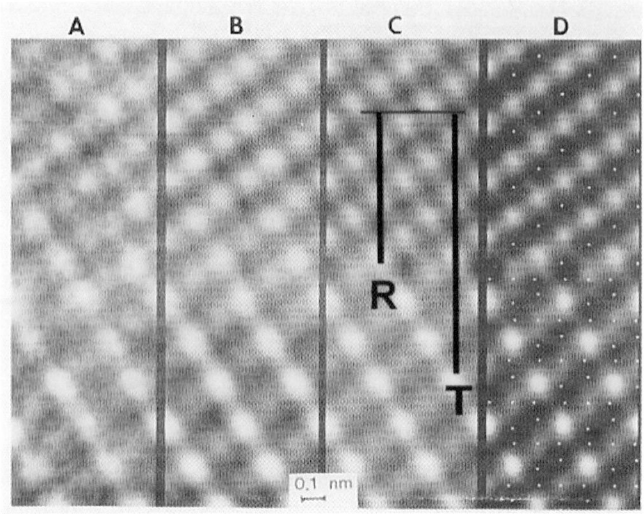

Fig. 3.12. Interface between Nb (*top*) and sapphire (*bottom*) after adjusting imaging parameters, translation distance T, and relaxation R of the terminating Nb layer. (**a**) Experimental original, (**b**) experimental filtered image, (**c**) best fit after iterative digital image matching, (**d**) structure model superimposed on "first guess" simulation. See Sect. 4.2 for details

suitable quality is often larger (owing to repetitive segments) than the simulated image area (one segment only to avoid computational redundancies). Filtering will extract the structure information from the big area and concentrate it as an average into a small area (Sect. 3.4). Figure 3.12a–c shows the improved quality of an image match between simulation and experiment, with and without application of a Fourier filter.

3.4 Iterative Digital Image Matching (IDIM)

In the general case of structure retrieval, for crystal defect cores for example, none of the fast techniques of direct image interpretation described above is applicable. On the other hand, the multi-step image formation process (Fig. 3.1) is too complicated to be simply inverted. A solution to this situation is *refinement*: given a first guess of the unknown structure, images can be simulated iteratively, with the best matching image then corresponding to the best structural solution (Fig. 3.13). This method is commonly used for the retrieval of crystal structures of bulk materials in crystallography by X-ray diffraction. In our case of high-resolution electron microscopy, the diffraction pattern is replaced by the HRTEM image intensity, and the atom coordinates of a perfect unit cell of a crystal are replaced by projected coordinates of atom columns in a defect. This strategy obviously allows a variety of individual mathematical modules to be combined flexibly.

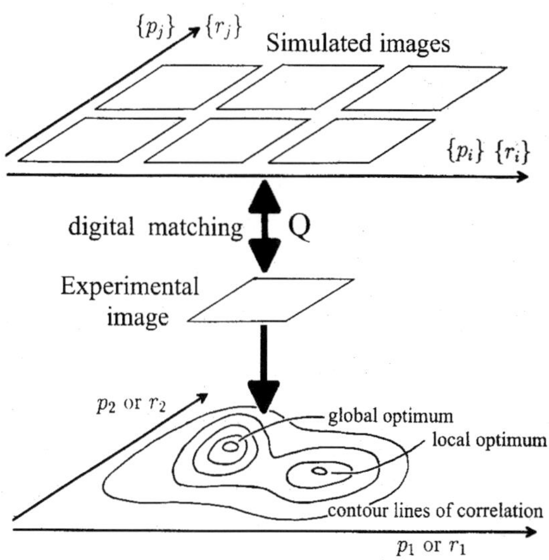

Fig. 3.13. Basic algorithm of iterative digital image matching: A series of structure models is converted into simulated images. These are evaluated against the experiment to obtain a map of fit-quality, which can be searched for the best-fitting parameter set

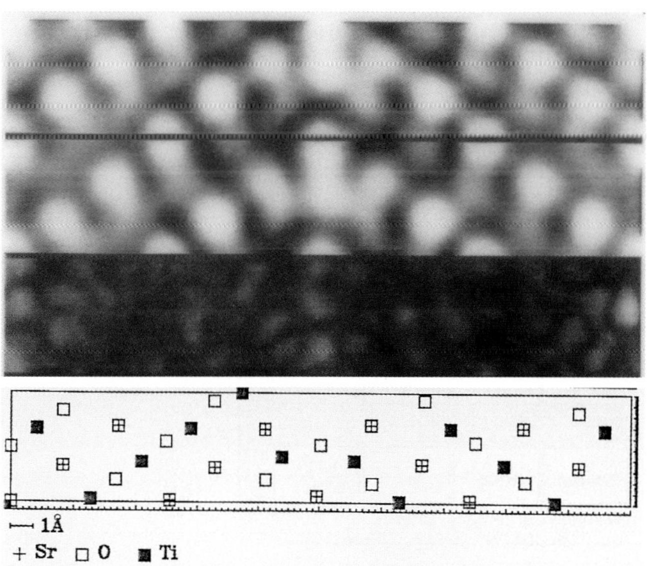

Fig. 3.14. Grain boundary ($\Sigma 3$) in $SrTiO_3$. From top: experimental image, best fit simulation, relative difference image (contrast enhanced for better visibility), structure model corresponding to best fit. From [31]

How refinement of a few free parameters in image simulation (optical data, atom layer positions $\boldsymbol{R}$, and half-crystal distance $\boldsymbol{T}$) leads to a significant improvement of image match between simulation and experiment is shown in Fig. 3.12b–d. Figure 3.14 shows a second application for grain boundary refinement comprising many individual atom columns. The resulting image match is exceptionally excellent (no excessive image difference at the boundary compared to the neighbourhood) since the specimen was very clean and the alignment was almost tilt-free.

3.4.1 Algorithms and Modules of Iterative Refinement

Behind the simple scheme of Fig. 3.13 there are several levels of choice betweeen alternative algorithms. Some decisions might be of general validity, while others vary from application to application. For convenience we refer in this section to one particular structure refinement software package, IDIM [3] (for *iterative digital image matching*), see Fig. 3.15. The discussion of its hierarchical and modular structure will start with the innermost switches and continues out to the main choice of structural topics.

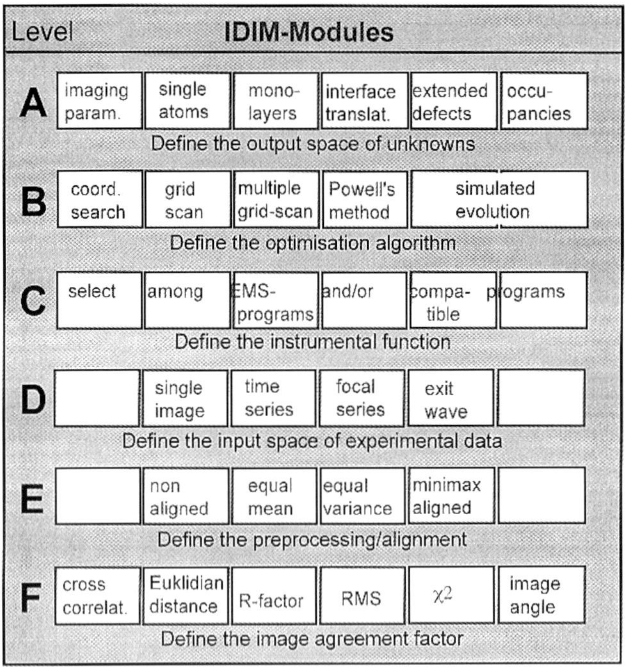

Fig. 3.15. Architecture of the IDIM-programme package ("iterative digital image matching")

Level F: Image-Agreement Factor. The bottom level of choice within Q-HRTEM and structure refinement is between the various formulae from which the similarity of simulated and experimental data is calculated. A metric, $D = ||\boldsymbol{E} - \boldsymbol{S}||$, needs to be defined on the experimental ($\boldsymbol{E}$) and simulated ($\boldsymbol{S}$) images, both being expressed here as vectors. Absolute or square distance are the most obvious choices, but arctan or ln-based metrics should not be excluded. Different refined structures result for different metrics [8]. This effect is not always significant but needs attention where the final match is far from perfect. The choice within a group of derived measures, internally sharing the same metric, is less severe; for the square metric for example, these might be: Euclidean distance, cross-correlation (xcf), RMS-value, normalised Euclidean distance (ned), χ^2-analysis (see [32,3] for details). The following conversion relations exist between ned, xcf and χ^2:

$$\mathrm{ned}(\boldsymbol{E}, \boldsymbol{S})^2 = \frac{\sum (E_{jk} - S_{jk})^2}{|\boldsymbol{E}||\boldsymbol{S}|} \tag{3.18}$$

$$= \frac{(\Delta A)^2}{|\mathbf{E}||\mathbf{S}|} + \frac{(\Delta s)^2}{|\boldsymbol{E}||\boldsymbol{S}|} + \frac{2 \cdot s_e s_s}{|\boldsymbol{E}||\boldsymbol{S}|} \mathrm{xcf}(\boldsymbol{E}, \boldsymbol{S}) \tag{3.19}$$

$$\mathrm{xcf}(\boldsymbol{E}, \boldsymbol{S}) = \left(\sum (E_{jk} - \overline{\boldsymbol{E}}) \cdot (S_{jk} - \overline{\boldsymbol{S}}) \right) / s_s s_e \,. \tag{3.20}$$

We have introduced s for the standard deviations, $\Delta s = s_e - s_s$ for their difference, $\Delta A = \overline{\boldsymbol{E}} - \overline{\boldsymbol{S}}$ for the difference in average image intensity. On the other hand,

$$N_{\mathrm{eff}} \cdot \chi^2(\boldsymbol{E}, \boldsymbol{S}) \propto (\boldsymbol{E} - \boldsymbol{S})^T (\underline{\sigma}^2)^{-1} (\boldsymbol{E} - \boldsymbol{S}) \approx \sum (E_{jk} - S_{jk})^2 / \sigma_{ij}^2 \tag{3.21}$$

$$\approx \frac{|\boldsymbol{E}||\boldsymbol{S}|}{\sigma_0^2} \mathrm{ned}(\boldsymbol{E}, \boldsymbol{S})^2 \,. \tag{3.22}$$

The matrix $\underline{\sigma}$ describes a noise distribution function with noise correlation between neighbouring pixels [33]; the first "≈" holds for uncorrelated noise and the second "≈" holds for spatially constant noise. N_{eff} is the effective number of pixels (reduced by the number of variables). As can be seen, xcf measures differences in pattern only (local distribution of contrast) and is invariant under differences in image brightness and overall contrast, often observed due to systematic errors (see level E below). On the other hand, ned measures the sum of differences in brightness, contrast and pattern [34]. A χ^2 value as image-agreement factor has a separate meaning (and is worth the effort) only if the information about the noise, to be entered as σ, is accurate and if the noise is additive to the signal. According to probability theory it then expresses the negative logarithm of the likelihood that an image difference $(\boldsymbol{E} - \boldsymbol{S})^2$ might be caused purely by noise. If χ^2 approaches its minimum allowed value of one, the image-matching process is complete, since no further improvement by refinement of parameters is possible. In the presence of systematic errors, however, the χ^2-value rapidly ceases to be useful.

Level E: Preprocessing of Images Before Matching. Optional noise suppression techniques may be applied, motivated in Sect. 3.3.8, or contrast alignment or symmetry operations (mirror or translational averaging) may be implemented. In any case, the aim of preprocessing is to accentuate on the essential information of a micrograph while some other information (average absorption, overall contrast, noise characteristics and symmetry deviations) is sacrificed. The choice of preprocessing will depend on individual properties of applications [3]. The more HRTEM-artefacts (misalignment, amorphous layers) or systematic errors are safely identified, the more useful will the preprocessing be to safe and separate the remaining undamaged information.

A mismatch of the overall contrast of HRTEM-images is frequently observed. Several authors have found that experimental images are weaker by a factor of around 3 in contrast (detailed findings vary from 1.5 to 9) than simulations that otherwise match the pattern ("Stobbs-factor" [34,35]). The origin of this factor remains unidentified. However, many issues create a difference between the real world and the ideal simulated world, including recording-medium resolution, drift and vibration of specimens, electrical fluctuations (not properly covered in the first-order envelope theory). A second important group consists of specimen irregularities, comprising strain, point defects, amorphous layers, phonon scattering, and atom irradiation displacements (see [36] for a survey).

Whenever the normalised pattern can be correctly reproduced by simulations, the author believes it is better to deliberately remove the Stobbs-factor by contrast normalization, rather than including absolute (and knowingly false) contrast into a fit programme. The reverse statement, of course, applies for research into the problem of contrast mismatch itself.

Level D: Set of Images as Input Data. Iterative structure refinement is normally applied to a single (often Scherzer-focus) micrograph. The precision of the refined structure data, the signal-to-noise ratio, and the security against false local optima can be enhanced significantly by employing image series (time or focal series), i.e. matching simulated against experimental arrays of images. Also, the use of a complex exit wave function, reconstructed by means of holographic techniques to provide input data is expected to enhance the reliability as well as providing a better start-up model for subsequent structure refinement.

Levels A and C: Instrumental Function and Output Space. The program architecture of iterative structure retrieval relies on a figure of merit (here: the image agreement factor $\mathrm{iaf}(\boldsymbol{E}, \boldsymbol{S})$) which is calculated as a function call out of an iterative loop on the main-level. While $\mathrm{iaf}(\boldsymbol{E}, \boldsymbol{S})$ is given by the formulae discussed under level F, and the experimental image $\boldsymbol{E}$ is unchanged, the simulated image is itself subject to a function-call $S = F_{\mathrm{I}}(\{a_i\})$.

This function F_I is called the instrumental function, which relates the simulation of the experimental data (the HRTEM image) to a parameterized model $\{a_i\} = \{r_{xi}, r_{yi}, c_i, Z_i, p_i\}$. The parameter vector is split into projected atom column coordinates $\boldsymbol{r}$, concentration factors (occupancies) $\boldsymbol{c}$ and atomic numbers (element symbols) $\boldsymbol{Z}$, as well as some variable image formation parameters $\boldsymbol{p}$, describing the specimen geometry (thickness and deviation from zone axis), the microscope alignment (focus, apertures, deviation from optic axis), and the recording conditions (modulation transfer function of film, or CCD-camera, the drift and vibration conditions). Many instruments in light-optics (microscopes and telescopes) can be represented approximately as a linear system, and the instrumental function F_I is then a closed expression with which the input image is convoluted. In HRTEM, owing to dynamical diffraction and the partially coherent, nonlinear character of image formation in phase contrast conditions, the situation is far more complicated: each function call F_I comprises a calculation of the projected scattering potential of a thin slice of the object model, followed by calculation of the propagation of a plane wave through the specimen, and finally by numerical calculation of the image intensity using the contrast-transfer function of the objective lens within the framework of transmission cross coefficients (Fig. 3.1).

Each HRTEM simulation software package (available for many platforms of computers from commercial sources or from several research institutes) generates F_I. While the IDIM package is relying on EMS [7], other choices of image simulation software (e.g. SHRLI, SEMPER, MACTEMPAS) may equally well be used.

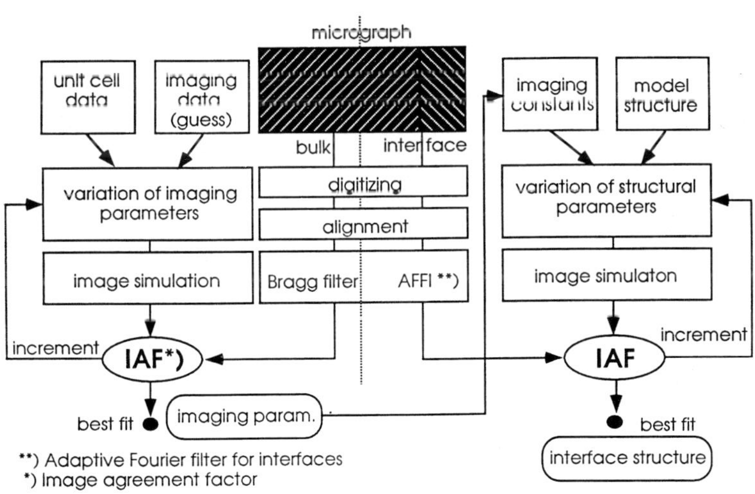

Fig. 3.16. Flow diagram for IDIM in the study of interface structures, shown as two consecutive loops of refinement for bulk and interface regions. The application of noise suppression (AFFI) is optional and is recommended for periodic interfaces

Within the output space of unknown parameters to be determined, only a subset is normally chosen as variable fit parameters. A recommended strategy is to use segments of perfect crystal lattice of known structure, often found in the vicinity of crystal defects, to fit the simulation parameters p_i; the atom coordinates r_i are then fitted using constant simulation parameters. This scheme is summarized in Fig. 3.16. Alternatively, or better still in addition, parameters among the p_i such as focus or specimen thickness should be checked by independent techniques: EELS plasmon spectra, or tilted interface projections for the thickness, or Fourier transforms of the image intensity of an amorphized specimen region (so-called diffractogram) for the focus value.

Level B: Optimization Algorithm. Finally, given the input data (images), output data (fit-parameters), and instrumental function F_{I}, we have to select the numerical optimizer that will govern the progress on a quality-scale determined by the image agreement factor iaf(E,S). Choices from textbooks on numerical mathematics [37,38] can be classified into three groups:

(i) Trivial but nonetheless useful techniques like "coordinate search" (fitting one parameter after another, Gauss–Seidel method) and "grid-scan techniques" (in which the whole N-dimensional parameter space is sampled on a grid with constant mesh-width, normally applicable for $N < 4$ only, see Fig. 3.17).

(ii) Downhill techniques, heading as quickly as possible for the next *local* optimum, e.g. non-linear least squares (Levenberg–Marquard), non-linear simplex method, Powell's method or conjugate gradients.

(iii) Finally a more recent group of algorithms, called "global optimizers" try to explore a large proportion of the N-dimensional configuration space at a lower efficiency and speed, but with the advantage of jumping back out of "suboptimal" (metastable) local optima. Examples are simulated annealing, simulated evolution and genetic algorithms [39].

Each of the three groups (i)–(iii) has its own domain of preference, group (i) algorithms are suitable for test calculations, generation of instructive dis-

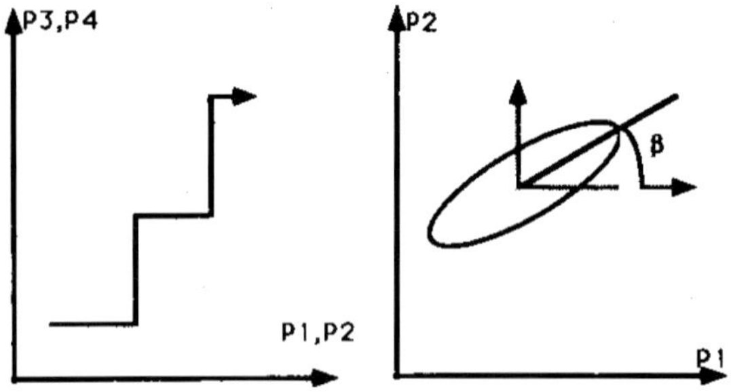

Fig. 3.17. Optimization by combined coordinate search and two-dimensional grid scan (iterated)

plays, preliminary focus–thickness matching and some simple low-dimensional interface problems (see applications). Group (ii) is the best choice for a complex high-dimensional structure-determination problem when the initial structure is known to lie close to the relaxed final solution. Global techniques of group (iii) are appropriate for all remaining cases of applications. Combinations are also feasible, e.g. starting with a quick local run to find a first optimum, then continuing with a coarse and wide global search to check for remote "better" local optimum positions, and finally returning to the global optimum and refining it down to ultimate precision with another local optimiser run.

3.4.2 Applications in Interface and Dislocation Science

Numerous results have been obtained meanwhile with the rather young discipline of automatic structure refinement. Digital difference images were originally used to distinguish between a few discrete structure models, e.g. dislocation cores [40,41]. Iteration was employed for the first time to the problem of focus–thickness matching using grid-scan or non-linear least-squares optimization [42,43]. A comparison of real-space versus reciprocal-space formulae for image agreement is given in [44].

A niobium–sapphire heterointerface problem was treated by the grid-scan technique in three separate steps. A pairwise imaging condition refinement was followed by a two-dimensional grid-scan to determine the translation vector (which defines the distance between the two lattices apart from interface relaxations) and, finally, the relaxation of the terminating monolayer of niobium was measured by another two-dimensional grid scan [45,46,8].

Multi-dimensional problems of atom-column positions in extended grain boundaries were solved using non-linear simplex or non-linear least squares [47,48,18,49]. Further structure refinement for a one-dimensional degree of freedom in structure has been applied to an oxygen position in a superconductor structure [50]. Global optimization by simulated evolution was employed for refinement of multiple imaging parameters and structural data at a copper–sapphire interface [51]. Further applications of structure refinement comprise grain boundaries in metals such as aluminum [52] and NiAl [53] and also in oxide-ceramics (Al_2O_3 [54], and $SrTiO_3$ [55]).

In the following we present a model application for each of the basic optimization techniques, introduced as (i), (ii), (iii) in Sect. 3.4.1 above.

Hetero-Interfaces and Translation States. Whenever the positions of contrast maxima/minima and atom columns disagree, a careful determination of the imaging condition is essential. We consider the terminating monolayer of a niobium thin film adjacent to the sapphire substrate crystal (Fig. 3.5). Errors in the imaging condition δp are likely to be propagated to the atom

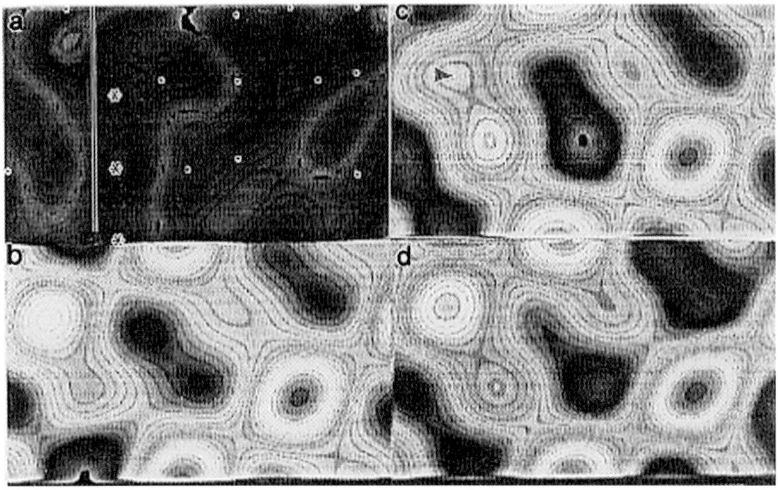

Fig. 3.18. Interface core area of Fig. 3.12 in contour line representation; left: Nb, right: sapphire. (**a**) atom positions on image difference (line marks the chemical interface); (**b**) experimental image; (**c**) simulated image without terminating oxygen layer (marked * in (a)); (**d**) simulated image with oxygen layer. The arrow in (c) points to a mismatching detail. From [8]

positions δr_i

$$(\delta r_i)^2 = (J_{ij}\delta p_j)^2 = \sum_j (\partial r_i/\partial p_j)^2(\delta p_j)^2 \tag{3.23}$$

because the Jacobian matrix J_{ij} does not vanish. On the other hand, hetero-interfaces offer the great advantage of providing twice as much information about the imaging condition because each perfect crystal lattice close to the defect can be investigated separately. A typical set of unknowns comprises the following eight (or even more) parameters p_i: focus, thickness, crystal tilt (x, y), beam tilt (x, y), and modulation transfer, including drift and mechanical vibration (x, y, angle α). Extra parameters can be added such as two-fold and three-fold astigmatism, coma, and partial coherence. Spherical aberration and aperture size are better determined from independent measurements. A simple cubic lattice does not offer enough linearly independent beams to determine all of these parameters unambiguously. For hetero-interfaces, the ambiguities are much reduced, leaving typically between one and three possible solutions on the focus–thickness map, among which often just one is reasonable.

Figure 3.12d displays the geometry and structure of the model system for this application, a Nb-Al_2O_3 interface [8]. The processing of the experimental image (a) into a filtered image (b) and its match to a multi-parameter fitted simulation (c) is demonstrated. The above-mentioned set of eight unknown imaging parameters was refined in a simple Gauss-Seidel type of coordinate

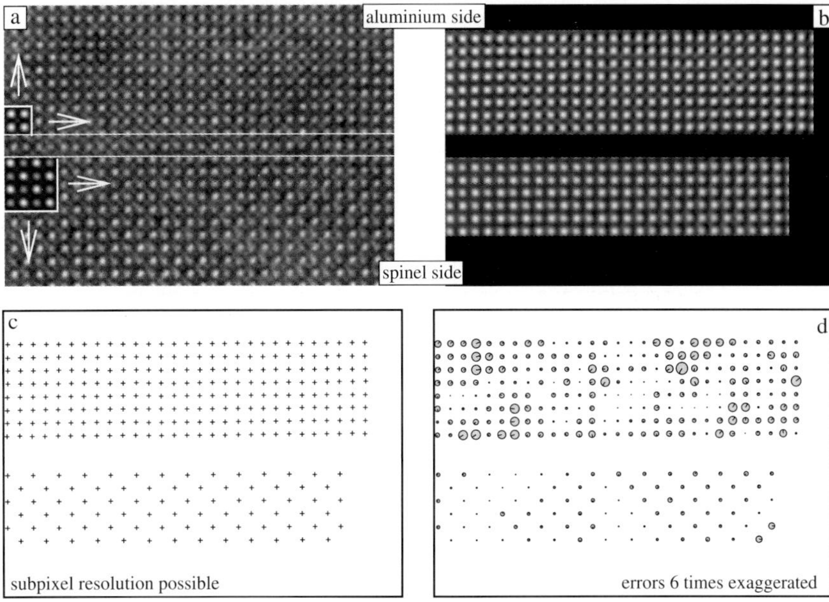

Fig. 3.19. (**a**) Al-$MgAl_2O_4$ interface with square templates for lattice measurement (recorded with the Stuttgart JEM-ARM1250, by R. Schweinfest). (**b**) Result after cross-correlation of the image with the templates. (**c**) Peak detection on the correlation map of (c). (**d**) Spatially resolved errors (deviation from a cubic reference lattice). From [56]

pair-wise search (Fig. 3.17). The further refinement comprised the translation vector $\boldsymbol{t}(x, y)$ and the relaxation $\boldsymbol{r}_1(x, y)$ of the first Nb monolayer. Figure 3.18 shows the final match at high magnification in contour line representation. To show the high sensitivity of the method, an alternative (worse) match with an oxygen-monolayer deficiency (marked as * in (a)) in the interface core is attached.

Although $\boldsymbol{t}, \boldsymbol{r}_i$ describe structural and p_i optical parameters, it is logically more sensible to consider $\boldsymbol{t}, p_i$ as one subspace of optimization and $\boldsymbol{r}_i$ as the other. This is easily understood by considering specimen tilt θ, as a free parameter, which couples strongly to the local mean lattice position and thus to $\boldsymbol{t}$, leaving us with $J_{\boldsymbol{t},\theta} > 0$ and $J_{\boldsymbol{t},r_i} = 0$ (by definition of $\boldsymbol{t}$).

An appropriate representation of the dependence of translation states $\boldsymbol{t}$ on the set of optical HRTEM-parameters has been proposed by Schweinfest et al. [56]. Figure 3.19a shows the coherent (misfit-free) interface between Al and $MgAl_2O_4$. Since no significant relaxations of individual atoms were found, the iterative structure-refinement problem collapses to an assessment of the multi-dimensional parameter-set $(\boldsymbol{t}, \{p_i\})$. As before, focus, thickness, specimen tilt and three-fold astigmatism A_3 are grouped into the vector $\boldsymbol{p}$. A projection of the distribution of individual trial sets (t_k, p_{ik}), for $k =$

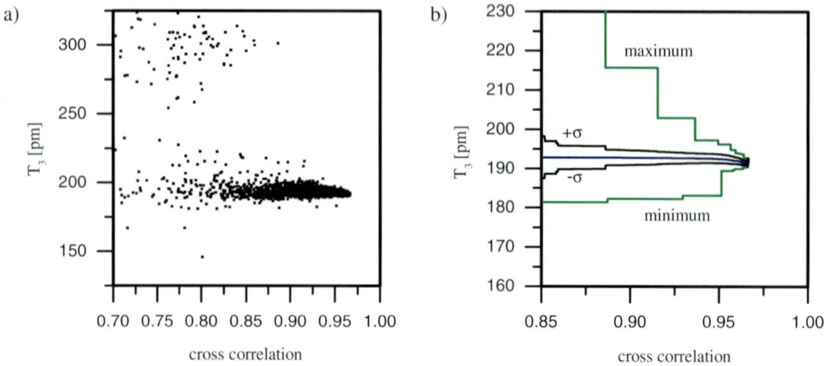

Fig. 3.20. (**a**) Point cloud showing distribution of trial translation states for the simulated Al/spinel interface depending on various optical imaging conditions assumed. Cross-correlation is used as the fit-quality criterion. (**b**) Envelopes constructed from (a). From [56]

1, 1000 has been recorded using the simulated evolution algorithm of IDIM [57]. From Fig. 3.20, a projection-width of ± 4 pm resulted for the error on $\boldsymbol{t}$, significantly lower than the estimate given by a two-dimensional parameter pair $(J_{\boldsymbol{t},A_3})$ [58]. The reduced influence of a single parameter that couples strongly to other parameters comes from a failure of square ("Gaussian") error propagation according to (3.23). This equation was originally intended to combine physically independent errors on individually measured input quantities. During multi-dimensional optimization, the parameters do not, however, fluctuate randomly, instead they are driven by the optimization quality-criterion towards mutual compensation.

Grain Boundary Structures by Local Multi-Dimensional Optimization. The model system is a $\Sigma 5$ grain boundary in NiAl [53]. The configuration space of this optimization problem consists of 12 projected atom column positions plus the translation vector. Nonlinear simplex [18] optimization has been applied to arrive at the best-fit structure model of Fig. 3.21, slightly asymmetric at the interface and with a wider translation state than the theoretical prediction. A local simplex algorithm could be applied here since the distance between the start-up model (obtained either from a symmetric coincidence-site-lattice construction or from a molecular statics refinement) and the best-fit solution is close enough to reduce the risk of trapping in other local optima.

Dislocation Structure Refinement by Global Optimization. After interface structures viewed in cross-section, dislocations viewed along their line direction form the second major group of applications for structure refine-

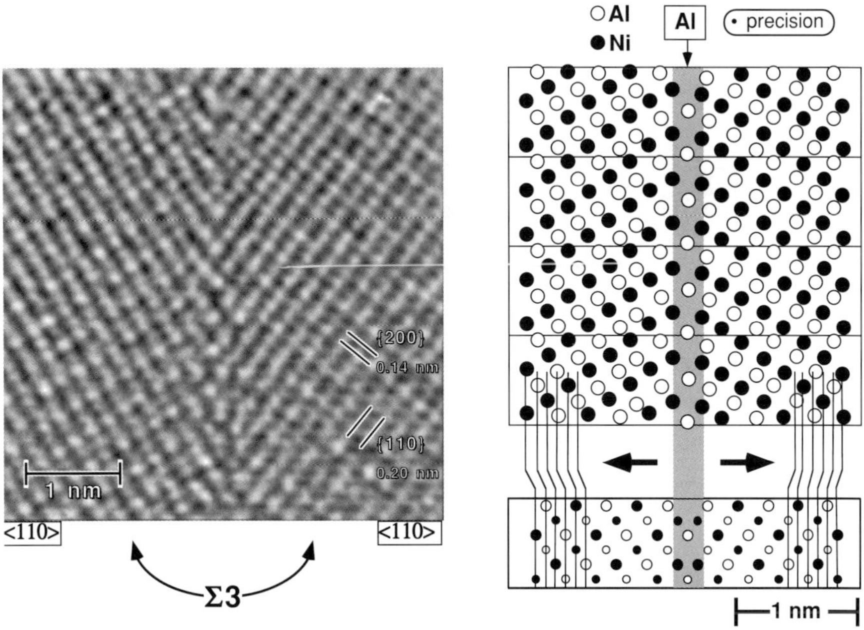

Fig. 3.21. *Left:* HRTEM image of a $\Sigma 5$ grain boundary in NiAl. Recorded with the Stuttgart JEM-ARM1250 in order to resolve the Ni and Al columns at a distance of 0.14 nm. *Right:* Structure model behind best-fit simulation (top) and theoretical prediction (bottom) from molecular statics simulation. From [53]

ment in HRTEM. The required supercell size for dislocations to deal with the periodic repetition requirement of FFT processes, is substantially larger than for short-periodic interface segments. Along with this large increase in the number of atom columns entering the simulation, there is also an increase in the number of *unknown* atom column positions, to be entered as fit-parameters. Furthermore, the signal-to-noise ratio can normally not be improved by applying Fourier filters or averaging, as it can with short-periodic interfaces. The model system is a bulk dislocation in niobium of Burgers vector (1/2)[111] and line-direction [100] (Fig. 3.22). It was found at about 5 nm from an interface between niobium and sapphire, far enough to consider it as a bulk dislocation. A reasonable strategy for the refinement of the structure of the visible edge-component must take into account the expected radial dependency of shifts R between contrast maxima and atom positions. The highly symmetric bulk-regions surrounding the dislocation core here yield $R = 0$, while R increases monotonically towards the dislocation core. A subdivision into segments I, II, and III as shown in Fig. 3.22a is therefore appropriate. In region I, positions are directly interpreted following the procedures of Sect. 3.3.3. In region II, a fast iterative structure-refinement is preferred. A

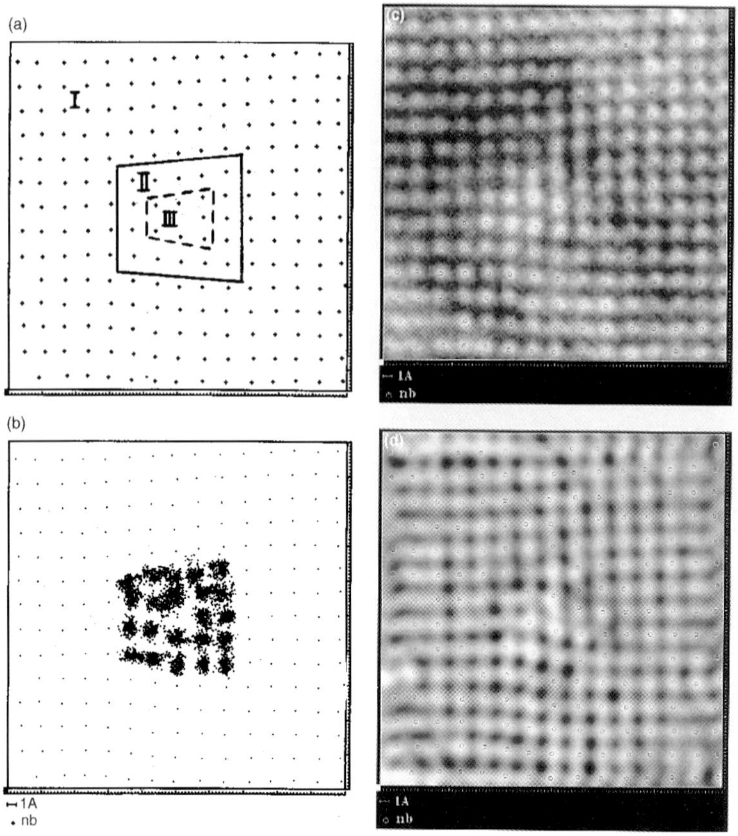

Fig. 3.22. Dislocation in niobium with line direction [100] and Burgers vector along [111]. (**a**) Supercell used for the simulation and regions I–III for subsequent iterative structure refinement. (**b**) Distribution of trial structures for a run of simulated evolution on the segment II of (a). (**c**) Experimental image (Stuttgart JEM-ARM-1250). (**d**) Best-fit simulation (from [3])

global optimization by simulated evolution is finally applied to the core atoms in region III. The application to the outer regions is in practice conducted including the innermost regions, thereby furnishing a good initial guess for the next refinement step. The imaging condition was refined independently in a previous simulated evolution run, using in parallel segments of perfect niobium and perfect sapphire found in the vicinity.

Three-Dimensional Effects. Dislocations are also good examples to demonstrate the basic requirement of straight atom columns with projected coordinates (r_x, r_y) in the refinement process, as assumed so far. After cross-sectional thin-foil preparation for HRTEM, the asymmetric stress of an edge-dislocation structure (high-stress region on the side of the extra half-plane) will interact with the free surfaces and lead to column bending near the surfaces. This in turn creates non-planar strain fields and a true 3D-structure. The 2D-structure refinement is thus to be interpreted as giving the effective projected structural coordinates, which approximate the straight central portions of an atom column with increasing specimen thickness. In Fig. 3.22c, the result of near-surface column bending can be seen as a region of excess brightness at the lower-left of the core, not included in a 2D-simulation. For semiconductor heterosystems, such an effect has been discussed by Bierwolf et al. [15].

3.4.3 Probability Calculus and Precision Estimation

Atom coordinates determined during iterative refinement, often printed with three digits after the decimal point on a nm scale, are worthless unless the "error bars" (or in the case of projected structures, *confidence circles*) are given as well. Minimization of these circles requires first of all experimental efforts to clean specimens, accurate microscope characterization and alignment, and selection of a suitable problem (good projection symmetry, not too many unresolved atomic columns per unit area). To a far lesser extent, inaccuracies of routine HRTEM simulation codes may contribute to the error.

The definition of confidence circles is, however, a different question. This has been tackled in several ways [18,8,51,52]. Here, only a brief survey of recent concepts is given [59,60]:

Jaynes' Entropy, Likelihood, and Bayesian Inference. "Maximum – a posteriori" (MAP) theory within the framework of Bayesian estimation (e.g. [61]) tries to answer the question: *What is the probability that possible sets of object parameters will emerge from the given experimental data?* (Fig. 3.23). Using $\boldsymbol{E}$ as the experimental data set (e.g. the HRTEM micrograph) and $\boldsymbol{a}$

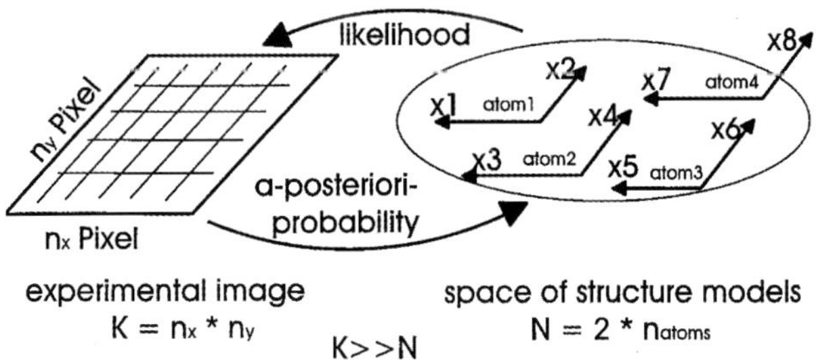

Fig. 3.23. Q-HRTEM regarded as an experiment that maps the input space (digital image) to the digital space of structure models (and vice versa)

as formal vector of object parameters (e.g. atom coordinates of an unknown defect) we have:

$$p(\boldsymbol{a}|\boldsymbol{E}) = \frac{q(\boldsymbol{E}|\boldsymbol{a}) \cdot t(\boldsymbol{a})}{p(\boldsymbol{E})}, \tag{3.24}$$

or in short notation: $p(\boldsymbol{a}) = q(\boldsymbol{a}) \cdot t(\boldsymbol{a}) / \int d\boldsymbol{a} q(\boldsymbol{a}) \cdot t(\boldsymbol{a})$. The probability $q(\boldsymbol{a})$ for $\boldsymbol{E}$ to result from a given $\boldsymbol{a}$, is known as the "likelihood" (Fig. 3.23). All a-priori knowledge is collected in $t(\boldsymbol{a})$. Classical likelihood estimation (Fig. 3.23), in the sense of a χ^2-fit, is then a special case of (3.24) with settings $t = 1$ and $q = \text{const} \cdot \exp(-\chi^2)$. For HRTEM work not based on either of the assumptions behind χ^2 or image entropy [8,51,31] we attempt to develop a link to (3.24) and extend its inherent derivation of error bars to the derivation of continuous density plots of $p(\boldsymbol{a})$.

Derivation of $p(\boldsymbol{a})$ from Jaynes' Equations. We now define an entropy in configuration space (as opposed to image entropy) in terms of $p(\boldsymbol{a})$ and subsequently maximize it subject to linear constraints [62]:

$$\delta/\delta p \left\{ \int d\boldsymbol{a} p(\boldsymbol{a}) \ln(p(\boldsymbol{a})) - \lambda_0 N - \sum_j \lambda_j [C_j(p) - C_{j0}] \right\} =: 0 \tag{3.25}$$

$$N = \left[\int d\boldsymbol{a}(\boldsymbol{a}) - 1 \right]. \tag{3.26}$$

The constraints are included by the technique of Lagrange parameters λ_j applied to a functional C_j of p being equal to a constant (see below). The derivative in variation calculus is written as δ. Neither (3.24) nor (3.25) requires the best fitting parameter-vector $\boldsymbol{a} = \boldsymbol{a}_0$ to be found. The unknown is the whole function p. We proceed from (3.25) for two complementary cases.

The Case of Known Likelihood. The likelihood $q(\boldsymbol{a})$ of (3.24) can be considered as available only if the entire noise-distribution statistics is known. Furthermore χ^2-analysis is limited to *additive noise* and systematic errors are assumed to be absent. There are two specific constraints for q and t of (3.25):

$$\frac{\delta}{\delta p} \left\{ \int d\boldsymbol{a} p(\boldsymbol{a}) \ln(p(\boldsymbol{a})) - \lambda_q (C_q - C_{01}) - \lambda_t (C_t - C_{02}) - \lambda_0 N \right\} =: 0. \tag{3.27}$$

Both constraints are identified as expectation values:

$$C_q = \langle Q \rangle = \int d\boldsymbol{a} p(\boldsymbol{a}) Q(\boldsymbol{a}); \quad C_t = \langle T \rangle = \int d\boldsymbol{a} p(\boldsymbol{a}) T(\boldsymbol{a}) \tag{3.28}$$

The quantity Q of (3.28) is related to q of (3.24) by $q(\boldsymbol{a}) = \exp[-\lambda_q Q(\boldsymbol{a})]$, and similarly $t(\boldsymbol{a}) = \exp[-\lambda_t T(\boldsymbol{a})]$. Making use of the intermediate result

$p = \exp[-(\lambda_0 + \lambda_q Q + \lambda_t T)]$, in which $\lambda_0 = \ln(\int d\boldsymbol{a} q t)$, (3.27) then simply reduces to (3.24), using the standard reasoning of statistical thermodynamics [62,63]. This setting guarantees consistency with the Bayesian logic.

The Case of Unknown Likelihood. If $q(\boldsymbol{a})$ and $Q(\boldsymbol{a})$ are unavailable because the noise statistic is unknown, we can still calculate a deviation D [e.g. the square difference $(\boldsymbol{E}-\boldsymbol{S})^2$] between experimental data and simulated data. It is again reasonable to weight the a-posteriori function $p(\boldsymbol{a})$ by D, which gives us a first constraint for (3.25).

$$C_d = \langle D \rangle = \int d\boldsymbol{a}\; p(\boldsymbol{a}) D(\boldsymbol{a})\,. \tag{3.29}$$

However, the width of the peaks of the distribution p, which will subsequently give the error bars, become arbitrarily dependent on the particular choice of difference measure (linear, quadratic, ...) for D. More constraints are needed therefore. The deviation measure D could contain free parameters c_i as in $D = [(\boldsymbol{E} - \boldsymbol{S})^2 - c_1]/c_2$; these would be adapted to meet constraints.

A promising suggestion is to use the second moment of the probability distribution (width W) in a particular area $\Delta\boldsymbol{a}$ of parameter space for calibration (an area in which the solution is known, e.g. in perfect crystals); written as a constraint, this has the form:

$$C_2 = \lambda_2[(1/\Delta\boldsymbol{a}) \int_{\Delta\boldsymbol{a}} (\boldsymbol{a} - \langle\boldsymbol{a}\rangle)^2 p(\boldsymbol{a}) d\boldsymbol{a} - W_{\Delta\boldsymbol{a}}]\,. \tag{3.30}$$

If we define the variance $V(\boldsymbol{a}) = (\boldsymbol{a} - \langle\boldsymbol{a}\rangle)^2$, we see by comparison of (3.28) and (3.30) that $V(\boldsymbol{a})$ now plays the role of the *a-priori* distribution $T(\boldsymbol{a})$. However, $V(\boldsymbol{a})$ is normally not known beforehand.

A Recipe for Practical Applications: Apart from the very powerful technique of printing point-cloud maps for structural solutions corresponding to best-fit conditions for various imaging parameters (Fig. 3.19), there is another highly suitable technique for noise-induced uncertainties in atom coordinates: a large number of copies of the true experimental noise patterns, e.g. at an amorphous edge, varying according to the true noise statistics (however complicated) are added to the image of a structural unit cell. Alternatively, in a perfect crystal region with many identical unit cells this is in fact done by nature itself. Refinement of the bulk crystal coordinates (assumed to be unknown for the moment) for all noise patterns is then a powerful tool by which to obtain complete probability-density functions for the coordinates in terms of histograms of refinement results (Monte-Carlo error estimation [37]).

Figure 3.24 shows this technique applied to the $\Sigma 3$ grain boundary in strontium titanate of Fig. 3.14. Evaluation of 36 unit cells of undisturbed

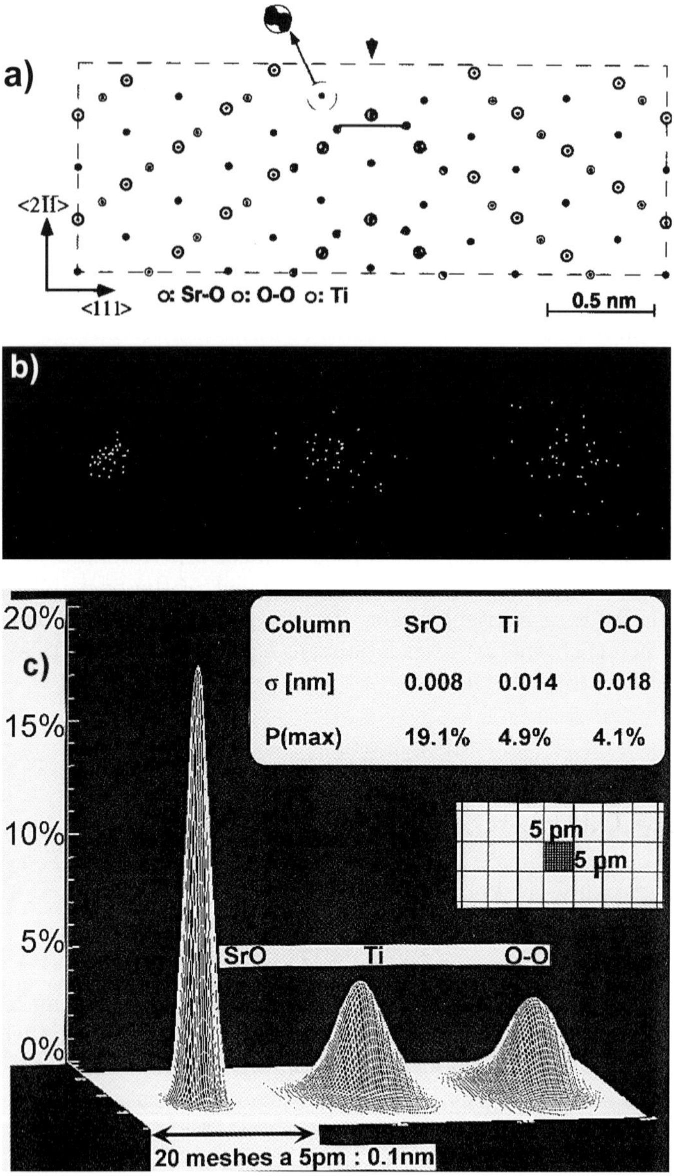

Fig. 3.24. Determination of confidence circles for atom positions in a grain boundary. (**a**) Structure model of $\Sigma 3$ grain boundary in $SrTiO_3$, redrawn from Fig. 3.14 with circles now indicating the degree of confidence for every atom position; (**b**) point cloud of fitted atom distribution using multiple unit cells away from the interface (fluctuations due to noise). Three atom-column compositions exist (from left: Sr-O, Ti, O-O); (**c**) conversion of point clouds to probability peaks by means of a Parzen technique. From [31,60]

material (excluding the boundary region) was possible using

$$\{\boldsymbol{E}^l\} = \{\langle \boldsymbol{E} \rangle + \boldsymbol{N}^l\}\,, \tag{3.31}$$

where l runs from 1 to 36.

A statistical sample of "best fit" positions can then be plotted after performing of structure refinement on all 36 segments in parallel; the result, printed as a point cloud in Fig. 3.24b, represents

$$\boldsymbol{x}^j = \langle \boldsymbol{x}^j \rangle + \delta\boldsymbol{x}^j = \{\boldsymbol{x}_{\text{true}} - \delta\boldsymbol{x}_{\text{S}}^j\} + \{\delta\boldsymbol{x}_{\text{N}}^j\}\,. \tag{3.32}$$

Here we have split off a systematic error (index S), defined by the deviation of the sample average from the ideal (true) crystal structure, and a noise error (index N), defined by the fluctuations around the true structure.

Postprocessing of the point-cloud by a Parzen technique (convolution of every delta-peak with a Gaussian kernel, subsequent integration and normalization) gives peaks of the probability-density distribution for every column composition (Fig. 3.24c). After definition of a "grid of interest" (here: $5\,\text{pm} \times 5\,\text{pm}$) the densities can be converted to "percentage"-probabilities. The standard deviations of the peaks give us the radii of the confidence circles (Sr-O: $\sigma = 0.008\,\text{nm}$; Ti: $\sigma = 0.014\,\text{nm}$; O-O: $\sigma = 0.018\,\text{nm}$) [31]. The peak-top probabilities are also printed in Fig. 3.24, but are restricted in meaning to the basic grid chosen [60]. More details on error assessment including the interface can be found in [31,64,60].

3.5 HRTEM-Based Structure Determination Techniques

A variety of important approaches to the computer-controlled interpretation of HRTEM-micrographs have been developed over the last few years, which could not be covered in detail in this chapter. They are briefly summarized here.

3.5.1 Classification of Techniques

From the previous sections it clearly follows that the choice of an algorithm from which the underlying object structure can be derived from an HRTEM image has to take into account

- the object under study (its thickness and sharpness of defects),
- the microscope alignment and microscope technology (coherence, contrast-transfer function),
- the kind of information introduced into a structure-retrieval problem,
- the ratio of available experimental data to unknowns in the structure-determination problem,
- the treatment of noise and systematic error sources.

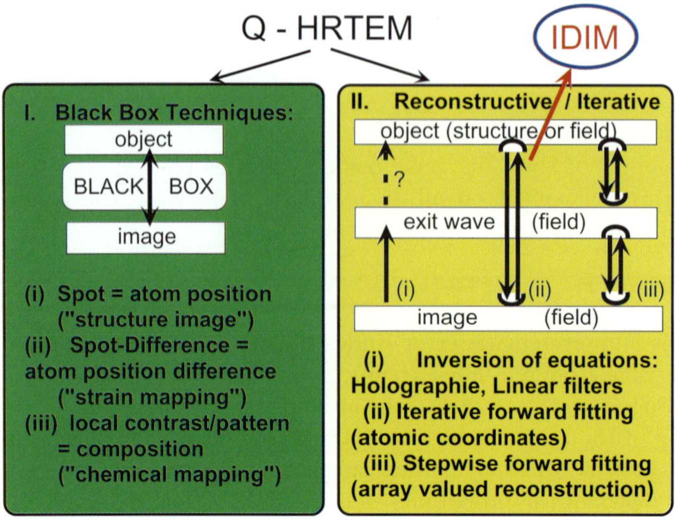

Fig. 3.25. Classification of strategies according to their direct or indirect direction of inference

Speed, ease-of-use, memory requirements, a-priori knowledge and/or the need for an initial guess for the structure solution are further important criteria.

Two ways of classifying structure-retrieval algorithms seem to be reasonable: the first scheme (Fig. 3.25) discriminates between "black box" approaches, techniques that use image simulation in forward direction (object $\longrightarrow$ image), and, true reconstruction algorithms that try to invert stepwise the image-formation theory.

A second scheme discriminates between the input and output planes within the stepwise image formation process, irrespective of the iterative or reconstructive nature of the algorithm. This second scheme is illustrated in Fig. 3.26 and collects various algorithms from the literature.

3.5.2 List of Techniques Not Covered in This Book

1. **Chemical Mapping:**
   Input: single HRTEM image; output: chemical concentration profiles; application range: coherent diffuse heterointerfaces; level: fully quantitative; type: black box; restrictions: chemically sensitive reflections [65]
   This technique works on digitized HRTEM images of diffuse but coherent interfaces. All atom coordinates are known beforehand. The stoichiometry of the crystals away from the interface must be assessed. Then, for a subset of materials and viewing directions, partially forbidden reflections exist (allowed in one lattice, forbidden in the other), the strengths of which are proportional to the concentration in the mixed area.

2. **"Quantitem", Local Fourier Analysis, Neural and Fuzzy Analysis:**
   Input: single HRTEM image; output: chemical concentration profiles; application range: coherent diffuse heterointerfaces; level: fully quantitative; type: black box; restrictions: linear-to-nonlinear pattern ratio must be proportional to concentration [66–68].
   These generalizations of (1) use the ratio between subsets of lattice-fringe-forming beams instead of forbidden reflections. The variants differ in whether they operate in real space (pattern recognition), or reciprocal space (unit-cell-wise spectrum) or whether they use floating-point data or robust classes of data-values. The application range is wider than (1) but is still restricted to known specimen geometries and structures and to orientations with image formation dominated by a few Bloch-waves [69].

3. **Nonlinear Focal Series Reconstruction:**
   Input: HRTEM focus series; output: exit-wavefunction; application range: all materials and defects; level: first part of structure solution (often sufficient); type: reconstructive (iterative); restrictions: precise knowledge of imaging conditions required.
   Inversion of the imaging by the lens (contrast-transfer function) has long been based on the linear approximation [70]. Such an approximate solution can nowadays be used as starting-point for more elaborate iterative schemes, in which an estimate of the complex exitwave emerging from the specimen is successively improved by matching the recalculated images (focal series) to the experimental focal series [71–73]. In this way, partial coherence and non linear interferences can easily be incorporated. Maximum likelihood is most common as the optimizer module.

4. **Off-Axis Holography with a Biprism:**
   Input: HRTEM hologram; output: exit wavefunction; application range: all materials and defects, level: first part of structure solution (often sufficient); type: reconstructive (analytical); restrictions: precise knowledge of imaging condition required, nearby hole in specimen [74].
   With the same objective as (3), off-axis holography restores the exit-wavefunction by first recovering the lost phase of the image amplitude from the hologram and then compensating the aberrations by deconvolution of the contrast-transfer function. The inherent linearity and speed and suppresion of inelastic scattering and speed are distinct advantages for a real-time reconstruction. For practical reasons owing to limited knowledge of the values of the aberration coefficients, iterative schemes have also been proposed [75].

5. **High-Angle Annular Dark-Field STEM:**
   Input: dark-field STEM image; output: map of high-angle scattering potential; application range: all materials and defects (except very light ele-

ments); level: structure solution (full or partial); type: direct or iterative (maximum entropy); restrictions: precise knowledge of probe function required, range of vailidity of the incoherent approximation [76,77].
High-resolution lattice images are also formed in the STEM as we know from the reciprocity theorem. While bright-field phase contrast provides little advantage over HRTEM in the TEM, an annular dark-field detector can be used in STEM. Integrating over the high-angle scattered electrons, this detector picks out those parts of the wavefront that appear to correspond most closely to an incoherent scattering mechanism. To such an approximation, the image-forming process then collapses to a linear system problem. Its solution, which can be obtained by deconvolution, Wiener filtering, maximum entropy or likelihood fitting, provides the scattering potential of the specimen for the high-angle regime. This is a sharply peaked distribution, which for perfect orientation, straight atom columns and low-noise conditions often directly reveals the atomic structure. The most striking advantages of the technique are the high sensitivity to the chemistry (a consequence of the absense of conflicting unknown imaging parameters) and the occupancies (number and density of atom columns in a completely unknown structure). Otherwise, for less peaked solutions, structure models may again have to be fitted to the continuous potential function, or in the case of residual coherence, to the image function.

Iterative techniques for retrieving structure information from electron-diffraction patterns [78] or chemical bonding information from convergent-beam diffraction disks [79] are beyond the scope of this chapter but are included in Fig. 3.26.

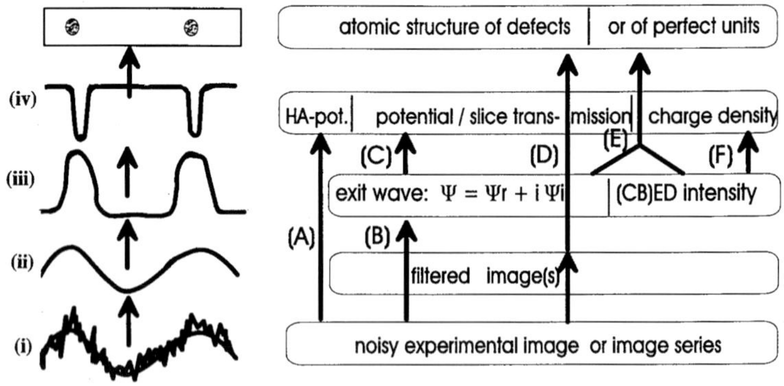

Fig. 3.26. Classification of strategies according to the input and output data used. Left: stepwise structure retrieval. Right: direction and levels of inference of several iterative techniques: (A) [76]; (B) [72]; (C) [80]; (D) [3]; (E) [78]; (F) [79]

3.5.3 A Practical Guide

The flow diagram of Fig. 3.27 is intended to assist the selection of the best suitable technique. Strain mapping and chemical mapping are grouped as "black box" techniques. Black box approaches can always be replaced by full simulation techniques, if there is any doubt about their validity. Black box techniques are essential *short cuts* to avoid redundant repetition of the lengthy image-formation calculation for one and the same *basic* imaging condition. Black boxes are faster by several orders of magnitude, can operate on much larger areas of the micrograph, and can be shown to exploit virtually *all* existing information in the micrograph (for suitable applications) concerning the output strain or concentration data.

The disturbance must be weak (weak strain or diffuse interfaces) in order to extrapolate the pattern from an undisturbed reference area. The relation between image pattern difference and strain/concentration must be shown to be linear. If it is nonlinear, a calibration curve needs to be recorded by full multislice techniques. This might still be worth the effort. Whatever the structural parameter we wish to measure by black-box techniques, it must be proved that it is the *leading* parameter in driving lateral pattern changes, and is thus significantly stronger than any systematic errors (e.g. spatially varying beam tilt) and spatially varying noise.

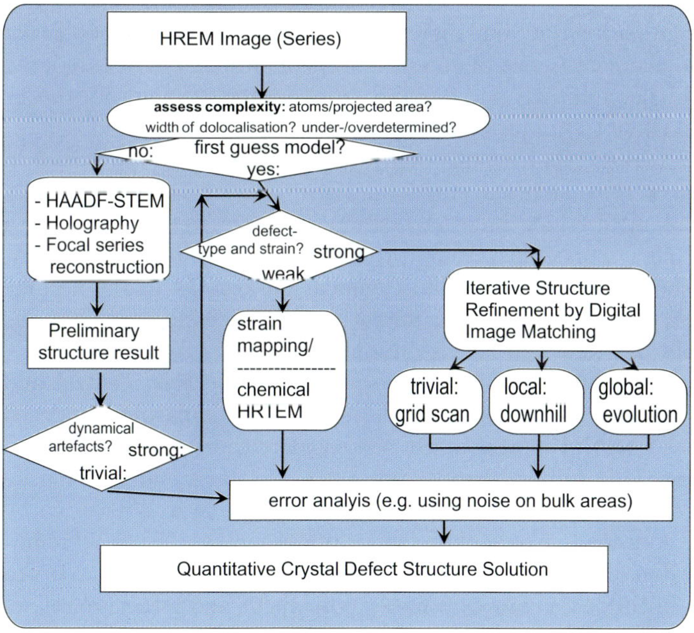

Fig. 3.27. Decision tree for structure determination strategies

The other choice on the flow diagram is between reconstruction by inversion and refinement by iterative simulation. Inversion has the advantage of being faster and not requiring an initial guess of the solution. Iteration has the advantage of being more flexible concerning the available input images; it provides the full precision of image simulation in the forward direction without any further approximation and it provides statistical information about the confidence circle (if not a complete probability distribution) of the parameters estimated. Since HRTEM image formation is a stepwise process, iterative techniques can be applied for individual steps and subsequently combined to form a quasi-inverse technique. This applies also for uninvertible (ill-posed) problems of dynamical diffraction [81]. The sequential combination of a full (but approximate) inverse technique with a precise iterative technique is particularly effective in cases in which the inverse technique delivers the initial structure for the iterative refinement (see Fig. 3.27). In such a serial combination of reconstructive structure solution and iterative structure refinement, all the advantages of the individual techniques can be exploited at once.

3.6 Conclusions and Outlook

Q-HRTEM has been defined in various ways in the past. Progress in hardware (e.g. replacement of film-plates by CCD-cameras) is accompanied by changes in the aim of HRTEM. The aim of Q-HRTEM is a final result that is not so much an image to look at, but more of crystallographic numbers (atom coordinates, displacement vector plots, concentration profiles) to be tabulated along with confidence intervals and possibly compared with quantum theory predictions. Will these techniques render non-quantitative HRTEM obsolete? "Paper-less microscopy", which links digitally recorded image intensities in the computer directly to structure retrieval algorithms without any images in between to look at, has now indeed become possible. However, just as with the "paper-less office", once mistakenly promised by the computer-industry, this might not be desirable. Certainly, chemical darkrooms might become obsolete, replaced by scanners, CCD-cameras, image-processing workstations and photo-quality printers. But this is only one aspect.

The early promise of Q-HRTEM to be perfectly objective, neutral and simply "better" than a human observer, especially in comparing simulated and experimental HRTEM images, must be considered (after 10 years of experience) only partially fulfilled: any significant (i.e. robust and systematic-error-tolerant) local image difference (e.g. Fig. 3.18) can be well detected by eye in a suitable contour-line or pseudo-colour representation. Even more important, supervision of automatic refinement by an experienced user seems to be unavoidable, both for imaging-parameter refinement and structure refinement. The behaviour of a particular image-agreement factor, the preferences and risks of numerical optimizers, the propagation and amplification of systematic errors, should ideally be assessed in test calculations for every new materials system under study. This can only be done by visual sampling of the progress (during image matching and modifying the structure model) at intermediate time-intervals during a refinement towards the best fit.

On the other hand, there is evidence that Q-HRTEM is indeed the only viable choice for every non-trivial structure retrieval attempts in HRTEM at < 0.01 nm precision. This argument relies on the sheer computing-power needed and on the fact that a human observer would never be able to conduct and control thousands of trial simulations, thousands of image matches, hundreds of spot-measurements with sub-pixel interpolation or examinations of basic pattern-contributions in chemical interfaces. The quantification revolution thus relies on factors such as ultimate mathematical precision and almost unlimited repetition of similar subroutines [both factors ever since attributed to "computer-intelligence"].

Apart from this progress in precision, *qualitative* progress must also be attributed to Q-HRTEM. New types of image interpretation have become possible, previously not even thought of:

1. Decision between a few discrete structural alternatives: for every alternative in discrete numbers of atoms or their chemistry at an atom site in question (e.g. a dislocation core), a complete refinement of all neighbouring atom positions is now routinely possible. Only then does a comparison of the discrete best matches within each group of structure models make real sense.
2. Falsification of hypotheses: a theoretically predicted structure model can be rejected if it does not match the experiment, only *after having scanned the whole reasonable range of imaging parameters* (as long as these are not fixed precisely enough by independent measurements).
3. Prediction of visibilites of details: The visibility of low contrast details, such as vacancies, impurities, and very faint relaxations, is often insufficiently estimated from one HRTEM simulation only. Q-HRTEM instead can automatically search for the most sensitive imaging condition and also detect ambiguity traps. Two structures may look identical at one condition and well distinct at another. Only an automatic multi-parameter scan for imaging-parameters can give a credible answer or prediction for the true capabilities of HRTEM to distinguish faint structural differences.

A Glance into the Future. The development of Q-HRTEM is far from complete, future trends will probably involve (see also Chap. 6):

- *Further extension of the resolution limit.* The information resolution for lattic fringe imaging has already reached about 0.1 nm for high-coherence field-emission guns at 200-300 kV. Monochromation and further stabilization by new compact instrument design might soon allow to improve contrast transfer envelopes to resolve lattice fringes in linear mode down to 0.07 nm and non-linearly down to 0.05 nm.

Concerning the point resolution, C_s-correction (at 200 kV) will result in simplification of structure determination similar to the use of high voltage (above 1 MV). In the sense of Sect. 3.5 and Fig. 3.22, the boundary between black-box and full dynamic/non-linear techniques will then shift towards direct structure retrieval. Also, the number of focal values required for non-ambiguous determination of imaging condition and defect structures will be reduced. However, for accurate data matching between simulation and experiment, neither high-voltage nor C_s-correction will obviate the need for dynamic plus nonlinear simulations. The failure of the weak-phase approximation and linear imaging approximation is inherent in the electron scattering of even single (heavier) atoms and in the recording of intensities (rather than amplitudes) under partial coherent illumination, neither of which circumstances is changed by the foreseen techniques.

- *Extension of structure determination to three-dimensional refinement.* This is achievable by using images of multiple zone axes (discrete angle tomography of crystal defects). Unfortunately high tilt-angles require large pole-piece gaps, not compatible with the requirement of high point-resolution. The availability of Cs-correction will resolve that trade-off conflict in the future (as with HV-HRTEM).
- *Techniques linking HRTEM numerically to other disciplines:* analytical TEM and highly spatially localised diffraction are ready to be combined with HRTEM, since all these techniques can now be simultaneously exploited in a modern FEG-TEM/STEM with CCD-camera recording and remote control of most microscope parameters and switches.
  More generally, combinations of techniques uniting TEM- and non-TEM-based knowledge will certainly emerge and grow. It is common sense in higher statistics and probability theory of physical measurements to link "soft data" (any fuzzy, assumptions-based, a-priori knowledge or published data) to the "hard" data given by the HRTEM experiment and its subsequent joint processing and matching to theoretical predictions.

Any researcher in the area of nanoscale structural characterization who has once set up and experienced the power of Q-HRTEM will certainly not want to abandon it in his future research.

References

1. Spence J.C.H. (1988) Experimental High Resolution Electron Microscopy. Oxford University Press, New York
2. Ibers J.A., Hamilton W.C. (Eds) (1974) International Tables for X-Ray Crystallography, Vol. IV. Kynoch, Birmingham.
3. Möbus G., Schweinfest R., Gemming T., Wagner T.,Rühle R. (1998) Iterative structure retrieval techniques: A comparative study and a modular program package. J Microsc, **190**, 109–130

4. Gemming T., Möbus G., Exner M., Rühle M. (1998) Ab-initio high resolution electron microscopy: A case study of sapphire. J Microsc, **190**, 89–98
5. Möbus G., Gemming T., Gumbsch P. (1998) Influence of phonon scattering on HRTEM-images. Acta Cryst A, **54**, 83–90
6. Ishizuka K. (1980) Contrast transfer of crystal images in TEM. Ultramicroscopy, **5**, 55–65
7. Stadelmann P. (1987) EMS a software package for electron diffraction analysis and HREM image simulation in materials science. Ultramicroscopy, **21**, 131–145
8. Möbus G., Rühle M. (1994) Structure determination of metal-ceramic interfaces by numerical contrast evaluation of HRTEM-micrographs. Ultramicroscopy, **56**, 54–70
9. Marks L.D. (1985) Image localisation. Ultramicroscopy, **18**, 33–38
10. Coene W., Jansen A.J.E.M. (1992) Image delocalisation in HRTEM. Scann Micr Suppl, **6**, 379–403
11. Lichte H. (1991) Optimum focus for taking holograms. Ultramicroscopy, **38**, 13–22
12. Decaro L., Giuffrida A., Carlino E., Tapfer L. (1995) Elastic stress relaxation in HRTEM specimens of strained semiconductor heterostructures. Microsc Microanal Microstr, **6**, 465–472
13. Hÿtch M.J., Plamann T. (2000) Imaging conditions for reliable measurement of rapidly varying displacement and strain in HREM. Ultramicroscopy, **87**, 199–212
14. Hÿtch M.J., Plamann T. (2000) Effect of the objective lens on the measurement of rapidly varying displacement fields from HRTEM images. Proceed. EUREM Brno, Czechia, **1**, 119–120
15. Bierwolf R., Hohenstein M., Phillipp F., Brandt O., Crook G.E., Ploog K. (1993). Direct measurement of local lattic distortions in strained layer structures by HREM. Ultramicroscopy, **49**, 273–285
16. Seitz H., Ahlborn K., Seibt M., Schröter W. (1998) Sensitivity limits of strain mapping procedures using HREM. J Microscopy, **190**, 184–189
17. Wang S.Q. (1995) Atom: X-windows based software for quantitative analysis of atomic images. J Appl Cryst, **28**, 837–839
18. Hofmann D. and Ernst F. (1994) Quantitative high-resolution transmission electron microscopy of the incoherent $\Sigma 3(211)$ boundary in Cu. Ultramicroscopy, **53**, 205–221
19. Kilaas R., Paciornik S., Schwartz A.J., Tanner L.E. (1994) Quantitative analysis of atomic displacements in HRTEM images. J Comp Assist Microsc, **6**, 129–138
20. Bayle P., Thibault J. (1994) Quantitative HREM Study of [001] Au/Ni Multilayers Proceed. ICEM, Paris, **1**, 397–398
21. Robertson M.D., Currie J.E., Corbett J.M., Webb J.B. (1995) Determination of lattice strains in epitaxial layers in HRTEM. Ultramicroscopy, **58**, 175–184
22. Rosenauer A., Remmele T., Fischer U., Forster A., Gerthsen D. (1997) Strain determination in mismatched semiconductor heterostructures by the digital analysis of lattice images. Inst Phys conf series, Bristol, UK, **157**, 39–42
23. Hÿtch M.J. (1997) Analysis of variations in structure from HREM images by combination of real space and Fourier space information. Microsc Microanal Microstruct, **8**, 41–57

24. C.J.D. Hetherington C.J.D., Dahmen U. (1992) An optical moiré technique for the analysis of displacements in lattice images Scann Microsc Suppl, **6**, 405–414
25. Inkson B.J., Möbus G., Rühle M. (1997) Atomic-resolution electron microscopy of TiB_2 precipitates in an industrial TiAl alloy. MRS Symp Proc, Boston, **466**, 151–156
26. Möbus G., Wagner T. (1999) Direct versus iterative structure retrieval for a Cu/Ti misfit dislocation: A comparison of various 1Å HREM Technologies. J Microsc, **194**, 124–141
27. Paciornik S., Kilaas R., Dahmen U. (1993) Assessment of specimen noise in HREM images of simple structures. Ultramicroscopy, **50**, 255–262
28. Möbus G., Necker G., Rühle M. (1993) Adaptive Fourier filtering technique for quantitative evaluation of high resolution electron micrographs of interfaces. Ultramicroscopy, **49**, 46–65
29. Paciornik S., Kilaas R., Turner J., Dahmen U. (1996) A pattern recognition technique for the analysis of grain boundary structures by HREM. Ultramicroscopy, **62**, 15–27
30. Kilaas R., Gronsky R. (1985) The effect of amorphous surface layers on imaging of crystals in HRTEM. Ultramicroscopy, **16**, 193–201
31. Kienzle O., Ernst F., Möbus G. (1998) Reliability of atom column positions in a ternary system determined by quantitative HRTEM. J Micros, **190**, 144–158
32. Hillebrand R. (1995) Quantitative analysis of HREM images: Measures of similarity. phys stat sol (a), **150**, 65–76
33. Taupin D. (1988) Probabilities, data reduction and error analysis in the physical sciences. les editions de physique, Les Ulis Cedex, France
34. Hÿtch M.J., Stobbs W.M. (1994) Quantitative comparison of high resolution TEM images with image simulation. Ultramicroscopy, **53**, 191–205
35. Hÿtch M.J., Stobbs W.M. (1994) Quantitative criteria for the matching of simulations with experimental HREM images. Microsc Microanal Microstruct, **5**, 133–151
36. Boothroyd C.B. (1998) Why don't high-resolution simulations and images match? J Micros, **190**, 99–108
37. Press W.H. et al. (1992) Numerical Recipes, 2nd ed. Cambridge University Press, Cambridge, UK
38. Schwefel H-P. (1981) Numerical Optimization of Computer Models. J. Wiley, New York
39. Bäck T., Schwefel H-P. (1993) An overview of evolutionary algorithms for parameter optimization. Evolutionary Computation, **1**, 1–23
40. Smith A.R., Eyring L. (1982) Calculation, display and comparison of electron microscope images modelled and observed. Ultramicroscopy, **8**, 65–78
41. Barry J.C. (1989) Semiquantitative image matching in HRTEM. In W. Krakow and M. O'Keefe, Eds, Computer Simulation of Electron Microscope Diffraction and Images. The Minerals, Metals and Materials Society, Pennsylvania
42. Thust A., Urban K. (1992) Quantitative high-speed matching of high-resolution electron microscopy images. Ultramicroscopy, **45**, 23–42
43. King W.E., Campbell G.H. (1993) Determination of thickness and defocus by quantitative comparison of experimental and simulated high-resolution images. Ultramicroscopy, **51**, 128–135
44. Tang D., Kirkland A.I., Jefferson D.A. (1994) Optimization of high-resolution image simulations. Ultramicroscopy, **53**, 137–146

45. Möbus G. (1994) Optimierung der digitalen Kontrastauswertung hochaufgelöster elektronenmikroskopischer Aufnahmen innerer Grenzflächen. Dissertation, Universität Stuttgart, Germany
46. Möbus G., Gutekunst G., Mayer J., Rühle M. (1994) High precision iterative digital image matching and limitations of quantitative HRTEM. Proceed 13th Int Congr Electron Microscopy, Paris, France, **1**, 373–374
47. Hofmann D., Möbus G., Ernst F. (1992) Quantitative HRTEM of incoherent twin boundaries in copper. Proceed Xth Europ Congr Electron Microscopy, Granada, Spain, 513–514
48. King W.E., Campbell G.H. (1993) Quantitative HREM study of the atomic structure of the $\Sigma(310)/[001]$ symmetric tilt grain boundary in Nb. MRS Symp Proc, **295**, 83–88
49. King W.E., Campbell G.H. (1994) Quantitative HREM using non-linear least-squares methods. Ultramicroscopy, **56**, 46–53
50. Zhang H., Marks L.D., Wang Y.Y., Zhang H., Dravid V.P., Han P., Payne D.A. (1995) Structure of planar defects in $(Sr_{0.9}Ca_{0.3})_{1.1}CuO_2$ infinite-layer superconductors by quantitative high-resolution electron microscopy. Ultramicroscopy, **57**, 103–111
51. Möbus G., Dehm G. (1996) Retrieval of crystal defect structures from HREM images by simulated evolution: II experimental image evaluation. Ultramicroscopy, **65**, 217–228
52. King W.E., Campbell G.H., Foiles S.M., Cohen D., Hanson K.M. (1998) Quantitative HREM observation of the $\Sigma 11(113)/[\bar{1}10]$ grain-boundary structure in aluminium and comparison with atomistic simulation. J Microsc, **190**, 131–143
53. Nadarzinski K., Ernst F. (1996) The atomistic structure of a Sigma=3,(111) grain boundary in NiAl studied by quantitative HRTEM. Phil Mag A, **74**, 641–664
54. Höche T., Kenway P.R., Kleebe H-J., Rühle M., Morris P.M. (1994) High-resolution transmission electron microscopy studies of a near $\sigma 11$ grain boundary in α-alumina. J Amer Ceram Soc, **77**, 339–348
55. Kienzle O., Ernst F., Möbus G. (1998) Reliability of atom column positions in a ternary system determined by high-resolution transmission electron microscopy. J Microsc, **190**, 144–158
56. Schweinfest R., Ernst F., Wagner T., Rühle M. (1998) Quantitative HRTEM at the $Al/MgAl_2O_4$ interface. Proceed ICEM-14, Cancun/Mexico (IOP-publ., Bristol, UK), **1**, 635–636
57. Möbus G. (1996) Retrieval of crystal defect structures from HREM images by simulated evolution: I basic technique. Ultramicroscopy, **65**, 205–216
58. Merkle K.L., Csencsits R., Rynes K.L., Withrow J.P., Stadelmann P.A. (1993) The effect of the three-fold astigmatism on measurments of grain boundary volume expansion by HRTEM. J. Microscopy, **190**, 204–213
59. Möbus G., Kienzle O. (1999) Interface structure retrieval by HREM: From entropy maximisation to r-factor fits. In Kiely, C.J., Edt, Proceedings of EMAG 1999, Sheffield, 263–266. IOP, Bristol, UK
60. Möbus G. (2000) Probability Calculus for quantitative HREM. Part II: Entropy and Likelihood concepts. Ultramicroscopy, **85**, 199–213
61. Skilling J. (1998) Probabilistic data analysis: An introductory guide. J Microsc, **190**, 28–36
62. Jaynes E.T. (1957) Information theory and statistical mechanics. Phys Rev, **106**, 620–630

63. Buck B., Macaulay V.A. (1990) Maximum Entropie in Action. Oxford Science Publications, Oxford, UK
64. Möbus G., Kienzle O. (2000) Probability Calculus for quantitative HREM. Part I: Monte Carlo and Point Cloud Techniques. Ultramicroscopy, **85**, 183–213
65. Ourmazd A., Baumann F.H., Bode M., Kim Y. (1990) Quantitative chemical lattice imaging: theory and practice. Ultramicroscopy, **34**, 237–255
66. Schwander P., Kisielowski C., Seibt M., Baumann F.H., Kim Y.O., Ourmazd A. (1993) Mapping Projected Potential, Interfacial Roughness, and Composition in General Crystalline Solids by Quantiative Transmission Electron-Microscopy Phys Rev Lett , **71**, 4150–4153
67. Stenkamp D., Jäger W. (1993) Compositional and structural characterization of Si_xGe_{1-x} alloys and heterostructures by HRTEM. Ultramicroscopy, **50**, 321–354
68. Hillebrand R. (1998) Fuzzy logic approaches to the analysis of HREM images of III-V compounds. J Microscopy, **190**, 61–72
69. Stenkamp D. (1998) Detection and quantitative assessment of image aberrations from single HRTEM lattice images. J Microscopy, **190**, 194–203
70. Saxton W.O. (1978) Computer Techniques for Image Processing in Electron Microscopy. Academic Press, New York
71. Kirkland E.L. (1984) Improved high resolution image processing of bright field electron micrographs. Ultramicroscopy, **15**, 151–172
72. Coene W., Janssen A., Op de Beeck M., Van Dyck D. (1992) Phase Retrieval Through Focus Variation For Ultra-Resolution in Field-Emission Transmission Electron-Microscopy Phys Rev Lett, **69**, 3743–3746
73. Thust A., Coene W.M.J., Op de Beeck M., Van Dyck D. (1996) Focal-series reconstruction in HRTEM: Simulation studies on non-periodic objects. Ultramicroscopy, **64**, 211–230
74. Lichte H. (1986) Electron holography approaching atomic resolution. Ultramicroscopy, **20**, 283–304
75. Lehmann M. (2000) Determination and correction of the coherent wave aberration from a simple off-axis electron hologram by means of a genetic algorithm. Ultramicroscopy, **85**, 165–182
76. Pennycook S.J., Jesson D.E. High-resolution incoherent imaging of crystals. Phys Rev Lett, **64**, 938–941
77. Nellist P., Pennycook S. (1998) Accurate structure determination from image reconstruction in ADF STEM. J Microsc, **190**, 159–170
78. Jansen J., Tang D., Zandbergen H.W., Schenk H. (1998) A least-square procedure for accurate crystal structure refinement from dynamical electron diffraction patterns. Acta Cryst A, **54**, 91–101
79. Zuo J.M, Spence J.C.H. (1991) Automated structure factor refinement from convergent beam patterns. Ultramicroscopy, **35**, 185–196
80. Lentzen M., Urban K. (1996) Reconstruction of the projected crystal potential from a periodic high-resolution electron microscopy exit plane wave function. Ultramicroscopy, **62**, 89–102
81. Scheerschmidt K. (1998) Retrieval of object information by inverse problems in electron diffraction. J microscopy, **190**, 238–248

4 Quantitative Analytical Transmission Electron Microscopy

P. Kohler-Redlich and J. Mayer

4.1 Introduction

Over the past decade analytical transmission electron microscopy (ATEM) has experienced a more rapid growth than any other major TEM technique. The main reasons for this development are the growing interest in the wealth of information that can be revealed by electron energy loss spectroscopy (EELS) and the rapid spread of new instrumental developments, in particular field-emission guns and imaging energy filters. One of the trends in transmission electron microscopy is to consider a microscope not primarily as an instrument to obtain micrographs but as an experimental tool on which information from a sample can be obtained via various channels in parallel [1]. The channels are defined by the available detectors, such as two-dimensional detectors for imaging and diffraction, electron counting devices for STEM bright- and dark-field imaging, an electron energy-loss spectrometer and an energy-dispersive X-ray spectrometer (EDS). There are also many competing analytical or spectroscopic techniques (some of which are be discussed in other chapters of this book) that are better in terms of energy resolution, detection limits, error of absolute compositional quantification, angular dependence, retrieval of three-dimensional information and reduction of sample damage due to irradiation. However, none of them offers a spatial resolution comparable to the one obtainable on a TEM, and none of them offers all the other high resolution imaging and diffraction techniques mentioned above.

The two main analytical TEM techniques, EELS and EDS, are complimentary. Each technique has its distinct advantages and typical applications. In general EDS and EELS are commonly available on the same microscopes. The detection limits and the error limits of quantification strongly depend on the exact nature of the problem (i.e. the type of EELS ionization edges or X-ray lines used for the analysis) and the quality of the sample. A comparison of EELS and EDS techniques has been made by Leapman and Hunt [2]. Comprehensive reviews of the theory and the quantification of EDS spectra can be found in the literature and will not be covered in this chapter [3] which is focused on the application of EELS. The following points may explain why there is currently a special interest in EELS in the TEM:

1. The spatial resolution of EELS is inherently higher than that of EDS owing to the fact that the angular distribution of inelastic scattering

is strongly forward peaked (see the chapter by Müller and Rose). The use of apertures reduces the collection angle of the EELS spectrometer and limits thereby the negative effect of the broadening of the electron beam by multiple scattering events in the sample. Furthermore, signal delocalisation by X-ray fluorescence is not a problem in EELS analysis.
2. Typical acquisition times for EDS are one to two orders of magnitude longer than for EELS, which in particular has a higher sensitivity to lighter elements. In the case of heavier elements the difference in collection efficiency becomes less pronounced and the time needed to acquire chemical information of certain elements can become equal.
3. The sensitivity of the EELS spectrum to the density of unoccupied states, reflected in the near-edge fine structure (ELNES), makes it possible to study bonding, local coordination and local electronic properties of materials. One recent trend in ATEM is to compare ELNES data quantitatively with the results of band structure calculations.
4. In EELS, low energy-loss processes (in the range of 1 eV to several tenths of eV including inter- and intraband transitions and surface, interface and volume plasmons) can also be analysed, in this range they are complementary to optical and vacuum ultraviolet spectroscopy techniques [4]. In this energy range, EELS is frequently referred to as valence EELS (VEELS) and can in principle be used to derive the complex dielectric function $\epsilon = \epsilon_1 + i\epsilon_2$ of a material with high spatial resolution. It has also been used to study plasmon excitations of internal interfaces [5] and small particles and clusters [6].
5. The development of energy filters for TEMs gave rise to new techniques based on electron spectroscopic imaging (ESI) and diffraction (ESD). These techniques can be used to map the chemical composition (and also bonding information, as will be discussed later in this chapter) of a sample in a very efficient way by quantification of a limited number of energy-filtered images.

In addition to these developments in spectroscopy techniques and instrumentation, ATEM has greatly benefited from major progress in the capabilities of the basic instruments. In particular, the use of high brightness field-emission guns improves the spatial resolution achievable with a focused probe by more than an order of magnitude and the energy resolution by at least a factor of two, compared to ATEMs equipped with a thermal emitter. On a STEM with field-emission gun, the probe diameter can be reduced to the dimensions of the separation of neighbouring atomic columns in a crystal and the lattice becomes visible in an image produced with a scanned probe. If the signal from a high-angle annular dark-field (HAADF) detector is used, incoherent images result in which the intensity of each individual atomic column is proportional to a good approximation to the square of the mean atomic number of the atoms forming the column [7–10]. Although strictly speaking not an analytical technique, this so-called Z-contrast imaging has shown itself capable of delivering very important information in analytical studies and has opened the door to atomic-level spectroscopy and analysis. It has already been shown that EEL spectroscopy can be performed on an individual atomic column [11–15], creating exciting perspectives for future studies of interfaces and defects in materials. However, the geometrical restrictions on this column-resolved spectroscopy are severe and more generalized techniques have to be developed in parallel, in order to fulfil all the needs of analytical characterization in materials science. In the case of interface studies, spectrum-line profiling and the spatial-difference technique can reveal important information about the chemistry and bonding at the interface. In geometrically even more complex situations, two-dimensional mapping and spectroscopy in an energy-filtering TEM is the method of choice [16] and has had remarkable success because of the ease of use and the efficient data collection.

In the present chapter, the reader will first be introduced briefly to the basic principles of analytical transmission electron microscopy (ATEM) with special emphasis on electron energy-loss spectroscopy (EELS) and energy-filtering TEM. The quantification of spectra to obtain chemical information and the origin and interpretation of near-edge fine structures in EELS (ELNES) are discussed. Special attention will be given to the characterization of internal interfaces and the literature in this area will be reviewed. Selected examples of the application of ATEM in the investigation of internal interfaces will be given. These examples include both EELS in the dedicated scanning transmission electron microscope (Sect. 4.3) and energy-filtered TEM (ESI) (Sect. 4.4). Convergent beam electron diffraction (CBED) is not commonly thought of as an analytical TEM technique. The last section of this chapter (Sect. 4.5), however, will present the basic theory and an example of CBED since the improvements in CBED are closely related to the progress in energy-filtered TEM.

4.2 Basics of Electron Energy-Loss Spectroscopy (EELS)

4.2.1 Inelastic Scattering Processes

All analytical techniques in the TEM are based on the inelastic scattering of the fast beam electrons by the electrons of the atoms in the material investigated. The primary event in each case is the transfer of energy and momentum from the fast electron to a sample atom, thereby exciting the latter from its ground state to an excited state. The primary electron looses energy in this inelastic scattering event. The energy losses are characteristic of the element and the local coordination and bonding of the atoms in the sample. Measuring the energy-loss spectrum with appropriate detectors therefore allows the chemical composition and the bonding of materials to be analysed. This technique is called electron energy-loss spectroscopy (EELS). It should be noted that EELS also allows collective excitations of valence electrons to be investigated: in this case the characteristic energy-losses of surface, interface and volume plasmons are analysed. Valence-loss EELS (VEELS) has been particularly successful in the characterization of the electronic and optical properties of nanometre-sized particles (for example silicon spheres [6], carbon nanotubes [17] or carbon onions [18]) and as angle-resolved EELS in a stand-alone spectrometer [19]. Secondary events resulting from these inelastic scattering processes are the generation of high-energy X-ray radiation and Auger electrons, as a consequence of the de-excitation to the ground state. The analysis of the intensity of the characteristic X-rays lines is the basis of energy-dispersive X-ray spectroscopy (EDS), which is routinely used for quantitative elemental analysis in ATEM. Limited use has been made of the Auger electrons in TEM so far, although there have been attempts to equip a TEM with a hemispherical Auger detector [20].

A schematic EELS spectrum for a wide range of energy-losses ΔE is shown in Fig. 4.1, covering the peak of the elastically scattered electrons (zero-loss peak), the range of the valence losses (plasmons, inter- and intraband transitions), and the core ionization edges for higher energy losses. The high dynamic range of the spectra makes it difficult to record low- and core-loss spectra (i.e. the spectra containing the zero-loss up to the valence losses and the core ionization edges), in a single spectrum with good signal-to-noise ratio for the whole range of energy losses. Measurements with different

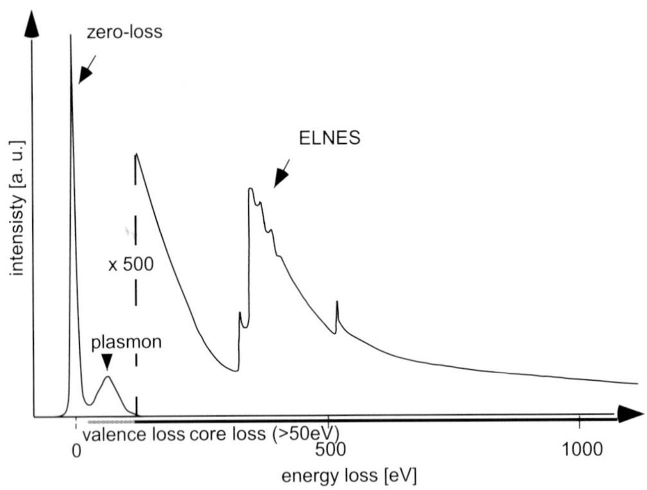

Fig. 4.1. Schematic EELS spectrum for a wide range of energy losses covering the peak of the elastically scattered electrons (zero-loss peak), the range of the valence losses (plasmons, inter- and intraband transitions), and the core-ionization edges. EELS spectra have a high dynamic range. Note that the ionization edges are superimposed on a high background signal

acquisition parameters are therefore performed to obtain optimized low-loss and core-loss spectra. The ionization edges are always superimposed on a background signal resulting from the tails of inelastic scattering processes with onset energies at lower energy losses. In order to use the ionization edges for chemical quantification, this background has to be removed by fitting an analytical function to a pre-edge region and extrapolating it to the energy-loss range containing the ionization edge. In most practical cases a function of the type

$$I(\Delta E, A, r) = AE^{-r} \tag{4.1}$$

where A and r are the fitting parameters, is used. The basic theory of EELS quantification methods, including the discussion of optimized signal-to-noise conditions for fitting and integration windows, the calculation of scattering cross-sections and the treatment of thickness effects (multiple scattering for thicker samples), has been reviewed by Egerton [21]. The more general book about transmission electron microscopy by Williams and Carter [3] is also recommended. Kothleitner and Hofer have analysed the dependence of the signal-to-noise in energy-filtering TEM on different types of ionization edges and the choice of quantification parameters [22].

A near edge structure (energy loss near edge structure, ELNES), the origin of which is schematically explained in Fig. 4.2, is always superimposed on

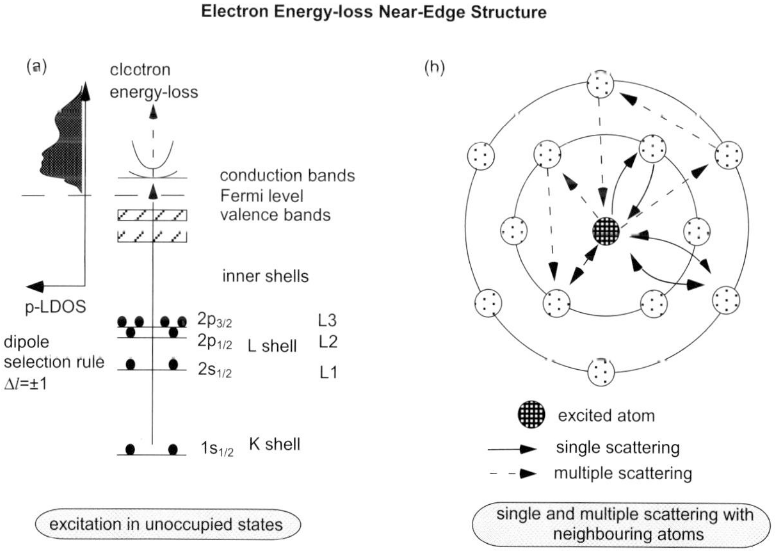

Fig. 4.2. Schematic representation of two models for the origin of electron energy-loss near-edge structures (ELNES) for the core-ionization edges. (**a**) Transition of strongly bound core electrons into unoccupied states, (**b**) multiple scattering description of ELNES

the ionization edges. Whereas the onset energy of the ionization edge and the area under the edge contain the information on the type and concentration of the chemical element (both depend mainly on the core level from which the electron is excited), the ELNES carries information about the chemical environment and hence the bonding of the excited atoms. The ELNES depends on the final state of the excited electron, which can be represented in two ways, as depicted in Fig. 4.2: in a band structure model by the local density of unoccupied states or in a real space model by an outgoing wave, which is multiply scattered by the surrounding atoms. Both pictures are fully equivalent within the appropriate limits and may be used alternatively in trying to understand the observed ELNES.

According to Fermi's golden rule, the intensity variation $I(E)$ in the near-edge region is proportional to the density of unoccupied (i.e. final) states $N_\mathrm{u}(E)$ and the square of the transition matrix element $M(E)$:

$$I(E) \propto N_\mathrm{u}(E) \mid M(E) \mid^2 \tag{4.2}$$

(see e.g. [23]). The transition matrix element is usually assumed to vary only slowly with energy and the ELNES hence provides a direct measure of the local density of unoccupied states. In addition, under normal experimental conditions, only allowed dipole transitions can be strongly excited, so that the actual measurement probes a site and angular-momentum-resolved partial density of states. In these conditions the ELNES is fully equivalent to the near-edge structure in X-ray absorption spectra (XANES), which has been exploited for the experimental and theoretical understanding of the observed ELNES in many different systems. The ELNES features of interest are usually found in the region between the edge onset and about 20–30 eV above the edge onset. Starting at about 50 eV above the edge onset, the fine structures are named EXELFS (extended energy-loss fine structures) and are again fully equivalent to the extended X-ray absorption fine structure (EXAFS). Both reflect the local arrangement of atomic scatterers and a quantitative analysis reveals information on the local-pair correlation function. Since the ELNES has received for more interest than EXELFS in the last few years, only the former will be considered in the examples discussed in this chapter.

4.2.2 Instrumentation: Dedicated Scanning and Energy-Filtering TEM

Figure 4.3 depicts the three-dimensional data space that has to be evaluated in a spatially resolved quantitative ATEM analysis [24]. The graphical representation suggests that the intensity distribution $I(x, y, \Delta E)$ can equally well be obtained by recording individual spectra or by employing the technique of electron spectroscopic imaging [25]. In the former case, which is illustrated in the examples of EELS in the dedicated STEM in this chapter, only a small spot on the specimen is illuminated and EEL spectra are recorded, whereas

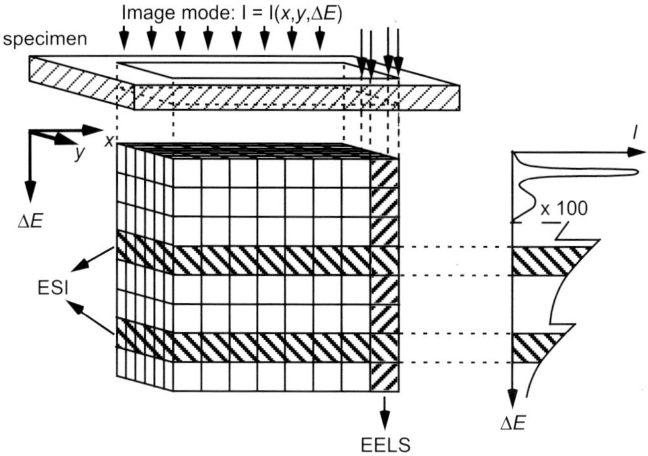

Fig. 4.3. Three-dimensional data space constituting the basis of spatially resolved quantitative TEM analysis

in the latter case a large specimen area is illuminated and filtered images are recorded with an energy-selecting window of finite width ΔE. The theory and the application of ESI will be discussed in Sect. 4.4. The different approaches of ATEM are closely related to the available instrumentation.

We distinguish serial and parallel EELS detectors, which allow individual spectra to be registered, and imaging filters, which allow images to be formed with electrons from a specific energy-loss window. Standard EELS detectors are fitted to the microscope below the viewing screen. Imaging filters can be placed either in-column or post-column. The most prominent examples of in column filters are the Henry-Castaing electrostatic and the omega-type magnetic filters [26–28]. Several designs of post-column energy filters as well as parallel EELS spectrometers have been developed by Krivanek [25,29]. The use of in-column and post-column imaging filters for ESI will be discussed in Sect. 4.4.

Progress in quantitative ATEM investigations of interfaces and grain boundaries has been largely dependent on the availability of fine electron probes. There are many microscopes available that are fitted with both an EELS spectrometer and an EDS detector and can be used for conventional, high resolution and analytical TEM. However, in high spatial resolution ATEM, dedicated scanning TEMs (the Vacuum Generators series VG HB-5, HB-501, HB-601 and HB-603) have been the equipment of choice for nearly two decades. A schematic diagram of a dedicated STEM is shown in Fig. 4.4. An electron beam of high coherence is generated by a field-emission source. The beam is focused by the objective lens into a sub-nanometre probe. Suitable coils allow the electron probe to be scanned across the sample. The

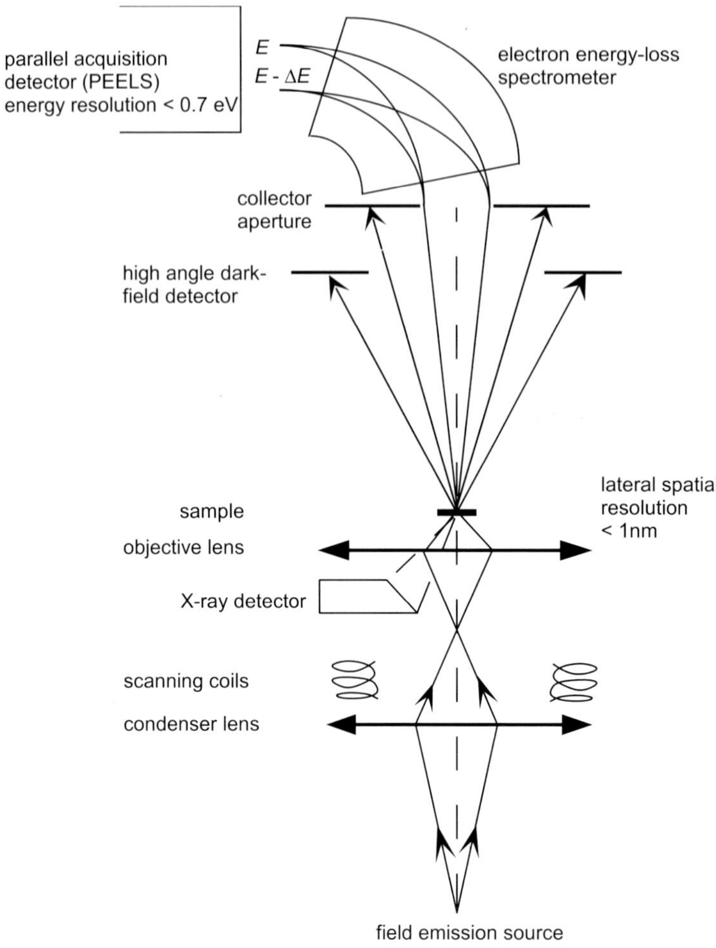

Fig. 4.4. Schematic diagram of a dedicated scanning transmission electron microscope as it is used for the ATEM studies of interfaces and grain boundaries

intensity of the transmitted electrons is recorded on several imaging and spectroscopic detectors.

- The bright-field detector is located in front of the spectrometer entrance and is inserted if bright-field imaging is desired.
- The annular dark-field (ADF) and the high-angle annular dark-field detectors (HAADF) are concentric detectors with characteristic inner and outer radii. Both detectors allow the simultaneous recording of images during the acquisition of spectral data with the EELS spectrometer since they leave the spectrometer entrance open. The HAADF detector collects electrons scattered through large angles. The specific geometry of the HAADF detector is the basis for Z-contrast imaging [7,30].

- The EDS detector is inserted from the side and can be used simultaneously with the ADF and the EELS spectrometer.
- The parallel EELS detector allows 1024 energy-loss channels to be recorded simultaneously on a photo diode or CCD array. The entrance apertures are used to control the acceptance angle of the spectrometer.

This type of dedicated STEM has been used for the interface studies given as examples in Sect. 4.3. Recently, the development of a spherical aberration corrector for the dedicated STEM by Krivanek [31] and high-efficiency CCD detectors for the EELS spectrometer have revitalized the interest in dedicated STEMs. On the TEM side, the recent development in instrumentation has now led to field-emission medium-voltage (200–300 keV) TEMs with energy-filter, annular dark-field detector and sub-nanometre probe forming capability [32]. These microscopes combine the potential of scanning and energy-filtering TEMs. As part of an initiative of the German Science Foundation (DFG) for further improvement in instrumentation, a Sub-Electronvolt–Sub-Ångström-Microscope (SESAM) is currently under development. The SESAM will be equipped with a monochromator for the electron gun [33] and an imaging filter of the mandoline-type [34]. This design will offer higher transmissivity of the imaging filter combined with a higher energy-resolution in EELS. A schematic drawing of the SESAM microscope is given in Fig. 4.5.

4.3 Investigation of Interfaces and Grain Boundaries

4.3.1 Experimental Techniques

The major experimental difficulty in interface studies using ATEM is to record the very weak signal reliably from the interface with good signal-to-noise ratio and subsequently to distinguish clearly the interface signal from signals arising from the adjacent bulk phases.

Various techniques to address this problem have been discussed in the literature. Using a focused probe in a STEM, the three main methods currently in use are:

1. Line scans of equidistant spot measurements across the interface.
2. Spot measurements with a probe focused on the interface. Using a probe with a diameter of 2 Å or less in combination with Z-contrast imaging makes it possible to perform atomic column-resolved EELS.
3. The spatial difference technique, which makes it possible to extract the interfacial signal from measurement areas with nm-size dimensions.

Each of these methods has distinct advantages and disadvantages and the best way to record the spectra will depend on the experimental situation and especially on the stability of the interface at the atomic level. Before discussing some individual examples, a brief survey of the main applications of the three techniques will be given.

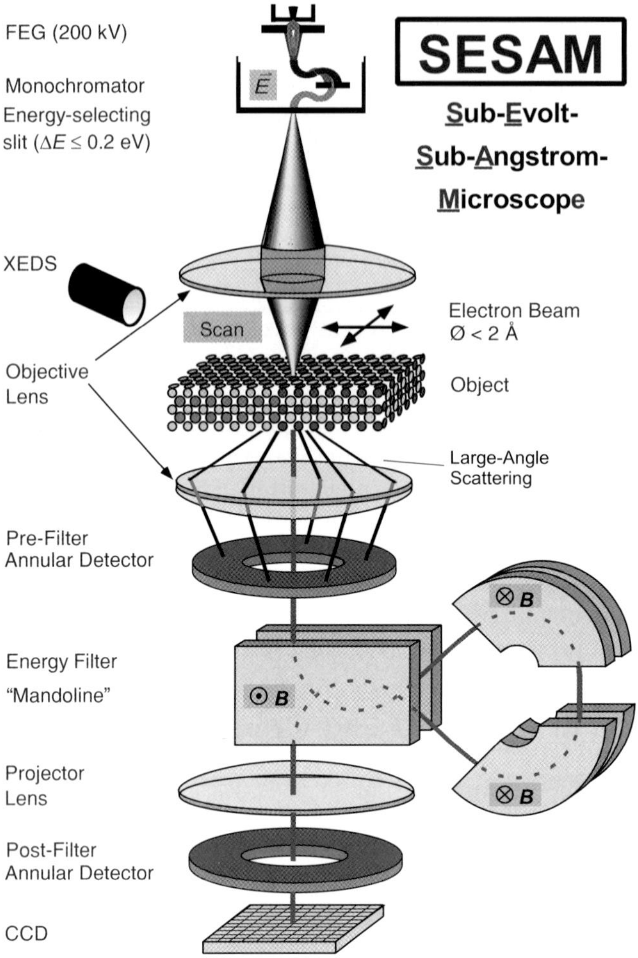

Fig. 4.5. Schematic design of the Sub-Electronvolt-Sub-Ångström-Microscope (SESAM) currently under development. The SESAM will be equipped with a monochromator for the electron gun and an imaging filter of the mandoline-type and is designed to be a next generation analytical TEM

In the line-scan method, pioneered by Colliex and coworkers [35–37], the electron probe is scanned across the region of interest (for example the interface) and a series of spectra recorded. Acquisition of line scans is the method of choice if gradual changes across an interface, resulting from interdiffusion for example, are to be investigated. The method has mainly been used to obtain profiles of the chemical composition of materials. However, the data can also be analysed with respect to changes of the ELNES at interfaces or to study valence losses (interface plasmons) at interfaces [5]. If interfacial ELNES is the aim of the analysis, the length of the line scan and the number of spectra recorded should be chosen such that the successive measuring spots (defined by the position increments and the effective probe diameter) overlap. A good spatial sampling should be reached. Even if only information on the interfacial layer itself is sought, experimental recording of a line scan has some advantages over a spot measurement: if the interface drifts during the measurement the line scan will nevertheless cross the interface in general. Inspection of the data then helps to identify the spectra containing ELNES components arising from the interface from among the larger data set. If the high-angle annular dark field signal (HAADF) is recorded simultaneously with the EELS spectra, this can facilitate the location of the exact position of the interface. From the line scan data gradual development of interfacial contributions in the ELNES across the interface can also be studied. Statistical analysis of the data is also facilitated since the number of data sets is larger. However, since the interfacial changes are usually very local, there are typically only a few spectra from a whole line scan that contain information about the interface. Spectra with longer integration times and hence good signal-to-noise ratios are sometimes difficult to record by this method. In the case of data sets obtained by line-scan ELNES measurements across interfaces, the use of statistical methods of analysing the data with the aid of the principal components seems to be a promising technique [38]. Although the main components that result from the analysis frequently do not have a direct physical meaning, careful interpretation can lead to largely unbiased determination of interfacial ELNES components. Artificial neural networks have been used by Duscher and Gatts [35] to extract interfacial components from EELS line scans. A line scan represents a set of spectra which has one spatial dimension. The coordinates of this data set are the energy loss ΔE and a spatial coordinate x. The spectral intensity $I(\Delta E, x)$ can therefore be visualized as a surface plot (see Fig. 4.6). Elemental quantification of the spectra leads to composition profiles $c_{\mathrm{A}}(x)$, $c_{\mathrm{B}}(x)$ of the elements A, B, ... present in the sample (see Fig. 4.6).

Spectra can also be recorded not only along one line scan but by scanning a whole region of interest with the focused electron probe in a raster pattern. This data set is usually called a *spectrum image* [24,40] and now represents the spectral intensity $I(\Delta E, x, y)$ with two spatial coordinates x and y. The method of recording this data cube (see Fig. 4.3) sequentially, spectrum by spectrum (spectrum imaging), is complementary to the method of electron spectroscopic imaging, which will be introduced in Sect. 4.4 of this chapter. The advantage of the spectrum image approach to *the filling of the data cube* is that a complete spectrum is available for every pixel of the image. This can be useful to improve background fitting or for the spatially resolved analysis of ELNES changes. The recording of the data can however be very time-consuming since the acquisition time increases as the square of the number of pixels along the sides of the region of interest. The availability of CCD detectors on parallel EELS spectrometers has improved the statistics of spectrum

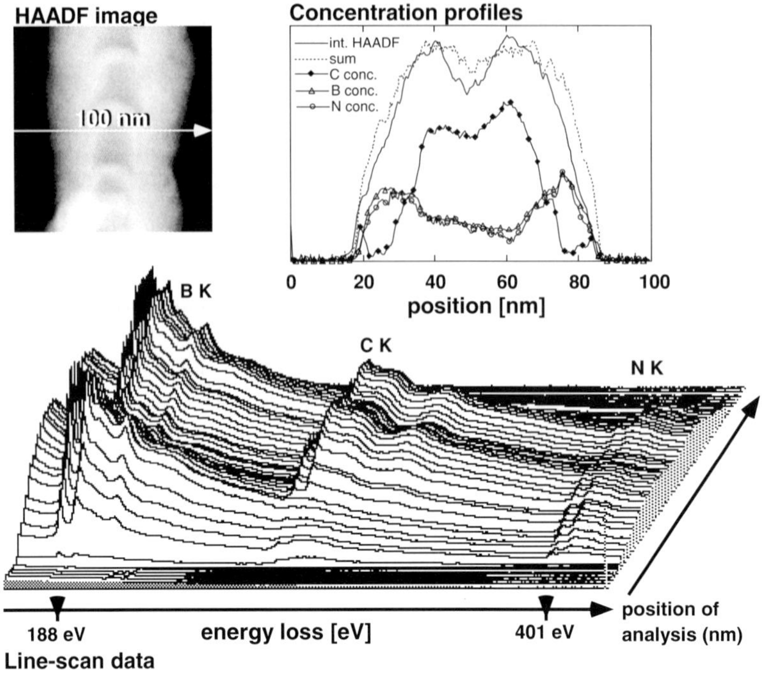

Fig. 4.6. Example for the data set obtained from EELS line scans, recorded on a VG HB501UX STEM. The series of spectra are displayed as a surface plot. The example shows a line scan across a B- and N-doped carbon nanotube [39]

acquisition and allows spectrum images to be recorded with much shorter acquisition times [37]. This development will make the spectrum image techniques feasible for larger regions of interest or higher sampling densities. As in the case of a line scan, the data can be used for elemental quantification. The result is not a chemical profile but a two-dimensional chemical map.

When the electron probe of the TEM can be reliably positioned on the interface for the whole acquisition time, such a spot measurement will necessarily contain a high contribution from the interface atoms. A measurement with sub-nanometre electron probes in dedicated scanning transmission electron microscopes (STEM) and TEM/STEMs in combination with high-angle annular dark-field imaging (Z-contrast imaging) constitutes the ultimate local interfacial analysis. A prerequisite for such measurements is that the atomic structure can be resolved while scanning the probe being used for the analysis across the interface. Pennycook and coworkers [7] were the first to demonstrate that this can be achieved in a dedicated STEM. A detailed investigation of the contrast mechanisms has been made by Silcox and coworkers [10,41] and Nellist and Pennycook [7] have demonstrated that sub-Ångström resolution can be achieved by HAADF-imaging. Z-contrast imaging has been applied to direct structure determination of grain boundaries [42]. It has also been shown that the electron probe can be positioned on individual atomic columns, which are imaged using the high-angle dark-field detector prior to locating the beam for spectroscopy [11]. One early example of near-atomic-level ELNES spectroscopy was given by Muller et al. [43] for the case of differently bonded carbon materials. In this study, it was possible to distinguish between graphite-like and diamond-like bonding with sub-nanometre resolution. The first reports on atomic-resolution chemical analysis using a dedicated STEM were given by Browning et al. [11]. The authors pointed out the benefits of simultaneous use of the high angle elastic scattering (Z-contrast) signal for imaging and the low angle inelastically scattered intensity for spectroscopy. Batson [14] showed that by using this scheme the local electronic structure in silicon-based semiconductors can be studied with atomic resolution.

Duscher et al. [12] give a survey of atomic-column-resolved EELS for different materials. Figure 4.7 shows their example of a Mn-doped 36.7° ⟨100⟩ tilt grain boundary in a $SrTiO_3$ bicrystal. In the Z-contrast image, the characteristic structural units of the grain boundary and the position of the electron probe for the respective measurements are marked. The experiments were performed with a 300 keV VG HB603 U and a 100 keV VG HB501 UX dedicated STEM. The spectra cover the energy-loss range in which the O-K edge (532 eV) and the Mn-L edge (640 eV) are visible. Spectra 5 and 6 recorded from atomic columns away from the interface show a strongly reduced inten-

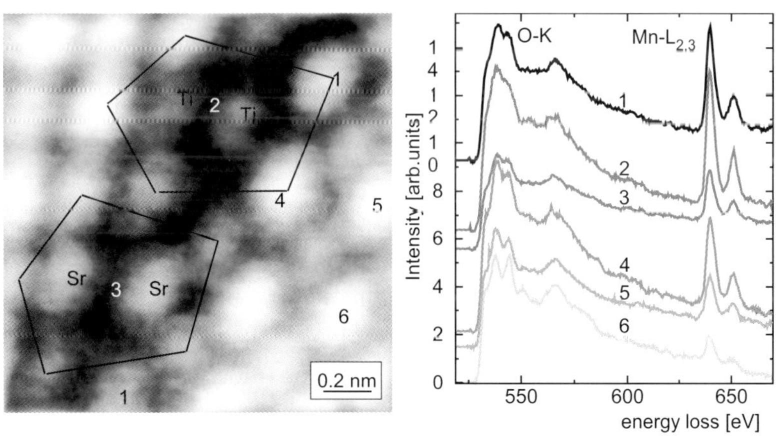

Fig. 4.7. Example of atomic-column-resolved EELS: Mn-doped 36.7° ⟨100⟩ tilt grain boundary in a $SrTiO_3$ bicrystal. In the Z-contrast image the characteristic structural units of the grain boundary and the position of the electron probe for the respective measurements are marked. The individual spectra recorded on individual columns cover the energy loss range in which the O-K edge (532 eV) and the Mn-L edge (640 eV) are visible. Note the changes in the O-K ELNES and the rapid changes in the Mn-L edge intensity for different atomic columns (after [12])

sity of the Mn edge intensity. Comparison of the Mn edge intensity for spectra 2 and 3 (both at the interface but at different locations) reveals a remarkable change, which suggests that the Mn is not homogeneously distributed in the grain boundary but is segregated at certain columns. There are also distinct changes visible in the O-K edge ELNES.

These results might imply that the local electronic structure of individual atomic columns could be analysed. However, the propagation of electron probes along individual atomic columns as a highly confined Bloch state is an area of ongoing research. It has been shown on the basis of Bloch wave calculations that coherent dynamical scattering of zone-axis-aligned crystals leads in good approximation to the formation of an incoherent structure image in the annular dark-field imaging mode [30]. The geometry of the annular detector leads to a dominance of the 1s-bound Bloch wave state. One important recent result is that the incoherent nature of the image formation process in HAADF-STEM is mainly based on the loss of the transverse coherence associated with the detector geometry. Phonon scattering (thermal diffuse scattering) only acts on the longitudinal coherence and has thereby mainly affects on the intensities observed. For quantitative interpretation of compositional changes the effect of phonon scattering has to be taken into account. Interference of the channelling probe along one column with neighbouring columns (crosstalk) becomes noticeable with increasing thickness and is caused by the dispersion of the eigenvalues of the Bloch states. Nellist and Pennycook [30] point out, however, that the dominant 1s-type Bloch states (selected by the detector geometry) show very little dispersion. Tunneling of intensity to neighbouring columns would therefore only occur for unrealistically thick samples. Duscher et al. [12] state that the resolution obtained in the Z-contrast image is not identical with the area analysed for the EELS analysis. The spatial shape of the electron probe is in general characterized by a sharp central peak but with extended tails.

In spatial difference measurements, the beam is scanned within a rectangular area (box) during acquisition of the spectra. These box measurements are repeated on the interface and in the bulk material on both sides. Early applications of the spatial difference method can be found in [44].

Müllejans and Bruley [45] gave a systematic explanation of the metho, which is presented schematically in Fig. 4.8. The rectangular scanning area is used to reduce electron irradiation damage and to ensure that the interface is in fact included and will remain within the measuring area during acquisition. The disadvantage is that the contribution of the atoms i at the interface $\sum_{n_{\mathrm{IF}}} I_{\mathrm{IF,i}}$ to the total signal I_{meas} measured while scanning the beam across the interface can become very small if larger box sizes are chosen. Even for boxes of a few nm width and length, the interfacial contribution can be as low as 10%. If we consider the interface signal $\sum_{n_{\mathrm{IF}}} I_{\mathrm{IF,i}}$ as the useful signal and the bulk contributions which have to be subtracted $n_{\mathrm{A}} I_{\mathrm{A}} + n_{\mathrm{B}} I_{\mathrm{B}}$ as the background, the box size should be carefully chosen in such a was that

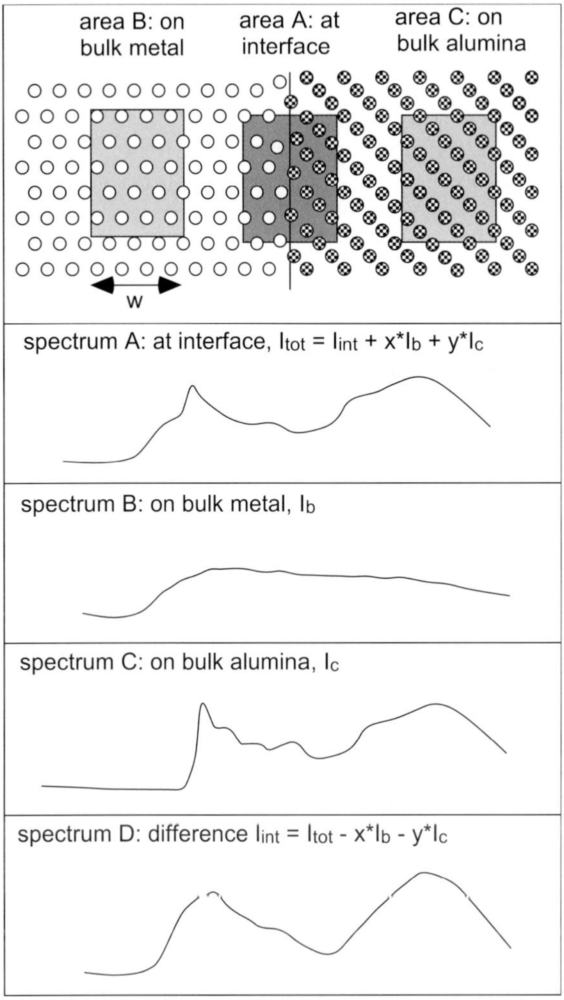

Fig. 4.8. Schematic explanation of the spatial difference method using rectangular boxes as measuring areas for the interface and the two adjacent bulk phases

the signal-to-background ratio will be high enough to ensure that a meaningful signal is obtained, compared to the statistical and systematic errors. The question of the signal-to-background ratio needed for spatial difference measurements has lead to some debate about the correct way of recording and interpreting interface-sensitive ELNES spectra. Muller has stressed the fact that core-level shifts influencing the interfacial spectra could easily lead to misinterpretations when using the spatial difference technique [46]. If the interface spectrum is slightly shifted in energy relative to the bulk spectra (assuming no further change in the ELNES for simplicity), the spatial difference will exhibit an interface-sensitive component. This component is, however,related only to the shift of the spectra. The shape of the resulting interface-sensitive signal cannot be interpreted in terms of ELNES, it is nothing else but the first derivative of the spectra from interface and bulk. Similarly, an artifact will be generated if the spectrometer experiences small energy instabilities during the recording of the bulk and the interface signals [47]. Hence, core-level shifts and the energy stability of the spectrometer have to be taken into account in the interpretation of spatial difference data. Owing to the high localization of the ionization process of the core electrons, one can regard the measured spectrum as a linear superposition of the excitations from all the individual atoms (or atomic columns) within the measured region. In the case of an interface (IF) between materials A and B, the measured spectrum I_{meas} is the sum of the spectra from all bulk columns of materials A and B, I_{A} and I_{B}, within the measured region, and the sum of all the spectra from the (in general inequivalent) interface columns, $I_{\mathrm{IF,i}}, i = 1 \ldots n_{\mathrm{IF}}$

$$I_{\mathrm{meas}} = \sum_{n_{\mathrm{A}}} I_{\mathrm{A}} + \sum_{n_{\mathrm{B}}} I_{\mathrm{B}} + \sum_{n_{\mathrm{IF}}} I_{\mathrm{IF,i}} = n_{\mathrm{A}} I_{\mathrm{A}} + n_{\mathrm{B}} I_{\mathrm{B}} + n_{\mathrm{IF}} \overline{I}_{\mathrm{IF}} \, . \tag{4.3}$$

Here n_{A} and n_{B} are the number of atomic columns of materials A and B in the measured area, n_{IF} is the number of interface columns in the measured area and $\overline{I}_{\mathrm{IF}}$ is the average interface spectrum calculated from all the $I_{\mathrm{IF,i}}$. For the sake of simplicity, we assume that the interface and bulk atomic columns are clearly distinguished. The spectral shapes of I_{A} and I_{B} can be derived from additional spectra recorded far from the interface in the corresponding bulk phases. It is not feasible in practice to normalize these spectra to the contribution of exactly one column. However, from the geometry of the measuring area at the interface and for the bulk spectra an estimate for the weights n_{A} and n_{B} can be deduced [48]:

$$I_{\mathrm{IF,total}} = n_{\mathrm{IF}} \overline{I}_{\mathrm{IF}} = I_{\mathrm{meas}} - n_{\mathrm{A}} I_{\mathrm{A}} - n_{\mathrm{B}} I_{\mathrm{B}} \, . \tag{4.4}$$

Another approach is to consider the weights n_{A} and n_{B} as free fitting parameters. Gu [49] has proposed a systematic method of determining the correct weights for the bulk subtraction in order to derive a spatial difference signal that is unbiased. The criteria for choosing the weights n_{A} and n_{B} are as follows: a set of parameters is chosen that minimizes the total intensity of the interface signal but avoids any obviously unphysical features in the derived interface signal (such as negative counts or reversal/inflection of minima or maxima of the ELNES).

In addition, the signal-to-background ratio and especially the visibility of the component from the interface atoms will depend sensitively on how different the shape of the interface ELNES is from that of the bulk ELNES. Very distinct differences facilitate the measurement. Close similarity between the bulk and interface spectra may easily hide spatial difference signals. An important advantage of improved energy resolution in future instrumentation (sub-eV microscopes) will be that minute difference between the ELNES from bulk phases and interface atoms will be more visible in cases in which they are at present smeared out by energy broadening. Theoretical predictions of interface ELNES spectra from the site and symmetry-projected local density of states will additionally help to identify interface components more easily. The interface contribution can also be enlarged if measuring boxes with a high aspect ratio, with the long axis of the rectangle parallel to the interfaces, are selected. An additional benefit is a more representative signal from a larger interface segment and a reduction of local electron dose.

4.3.2 Segregation at Grain Boundaries in Copper

Segregants at grain boundaries can have dramatic effects on the macroscopic electronic or mechanical properties of a material. Segregation of these dopants to grain boundaries or interfaces must be controlled in order to improve their properties or to avoid negative side effects. Certain alloying elements are known to habe a considerable effect on the mechanical properties of metals. In many cases, these effects are exploited to improve the properties of the alloy. On the other hand, certain unwanted contaminants can make metallic parts prone to brittle failure even at very low concentrations. A well-known example of this behaviour is the embrittlement of polycrystalline Cu by small concentrations of Bi or the embrittlement of Al by Ga. Enrichment of minor alloying elements becomes an issue owing to repeated recycling in the copper industry. The Bi effect on copper has been known for more than 100 years. A better understanding of the segregation behaviour and the microscopic mechanisms of embrittlement at the grain boundaries was the purpose of the investigations by Alber, Müllejans and Rühle [47], a summary of which will be given below. The first quantitative TEM/STEM investigation of Bi segregation in Cu by Baumann and Williams [50] on polycrystalline Cu(Bi) samples gave evidence for an anisotropy of the Bi segregation at different grain boundaries. In a subsequent study Michael and Williams [51] explained the time and temperature dependence of the Bi segregation at grain boundaries with the aid of McLean isotherms. In the special case of a $\Sigma 3$ (111)[110] grain boundary, no Bi segregation was detected. The progress of the work by Alber and coworkers [47]reported here lies in the preparation of specific interfaces. This made it possible to assess the Bi segregation quantitatively as a function of grain boundary orientation, temperature and total Bi concentration.

Bicrystals of copper with well-defined orientation relationship were grown by the Bridgman technique from high purity (99.999%) copper. The bicrystals were doped with Bi from the vapour phase to well-defined concentrations of Bi. The samples were then homogenized at 850°C for one week. Finally, annealing at different temperatures was used to reach the respective equilibrium segregation of Bi at the grain boundaries. Electron transparent samples for the TEM were prepared by spark cutting and subsequent electrochemical

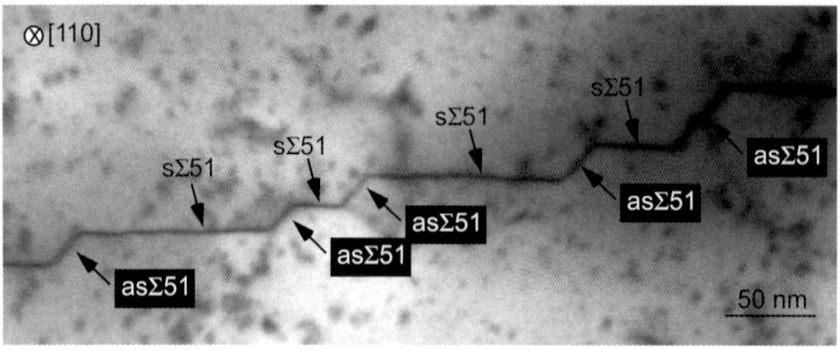

Fig. 4.9. STEM bright-field micrograph of a micro-faceted $\Sigma 5$ [110] grain boundary which has rearranged into symmetrical s$\Sigma 51$ and asymmetric as$\Sigma 51$ facets. The micrograph was recorded at the VG HB501UX STEM in [110] projection from a Cu(Bi) sample annealed at 700° [47]

polishing. The specimens were analysed with respect to their overall morphology and their orientation relationship by conventional TEM. This was necessary because some of the bicrystals did not exactly grow with the orientation of the seed crystals in the Bridgman furnace. Grain-boundary dissociation into two grain boundaries and faceting of the grain boundary were also observed. A micrograph of a microfaceted $\Sigma 51$ [110]grain-boundary is shown in Fig. 4.9. Bismuth segregation to individual facets of the grain boundary was measured quantitatively by EDS [47]. Bi segregation at the symmetrical facets of the $\Sigma 51$ [110]grain-boundary was determined to reach an equivalent of 0.09 ± 0.03 monolayers. The Bi concentration at the asymmetrical facet was below the detection limit of 0.03 monolayers. This result clearly shows that the segregation depends strongly on the grain boundary orientation and geometry. Alber and coworkers [47] analysed systematically the Bi segregation to $\Sigma 3$, $\Sigma 11$, $\Sigma 17$, $\Sigma 51$ and $\Sigma 5$ grain boundaries both at different annealing temperatures and Bi doping concentrations. The quantification allowed the investigated grain boundaries to be classified into those which show no detectable Bi-segregation ($\Sigma 3$, $\Sigma 11$) and those with Bi-segregation ($\Sigma 17$, $\Sigma 51$, $\Sigma 5$). For the latter grain boundaries, the segregation enthalpy could be determined from the EDS data: for the $\Sigma 51$ and $\Sigma 17$ the segregated Bi atoms at the grain boundary show an attractive interaction reflected in a decrease of segregation enthalpy for increasing Bi coverage of the grain boundary.

To obtain this EDS result an improved method for the quantification of the EDS data for grain-boundary segregation was developed by Alber and coworkers [52]. The authors showed that it is necessary to consider the broadening of the electron beam with increasing specimen thickness to avoid systematic errors in the Bi concentration at grain boundaries. This holds both for spot and box mode measurements. The effect of beam broadening leads to an effective scanned area greater than the area on which the electron beam rasters at the specimen entrance surface. The results of Bi segregation have to be corrected for different specimen thicknesses. This effect becomes notable for specimens thicknesses above 50 nm for the 100 keV microscopes used.

This example of a high resolution EDS investigation of a grain boundary clearly illustrates the advantage of using a highly focused electron beam in ATEM. The spatial resolution of EDS measurements is determined by the excitation volume in the sample, which is given by the probe size and the subsequent beam broadening. Similar to the situation in EELS, spot, line-scan or box acquisition modes should be chosen depending on the specific requirements of the experiment.

In the above-mentioned set of grain boundaries, the s$\Sigma 5$ always showed increased Bi segregation. The Bi-doped $\Sigma 5$ (013) [110]grain boundary (material 87 wt.-ppm Bi, annealed at 70°C, surface density of Bi at the grain boundary $\Gamma_{\mathrm{Bi}} = 2.9 \pm 0.5$ Bi atoms per nm^2) was therefore investigated in more detail with a view to determining the atomic positions of the segregated Bi atoms at the grain boundary. Figure 4.10 shows a conventional high-resolution phase-contrast image of the grain boundary (**a**) and a Z-contrast image recorded at the VG HB603 at ORNL (**b**). Figure 4.10c is the Z-contrast image of an undoped $\Sigma 5$ (013) [110]grain boundary. The image contrast of the HRTEM image cannot be directly interpreted. It was therefore not possible to determine the Bi positions from HRTEM images. New results obtained by Gemming and coworkers [53] show that quantitative image simulation and comparison with HRTEM images leads to results that match the Z-contrast results closely (**b**). A periodic sequence of brighter spots at the grain boundary is clearly visible in the Z-contrast image. This contrast feature is completely absent for the undoped grain boundary (**c**). Since the image intensity in Z-contrast is in a first approximation proportional to the atomic number of the scattering atoms, the brighter spots indicate that these columns are Bi-rich.

Owing to the embrittlement effect, the investigation of possible changes in the bonding of Cu atoms at the grain boundaries induced by the Bi segregation is of high interest. Changes in the bonding character of Cu atoms at the grain boundary should lead to changes of features in the ELNES of the grain boundary as compared to the bulk. The key problem in this kind of study is to record the weak interface signal, as discussed above. The ELNES of grain boundaries in polycrystalline Cu(Bi) samples was first investigated by Bruley et al. [54]. The Cu-$L_{2,3}$ edge ELNES of the grain boundary was found to differ from that of bulk copper. The grain-boundary-sensitive signal derived by the spatial difference method revealed similarities with the ELNES of oxidized Cu^{1+} (white lines). Bruley et al. [54] calculated a charge transfer of 0.15 electrons from the Cu to the segregated Bi atoms. In a subsequent article, the experimental results were compared with band-structure calculations of Cu–Bi clusters designed to model the grain-boundary situation [55].

Fig. 4.10. Bi-doped $\Sigma 5$ (013) [110]grain boundary (material 87 wt.-ppm Bi, annealed at 700°C, area density of Bi at the grain boundary $\Gamma_{\mathrm{Bi}} = 2.9 \pm 0.5$ Bi atoms per nm^2): (**a**) high-resolution phase-contrast image (ARM1250) of the grain boundary and a Z-contrast image recorded with the VG HB603 at ORNL, Tennessee, USA (**b**). (**c**) Z-contrast image of an undoped $\Sigma 5$ (013) [110]grain boundary [47]

In other studies, changes in the mechanical properties of metals or intermetallic alloys caused by segregants have been successfully explained by charge transfer. Muller et al. [56] have explained the ductilization of boron-doped Ni_3Al by charge transfer from the segregated B-atoms to the grain-boundary Ni atoms. However, Muller has also stated that great care has to be taken to interpret the grain boundary data derived from spatial difference methods [46] correctly: if the ionization edges for the grain boundary and the bulk are shifted in energy, the subtraction will lead to artifacts which – for example in the case of Cu – will resemble white lines. Such energy shifts can be caused by core-level shifts or instrumental instabilities. In the above example of ATEM experiments at well-defined grain boundaries, a detailed study of the ELNES changes at the Cu-$L_{2,3}$ edge revealed no significant change in the

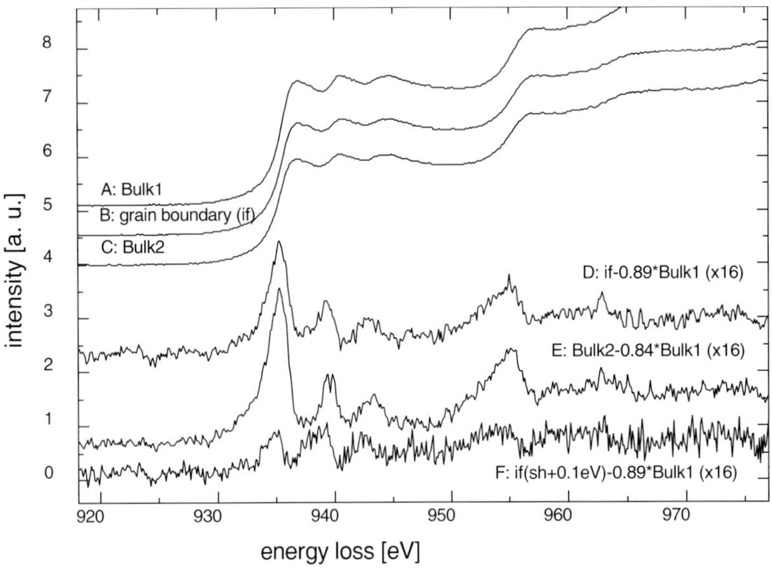

Fig. 4.11. Typical experiment to measure the grain boundary sensitive ELNES. Three spectra were recorded in the 2 adjacent grains (A,C) and at the grain boundary (B) of a Bi-doped $\Sigma 5$ (013)[110]grain boundary (material 85 wt.-ppm Bi, annealed at 700°C). The difference spectrum (D) shows two white lines at 935 and 939 eV which resemble the characteristic ELNES of oxidized Cu. Spectrum (E) shows an efficient test of the data. The two bulk spectra (A,C) are subtracted from each other after intensity normalization. The result of this operation (E) again shows white lines. This is a strong indicator that the difference spectrum is due to energy shifts and reflects no real bonding change at the grain boundary. Spectrum (F) on the other hand reveals that the resulting white lines from (D,E) are vanishing if one of the spectra is shifted by 0.1 eV before substraction [47]. No significant charge transfer could be detected in this experiment

ELNES or the electronic structure, as is evident from the results presented in Fig. 4.11.

4.3.3 Bonding at Metal-Ceramic Interfaces

Control of the properties of metal–ceramic interfaces is of increasing importance in applications of advanced materials [57]. Typical classes of materials in which those interfaces are present are metal-matrix composites (MMC), oxide-dispersion-strengthened metallic alloys, electric contacts to functional ceramics, thermal barrier coatings and other ceramic coatings on metals used to improve wear resistance or hardness and to obtain specific optical properties. The mechanical properties of such metal–ceramic interfaces are one important factor controlling the overall mechanical and functional behaviour of the material they are contained in [58]. It is now realized that many aspects of the behaviour of materials that are already found in practical applications are largely controlled by atomic-scale microstructural features at internal interfaces. Improvements of the performance of these materials or the control of production quality and reliability (which is of crucial importance for many industrial processes) depend strongly on the ability to characterize interfacial structures, and on a better understanding of the atomic bonding which determines their behaviour.

Metal-ceramic interfaces between α-Al_2O_3 and different metals (Al, Cu, Nb, Ni, Fe, NiAl) have been studied by many groups by combination of high-resolution transmission electron microscopy (HRTEM) and analytical TEM. The Cu/α-Al_2O_3 interface is considered as a technologically relevant model system. The results on this system given below have been obtained by our colleagues Ch. Scheu, G. Dehm and M. Rühle in cooperation with several coworkers. Their research has been chosen by the authors as one of the most complete examples of advanced grain boundary studies using EELS and ELNES. In first experiments the, bonding of Cu on the basal plane of α-Al_2O_3 has been investigated [59–62]. The main questions of interest were (i) what is the terminating layer of the sapphire crystal and (ii) what types of bonds develop between the atoms of this terminating layer and the Cu overlayer. This implies specifically the question of whether the Cu changes the oxidation state. In order to study interfacial cohesion, Ti interlayers of nominally 0.1 nm, 3 and 10 nm thickness were deposited prior to the growth of the Cu overlayer. Deposition of both Cu and Ti was performed by molecular beam epitaxy at 100 to 200°C under ultra-high vacuum conditions on the basal plane of sapphire. HRTEM showed that atomically sharp interfaces could be obtained as shown for the example of the pure Cu/α-Al_2O_3 interface in Fig. 4.12.

The presence of Ti at the interface with nominally 0.1 nm thickness was verified by EELS and EDS analysis. It was also possible to image the Ti interlayer directly with electron spectroscopic imaging (ESI) [63]. This demonstration of the typical detection limits of this technique is discussed later in this chapter.

In addition to the microstructural characterization, the mechanical properties of the interfaces were measured using hardness indentation techniques by Dehm et al. [64]. From the mechanical tests, the authors were able to show that the interfacial shear-strength of the Cu/α-Al_2O_3 interface was increased by the presence of a 0.7 nm or 3 nm thick Ti interlayer [64]. For the investigation of the bonding on the atomic scale, a Cu/0.1 nm Ti/α-Al_2O_3 system was prepared and compared with pure Cu/α-Al_2O_3. Changes of the local electronic environment of any of the elements present at the boundary (Cu, Ti, Al, O) should lead to changes in the near edge fine structure of the respective ionization edges. For the interpretation of such data, it is useful to refer to reference ELNES spectra for the corresponding element in different phases and compounds. Figure 4.13a collects the "fingerprints" for the

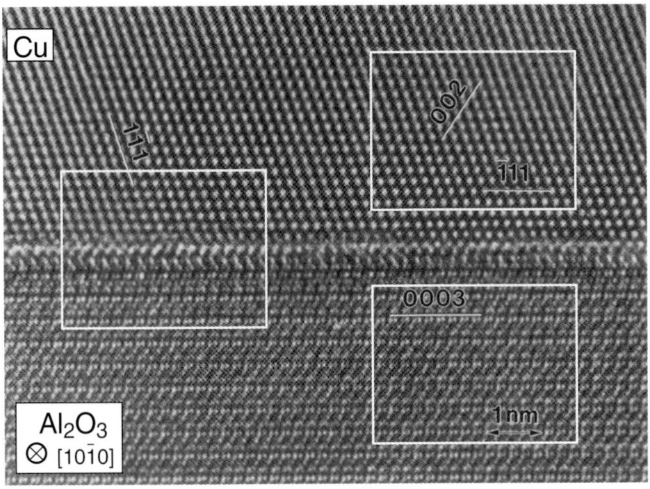

Fig. 4.12. High resolution TEM micrograph of the interface between Cu deposited on the basal plane of α-Al_2O_3 using molecular beam epitaxy [59]

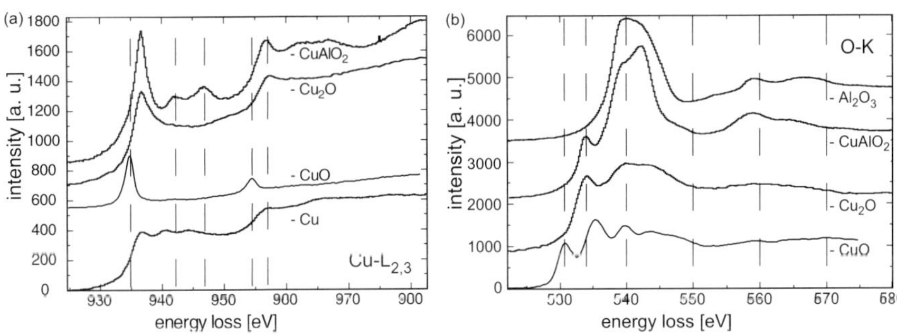

Fig. 4.13. ELNES reference spectra for the (**a**) Cu-$L_{2,3}$, (**b**) O-K edge

Cu-$L_{2,3}$ edge as pure metal (oxidation state 0), as cupric oxide (Cu_2O, ionization state +1), as cuprous oxide (CuO, ionization state +2) and as $CuAlO_2$ (Cu^{+1}). There is a chemical shift for the CuO with copper in the oxidation state +2 relative to the metallic copper or the compounds with Cu^{+1}. Partial transfer of electrons from the 3d band of Cu caused by oxide formation and hybridization leads to two characteristic sharp peaks (white lines) in the ELNES. In Fig. 4.13b the O-K ionization-edge ELNES for several phases is compared. There is also a chemical shift for CuO. ELNES reference spectra are compiled in [65] for major phases but are otherwise scattered throughout the literature.

Scheu and coworkers [60,61,66] investigated systematically the interface-sensitive ELNES of all the above-mentioned Cu/α-Al_2O_3 and Cu/Ti/α-Al_2O_3 interfaces. Figure 4.14 shows the result of a spatial difference analysis of the

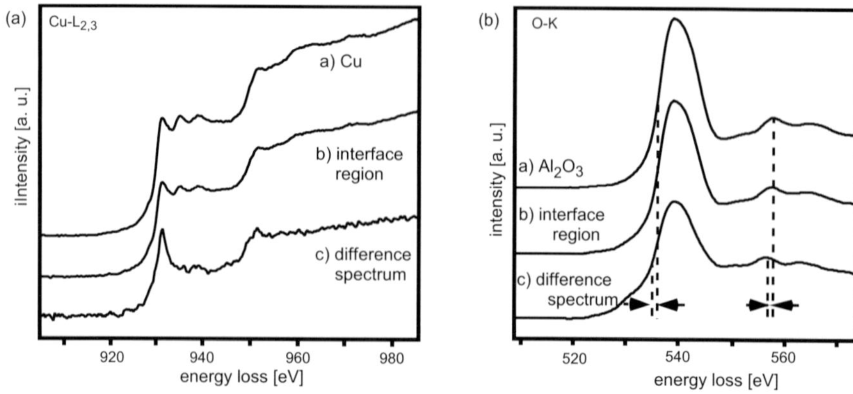

Fig. 4.14. Comparison of the interface-sensitive ELNES obtained by spatial difference for a Cu/sapphire interface grown by MBE (basal plane) with reference spectra for the (**a**) Cu-$L_{2,3}$ and (**b**) O-K edge. The Cu-$L_{2,3}$ edge reveals characteristic white lines in close correspondence with the reference spectrum for the Cu_2O bulk phase. The O-K signal is predominantly sapphire-like but shows a pre-edge feature, which indicates that O 2p-Cu 3d hybridised states have formed (from [60])

ELNES at the Cu/Al_2O_3 interface. The Cu film was deposited on the basal plane of the sapphire substrate by MBE. No spatial difference signal was obtained for the Al-$L_{2,3}$ edge, which indicates that the Al atoms are not involved in the interface bonding and remain in their octahedral oxygen coordination. The Cu-$L_{2,3}$ (Fig. 4.14a edge reveals characteristic white lines in close correspondence with the reference spectrum for the Cu_2O bulk phase. The Cu atoms at the interface are therefore expected to be in a nominal Cu^{+1} oxidation state. Rapid charge screening in the copper leads to a strong confinement of the effect of bond formation on the ELNES. It is therefore understandable that the Cu-$L_{2,3}$ edge can resemble the ELNES fingerprint of the bulk Cu_2O so closely. The O-K signal, however, is affected by several nearest neighbour atomic shells and is therefore much more influenced by the adjacent bulk alumina. The O-K edge (Fig. 4.14b) is hence predominantly Al_2O_3-like. The weak pre-edge feature indicates that O 2p-Cu 3d hybridised states have formed [60] and is consistent with the formation of Cu–O bonds of mixed ionic-covalent character. These spatial difference results show that the interface can be modelled as an O-terminated basal plane of alumina with the development of Cu–O bonds across the interface. This finding is in accordance with the quantitative HRTEM investigations performed in parallel [67].

Figure 4.15 compares the interface-sensitive ELNES obtained by spatial difference for the Cu/0.1 nm Ti/Al_2O_3 and the Cu/10 nm Ti/Al_2O_3 systems with bulk Ti and TiO_2. The TiO_2 edge shows a characteristic splitting of the L_2 and L_3 lines. This crystal field splitting is very sensitive to the local coordination of the Ti atoms. It does not exist in metallic Ti and is not present in the two interface ELNES spectra. This result was interpreted by

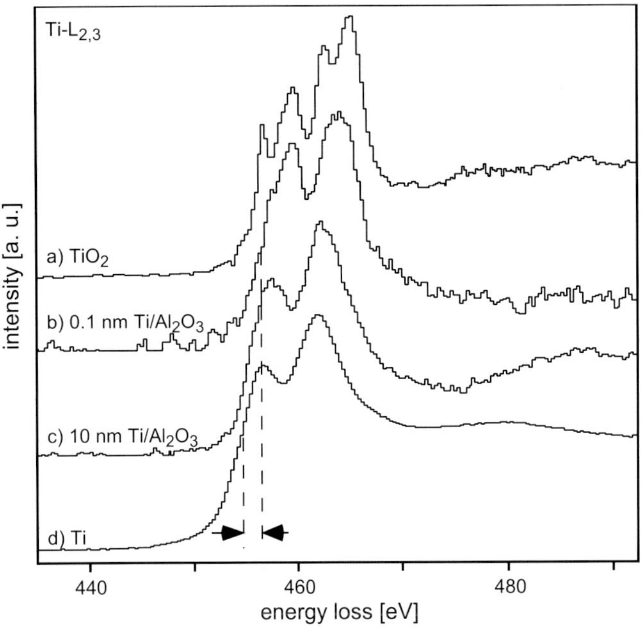

Fig. 4.15. Comparison of reference spectra for pure TiO_2 (**a**) and pure Ti (**d**) with the interface sensitive ELNES obtained by spatial difference for a Cu/0.1 nm Ti/Al_2O_3 (**b**) and a Cu/10 nm Ti/Al_2O_3 (**c**) interface. The Ti-$L_{2,3}$ edge in (**b**) is shifted by 1.6 eV ± 0.3 eV with respect to metallic Ti (**d**) and by 0.4 eV ± 0.2 eV as compared to the thicker Ti film (**c**) (from [61])

Scheu and coworkers as a change in the local coordination of the Ti atoms at the interface as compared to bulk TiO_2. The sign of metallic Ti bonding is a chemical shift to lower energy loss relative to TiO_2. The Ti-$L_{2,3}$ edge for the thin Ti interlayer is shifted by 1.6 eV ± 0.3 eV with respect to metallic Ti and is therefore very close to the energy onset of TiO_2. There is a shift of 0.4 eV ± 0.2 eV as compared to the thicker Ti film. Scheu [61] has interpreted this result as indication of a nominal Ti^{4+} oxidation state for the first atomic plane of the thin Ti layer. For the thicker Ti layers, the interface Ti atoms are in a mixed Ti^{3+}/Ti^{4+} oxidation state. This interpretation is further supported [61] by an analysis of the white line branching ratio, that is the intensity ratio of the L_3/L_2 lines, which has been frequently used to measure oxidation states of transition metal atoms.

It is increasingly realized that the bonding at atomically sharp interfaces depends sensitively on the orientation of the interfaces, their pretreatment before bonding or growth and the possible formation of reaction layers. The interfacial chemistry and the bonding between Cu and single crystals of α-Al_2O_3 have been systematically investigated as a function of the oxygen partial pressure (in the range between 10 to 10^{-10} Pa) during interface formation

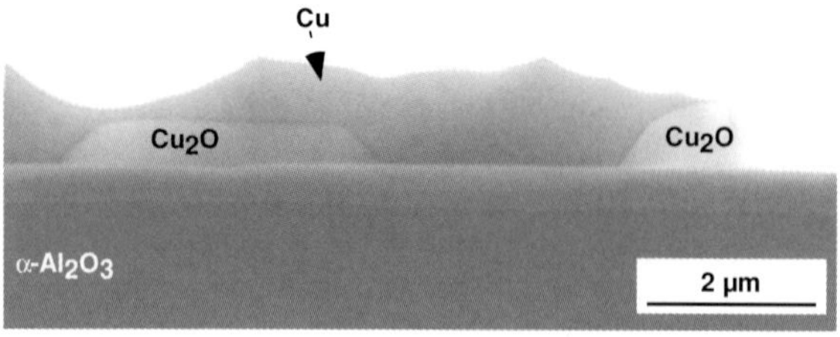

Fig. 4.16. Bright field micrograph of the interface between Cu and the R-plane of α-Al_2O_3 prepared by melting the Cu at an oxygen partial pressure of 10^{-5} bar on the sapphire substrate (VG HB501UX STEM). Copper oxide islands are formed at the interface [68]

both for the basal (B-) (0001) plane of α-Al_2O_3 and the rhombohedral $(10\bar{1}2)$ plane of α-Al_2O_3 by Alber et al. [68]. The interfaces were prepared by the sessile drop method on sapphire plates of the respective orientation. It should be noted that this method is different from the MBE Cu deposition used for the interfaces discussed above. The resulting interface bonding may therefore not be comparable. One example illustrating the microstructure of the interfaces obtained is given in Fig. 4.16. For the samples prepared at lower oxygen partial pressures, the cohesion of the interfaces was so low that TEM specimen preparation was not possible.

The electronic structure of both types of interfaces was studied by high spatial resolution EELS using the spatial difference method in order to separate the interface-sensitive component. In the example discussed here, EELS spectra in the energy ranges of the four accessible ionization edges Cu-$L_{2,3}$, O-K, Al-$L_{2,3}$ and Cu-$M_{2,3}$ were recorded for samples prepared at different oxygen partial pressures for both the basal and the rhombohedral plane. In Fig. 4.17 the results of the spatial difference analysis for the rhombohedral plane ($p_O = 10^{-5}$ bar) are given. The Cu-$L_{2,3}$ shows a clear spatial difference signal, the spectral shape of which is characterized by the L_2 and L_3 white lines. The overall ELNES resembles that of Cu_2O, which is, however, shifted by 1.5 eV to higher energies as compared to a Cu_2O reference. The main peak of the interface-specific signal for O-K is broadened. In the energy range of the overlapping edges Al-$L_{2,3}$ and Cu-$M_{2,3}$, a featureless interface signal has been obtained. These results were reproduced for all other samples grown at different oxygen partial pressures on the rhombohedral plane. The basal planes, however, showed no interface sensitive component for the Cu-$L_{2,3}$ and a rather broad IF spectrum for the O-K edge. This result is different from the finding of Scheu et al. [60] for MBE-grown copper films on the basal plane of sapphire.

The differences in the ELNES for the sessile-drop-prepared basal and rhombohedral plane interfaces indicate a distinct difference in the bonding mechanism over the interface. In [68], this difference was qualitatively explained by the charged nature of the rhombohedral surface and the formation

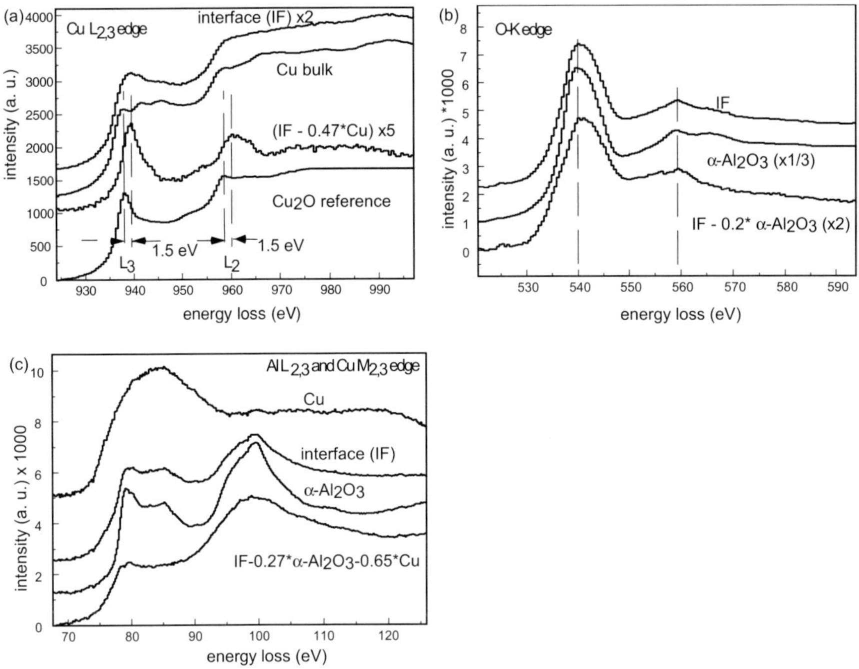

Fig. 4.17. Spatially resolved ELNES of the three available ionization edges for the Cu/α-Al_2O_3 interface (R-plane, oxygen partial pressure 10^{-5} bar for (**a**) and 10^{-7} bar for (**b**) and (**c**)). The interface characteristic spectral component is calculated by the spatial difference method from the interface spectrum (IF) and the respective bulk spectra. (**a**) Cu-$L_{2,3}$: The spatial difference signal shows a spectral shape with L_2 and L_3 white lines, which are shifted by 1.5 eV to higher energies relative to a Cu_2O reference. (**b**) O-K: The main peak of the interface specific signal is broadened. (**c**) Al-$L_{2,3}$/Cu-$M_{2,3}$: Note the non-vanishing interface signal

of ionic bonds whereas the basal plane is known to reconstruct to be neutral. The interpretation of the shape of the Cu-$L_{2,3}$ ELNES signal for the rhombohedral plane is not straightforward. The IF signal shows an increase in the so-called white-line intensities relative to the bulk. The observed chemical shift of 1.5 eV makes it difficult to interpret the interface signal as a fingerprint of a Cu_2O-like electronic structure, corresponding to Cu^{1+} atoms at the interface, which could also be interpreted as a direct bonding of oxygen atoms from the sapphire to the terminal layer of copper. Scheu and coworkers have recently found that the ELNES is identical with that of a Cu–Al intermetallic compound [69]. More generally, an increase of the white-line intensity corresponds to a reduction in the occupancy of the Cu 3d-orbitals and a transfer of electrons away from the Cu. This corresponds in the chemical sense to an oxidation, but the formation of intermetallic bonding could have a similar effect. It is not possible to identify straightforwardly the IF components for the other ELNES edge reference fingerprints. The example shows that spatially resolved ELNES analysis can at present yield some qualitative information on the bonding at internal interfaces. This is already very important for a better understanding of atomic effects on macroscopic properties, the wetting behaviour of and the cohesion at metal-ceramic interfaces for example. The potential of this technique has not yet been even partially exploited: improvements in terms of reducing the energy spread of the primary electrons, increasing the brightness of electron sources, using optimized energy filters and EELS detectors and the approach to 0.1 nm resolution with very fine electron probes will certainly help to record experimental data with better energy resolution and signal-to-noise ratio in shorter times. The other important development is related to the application of modern band-structure calculation methods to identify low-energy interface configurations and to the possibility of calculating site and symmetry-projected density of states of bulk crystals and interface configurations (using supercells) in a quantitative manner. Theoretical calculations will then become an independent additional technique to be combined with TEM methods. This has been shown in a recent study of the rhombohedral twin boundary in α-Al_2O_3 by Nufer et al. [70]. In this study, the results from quantitative HRTEM, spatially resolved EELS and density functional calculations of the structure and the ELNES led to a consistent interpretation and unambiguous structural model of this twin boundary. An important result of this project was also the demonstration that the preparation of the TEM samples critically influences the possibility to record high quality data. Only after optimization of several specimen-preparation steps could the deleterious effect of surface damage during ion-beam thinning be reduced to the point at which α-Al_2O_3 could be recorded without immediate electron irradiation damage. After optimization of the specimen quality, reasonably long acquisition times were possible without the appearance of a pre-peak in the O-K edge, which is a direct indication of radiation damage in α-Al_2O_3.

The examples given above highlight the information that can be obtained about the bonding at internal interfaces by using high spatial resolution EELS. The detailed investigations of the Cu/α-Al_2O_3 interface have revealed that the bonding across atomically sharp interfaces can be influenced by several factors. These include the choice of the substrate orientation, the substrate pretreatment (not discussed in this contribution), the nature of the interlayers (for example Ti) introduced to increase cohesion and the deposition method for the metallic overlayer. It should be noted that earlier investigations of Nb/alumina interfaces prepared by diffusion bonding [71] or the Fe or Ni/alumina interface prepared by hot isostatic pressing [72] have revealed the formation of an Al-terminated interface. Oxygen loading of Cu films and subsequent annealing treatments were found to result in extended reaction layers at the interface to the sapphire [73], which also increased the interfacial shear strength. A better understanding of the atomic processes can help us to understand the properties of interface-controlled materials, to avoid failure and to design interfacial bonding in a more controlled manner. The combination of high spatial resolution EELS with theoretical calculations will challenge our concepts about the local changes in electronic structure at atomic discontinuities in materials. Improved instrumentation – already close to the market place – will open up new perspectives for such research activities in the coming years.

4.4 Energy-Filtering Transmission Electron Microscopy

In analytical TEM, the main advantage associated with the new energy-filtering techniques, summarized under the keyword "electron spectroscopic imaging" (ESI), is the fast two-dimensional data acquisition. In all the examples given above, the three-dimensional data-space of the spatially resolved energy-losses, $I(x, y, \Delta E)$ (see Fig. 4.3) has been sequentially filled by recording individual spectra. High spatial resolution could be obtained only by using a field-emission TEM capable of producing a very narrow electron probe. The significance of filling the data cube with information by acquiring spectra for a two-dimensional array of probe-positions was first pointed out by Jeanguillaume and Colliex [24]. The resulting intensity distribution is named a "spectrum-image". Jeanguillaume and Colliex [24] list a number of ways of analysing the data-space and also give references to a number of earlier attempts to fill the data cube by means of energy-filtering TEMs equipped with a Castaing–Henry type of energy filter. An energy-filtering TEM with this type of energy filter was also used by Reimer and coworkers [74,75] in their pioneering work on the operation modes of an EFTEM. However, the rapid spread of energy filters in analytical microscopy only started when the omega-type energy filters [26–28] and the post-column energy filter became commercially available [29]. Early work with these energy filters concentrated on applications in elemental mapping and numerous examples can be found in literature, e.g. the work of Hofer and coworkers [76,77], Crozier [78] and Mayer and coworkers [79], to name just a few examples. A concise survey of the background and the different operating modes of EFTEM is given in the book edited by Reimer [16]. The analysis of spectra extracted from a series of energy-filtered images was pioneered by Körtje [80], Martin and coworkers [81] and Beckers et al. [82]. The first attempts to analyse such spectra quantitatively using standard EELS techniques were made by Beckers et al. [83] and by Mayer et al. [84]. Martin et al. [85] and Mayer and Plitzko [86] showed that information on the ELNES can also be extracted from a series of ESI images, even on the nanometre scale. Recently, Midgley and coworkers [87] demonstrated that full spectra extending from the zero-loss up to the 1–2 keV energy-loss range can be retrieved from extended series of ESI-images. In all these applications it is demonstrated that the use of electron spectroscopic imaging (ESI) in an energy-filtered TEM allows EELS data to be recorded from a large specimen area with a limited number of measurements. Any quantitative analysis of the resulting intensity distribution $I(x, y, \Delta E)$ should be independent of the sequence in which the data are acquired. In the following, we shall show how one can employ standard EELS programs, which are in common use and well tested, for the analysis of ESI series. The main advantages of the ESI technique are that a large number of pixels (typically 1024×1024, or 512×512, using 2-fold binning) are recorded in parallel during each image acquisition and that the total number of ESI images and the energy window-width can be easily adapted to the requirements of the experimental analysis. EEL spectra can subsequently be extracted from a series of energy-filtered images for any individual pixel of the image or, by integrating over the corresponding pixels, for a region of interest on the sample.

4.4.1 Basic Equations for Quantification

After background subtraction, the integrated intensity under an inner-shell loss edge is given by

$$I_{\mathrm{K}}(\Delta) = N_{\mathrm{a}} \sigma_{\mathrm{K}}^{\mathrm{a}}(\Delta) I_{\mathrm{L}}(\Delta) \tag{4.5}$$

where N_{a} is the number of atoms of the element 'a' per unit area, $\sigma_{\mathrm{K}}^{\mathrm{a}}(\Delta)$ the inelastic scattering cross-section for the K-edge of element 'a' in a spectral integration interval of width $\Delta = [\Delta E_1, \Delta E_2]$ and $I_{\mathrm{L}}(\Delta)$ the intensity in the low-loss region integrated up to the energy loss Δ [21]. Choosing the starting point of the integration window 10-20 eV above the edge onset has the advantage that the first strong oscillations of the near-edge structure are omitted. Calculated inelastic scattering cross-sections $\sigma_{\mathrm{K}}(\Delta)$ then describe the actual edge shape more accurately. From (4.5) the area density N_{a} (atoms/nm^2) can be calculated. N_{a} is related to the specimen thickness t by

$$t = \frac{N_{\mathrm{a}}}{n_{\mathrm{a}}} \tag{4.6}$$

where n_{a} is the number of atoms per unit volume (atoms/nm^3) of element 'a'. The atomic density n_{a} of element 'a' in a compound $\mathrm{a}_x\mathrm{b}_y(\mathrm{c}_z\ldots)$ is related to the mass density ρ of the compound by

$$n_{\mathrm{a}} = x \cdot n \tag{4.7}$$

and

$$n = \frac{\rho N_{\mathrm{A}}}{A} \tag{4.8}$$

where n is the number of formula units per unit volume, A the molar weight in g and $N_{\mathrm{A}} = (6.023 \times 10^{23})$ is Avogadro's number.

If only concentration *ratios* between two different elements 'a' and 'b' are to be determined, the low-loss intensity $I_{\mathrm{L}}(\Delta)$ does not have to be measured and the result is determined by the ratio of the intensities under the core-loss edges after background subtraction

$$\frac{N_a}{N_b} = \frac{\sigma_K^b(\Delta) I_K^a(\Delta)}{\sigma_K^a(\Delta) I_K^b(\Delta)} . \tag{4.9}$$

The total specimen thickness t can be computed from the inelastic mean-free-path length λ for inelastic electron scattering with energy losses smaller than an integration limit ΔE_{int} [21]

$$t = \lambda \ln \left(I_{tot}/I_0 \right) , \tag{4.10}$$

where I_{tot} is the total intensity integrated up to ΔE_{int} and I_0 is the intensity in the zero-loss peak only. For an accurate quantification, the integration limit ΔE_{int} should be chosen in the range of 150 to 200 eV.

4.4.2 Elemental Distribution Images

As discussed in Sect. 4.2, inner-shell ionization of specimen atoms by the beam electrons leads to characteristic edges in the energy-loss spectrum. The onset energy of each edge can be used to identify the corresponding atomic species. The concentration of an element can be determined from an EELS spectrum if the pre-edge background is extrapolated and subtracted from the signal above the edge. In ESI images this has to be done for each individual pixel. An efficient method for background subtraction, the three-window technique [88], is illustrated in Fig. 4.18. Two ESI images are acquired in the background region before the edge and the extrapolated background is

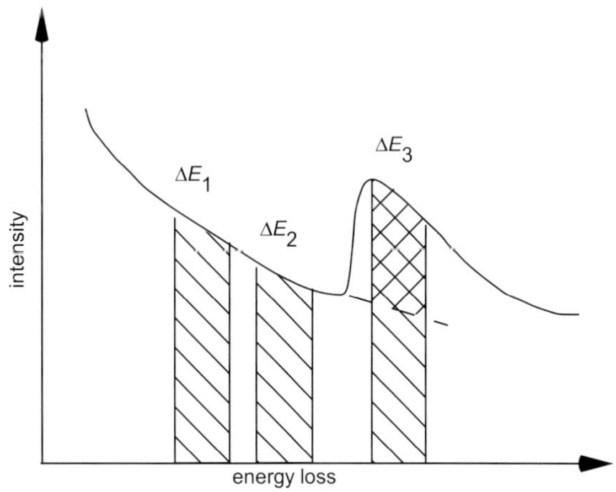

Fig. 4.18. Illustration of the three-window technique for background subtraction in ESI quantification. The result is an elemental distribution image for the given element

then subtracted from the ESI image containing the signal above the edge. As a result, the difference image contains intensity only in regions where the corresponding element is present in the sample and thus maps the distribution of this element. In the three-window technique, however, the intensity in the difference image depends not only on the concentration of the element but may also vary with thickness or with the Bragg orientation of crystalline grains. The latter causes artefacts usually referred to as preservation of elastic scattering contrast, a detailed discussion of which can be found in Hofer et al. [89].

Owing to the low intensity of the individual ESI images (up to a factor of 100 to 1000 less than the corresponding bright-field image), the difference image will also contain considerable noise, which makes it impossible to detect elements in very small concentrations (below 1 at.%). The noise in the elemental distribution images can be reduced by special image processing techniques, which may, however, result in a loss of resolution.

In the three-window technique, the optimum position of the energy windows and their width depends on several parameters [22]: (i) the intensity in the energy-loss spectrum, which decreases strongly with increasing energy loss, (ii) the shape of the edge, which only shows a sharp onset for the light elements, and (iii) the width of the unstructured background region before the edge.

In most cases, the distribution of several elements has to be studied in one sample area. The resulting elemental distribution images can be combined in one image by using different colours for each element and overlaying the individual images. If two or more of the elements under investigation are present in one sample area, mixed colours will occur. Mixed colours thus reveal important information about the occurrence of phases that contain more than one of the elements under investigation.

A straightforward quantitative analysis of elemental distribution images obtained with the three-window technique is possible if concentration ratios between two different elements have to be determined. Crozier [78], Bentley et al. [90] and Hofer et al. [77] have applied a simple division of the two corresponding elemental maps with subsequent normalization by the ratios of the core-loss scattering cross-section to produce quantitative two-dimensional maps of concentration ratios.

4.4.3 Noise Statistics

The noise statistics in an elemental map is governed by the electron counting statistics in each pixel of the CCD camera that is used as a detector. The number of counts N_{ij} in pixel (i, j) is related to the number of electrons via the conversion efficiency c. For most CCD cameras, c is close to unity and thus the deviation from unity will not be taken into account in the following. Additional noise introduced by the CCD camera can be described by its detection quantum efficiency (DQE). The spreading of intensity into neighbouring pixels is described by the point-spread function (PSF) [91].

For Poisson statistics, the variance of the signal is equal to its mean value [92]:

$$\sigma^2\left(N_{ij}\right) = \left\langle \left(N_{ij} - \langle N_{ij}\rangle\right)^2 \right\rangle = \langle N_{ij}\rangle \,. \tag{4.11}$$

Equation (4.11) holds for all three individual images from which the elemental map is calculated. Subtracting the extrapolated background yields the intensity in the elemental distribution image: $N_{ij}^{\mathrm{S}} = N_{ij}^{3} - N_{ij}^{\mathrm{B}}$. The noise in the signal N_{ij}^{S} is composed of the Poisson noise N_{ij}^{3} in image three and the noise of the extrapolated background N_{ij}^{B} and may be amplified by the DQE of the CCD camera

$$\sigma^2\left(N_{ij}^{\mathrm{S}}\right) = \frac{1}{\mathrm{DQE}}\left(\sigma^2\left(N_{ij}^{3}\right) + \sigma^2\left(N_{ij}^{\mathrm{B}}\right)\right) \,. \tag{4.12}$$

If the extrapolation region is much smaller compared to the energy loss, (4.12) can be rewritten by introducing a parameter h [92]:

$$\sigma^2\left(N_{ij}^{\mathrm{S}}\right) = \frac{1}{\mathrm{DQE}}\left(\langle N_{ij}^{3}\rangle + h\langle N_{ij}^{\mathrm{B}}\rangle\right) \,. \tag{4.13}$$

The parameter h is a measure of the additional noise that is introduced by the background extrapolation procedure. For the case of equidistant energy windows ($\Delta E_2 - \Delta E_1 = \Delta E_3 - \Delta E_2$), a setting used in most cases, a value of $h = 6$ is typically obtained. Owing to the background extrapolation, the variance in N_{ij}^{S} is increased by a factor of 3.5 to 6 (compared to the variance of the third image N_{ij}^{3}, for an edge with $N_{ij}^{\mathrm{S}} = N_{ij}^{\mathrm{B}}$ and with $N_{ij}^{\mathrm{S}} < N_{ij}^{\mathrm{B}}$, respectively. The signal-to-noise ratio in the elemental distribution image is given by:

$$(\mathrm{S/N})_{ij} = \frac{\langle N_{ij}^{\mathrm{S}}\rangle}{\sqrt{\mathrm{var}\{N_{ij}^{\mathrm{S}}\}}} = \frac{\langle N_{ij}^{\mathrm{S}}\rangle\sqrt{\mathrm{DQE}}}{\sqrt{\langle N_{ij}^{3}\rangle + h\langle N_{ij}^{\mathrm{B}}\rangle}} \,. \tag{4.14}$$

A signal can be distinguished from noise if the criterion $\mathrm{S/N} \geq 5$ is fulfilled. In order to maximize the signal and hence to approach the theoretical detection limit, it is essential to choose the imaging parameters carefully [93]. In many cases the S/N ratio can be increased by a summation over several pixel, for example by integration along a straight boundary.

4.4.4 Detection Limits

Owing to the smallness of thus cross-sections for inner-shell excitations, the detection limit is governed by the S/N ratio. The signal intensity in the final elemental distribution image indicating the presence of element S is given by [21]

$$I_\mathrm{S} = \frac{1}{e} j_0 n_\mathrm{S} \sigma_\mathrm{S} \tau \,, \tag{4.15}$$

where j_0 is the current density of the incident beam, n_S the number of atoms per unit area of the element under consideration, τ the integration time (the exposure time of the image) and $\sigma_\mathrm{S} = \sigma_\mathrm{S}(\Delta E, \Delta E, Q_c, q_0)$ the integrated inelastic scattering cross-section for the chosen energy window (energy loss ΔE, window width ΔE), and illumination and objective aperture angles (Q_c, q_0). The intensity I_S is superimposed on a background intensity

$$I_\mathrm{B} = \frac{1}{e} j_0 \left(n_\mathrm{S} \sigma_\mathrm{S}^\mathrm{B} + \sum_x n_x \sigma_x^\mathrm{B} \right) \tau \tag{4.16}$$

to which the element S and all other elements contribute according to their density n_x. The two images acquired at energy losses ΔE_1 and ΔE_2 below the edge show only this background intensity. The intensities in the ESI images acquired around the inner-shell loss edges are a factor of 100 to 10000 times lower than the intensities in normal bright-field images. A power law approximation $I_\mathrm{B} = A \Delta E^{-r}$ is used to extrapolate the background and to subtract it from the intensities $I_3 = I_\mathrm{S} + I_\mathrm{B}$ in the third image in order to reveal I_S.

The noise statistics discussed above governs the detection limits for a given element in its distribution image. In the following, we discuss the detection

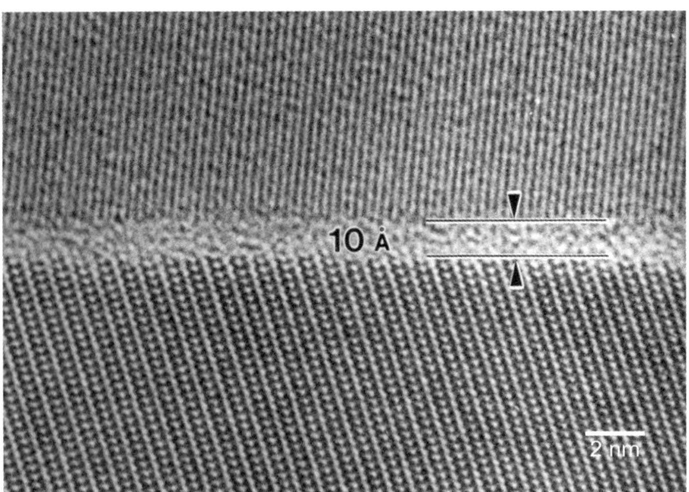

Fig. 4.19. A high resolution image of a characteristic sample area of the Si_3N_4 ceramic material with an amorphous oxide grain-boundary film, 1.0 nm thick

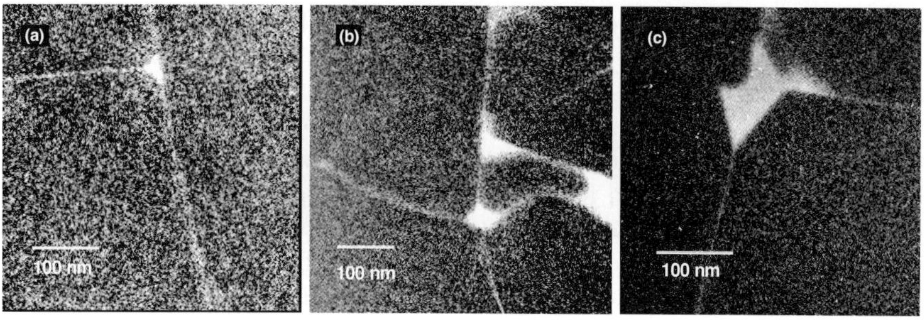

Fig. 4.20. Oxygen distribution images for (**a**) the material with 0.7 nm film thickness obtained on the Zeiss EM 912 Omega, (**b**) the 1.0 nm film thickness material on the Philips CM 20 with GIF and (**c**) the 1.5 nm material on the JEOL ARM1250, which is equipped with a high voltage imaging filter

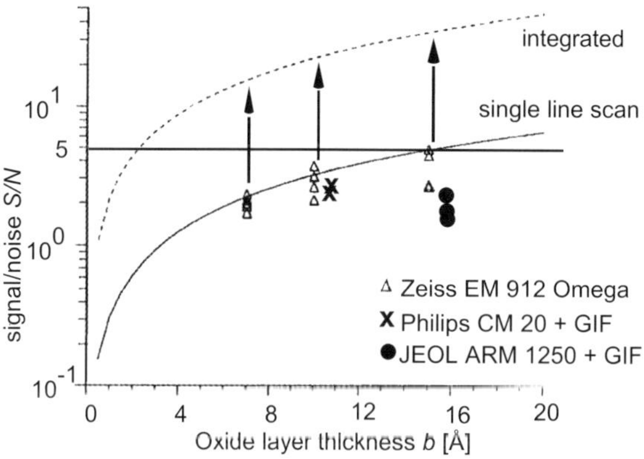

Fig. 4.21. S/N ratio determined from a representative number of grain boundaries in each system on three different TEMs equipped with energy filters, and comparison with theoretical prediction for a single line scan and for an integrated line scan (50 pixels parallel to boundary) [93]

limits for the example of interfacial segregation layers in Si_3N_4 ceramics. For these ceramics it has been predicted theoretically [94] and proven experimentally [95,96] that, during sintering with oxide additives, an amorphous grain-boundary film forms, which possesses an equilibrium thickness. The film thickness is characteristic for a given material and can be controlled by doping with small amounts of Ca [97]. In the following example, materials with an equilibrium film thickness of 0.7, 1.0, and 1.5 nm have been studied. A high-resolution image of a characteristic sample area of the material with 1.0 nm film thickness is shown in Fig. 4.19. Elemental distribution images for oxygen acquired on the Zeiss EM 912 Omega revealed the presence of such grain-boundary films for all three materials. Figure 4.20a depicts an oxygen distribution image for the material with 0.7 nm film thickness. In order to extend the systematic studies to microscopes with different accelerating voltages and different energy filters, the material with 1.0 nm film thickness was investigated on a Philips CM 20 fitted with a GIF (Fig. 4.20b) and the material with 1.5 nm film thickness on the JEOL ARM1250, which is equipped with a high-voltage imaging filter (GIF, Gatan, Pleasanton) (Fig. 4.20c). The S/N ratio was determined for a number of grain boundaries in each system and the results are shown in Fig. 4.21. The S/N ratio was obtained by evaluating line profiles parallel to the boundary according to (4.8) [93].

The results obtained using the Zeiss EM 912 Omega are in good agreement with the results of theoretical calculations for the given sample geometry [93]. In these simulations the S/N ratio in the final elemental distribution image was calculated according to (4.14) using the appropriate inelastic scattering cross-sections for the background before and the signal above the edge (4.15), the optimized imaging conditions being determined for the microscope and the response function (DQE, PSF) of the CCD camera. Figure 4.21 shows the resulting S/N ratio as a function of the thickness of the grain boundary film [93]. Considering that a signal can be discriminated from noise if the S/N ratio is larger than about 5, Fig. 4.21 indicates that the intergranular film can be identified in an individual line profile only if the width is larger than 2 nm. The S/N ratio can be improved by integrating parallel to the boundary (integration over 100 pixels increases the S/N ratio by a factor of 10). The theoretical curve (Fig. 4.21) predicts, that, in this case, even one monolayer of oxide should be detectable. The effect of increasing the high voltage up to 1.25 MeV, i.e. by a factor of ten compared to the Zeiss EM 912, is indicated by the corresponding data points in Fig. 4.21.

With increasing accelerating voltage of the beam electrons, the inelastic scattering cross-sections σ_s decrease. This can be compensated for by using a greater specimen thickness, which furthermore reduces the detrimental influence of surface oxide layers and contamination. This explains why the results obtained on the CM 20 at 200 keV are slightly better than the results obtained at 120 keV on the Zeiss EM 912. However, at 1250 keV the influence of the decreasing scattering cross-sections could no longer be compensated for and furthermore the efficiency of the detection system decreases, which explains the lower S/N ratio found experimentally. No difference between the two different filter concepts could be detected; neither system deteriorates the S/N ratio or the resolution in the elemental distribution images.

4.4.5 Resolution Limits

The resolution limit in ESI images, or in the elemental distribution images derived from them, are controlled by a number of factors: the ultimate limit is defined by the aberrations of the electron optical elements of the instrument and is referred to as the instrumental resolution limit. However, the resolution is also degraded by the delocalisation of the inelastic scattering process. Newer calculations show that this contribution is small and can be ignored for inner-shell loss edges with energy losses of 100 eV and higher [98]. In many cases, the dominant factor arises from the statistical nature of the inelastic scattering processes and the weak signal resulting from the small inelastic scattering cross-sections. Thus, structures close to the instrumental resolution limit are not visible in the images because of the poor signal/noise ratio. Taking this into consideration, one can define an object-related resolution limit, which can easily be a factor of two to five times worse than the instrumental resolution limit [99].

In the following, we first discuss the instrumental resolution limit. The most important imperfections of the lenses of a TEM result in a degradation of the resolution by spherical aberration

$$d_{\mathrm{s}} = 0.5 C_{\mathrm{s}} \theta^3 M \tag{4.17}$$

and chromatic aberration

$$d_{\mathrm{c}} = 0.5 C_{\mathrm{c}} \frac{\Delta E}{E_0} \theta M \,, \tag{4.18}$$

where M is the magnification in the image plane and $M = 1$ refers to the case of denoting the smallest distances which can be resolved in the object plane. Both limitations depend on the scattering angle θ. Since the beam divergence diminishes with increasing magnification, only the first image forming lens needs to be considered, i.e. only the objective lens is relevant. The degradation in resolution caused by these two aberrations can be reduced by limiting the acceptance angle with an objective aperture, but this in turn limits the resolution by forming diffraction discs with a diameter

$$d_{\mathrm{b}} = 0.6 \frac{\lambda}{\theta} M \,. \tag{4.19}$$

The instrumental resolution limit for present instruments is entirely governed by the properties and the aperture limit of the objective lens. The higher order aberrations of the energy filter itself do not limit the resolution. The primary magnification can always be set to a value at which the distance between two object points resolvable by the objective lens can be imaged through the filter. The only restriction then results from the total number of independent pixels that can be transferred through the filter. For currently existing filters this number is much larger than the pixel numbers of the detectors that are used, e.g. the CCD-cameras, and thus does not impose any serious restrictions.

In contrast to these instrumental limitations, which may become less severe with future instrumental developments, the delocalisation of the inelastic scattering event imposes a physical limit on the achievable resolution. The delocalisation is a consequence of the quantum mechanical uncertainty principle, and in a simple approximation can be described by [21]

$$d_{\mathrm{del}} = 0.5 \frac{\lambda}{\theta_E^{3/4}} M \,. \tag{4.20}$$

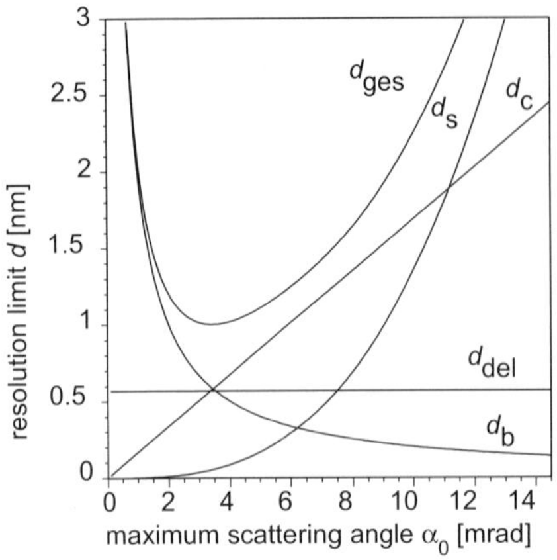

Fig. 4.22. The dependence of the resolution limit in ESI images on the maximum scattering angle that is allowed to contribute to the image formation. The individual components are explained in the text

Combining all the contributions in an incoherent approximation, one can compute a diagram giving the dependence of the resolution limit

$$d_{\mathrm{ges}} = \sqrt{d_{\mathrm{c}}^2 + d_{\mathrm{s}}^2 + d_{\mathrm{b}}^2 + d_{\mathrm{del}}^2} \tag{4.21}$$

on the maximum scattering angle that is allowed to contribute to the image formation. The result is shown in Fig. 4.22 for the Zeiss EM 912 Omega.

An example showing that the predicted resolution can actually be approached in favourable cases is shown in Fig. 4.23 for the case of a Si–Ge multilayer heterostructure. The Si–Ge layers possess a periodicity of 1.5 nm, which is just above the resolution limit predicted in Fig. 4.22. Figure 4.23 depicts a Si elemental distribution image, which was obtained on the Zeiss EM 912 Omega quantified from the Si-L edge located at $\Delta E = 99$ eV. Details of these studies are given by Jäger and Mayer [100]. In many areas, Fig. 4.22 shows clearly that the 1.5 nm periodicity of the Si layers can be resolved, although the heterostructure does not show a great perfection because it was not grown on a lattice-matched gradient buffer substrate.

The resolution limit can be improved by using higher accelerating voltages and lower C_{s} and C_{c} objective lenses. Figure 4.24 shows a comparison between predicted values for the Zeiss EM 912 Omega, a 200 keV FEG instrument and a 1.25 MeV high-voltage microscope. The delocalisation of the inelastic scattering event has not been taken into account for the curves shown in Fig. 4.24. This seems to be justified by newer calculations by Pennycook et al. [98], who found that even for low energy losses around 100 eV the delocalisation is only of the order of 0.1 nm. Freitag and Mader [101] present experimental evidence that element specific imaging is possible with

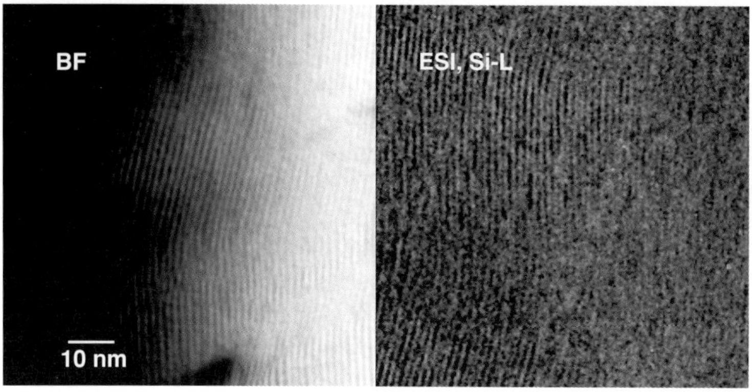

Fig. 4.23. Si elemental distribution image from SiGe heterostructure with 1.5 nm periodicity which was obtained on the Zeiss EM 912 Omega using the Si-L edge signal [100]

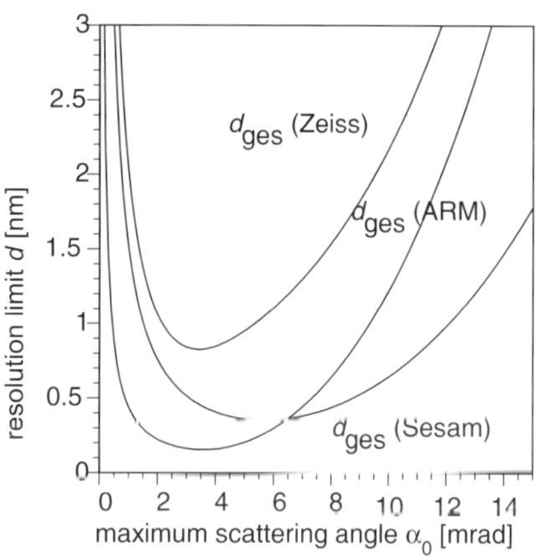

Fig. 4.24. Comparison of the theoretical resolution limits of the Zeiss EM 912 Omega, a 200 keV FEG instrument and a 1.25 MeV high-voltage microscope. The delocalisation of the inelastic scattering event has not been taken into account for the curves shown, which is a good approximation for inner-shell loss edges with $\Delta E > 100\,\text{eV}$

a resolution of 0.4 nm on a 300 keV field-emission gun (FEG) instrument in jump-ratio images obtained with the B–K edge.

The instrumental resolution limit introduced above is a measure of the smallest object distance $d_{\min}$ that can be resolved with appreciable contrast. The common definition is based on two point scatterers. These can be resolved individually for a minimum distance $d_{\min}$ for which the intensity distribution in the final image plane still exhibits two maxima, as judged e.g. by Raleigh's criterion. Since the broadening of the intensity distribution emerging from one point scatterer reflects the influence of all the aberrations discussed above, the analysis of an inelastic scattering experiment with randomly distributed point scatterers would be the best means to assess the instrumental resolution limit. However, in elemental maps, the signal is so low that an analysis of individual pairs of point scatterers is impossible close to the resolution limit because of the dominant role of statistical noise. Therefore, Golla and Kohl [99] suggested that the method of Young's fringes [102] should be used for the analysis of the signal of a random distribution of inelastic scatterers. The basic problem is to find a way to distinguish between the signal and the noise in a Fourier transformed image of the inelastically scattered intensities. This is not a trivial task since the statistical noise does not form a constant background in the Fourier transformed image, but also shows a decay at higher q-values, which is mainly caused by the properties of the detection system, typically a CCD camera, employed [91]. We have to find the actual signal contribution, which is superimposed on this background and smoothly fades out towards the resolution limit.

The method of Young's fringes resolves this problem by overlaying a well-defined modulation on the signal part of the Fourier transform, which for statistical reasons does not occur in the noise part. Such a modulation with arbitrary wavelength in the Fourier transform can be produced by superimposing two different pictures of the same specimen area with a small lateral displacement. In the Fourier transform of the sum image, a cosine modulation will occur with a wavelength that is inversely proportional to the displacement. The amplitude of the modulation is governed by the intensity distribution in the signal part, which is assumed to be the same for both images. One problem is that any fixed pattern on the detector would also contribute in the same way and careful gain normalization is therefore required. Any statistical noise resulting from the incident electrons or the detection system, however, does not contribute to the amplitude of the modulation. The intensity distribution $S(q)$ of the system of Young's fringes can thus be used as a measure of the range of reproducible information in Fourier space. $S(q)$ is given by [99]

$$S(q) = \alpha \cos\left(q\frac{a}{2}\right) \tilde{I}(q) , \tag{4.22}$$

where a is the lateral displacement of the two images and $\tilde{I}(q)$ is the Fourier transform of the signal part of the images. The analysis is then based on the fringe extension in reciprocal space, which requires determination of the q-value for which the maxima or minima in the fringes can no longer be distinguished from the background. The reciprocal value of this limiting spatial frequency $q_{\max}$ is equal to the object resolution limit as defined above. The envelope of the cosine modulation can be considered as the contrast transfer function for the inelastic scattering signal, which because of the incoherent nature of the inelastic scattering event is entirely formed by amplitude contrast.

Fig. 4.25. Young's fringe pattern obtained from an amorphous carbon film used as test specimen, which exhibits a large range of spatial frequencies owing to modulations in the density of the atomic arrangement and statistical thickness fluctuations

An example from the work by Golla and Kohl [99] is shown in Fig. 4.25. An amorphous carbon film was used as test specimen, which exhibits a large range of spatial frequencies owing to modulations in the density of the atomic arrangement and statistical thickness fluctuations. ESI images were recorded for low energy losses in the range of 50 to 80 eV. Except for the lowest energy loss, the resolution limit was in the range of 2 to 4 nm and was shown to be a function of the energy loss, the slit width and the acceptance angle. In general, the theoretically predicted instrumental resolution limit could not be reached in the study of an amorphous arrangement of inelastic scattering centres. The main reason for this is that only very thin specimens exhibit a large enough fluctuation in the density of the scatterers, but these specimens in turn produce a signal that is too small to give a significant result in the Fourier transform.

In the literature most of the attempts to assess the resolution limit in elemental distribution images were therefore made with a periodic arrangement of layers. Besides the example by Jäger and Mayer [100], in which periodic Si-Ge heterostructures have been characterized, other authors have investigated the modulation of a chemical element within a unit cell for crystals with large unit cell and defects in these crystals. Experimental examples have been given by Hashimoto et al. [103], Mayer et al. [104] and Freitag and Mader [101]. In these cases the low signal intensities for each individual atomic column can be compensated for by integrating along the lattice planes or by applying Fourier techniques for noise reduction.

The main problem in the study of crystalline materials is that, unlike in the case of amorphous materials, the intensity distribution of the incident beam is strongly modified by the dynamical interaction of the beam electrons with the crystals. The amplitude modulation of the electron wave in the Bloch wave field within the crystal not only influences the probability of an inelastic scattering event for a given column but also influences the propagation of the electrons through the crystal after the incoherent inelastic scattering event. In other words, an amplitude modulation that reflects the crystal periodicity will frequently be superimposed on the inelastically scattered waves in the course of their propagation through the crystal. This is commonly referred to as the preservation of elastic scattering contrast in ESI images. It has been shown above that the resolution that can be obtained on modern EFTEMs approaches atomic dimensions. We want discuss whether one can obtain reliable information on the distribution of an element within a unit cell despite the fact that elastic scattering contrast is preserved.

4.4.6 Preservation of Elastic Scattering Contrast

The inelastic scattering event may transfer the electrons from one Bloch wave field into another, i.e. interband scattering may occur. However, in most cases, the inelastic scattering angles are much smaller than the elastic ones and intraband scattering dominates. In this case the shape of the Bloch wave field is unaltered and the intensity distribution at the exit surface resembles closely the one for elastic scattering only, apart from the fact that the electrons have a somewhat reduced energy. This preservation of elastic scattering contrast in inelastically filtered images has been studied theoretically by Stallknecht and Kohl [105]. They computed lattice-fringe images for GaAs, using Bloch wave theory for the description of the elastic scattering and an exciton-like model for the description of the final states of the crystal after the inelastic scattering event. The inelastic scattering event itself is described using first-order perturbation theory.

In their work, Stallknecht and Kohl calculated thickness series of energy-loss images and compared them with zero-loss images. The computed im-

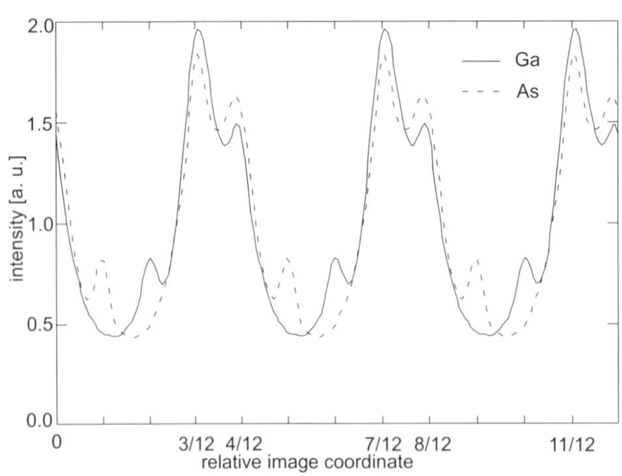

Fig. 4.26. Inelastic image contrast simulated for the Ga or As-L edge elemental distribution images of GaAs with 222 Å thickness [105]

ages describe the expected signal after background subtraction, because in the computations only the Ga-L or As-L edge were taken into consideration. GaAs was chosen since Ga and As have very similar elastic scattering amplitudes making their elastic diffraction behaviour almost identical. Furthermore, the GaAs-crystal can be oriented in such a way that the lattice planes parallel to the beam contain alternately Ga or As atoms exclusively. In an inelastic image simulated for the Ga or As-L edges, only the corresponding set of lattice planes should thus lead to a maximum in the intensity distribution. Figure 4.26 shows computed line scans for the two inner-shell loss images obtained for a crystal thickness of 222 Å.

In intensity distributions calculated for a thickness of 10 Å the peaks are still correlated with the sites of the lattice planes. In this case the intensity distribution in the inelastic images would thus clearly indicate the atomic positions of the corresponding species. However, in an actual experiment, the signal obtained at such a specimen thickness would be much too weak for the analysis and it is also very doubtful weather such a small specimen thickness could successfully be prepared and studied in a TEM.

At a specimen thickness of 222 Å, which corresponds to the lower limit of thickness used in actual ESI experiments, the situation has already changed completely. Intensity redistribution in the Bloch wave field after the inelastic scattering process has led to additional intensity maxima on the planes of the unlike atoms. In the Ga line scans, there appears a second peak at the position of the As atoms and vice versa. Note, that in the Ga line scan the peak with the maximum intensity is not located at the Ga position but at the position of the As lattice planes. Furthermore, in both cases, a small additional peak has appeared, which cannot be associated with any atomic

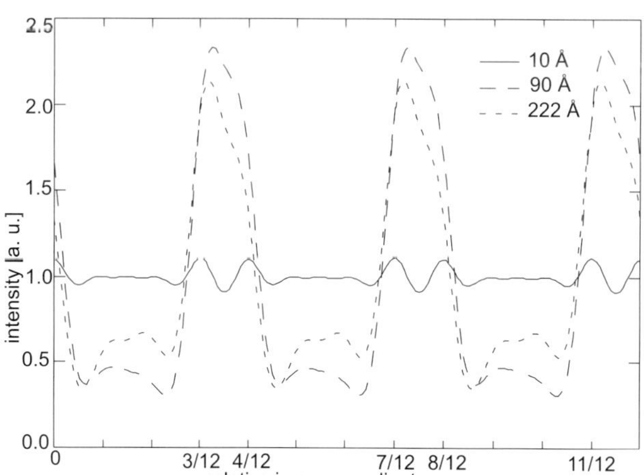

Fig. 4.27. Computed line scans for elastic images of a GaAs crystal for three different specimen thicknesses

position within the unit cell. A comparison with computed line scans for elastic images obtained at the same specimen thickness (Fig. 4.27) indicates that the intensity distribution in the inner-shell loss images is mainly dictated by the elastic scattering process. Therefore, a direct interpretation of the inner shell loss images would actually lead to a wrong result and has to be performed with caution. As in the case of HRTEM images, simulation of image contrast will play a more and more important role in future investigations of elemental distributions at or near the atomic scale.

4.4.7 Relativistic Intensity Distribution

Another factor that has to be taken into account in high-resolution elemental maps is that relativistic effects start to influence the intensity distribution at electron energies of 200 keV or higher. The difference between the relativistic and non-relativistic cases has been discussed in the work of Knippelmeyer and Kohl [106], which will briefly be summarized in the following. The calculations were based on a fully relativistic theory including retardation and magnetic interaction. Inelastic images of single atoms were calculated in order to study the basic effects. Figure 4.28 shows the intensity distribution in the image of a single atom for three different accelerating voltages. A noticeable

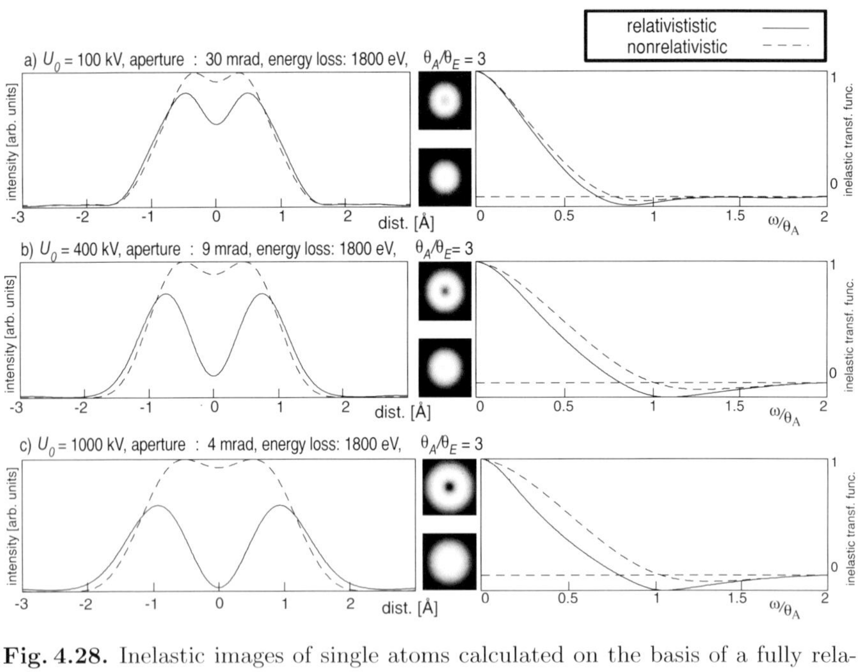

Fig. 4.28. Inelastic images of single atoms calculated on the basis of a fully relativistic theory including retardation and magnetic interaction in order to study the basic effects for three different accelerating voltages

effect is already obtained at an accelerating voltage of 100 keV. The main change resulting from the relativistic effects is a minimum in the intensity distribution at the centre position of the atoms, which approaches zero for higher electron energies. In addition, a widening of the intensity distribution occurs, which partly compensates for the loss of intensity in the center. This explains why the relativistic scattering cross-sections deviate only very little from their non-relativistic values [106].

From the changes observed in the image of a single atom, two important consequences of the relativistic effects can be expected for the intensity distribution in high-resolution elemental maps:

1. Because of the broadening of the intensity profiles, the resolution will be poorer in the relativistic than in the non-relativistic case.
2. The dip in the intensity at the position of the atom centre may lead to artifacts in elemental maps acquired at high electron energies and high spatial resolution.

The changes in image contrast can be related to changes in the contrast transfer function for the inelastically scattered intensities. If lens aberrations and an aperture limit imposed by the acceptance angle of the objective aperture are included, the relativistic changes are somewhat reduced. This can be ex-

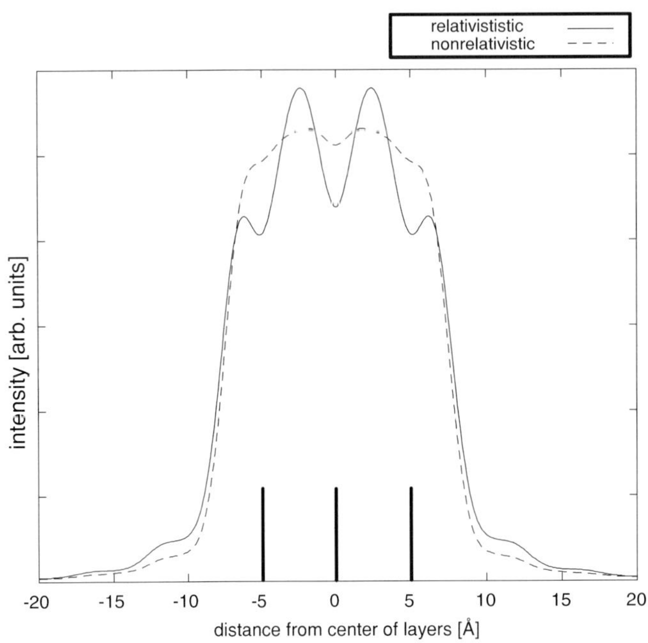

Fig. 4.29. Simulation of the contrast in an oxygen distribution image for a structure consisting of three oxygen layers with a distance of 5 Å showing distinct differences between the relativistic and the non-relativistic case

plained by the degradation in resolution. However, the relativistic effects still lead to significant changes of the contrast transfer function for accelerating voltages of 200 to 300 keV and beyond.

In order to demonstrate the influence of the relativistic effects on highly resolved elemental maps of systems of practical interest, Knippelmeyer and Kohl [106] have simulated the changes for a structure consisting of three oxygen layers at a distance of 5 Å. The results show distinct differences between the relativistic and the non-relativistic case (Fig. 4.29). In the relativistic simulation, the superposition of the single atom intensity distribution (shown in Fig. 4.28) leads to intensity maxima that are located in the middle between the atomic layers and minima at the location of the planes. These artifacts would lead to a misinterpretation of the layering in the sample if no relativistic simulation of the image contrast were performed.

4.4.8 Quantitative Analysis of ESI Series

The previous sections on energy filtering were concerned with the applications and artifacts of elemental distributions images obtained with the three-window technique. We will now focus our attention on the additional information that can be obtained by acquiring whole series of ESI images. The energy-loss spectra extracted from an ESI series with n images can be visualized graphically in several different ways (Fig. 4.30). The data are obtained as intensities $I(\Delta E)$ integrated over the energy window ΔE defined by the slit aperture. A simple plot would consist of a series of n data points, which give the integrated intensities at the centre positions ΔE_i ($i = 1 \ldots n$) of the corresponding energy windows (Fig. 4.30). Most analysis programs use a bar representation showing the intensities in steps with a width that corresponds to the energy increment. In the following, we will use this type of representation for the low loss spectra. For the spectra in the core-loss region, we use linear interpolation between the individual data points (Fig. 4.30). The resulting spectra resemble very closely the spectra which would be obtained

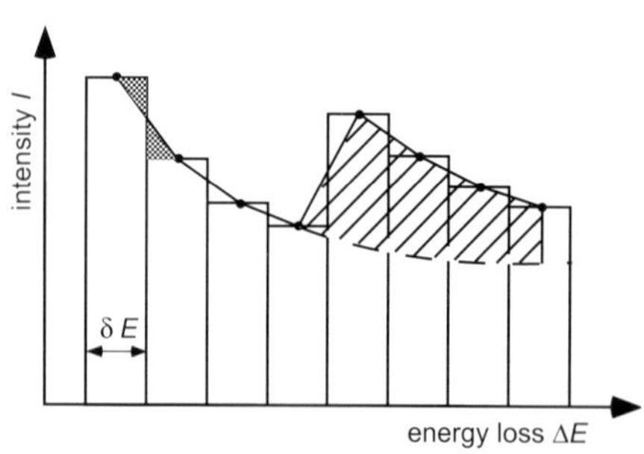

Fig. 4.30. Energy-loss spectra extracted from an ESI series with n images can be visualized graphically in several different ways

with a parallel EELS (PEELS) detector with much higher sampling frequency and can be treated with all currently available programs for quantification.

It can easily be shown that summing (or integrating) over all data points produced by linear interpolation exactly reproduces the original intensities, as long as the integration extends from one original data point to any other. Graphically this can also be seen from the equality of the two hatched triangles in Fig. 4.30.

The background extrapolation and subtraction is performed via a power-law background fit. We have found that, using the spectra obtained by linear interpolation, very accurate background fits can be obtained. However, it should be kept in mind that the linear interpolation is only an approximation (as is the power-law background fit). For higher accuracy, modelling of the exact functional dependence of the intensity variation for windows with a finite width ΔE is required.

As an experimental example, we describe investigations performed in the materials system Al_2O_3–Ti–Cu, which has already been introduced in Sect. 4.3.3 [63]. The thin interlayer of titanium is incorporated between an Al_2O_3 substrate and a Cu metallization layer to enhance the adhesion of the copper on the sapphire substrate [67].

In the following, we discuss the quantification of the number of atoms per unit area of the Ti interface layer with nominally 1 nm thickness. In the experiment, a series of 30 ESI images in the energy-loss range between 380 and 670 eV were acquired. This range includes the Ti-$L_{2,3}$ edge (456 eV) and the O-K edge (535 eV). The slit-width was calibrated to be 10 eV and every image was acquired with 10 s exposure time and twofold binning of the CCD camera pixels. Figure 4.31a shows one ESI image from the series around the

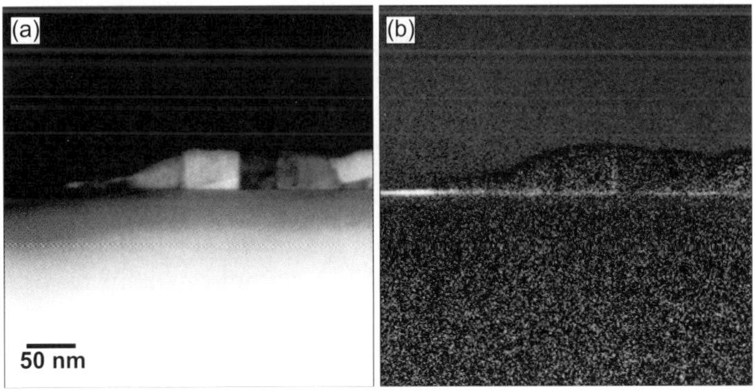

Fig. 4.31. (**a**) ESI image at 380 eV loss showing the microstructure of the Cu-Ti-Al_2O_3 layer system. The image is the first of an ESI series of 30 images. The polycrystalline copper can clearly be identified, whereas at the given energy loss, no contrast from the Ti layer can be observed. (**b**) The titanium distribution image clearly shows the presence and the thickness variations of the nominally 1 nm thick Ti-film

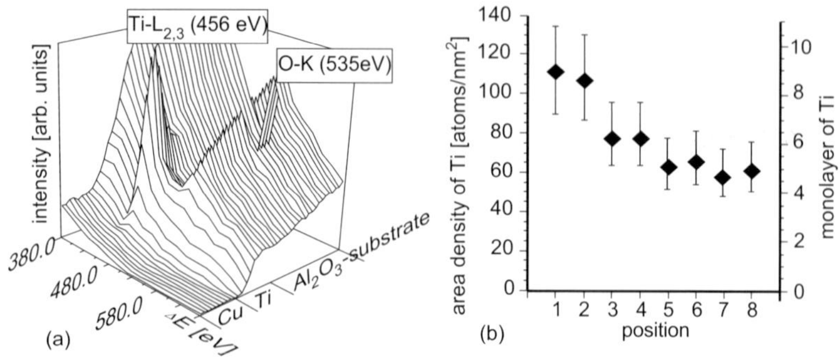

Fig. 4.32. (**a**) Spectrum line profile across the interface in the thin area on the left side of Fig. 4.31. The signal is integrated in areas of 1×50 pixels for each individual spectrum, which corresponds to $1.5 \times 75\,\text{nm}^2$ on the specimen. (**b**) Absolute concentrations in Ti-atoms/nm^2 determined as a function of the position along the interface

Ti-L edge, which illustrates the distribution of titanium in the sample. The Bragg contrast of individual grains in the polycrystalline copper is clearly visible and in the left part of the image the copper was removed during the ion milling of the sample. Three appropriate images of the ESI series before and after the Ti-L edge were used to calculate a titanium elemental map, which is depicted in Fig. 4.31b. The Ti distribution seems to be very homogeneous, except in the area where the copper has been removed. In this region an increase in intensity can be seen.

Line-profile analysis across the interface was performed by integrating the signal parallel to the interface in areas of 1×50 pixels, which corresponds to $1.5 \times 75\,\text{nm}^2$ on the specimen. Figure 4.32a shows an example of these compositional line profiles. In the case of this profile the selected area was positioned in the left part of the image, where an intensity increase in the Ti layer can be seen. The titanium and the oxygen edges are clearly visible. The apparent width of the Ti-layer is larger than its nominal width because of the combined effect of a slight inclination of the film, the focus spread caused by chromatic aberration and the blurring caused by the point spread in the CCD camera. Some overlap between the extension of the Ti peak and the extension of the O signal from the Al_2O_3 substrate can be seen. This may be indicative of a partial oxidation of the Ti; however, a slight inclination of the film and the blurring caused by the mechanisms discussed above certainly also contribute to the oxygen signal in the Ti layer. Furthermore, the lack of the overlaying Cu layer may be an additional reason for the oxidation of the titanium found in this position. All experimental results were confirmed by focused-probe PEELS investigations on a VG HB 501 dedicated STEM [61].

For the quantitative analysis of the Ti signal according to (4.5), the low-loss intensity $I_{\mathrm{L}}(\Delta)$ has to be known. Thus, in a second experiment, information on the EEL spectrum is recorded in the low-loss region up to typically 150 eV, including the zero-loss peak. With the help of an EFTEM, the required information can be obtained from the whole sample area imaged in a single series of ESI images in an energy range from the zero-loss to energy losses of about 150 eV. The typical slit width and step width are 10 eV and the exposure times are only a few seconds per image.

By integration over the Ti signal in the line profile and using the low-loss intensities, the area density N_a of the titanium atoms forming the interlayer can be determined in a first step. Using this number and the specimen thickness t, which can be determined from the low loss, we can then compute the volume density n_a of titanium atoms within the layer:

$$n_a = \frac{N_a}{t} . \quad (4.23)$$

In a next step, this can be converted into the number of atoms per unit area of the interface N_{int} which is obtained by integrating the signal across the interface

$$N_{int} = \int_{d_i} N_a \mathrm{d}x , \quad (4.24)$$

where d_i is the image width of the boundary layer and x is the coordinate perpendicular to the boundary. N_{int} is given in atoms/nm^2 in the interface plane. This can finally be converted into the thickness d (in nm) of the layer, which is given by

$$d = N_{int} \frac{A}{\rho N_A} , \quad (4.25)$$

where A is the molar weight of the element or compound.

After performing this in one location, the analysis is continued along the interface to determine the thickness variation, which is evident from the result in the elemental map (Fig. 4.31b). Spectra from eight different areas were analyzed and the result is plotted in terms of atoms per nm^2 in the interface plane as well as in the equivalent in monolayers in the diagram shown in Fig. 4.32b. In the larger area with homogeneous thickness of Ti, the value determined using (4.25) is 1.1 ± 0.3 nm, which corresponds to 4.7 ± 1.0 monolayers. These results are in good agreement with the expected values. The higher titanium concentration on the left side in Fig. 4.32b may be caused by variations during the MBE process, or by a possible accumulation of titanium during ion-beam thinning after the removal of the copper overlayer.

The example discussed above demonstrates that ESI series can be quantified not only in terms of relative concentrations but also in terms of absolute numbers of atoms per nm^2. The present investigations on thin-film systems clearly indicate that quantitative ESI analysis is applicable to segregation and precipitation in the monolayer range. In this case, drift determination and correction is very important to suppress artifacts and to maximize the signal-to-noise ratio. The main advantage of our ESI series technique over the well-known spectrum-imaging technique [93,94] is the short time in which two-dimensional analytical information can be obtained. The main advantage of the spectrum-imaging method is the much higher energy resolution in spectra obtained on a STEM with field-emission source. This illustrates that, if only the chemistry has to be analysed and time or number of pixels becomes a concern, then ESI will be the method of choice and results can be obtained even in the submonolayer range, as has also been demonstrated recently on an EFTEM with a FEG source [104].

4.4.9 Analysis of Near-Edge Fine Structure

The presence of an edge in the energy-loss spectrum not only reflects the occurrence of the corresponding element, as discussed in Sect. 4.2; the near-edge fine structure (ELNES) of the edge also contains information on the three-dimensional atomic coordination of this element in the sample. In a number of biological applications, it has already been shown that the ELNES can be retrieved from a series of ESI images [107,108]. In the application presented here, ESI image series are used to study the bonding across an interfacial layer only a few nanometres thick. As a model system, CVD-grown diamond films on Si substrates [109] were used. At the interface between the film and the substrate, an amorphous carbon layer is formed [110,111]. An analysis of the ELNES makes it possible to distinguish between the two different phases of carbon, namely diamond and amorphous carbon.

The ESI investigations were performed on a Zeiss EM 912 Omega operated at 120 keV. The ESI images and the reference EELS spectra were recorded on a Gatan 1024 × 1024 slow-scan CCD camera. The ESI images were recorded in the binning mode, in which 2 × 2 pixels are summed up into one effective pixel during the read-out process. For the high-resolution investigations a JEOL 4000 EX operated at 400 keV was used.

Figure 4.33 shows a high resolution micrograph of a typical interface area, which exhibits the amorphous film between the Si substrate and the diamond film and also considerable surface roughness. The amorphous film with a thickness of 1 to 4 nm could also be identified in tripod-polished TEM specimen, thus demonstrating that it is not an artifact produced during ion-beam thinning. The amorphous layer is present almost everywhere along the interface and there exist only a few isolated islands where the diamond film is in direct contact with the substrate. The areas with no detectable amorphous layer can be found for example at elevated surface steps and are thought to be the nucleation centres for the diamond film [112]. The diamond film subsequently overgrows the areas onto which amorphous carbon has been deposited during the initial deposition steps. Since the two phases of interest, diamond and amorphous carbon, are formed by the same chemical element, they cannot be distinguished from the presence of the carbon edge alone.

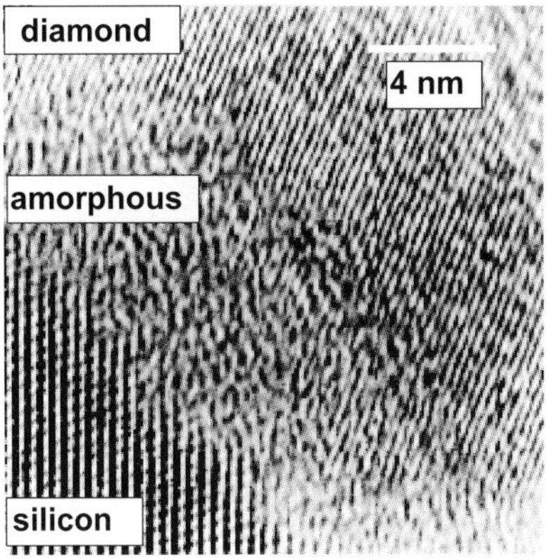

Fig. 4.33. High-resolution micrograph of a typical interface area exhibiting considerable surface roughness; an amorphous film of 1 to 4 nm thickness can be seen between the Si substrate and the diamond film

Rather, the ELNES has to be used in order to separate the areas in which the two different phases occur [113,114]. It should be kept in mind that EEL spectra could not directly be obtained from the 1 nm thick layer at the diamond/silicon interface on the Zeiss EM 912 Omega because the beam on an analytical TEM equipped with a tungsten or LaB_6 filament cannot be focused into a 1 nm probe. This would require the use of a FEG–STEM with its higher brightness for small probe diameters, as has been demonstrated for the same system by Muller et al. [43].

In order to reveal this difference in the ELNES by electron spectroscopic imaging, we have acquired a series of ESI images in a range of energies across the carbon K-edge. An energy window width of 5 eV was chosen, which is a compromise between the required energy resolution and maximizing the signal in each individual ESI image. The energy increment between the individual ESI images was set to 2 eV. We choose an energy increment that is smaller than the actual energy window width to make sure that one of the ESI images of the series is centred around the π^*-peak and another one around the σ^*-excitations. In total, the ESI series consisted of 20 images from $\Delta E = 265$ eV to $\Delta E = 303$ eV. The exposure time for each image was 10 seconds so that the total acquisition time for the whole series was 200 seconds. One image from the series is shown in Fig. 4.34a. This ESI image was acquired at $\Delta E = 283$ eV, where the strongest π^* signal is obtained. In this image a thin layer at the interface between the diamond film and the Si substrate is noticeably brighter than the surrounding phases. The schematic drawing of Fig. 4.34b illustrates the arrangement of the silicon substrate, the diamond film and the amorphous layer in between.

From the whole series of ESI images, information on the ELNES of the carbon K-edge can be retrieved for any given area in the image. The energy-

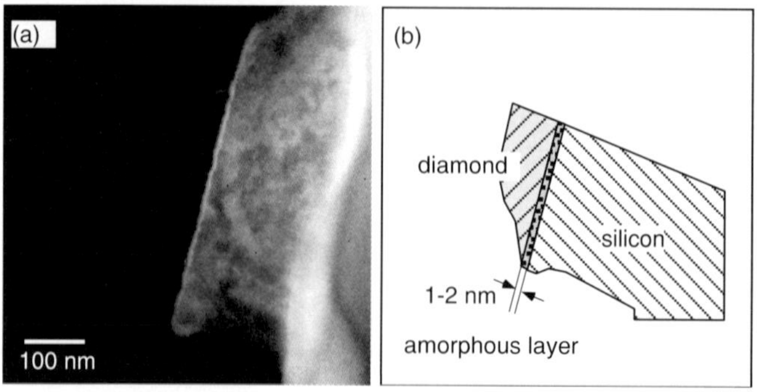

Fig. 4.34. ESI investigation of the interface between a Si substrate and a diamond film. (**a**) One image from the series acquired at $\Delta E = 283\,\mathrm{eV}$, where the strongest π^* signal is obtained. (**b**) Schematic drawing illustrating the arrangement of the silicon substrate, the diamond film and the amorphous layer in between

loss spectrum is obtained by simply extracting the intensity from the same area in the series of ESI images and plotting it as a function of the corresponding energy loss. Basically, this can be performed for each individual pixel in the images. However, the resulting spectra would be very noisy. In order to reduce the noise, the intensities were integrated over a certain area in the images. Prior to this, drift correction has to be applied to the individual ESI images in order to align the corresponding areas properly in the series of images. The magnitude of the drift correction can be determined either by cross-correlation or by visual inspection. From the drift-corrected series, we extracted the integrated intensities of a line profile with a length of 150 pixels and a width of one pixel, which was placed in the centre of the amorphous layer. Thereafter, the line profile was shifted parallel into the diamond layer. The resulting intensity data are plotted in Fig. 4.35. The carbon K-edge of the material forming the amorphous layer clearly shows a π^*-peak, whereas the edge from the diamond film shows an onset at about 4 eV higher energy losses and a more pronounced σ^*-peak. Qualitatively, the ELNES features reproduced in Fig. 4.35 are in good agreement with the shape of reference spectra [86], with an energy resolution which is reduced to about 5 eV, as defined by the slit width used for the ESI series. The carbon K-edge of the amorphous layer is superimposed on a much stronger background than the K-edge from the diamond film, which reflects both the increasing thickness towards the substrate and the amount of Si which is presumably dissolved into the amorphous layer.

The spectra in Fig. 4.35 show that ESI series obtained in an energy-filtering TEM make it possible to obtain two-dimensional information on the variation of the ELNES on a nanometre scale. Thus, ESI presents itself as an alternative approach to the standard EELS method, in which a fine probe is stepped across a sample and EEL spectra are recorded consecutively. ESI

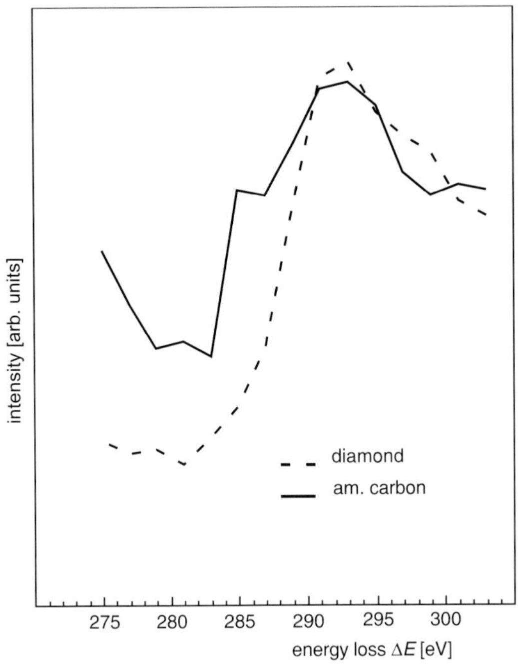

Fig. 4.35. Characteristic spectra for the diamond film and the amorphous interface layer reconstructed from the ESI image series revealing the characteristic ELNES features of the two different phases

makes it possible to obtain two-dimensional information in a much shorter time than in the scanning approach. In comparison, the main advantages of a dedicated STEM are the higher energy resolution of about 0.5 eV and the better spatial resolution in the range of 0.5 to 1 nm. However, using an EFTEM with a FEG emitter a similar spatial and energy resolution can be reached in ESI studies. Employing such an instrument thus will make it possible to select the most appropriate way to analyse the energy-loss space in each case – via PEELS acquisition in spot mode or via ESI series in the TEM imaging mode.

4.5 Quantitative Convergent Beam Electron Diffraction

Energy filtering has many different applications in electron diffraction, a survey of which can be found in [21]. In the context of the present paper we restrict the discussion to the quantitative analysis of zero-loss filtered convergent-beam electron diffraction (CBED) patterns.

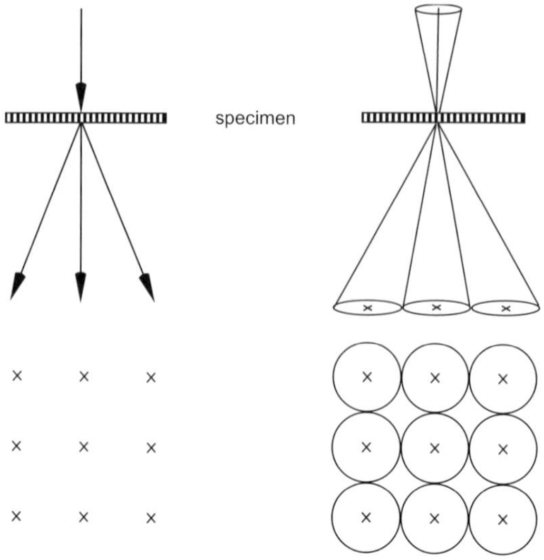

Fig. 4.36. Diagram illustrating the geometry of CBED pattern formation

4.5.1 Basic Principles of CBED

CBED patterns are obtained by projecting a demagnified image of the electron source onto the specimen. A convergent cone-shaped beam is thus formed, which stands upside-down on the surface of the sample (Fig. 4.36). The convergence angle is defined by the diameter of the condenser aperture (but can be varied electron optically within certain limits). This geometry of the illumination results in a continuous variation of the incident beam direction within the limits given by the surface of the cone. In the back focal plane of the objective lens, a pattern consisting of discs is formed rather than the spot pattern obtained for parallel illumination. The discs possess sharp edges and the diameter is given by the convergence angle. Each disc in the CBED pattern corresponds to one Bragg reflection $\boldsymbol{g}$ of the crystal.

The keypoint to the understanding of CBED is that for an incoherent electron source (tungsten, LaB_6) the continuous variation of incident beam directions is totally equivalent to successively varying the beam tilt for parallel illumination. For each incident beam direction within the illumination cone, one spot pattern is obtained in the back focal plane. The intensity of the reflections in this spot pattern is independent of the intensities in all the other spot patterns obtained for different incident beam directions. A CBED pattern can thus be understood as an array of individual spot patterns laid side by side within the boundaries given by the borders of the discs [115].

In the evaluation of CBED patterns, we have to distinguish between low-indexed reflections belonging to the zero-order Laue zone (ZOLZ) and high-indexed reflections belonging to the higher-order Laue zones (HOLZ). In contrast to the low-indexed reflections, scattering into the HOLZ-reflections involves large scattering angles and is only possible for a very narrow angular range. The scattering into the HOLZ reflections is mostly investigated by looking at the deficiency lines (the so-called HOLZ-lines) within the (000) disc. The arrangement of the HOLZ lines is very sensitive to small changes in accelerating voltage, local lattice parameter or composition. Thus, a quantitative analysis of the HOLZ line positions makes it possible to measure local lattice strain with high spatial resolution. Furthermore, the accuracy of the method greatly benefits from energy filtering the patterns, as has been demonstrated in a growing number of cases [116].

In the present discussion of recent work, we restrict ourselves to the analysis of the dynamical effects, which are most pronounced in the low-order reflections and make it possible to determine structure factors with high accuracy. The technique was first developed by Zuo and Spence [117] and was modified by us to make the most efficient use of the two-dimensional energy-filtered data that can be obtained on modern EFTEMs [118].

An alternative approach based on energy-filtered zone-axis CBED patterns has been proposed by Bird and Saunders [119]. The advantage of their technique is that several low-indexed structure factors can be refined simultaneously from one pattern while a disadvantage is that higher-indexed structure factors cannot be measured directly.

The intensity distribution in any CBED pattern depends on the incident beam direction, the crystal thickness and the structure factors of the reflections. This assumes that the crystal structure and the Debye–Waller factors at a given temperature are known. In the quantitative evaluation of CBED patterns, it is not possible to invert the diffracted intensities and directly extract the underlying parameters. Instead, a quantitative evaluation is always based on a comparison between experimental and simulated CBED patterns.

4.5.2 Determination of Bonding Charge Densities

For the simulation, an initial set of parameters has to be guessed and adjusted until a best fit is obtained. This can be achieved using computer-based refinement algorithms [117,118]. The refinement of the calculated line profiles with respect to the experimental data is performed via minimization of a function χ^2 [117], which is given by:

$$\chi^2 = \sum_{i=1}^{n} \frac{w_i \left(c f_i^{\text{theo}} - f_i^{\text{exp}}\right)}{\sigma_i^2} . \tag{4.26}$$

Here n is the number of pixels to be compared, w_i is a weighting factor which is normally set to unity, f_i^{theo} and f_i^{exp} are the theoretical and experimental intensities, respectively, and σ_i^2 is the experimental noise in data point i. There is only one scaling constant c which is used to scale the normalized theoretical intensities to the experimentally reached intensity level.

In the refinement, the whole set or a subset of the parameters described above is varied until a minimum of the function χ^2 is found. The refinement algorithms can be distinguished by whether they are able to leave local minima and to find the global minimum (global refinement) or whether they are liable to get trapped in local minima (local refinement).

In the experiment, two-dimensional energy-filtered CBED patterns are acquired with a slow-scan CCD camera. The sample is cooled to liquid-nitrogen temperature and tilted to a systematic orientation which is 5 to 10 degrees away from the next major zone axis. The specimen thickness is chosen in such a way that several but not too many Pendellösung oscillations are visible within the CBED discs.

The low-indexed reflections carry information on the bonding in the crystal. This information can be extracted from the structure factors by calculating the charge-density distribution in the crystal. The charge-density distribution is related to the electrostatic potential $V(\boldsymbol{r})$ via the Poisson equation:

$$\nabla^2 V(\boldsymbol{r}) = -\frac{|e|}{\varepsilon_0}\left(\rho_{\mathrm{n}}(\boldsymbol{r}) - \rho_{\mathrm{e}}(\boldsymbol{r})\right) . \tag{4.27}$$

Here the charge density can be separated into the contributions from the nucleus, ρ_{n}, and the contribution from the electrons, ρ_{e}. The former can be expressed as a point charge of magnitude Z_i at the lattice position $\boldsymbol{r}_i$ of the atom i. $V(\boldsymbol{r})$ can be calculated by summation over all Fourier coefficients V_g of the reciprocal lattice points g:

$$V(\boldsymbol{r}) = \sum_g V_g \exp(-2\pi i \boldsymbol{g}\cdot\boldsymbol{r}) . \tag{4.28}$$

From the total electronic charge density only the part containing information on the bonding in the crystal is of interest. The contribution from the inner shells can be assumed to be spherical and does not directly influence any of the properties of the crystal. The charges forming the covalent bonds in a crystal for example, can be composed of charges as small as 10^{-4} of the total electronic charge density. This contribution will thus not be visible in plots of the total charge density within a unit cell. The problem can be solved by plotting the difference charge density, which will subsequently be referred to as the 'bonding charge density' [91]:

$$\Delta\rho(\boldsymbol{r}) = \rho_{\mathrm{e}}(\boldsymbol{r}) - \rho_{\mathrm{a}}(\boldsymbol{r}) \tag{4.29}$$

where $\rho_{\mathrm{e}}(\boldsymbol{r})$ is the charge density of the real crystal, which has been measured experimentally, and $\rho_{\mathrm{a}}(\boldsymbol{r})$ is the charge density of a hypothetical crystal with neutral spherical atoms at the lattice sites. Such a difference charge density map will contain areas of positive and negative charge density, indicating both accumulation and depletion of electronic charges. In the map, covalent bonds can be identified by an accumulation of charges between two atoms and the strength of the bond can be estimated by integrating over the three-dimensional area with a surplus of electronic charges. The presence of ionic bonds can be inferred from a positive or negative charge balance at the atom positions. For the calculation of the charge density $\rho_a(\boldsymbol{r})$ the X-ray scattering amplitudes $f_i^X(g)$ tabulated in the International Tables for X-ray Crystallography are used. From these the X-ray structure factors $F^X(g)$ are calculated:

$$F^X(g) = \sum_i f_i^X(g) \exp(-B_i g^2/4) \exp(-2\pi i \boldsymbol{g} \cdot \boldsymbol{r}_i) , \tag{4.30}$$

where B_i is the Debye–Waller factor for species i.

By definition, the X-ray structure factors are the Fourier coefficients of the electronic charge density $\rho_e(\boldsymbol{r})$:

$$\rho_e(\boldsymbol{r}) = \frac{1}{\Omega} \sum_g F^X(g) \exp(-2\pi i \boldsymbol{g} \cdot \boldsymbol{r}_i) , \tag{4.31}$$

where Ω is the unit cell volume.

The actual Fourier summation for the determination of the charge density distributions is thus best performed in terms of the X-ray structure factors. Hence, the electron scattering structure factors

$$U_g^c = \frac{2me}{h^2} V_g , \tag{4.32}$$

which are proportional to the Fourier coefficients V_g of the crystal potential, have to be converted into X-ray structure factors. Here m is the relativistic electron mass and h is Planck's constant. The relationship between both is given by the Mott–Bethe relationship which can be derived from the Poisson equation (4.27):

$$V_g = \frac{|e|}{4\pi^3 \varepsilon_0 \Omega} \sum_i \frac{Z_i - f_i^X(g)}{g^2} \exp(-B_i g^2/4) \exp(-2\pi i \boldsymbol{g} \cdot \boldsymbol{r}_i) \tag{4.33}$$

for atoms of atomic number Z_i at position r_i.

For the determination of the charge density distribution (4.33) has to be inverted using (4.32) which yields

$$F^X(g) = \sum_i Z_i \exp(-B_i g^2/4) \exp(-2\pi i \boldsymbol{g} \cdot \boldsymbol{r}) - \frac{2\pi^2 \varepsilon_0 h^2 g^2 \Omega}{\gamma m_e e^2} U_g^c , \tag{4.34}$$

where γ is the relativistic constant.

With the help of this equation we can convert each individual electron structure factor (or Fourier coefficient of the crystal potential) into its corresponding X-ray structure factor (i.e. Fourier coefficient of the electronic charge density). The bonding charge density can thus be obtained by building the Fourier sum over the differences between the X-ray structure factors derived from the experimental U_g and the X-ray structure factors computed for a crystal with neutral spherical atoms:

$$\Delta\rho(\boldsymbol{r}) = \frac{1}{\Omega}\sum_{g}\left[F^{\mathrm{X}}_{\mathrm{exp}}(g) - F^{\mathrm{X}}_{\mathrm{at}}(g)\right]\exp\left(-2\pi i\boldsymbol{g}\cdot\boldsymbol{r}_i\right) . \tag{4.35}$$

4.5.3 Bonding Charge Density of NiAl

The procedure outlined above has been applied by a number of different groups to various materials [116]. The first intermetallic phase investigated was TiAl, for which bonding effects have been studied in the pure material [120] as well as in the Mn-doped material [121].

Here, as an example, a study on the bonding charge density in the intermetallic phase NiAl is presented. In equilibrium, NiAl forms the B2 phase with the CsCl structure. There are two atoms in the unit cell, the Ni atom at (0,0,0) and the Al atom at (1/2, 1/2, 1/2). NiAl shows very little ductility at room temperature, which is a bulk property found even in single crystals. Formally, this is a consequence of the [100] glide system, which is activated at room temperature in NiAl [116]. Physically, the choice of the activated glide system is certainly influenced by the details of the interatomic bonding which motivated the studies of the charge density distribution by CBED.

The first electron diffraction experiments on NiAl were carried out by Fox and coworkers [122] using the critical-voltage technique. They determined the four lowest order structure factors with an accuracy of 0.4 to 2.1%. Menon and Fox [123] measured the 12 lowest structure factors up to 222 by X-ray powder diffraction. As pointed out by these authors the accuracy was 0.4 to 1.2% which is comparable to the change in structure factors caused by bonding effects. Hence, there is clearly a need for more accurate measurements. Theoretical calculations have been performed for NiAl by Lu et al. [124] using a full potential LAPW band-structure code. They found a dominating ionic nature of the bonding and a weak covalent component. This is in qualitative agreement with the results given in [122]. In contrast, the calculations of Schultz and Davenport [125] indicated a more covalent bonding type.

In the experiment, CBED patterns were recorded on a Zeiss EM 912 Omega at a nominal accelerating voltage of 120 keV using a double-tilt liquid-nitrogen holder at a temperature of 110 K. The patterns were recorded on a 1024×1024 slow scan CCD-camera with an energy window of 10 eV width centred around the zero-loss peak. The point-spread function of the camera was measured by Weickenmeier et al. [91] and was deconvoluted from the patterns. Further details of the experiment are given by Nüchter et al. [126].

For the measurement of the individual structure factors, the specimen was tilted to a systematic-row orientation in which normally two reflections of interest are excited in Bragg condition within the boundaries defined by

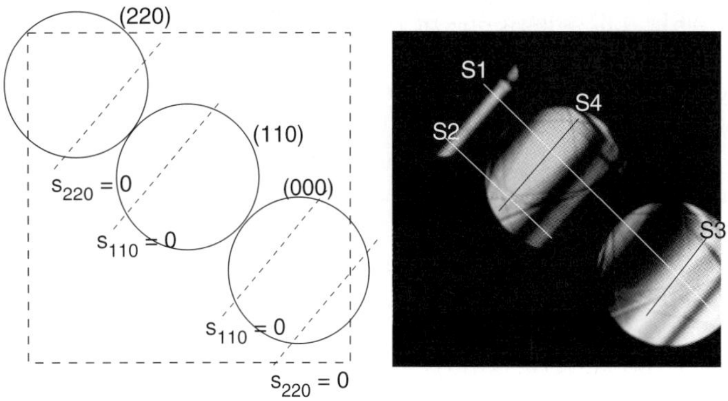

Fig. 4.37. CBED pattern used for the determination of the 110 and 220 structure factors of NiAl

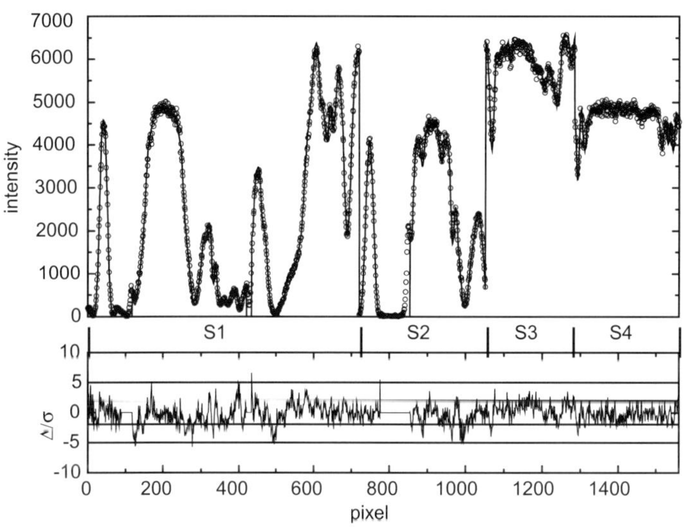

Fig. 4.38. Refinement of the simulated intensities by fitting the 110 and 220 structure factors, the corresponding absorption factors, the incident beam direction and relative magnification of the pattern, a possible anisotropic magnification, the specimen thickness, and the magnitude of a constant background resulting from thermal diffuse scattering. The simulated best fit (solid line) to the experimental data (circles) and the difference between experiment and best fit (bottom) scaled by the standard deviation σ of the experimental data are plotted

the discs. The full beam-convergence angle was set to slightly less than twice the Bragg angle of the particular systematic row, so that the CBED discs just do not overlap. In most cases, it was possible to have two reflections of interest in the Bragg position, allowing two structure factors to be fitted at the same time from one pattern.

As an example, Fig. 4.37 shows one of the CBED patterns used for the determination of the 110 and 220 structure factors. The geometry of the pattern and the Bragg positions of the individual reflections are schematically indicated on the left-hand side. On the right hand side, the experimental CBED is shown together with the positions of the line scans along which the data used for the analysis were extracted. The experimental intensities obtained for the line scans are put together in one diagram as a series of data points, as shown in Fig. 4.38. In the refinement of the simulated intensities, the following parameters were fitted: the 110 and 220 structure factors, the corresponding absorption factors, the incident beam direction and relative magnification of the pattern, a possible anisotropic magnification, the specimen thickness, and the magnitude of a constant background resulting from thermal diffuse scattering. The simulated best fit to the experimental data is also plotted in Fig. 4.38. At the bottom of the diagram, the difference between experiment and best fit scaled by the standard deviation σ of the experimental data is plotted. Here, σ is given by the Poisson noise resulting from the statistical nature of the electron intensities and an additional noise contribution from the CCD detector. In such a difference plot, only normalized differences above a value of 5 are statistically significant [116] and it can be seen that, for the bulk of the data, the difference stays well below this level.

A list of the reflections investigated and structure factors obtained in the refinement is given in Table 4.1 along with the experimental error. Each indi-

Table 4.1. Mean values of the measured structure factors (column 3) and difference from the reference (column 4). The standard deviations of the experimental values are given in column 5 and the estimated error of the reference in column 6. Absolute errors $\sigma_{\text{ges}} = \sqrt{\sigma(F^{\text{X}}_{\text{exp}})^2 + \sigma(F^{\text{X}}_{\text{ref}})^2}$ are given in the last column

No.	hkl	$F^{\text{X}}_{\text{exp}}$	$F^{\text{X}}_{\text{exp}} - F^{\text{X}}_{\text{ref}}$	$\sigma(F^{\text{X}}_{\text{exp}})$	$\sigma(F^{\text{X}}_{\text{ref}})$	σ_{ges}
1	100	13.64	+0.21	0.01	0.01	0.01
2	110	28.43	−0.16	0.02	0.02	0.03
3	111	10.60	+0.02	0.04	0.03	0.05
4	200	23.57	−0.22	0.04	0.04	0.06
5	210	8.76	+0.06	0.02	0.04	0.04
6	211	20.28	−0.04	0.04	0.05	0.06
7	220	17.79	−0.02	0.04	0.06	0.07
8	221	6.89	+0.35	0.05	0.06	0.08
9	300	6.85	+0.29	0.06	0.06	0.08
10	310	16.1	+0.6	0.2	0.06	0.2
11	311	6.23	+0.34	0.05	0.06	0.08
12	222	15.4	+1.2	0.5	0.07	0.6

vidual structure factor was measured several times using different specimens and specimen thicknesses to test the reliability and to obtain a measure of the statistical error. A more detailed list of the results and an error discussion can be found in Nüchter et al. [126].

The bonding charge density of NiAl was constructed from the refined values of the structure factors of the reflections 100, 110, 111, 200, 210, and 211. As discussed above, the experimental U_g were transferred into the corresponding X-ray structure factors using (4.33) and the structure factors of a crystal with neutral spherical atoms (calculated from (4.30)) were subtracted. The resulting Fourier coefficients of the bonding charge density are listed in Table 4.1. From these Fourier coefficients, the bonding charge density can be synthesized in three dimensions for the whole unit cell or for any section through the unit cell. Here, we restrict ourselves to the central (110) section through the unit cell, which contains both types of atoms. The result is plotted in Fig. 4.39. Substantial charge redistribution compared to neutral spherical atoms is revealed and can be analysed in more detail. However, one has to bear in mind that the bonding charge density has been constructed from a Fourier series with a rather limited number of Fourier coefficients. The spatial resolution is therefore limited and details of the charge density in the region of the atom cores may be obscured. Nonetheless, some details of the charge density distribution can be interpreted in terms of the usual, but somewhat arbitrary, division into "covalent", "ionic" and "metallic" bonding. In Fig. 4.39 three characteristic features can be recognized. Most importantly, we notice a concentration of electrons along the nearest neighbor (NN) axis of Ni and Al. Within a sphere with its centre half way along the NN axis and with a radius equal to 0.26 times the NN distance, 0.024 electrons are accumulated, forming a directional contribution to the bonding. This can be interpreted as partial formation of a covalent bond (note that there are eight such bonds per atom).

Another observation is that electrons are drawn away from the core regions, in particular from the Al core. In order to quantify this charge transfer, we integrate the bond charge density within the sphere with negative difference charge density around the Ni or Al nuclei. For Al, the integration radius is 0.38 times the NN distance and the loss of electrons is 0.23. For Ni we find a lack of only 0.02 electrons in an integration radius of 0.24 of the NN distance. Thus both atoms are equally charged and do not have opposite ionicities, which can be understood from the fact that they have to donate their electrons for the formation of covalent or metallic contributions to the bonding in the crystal.

The values given above are an example of the type of quantitative information that can be extracted from the bonding charge densities. However, such results must be regarded with some caution because of the limited number of Fourier coefficients, as indicated above. A comparison with theoretical

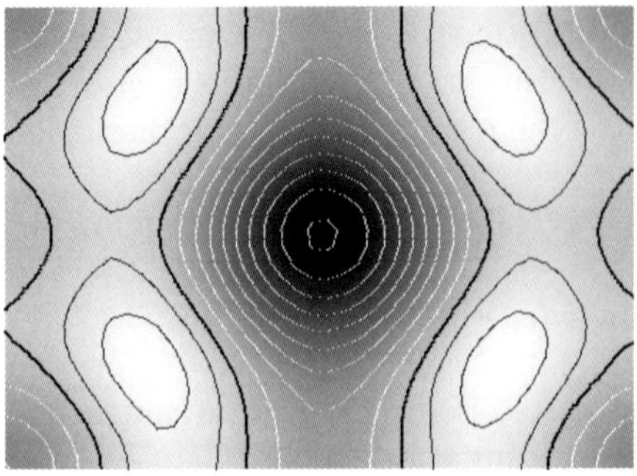

Fig. 4.39. Bond charge density constructed from the coefficients 100, 110, 111, 200, 210 and 211. Shown here is a (110) section through the unit cell. The Ni atoms are located at the corners and the Al atom is in the centre of the cell. Contour line spacing is 0.02 electrons Å^3. The black solid lines indicate zero electron density, the white lines positive (depletion of electronic charge) and the black lines negative charge density, respectively

calculations by Elsässer et al. indicates that full convergence is not reached with the number of Fourier coefficients used to construct Fig. 4.39.

In conclusion, the work discussed above illustrates that the quantitative analysis of CBED patterns makes it possible to determine structure factor amplitudes with an accuracy that can reach 0.1% (after conversion to X-ray structure factors). In studies of BeO, Zuo et al. [127] have shown that structure factor phases can be measured with an accuracy of 1¡ by analysing three-phase invariants. Convergent-beam electron diffraction thus allows one to reach an accuracy that comes very close to that reached with X-ray methods or even surpasses it, as in the case of the phase measurement.

Acknowledgements

We thank C. Scheu, G. Dehm and M. Rühle for helpful discussions and U. Eigenthaler and S. Krämer for assistance in preparing the manuscript.

References

1. Jeanguillaume, C. Colliex C. (1992) New STEM multisignal imaging modes, made accessible through the evaluation of detection efficiencies. Ultramicroscopy **45**:205-217
2. Leapman R.D., Hunt J.A. (1991) Comparison of detection limits for EELS and EDXS. Microsc Microanal Microstruct **2**:231-244
3. Williams D.B., Carter C.B. (1996) Transmission Electron Microscopy-A Textbook for Materials Science, Plenum Press, New York/London
4. Dorneich A.D., French R.H., Müllejans H., Loughin S., Rühle M. (1998) Quantitative analysis of valence electron energy-loss spectra of aluminium nitride. J Microscopy **191**:286-296

5. Moreau, P., Brun, N., Walsh, C.A., Colliex, C., Howie, A. (1997) Relativistic Effects in Electron Energy-loss-spectroscopy observations of the Si/SiO_2 interface plasmon peak. Phys Rev B **56**:6774-6781
6. Ugarte D., Colliex C., Trebbia P. (1992) Surface- and interface-plasmon modes on small semiconducting spheres. Phys Rev B **45**:4332-4343
7. Nellist P.D., Pennycook S.J. (1998) Sub-Ångström resolution by underfocused incoherent transmission electron microscopy. Phys Rev Lett **81**:4156-4159
8. Pennycook S.J., Boatner L.A. (1988) Chemically sensitive structure-imaging with a scanning transmission electron microscope. Nature **336**:565-567
9. Pennycook S.J., Jesson, D.E. (1990) High-resolution incoherent imaging of crystals. Phys Rev Lett **64**:938-941
10. Silcox J., Xu P., Loane R.F. (1992) Resolution limits in annular dark field STEM. Ultramicroscopy **47**:173-186
11. Browning N.D., Chisholm M.F., Pennycook S.J. (1993) Atomic-resolution chemical analysis using a scanning transmission electron microscope. Nature **366**:143-146
12. Duscher G., Browning N.D., Pennycook S.J. (1999) Atomic column resolved electron energy-loss spectroscopy. phys stat sol (a) **166**:327–342
13. Browning N.D., Pennycook S.J. (1995) Atomic-resolution electron energy-loss spectroscopy in the scanning transmission electron microscope. J Microscopy **180**:230-237
14. Batson P.E. (1996) Atomic resolution electronic structure in silicon-based semiconductors. J Electron Microscopy **45**:51-58
15. Muller D.A., Shashkov D.A., Benedek R., Yang L.H., Silcox J., Seidman D.N. (1998) Phys Rev Lett **80**:4741-4744
16. Reimer L. (1995) Energy Filtering Transmission Electron Microscopy Springer, Berlin
17. Stöckli T., Bonard J.M., Stadelmann P.A., Chatelain A. (1997) EELS investigation of plasmon excitations in aluminum nanospheres and carbon nanotubes, Z Physik D **40**:425-428
18. Stöckli T., Bonard J. M., Chatelain A., Wang Z. L., Stadelmann P. (1998) Plasmon excitations in graphitic carbon spheres, Phys Rev B **57**:15599-15612
19. Fink J. (1989) Recent developments in energy-loss spectroscopy. In: Advances in Physics and Electron Physics, Academic Press, London **75**:121-232
20. Kruit P., Venables J.A. (1988) High-spatial-resolution surface-sensitive electron spectroscopy using a magnetic parallelizer. Ultramicroscopy **25**:183-194
21. Egerton R. (1996) Electron Energy-loss Spectroscopy in the Transmission Electron Microscope. 2nd edition, Plenum Press, New York/London
22. Kothleitner G., Hofer F. (1998) Optimization of the signal to noise ratio in EFTEM elemental maps with regard to different ionization edge types. Micron **29**:349-357
23. Inokuti M. (1971) Inelastic collisions of fast charged particles with atoms and molecules: the Bethe theory revisited. Rev Mod Phys **43**:297-347
24. Jeanguillaume C., Colliex C. (1989) Spectrum-image: The next step in EELS digital acquisition and processing. Ultramicroscopy **28**:252-257
25. Krivanek O.L., Gubbens A.J., Dellby N. (1991) Developments in EELS instrumentation for spectroscopy and imaging. Microsc Microanal Microstr **2**:315-312
26. Lanio, S. (1986) High-resolution imaging magnetic energy filter with simple structure. Optik **73**:99-107

27. Probst W., Benner G., Bihr J., Weimer E. (1993) An "Omega" energy filtering TEM – principles and applications. Adv Mater **5**:297-300
28. Tanaka M., Tsuda K., Terauchi M., Tsuno K., Kaneyama T., Honda T., Ishida M. (1999) A new 200 kV Omega-filter electron microscope. J Microscopy **194**:219-227
29. Krivanek O.L., Gubbens A.J., Dellby N., Meyer C.E. (1992) Design and first applications of a post-column imaging filter. Microsc Microanal Microstruct **3**:187-199
30. Nellist P.D., Pennycook S.J. (1999) Incoherent imaging using dynamically scattered coherent electrons. Ultramicroscopy **78**:111-124
31. Krivanek O.L., Dellby N., Lupini A.R. (1999) Towards sub-Åelectron beams. Ultramicroscopy **78**:1-11
32. James E.M., Browning N.D. (1999) Practical aspects of atomic resolution imaging and analysis in STEM. Ultramicroscopy **78**:125-139
33. Rose H. (1999) Prospects for realizing a sub-Å sub-eV resolution EFTEM. Ultramicroscopy **78**:13-25
34. Uhlemann S., Rose H. (1994) The MANDOLINE-filter – a new high-performance imaging filter for sub-eV EFTEM. Optik **96**:163-178
35. Gatts C., Duscher G., Müllejans H., Rühle M. (1995) Analyzing line scan profiles with neural pattern recognition. Ultramicroscopy **59**:229-240
36. Colliex C., Tencé M., Lefevre E., Mory C., Gu H., Bouchet D.,Jeanguillaume C. (1994) Electron energy-loss spectrometry mapping. Microchim Acta **112**:71-87
37. Tencé M., Quartuccio M., Colliex, C. (1995) PEELS compositional profiling and mapping at nanometer spatial resolution. Ultramicroscopy **58**:42-54
38. Bonnet N., Brun N., Colliex C. (1999) Extracting information from sequences of spatially resolved EELS spectra using multivariate statistical analysis. Ultramicroscopy **77**:97-112
39. Redlich P., Carroll D.L., Ajayan P.M. High spatial resolution imaging and spectroscopy in nanostructures. Curr Opinion Sol State Mater Sci **4**:325-336
40. Hunt J.A., Williams D.B. (1991) Electron energy-loss spectrum-imaging. Ultramicroscopy **38**:47-73
41. Shin D.H., Kirkland E.J., Silcox, J. (1989) Annular dark field electron microscope images with better than 2Å resolution at 100 kV. Appl Phys Lett **55**:2456-2458
42. McGibbon M.M., Browning N.D., Chisholm M.F., McGibbon A.J., Pennycook S.J., Ravikumar V., Dravid V.P. (1994) Direct determination of grain boundary atomic structure in $SrTiO_3$. Science **266**:102-104
43. Muller D.A., Tzou Y., Raj R., Silcox J. (1993) Mapping sp^2 and sp^3 states of carbon at sub-nanometre spatial resolution. Nature **366**:725-727
44. Berger S.D., Pennycook S.J. (1982) Detection of nitrogen at {100} platelets in diamond. Nature **298**:635-637
45. Müllejans H., Bruley J. (1995) Electron energy-loss near edge structure of internal interfaces by spatial difference spectroscopy. J Microsc **180**:12-21
46. Muller D.A. (1999) Why changes in bond lengths and cohesion lead to core-level shifts in metals, and consequences for the spatial difference Method. Ultramicroscopy **78**:163-174
47. Alber U. (1998) Ursachen der mechanischen Eigenschaften von Cu-Korngrenzen und Cu/α-Al_2O3-Grenzflächen. Dissertation, Universität Stuttgart, Stuttgart, Germany

48. Kienzle O. (1999) Atomistische Struktur und chemische Zusammensetzung innerer Grenzflächen von Strontiumtitanat. Dissertation, Universität Stuttgart, Stuttgart, Germany
49. Gu H., Ceh M., Stemmer S., Müllejans H., Rühle M. (1995) A quantitative approach for spatially-resolved electron energy-loss spectroscopy of grain boundaries and planar defects on a subnanometer scale. Ultramicroscopy **59**:215-227
50. Baumann S.R., Williams D.B. (1981) A STEM/x-ray microanalytical study of the equilibrium segregation of bismuth in copper. J Microscopy **123**:299-305
51. Michael J.R., Williams D.B. (1984) An analytical electron microscope study of the kinetics of the equilibrium segregation of bismuth in copper. Metall Mater Trans A. **15A**:99-105
52. Alber U., Müllejans H., Rühle M. (1997) Improved quantification of grain boundary segregation by EDS in a dedicated STEM. Ultramicroscopy **69**:105–116
53. Gemming T.: unpublished work (UPDATE IN PROOFS)
54. Bruley J., Keast, V.J., Williams, D.B. (1996) Measurement of the Localized Electronic Structure Associated with Bismuth Segregation to Copper Grain Boundaries. J. Phys. D: Appl. Phys., **29** 1730-1739.
55. Keast V.J., Bruley J., Rez P., MacLaren J.M., Williams D.B. (1997) Chemistry and bonding changes associated with the segregation of Bi to grain boundaries in Cu. Acta Metall Mater **6**:481-490
56. Muller D.A., Subramanian S., Batson P.E., Silcox J., Sass S.L. (1996) Structure, chemistry and bonding at grain boundaries in Ni_3Al-I. The role of boron in ductilizing grain boundaries. Acta metall mater **44**:1637-1645
57. Rühle M. (1996) Structure and composition of metal ceramic interfaces. J Eur Ceram Soc **16**:353-365
58. Rühle M., Evans A.G., Ashby M.F., Hirth J.P. (1990) Metal-Ceramic Interfaces. Pergamon Press, Oxford
59. Dehm G., Scheu C., Rühle M., Raj R. (1998) Growth and structure of Cu/Al_2O_3 and Cu/Ti/Al_2O_3 interfaces. Acta Mater **46**:759-772
60. Scheu C., Dehm G., Rühle M., Brydson, R. (1998) Electron-energy-loss spectroscopy studies of Cu-α-Al_2O_3 interfaces grown by molecular beam epitaxy. Phil Mag A **78**:439–465
61. Scheu C. (1996) Analytische Untersuchungen an Cu/Al_2O_3- und Cu/Ti/Al_2O_3 Grenzflächen, Dissertation, Universität Stuttgart, Stuttgart, Germany
62. Dehm G. (1995) Struktur, Zusammensetzung und mechanische Eigenschaften von Cu/Al_2O_3 und Cu/Ti/Al_2O_3 Grenzflächen, Dissertation, Universität Stuttgart, Stuttgart, Germany
63. Plitzko J.M., Mayer J. (1999) Quantitative thin film analysis by energy filtering transmission electron microscopy. Ultramicroscopy **78**:207-219
64. Dehm G., Rühle M., Conway D., Raj R. (1997) A microindentation method for estimating interfacial shear strength and its use in studying the influence of titanium transition layers on the interface strength of epitaxial copper films on sapphire. Acta Mater **45**:489-499
65. Ahn C.C., Krivanek O.L. (1983) EELS Atlas, Center for Solid State Science, Arizona State University, Tempe, Arizona.
66. Scheu C., Dehm G., Müllejans H., Brydson R., Rühle M. (1995) Electron energy-loss spectroscopy of metal-alumina interfaces. Microsc Microanal Microstruct **6**:19-31

67. Dehm G., Scheu C., Möbus G., Brydson R., Rühle M. (1997) Synthesis of analytical and high-resolution transmission electron microscopy to determine the interface structure of Cu/Al_2O_3. Ultramicroscopy **67**:207-217
68. Alber U., Müllejans H., Rühle M. (1999) Wetting of copper on $\alpha - Al_2O_3$ surfaces depending on the orientation and oxygen partial pressure. Micron **30**:101-108
69. Scheu Ch., Stein W., Rühle M. (2000) Electron energy-loss near-edge structure studies of a $Cu/(11\bar{2}0)\alpha$-Al_2O_3 interface. phys stat sol (b) **222**:199-211
70. Nufer S., Marinopoulos A.G., Gemming T., Elsässer C., Kurtz W., Köstlmeier S., Rühle M. (2001) Quantitative atomic-scale analysis of interface structures: Transmission electron microscopy and local density functional theory. Phys Rev Lett **86**:5066-5069
71. Bruley J., Brydson R., Müllejans H., Mayer J., Gutekunst G., Mader W., Knauss D., Rühle M. (1994) Investigation of the chemistry and bonding at niobium-sapphire interfaces. J Mater Res **9**:2574–2583
72. Brydson R., Müllejans H., Bruley J., Trusty P., Sun X., Yeomans J., Rühle M. (1995) Spatially resolved electron energy-loss studies of metal-ceramic interfaces in transition metal/alumina cermets. J Microscopy **177**:369-386
73. Liedtke A. (1997) Einflußdes Sauerstoffgehalts in Kupfer auf die Reaktivität von diffusionsverschweißten Cu–Al_2O_3 Grenzflächen. Dissertation, Universität Stuttgart, Stuttgart, Germany
74. Reimer L., Fromm I., Rennekamp R. (1988) Operation modes of electron spectroscopic imaging and electron energy-loss spectroscopy in a transmission electron microscope. Ultramicroscopy **24**:339-354
75. Reimer L., Fromm I., Hirsch P., Plate U., Rennekamp R. (1992) Combination of EELS modes and electron spectroscopic imaging and diffraction in an energy filtering electron microscope. Ultramicrosc **46**:335-347
76. Hofer F., Warbichler P., Grogger W. (1995) Imaging of nanometer-sized precipitates in solids by electron spectroscopic Imaging. Ultramicroscopy **59**:15-31
77. Hofer F., Grogger W., Kothleitner G., and Warbichler P. (1997) Quantitative Analysis of EFTEM elemental distribution images. Ultramicroscopy **67**:83-103
78. Crozier P. A. (1995) Quantitative elemental mapping of materials by energy-filtered imaging. Ultramicroscopy **58**:157-174
79. Mayer J., Szabo D.V., Rühle M., Seher M., Riedel R. (1995) Polymer derived Si-based ceramics, Part II: microstructural characterisation by electron spectroscopic imaging. J Eur Ceram Soc **15**:717-727
80. Körtje K.-H. (1994) Image-EELS: simultaneous recording of multiple electron energy-loss spectra from series of electron spectroscopic images. J Microscopy **174**:149-159
81. Lavergne J.-L., Foa C., Bongrand P., Seux D., Martin J.-M. (1994) Application of recording and processing of energy-filtered image sequences for the elemental mapping of biological specimens: Imaging-Spectrum. J Microscopy **174**:195-206
82. Beckers A.L.D., De Bruijn W.C., Gelsema E.S., Cleton-Soeteman M.I., van Eijk H.G. (1994) Quantitative electron spectroscopic imaging in bio-medicine: methods for image acquisition, correction and analysis. J Microscopy **174**:171-182

83. Beckers A.L.D., Gelsema E.S., De Bruijn W.C., Cleton-Soeteman M.I., van Eijk H.G. (1996) Quantitative electron spectroscopic imaging in bio-medicine: evaluation and application. J Microscopy **183**:78-88
84. Mayer J., Eigenthaler U., Plitzko J.M., Dettenwanger F. (1997) Quantitative analysis of electron spectroscopic imaging (ESI) series. Micron **28**:361-370
85. Martin J.-M., Vacher B., Ponsonnet L., Dupuis V. (1996) Chemical bond mapping of carbon by image-spectrum EELS in the second-derivative mode. Ultramicroscopy **65**:229-238
86. Mayer J., Plitzko J.M. (1996) Mapping of ELNES on a nanometre scale by electron spectroscopic imaging. J Microscopy **183**:2-8
87. Thomas P.J., Midgley P.A. (1999) Image-Spectroscopy: New developments and applications. Microsc Microanal **5 (Suppl. 2)**:618-619
88. Jeanguillaume C., Trebbia P., Colliex C. (1978) About the use of electron energy-loss spectroscopy for chemical mapping of thin foils with high spatial resolution. Ultramicroscopy **3**:237-242
89. Hofer F., Warbichler P. (1996) Improved imaging of secondary phases in solids by energy-filtering TEM. Ultramicroscopy **63**:21-25
90. Bentley J., Hall E.L., Kenik E.A. (1995) Quantitative elemental concentrations by energy filtered imaging. In: G.W. Bailey, M.H. Ellisman, R.A. Hennigar and N.J. Zaluzec (Eds.) Microscopy and Microanalysis 1995, Jones and Begell (New York):268-269
91. Weickenmeier A.L., Nüchter W., Mayer J. (1995) Quantitative characterization of point spread function and detection quantum efficiency for a YAG scintillator slow scan CCD camera. Optik **99**:147-154
92. Berger A., Kohl H. (1992) Optimum imaging parameters for elemental mapping in an energy filtering transmission electron microscope. Optik **4**:175-193
93. Berger A., Mayer J., Kohl H. (1994) Detection limits in elemental distribution images produced by EFTEM: Case study of grain boundaries in Si_3N_4. Ultramicroscopy **55**:101-112
94. Clarke D. R. (1987) On the equilibrium thickness of intergranular glass phases in ceramic materials. J Am Ceram Soc **70**:15-22
95. Kleebe H.-J., Cinibulk M.K., Cannon R.M., Rühle M. (1993) Statistical analysis of the intergranular film thickness in silicon nitride ceramics. J Am Ceram Soc **76**:1969-1977
96. Cinibulk M.K., Kleebe H.-J., Schneider G.A., Rühle M. (1993) Amorphous intergranular films in silicon nitride ceramics quenched from high temperatures. J Am Ceram Soc **76**:2801-2808
97. Tanaka I., Kleebe H.-J., Cinibulk M.K., Bruley J., Clarke D.R., Rühle M. (1993) Calcium concentration dependence of the intergranular film thickness in silicon nitride. J Am Ceram Soc **76**:911-914
98. Rafferty B., Pennycook S.J. (1999) Towards atomic column-by-column spectroscopy. Ultramicroscopy **78**:141-151
99. Golla U., Kohl H. (1997) Theoretical and experimental investigations of resolution and detection limits in energy filtering electron microscopy. Micron **28**:397-406
100. Jäger W., Mayer J. (1995) Energy filtering transmission electron microscopy of Si_mGe_n superlattices and Si–Ge heterostructures – I. experimental results. Ultramicroscopy **59**:33-45
101. Freitag B., Mader W. (1999) Element specific imaging with high lateral resolution: An experimental study on layer structures. J Microscopy **194**:42-57

102. Frank J. (1975) A practical resolution criterion in optics and electron microscopy. Optik **43**:25-34
103. Endoh H., Hashimoto H., Makita Y. (1994) Theoretical and observed electron microscope images of impurity atoms in thin crystals formed by L-shell ionization electrons. Ultramicroscopy **56**:108-120
104. Mayer J., Matsumura S., Tomokiyo Y. (1998) First ESI experiments on the new JEOL 2010 FEF. J Electron Microscopy **47**:283-291
105. Stallknecht P., Kohl H. (1996) Computation and interpretation of contrast in crystal lattice images formed by inelastically scattered electrons in a transmission electron microscope. Ultramicroscopy **66**:261-275
106. Knippelmeyer R., Kohl H. (1999) Relativistic calculations of intensity distributions in elemental maps using contrast transfer functions. J Microscopy **194**:30-41
107. Lavergne J.-L., Foa C., Bongrand P., Seux D., Martin J.-M. (1994) Application of recording and processing of energy-filtered image sequences for the elemental mapping of biological specimens: Imaging-spectrum. J Microscopy **174**:195-206
108. Körtje K.-H. (1994) Image-EELS: Simultaneous recording of multiple electron energy-loss spectra from series of electron spectroscopic images. J Microscopy **174**:149-159
109. Rösler M., Zachai R., Füßer H.-J., Jiang X., Klage, C.-P. (1993) Structural properties of heteroepitaxial diamond on silicon. In: Proc. 2nd International Conference on the Applications of Diamond Films and Related Materials. Yoshikawa, M. Murakawa, M. (Eds), Tokyo,691-696
110. Stoner B.R., Ma G.-H.M., Wolter S.D., Glass J.T. (1992) Characterization of bias-enhanced nucleation of diamond on silicon by in vacuo surface analysis and transmission electron microscopy. Phys Rev B **45**:11067-11084
111. Tzou Y., Bruley J., Ernst F., Rühle M., Raj R. (1994) TEM study of the structure and chemistry of a diamond/silicon interface. J Mater Res **9**:1566-1572
112. Plitzko J., Rösler M., Nickel K.G. (1997) Heteroepitaxial growth of diamond thin films on silicon: Information transfer by epitaxial tilting. Diamond Rel Mater **6**:935-939
113. Egerton R.F., Whelan M.J. (1974) Electron energy-loss spectra of diamond, graphite and amorphous carbon. J Electron Spectrosc **3**:232-236
114. Berger S.D., McKenzie D.R., Martin P.J. (1988) EELS analysis of vacuum arc-deposited diamond-like films. Phil Mag Lett **57**:285-290
115. Spence J.C.H., Zuo J.M. (1992) Electron Microdiffraction. Plenum Press, New York
116. Mayer J., Deininger C., Reimer L. (1995) Electron Spectroscopic Diffraction. In: L. Reimer (Ed.) Energy Filtering Transmission Electron Microscopy. Springer Series in Optical Sciences **71**:291-345
117. Zuo J.M., Spence J.C.H. (1991) Automated structure factor measurement by convergent-beam electron diffraction. Ultramicroscopy **35**:185-196
118. Deininger C., Necker G., Mayer J. (1994) Determination of structure factors, lattice strains and accelerating voltage by energy filtered convergent beam electron diffraction. Ultramicroscopy **54**:15-30
119. Bird D.M., Saunders M. (1992) Sensitivity and accuracy of CBED pattern matching. Ultramicroscopy **45**:241-251

120. Swaminathan S., Jones I.P., Zaluzec N.J., Maher D.H., Fraser H.L. (1993) Experimental determination of low order structure factors in the intermetallic compound TiAl. Mater Sci Eng A**170**:227-235
121. Holmestad R., Zuo J.M., Spence J.C.H., Høier R, Horita Z. (1995) Effect of Mn doping on charge density in gamma-TiAl by quantitative convergent beam electron diffraction. Phil Mag A **72**:579-601
122. Fox A.G., Tabbernor M.A. (1991) The bonding charge density of bNiAl. Acta Metall Mater **39**:669-678
123. Menon E.S., Fox A.G. (1996) On the determination of the Debye-Waller factor and structure factors of NiAl by X-ray powder diffraction. Acta Mater **44**:2547-2555
124. Lu Z.W., Wei S.-H., Zunger A. (1992) Theory of bonding charge density in 'NiAl. Acta Metall Mater **40**:2155-2165
125. Davenport J.W., Schultz P.A. (1992) Bonding and brittleness in B2 structure 3d transition metal aluminides: ionic, directional, or does it make a difference? Scripta Metall Mater **27**:629-634
126. Nüchter W., Weickenmeier A.L., Mayer J. (1998) Determination of bonding charge density in NiAl by quantitative convergent beam electron diffraction. phys stat sol (a) **166**:367-379
127. Zuo J.M., Spence J.C.H., Downs J., Mayer J. (1993) Measurement of individual structure factor phases with one degree accuracy: the (002) in BeO studied by dynamical electron diffraction. Acta Cryst **A49**:422-429

5 Advances in Electron Optics

H. Rose

5.1 Fundamentals of Image Formation

The elucidation of the atomic structure of solids is a major goal of high-resolution transmission electron microscopy. The attainable resolution of all imaging microscopes is determined by the wavelength of the radiation employed (e.g. light, sound, charged particles) and the defects of the image-forming lenses. The resolution of microscopes that do not use lenses, such as the scanning tunneling microscope or the atomic force microscope, is not limited by diffraction. Unfortunately, these microscopes can only image the surface of the sample whereas detailed information about the atomic bulk structure is necessary for elucidating the properties of real solid objects.

Image formation in the transmission electron microscope (TEM) can only be described sufficiently accurately by taking into account the wave nature of the electron. The electron wave propagating from the source to the final image plane of the instrument traverses macroscopic electromagnetic fields in the regions in front of and behind the object and microscopic fields within the object.

The macroscopic fields do not vary appreciably over distances of several electron wavelengths. This behaviour differs from that of the microscopic fields produced by the atoms within the specimen. As a consequence, the propagation of the electrons through the static fields of the instrument can be described very accurately within the frame of geometrical optics, which considers the electrons as classical particles. In order to account for diffraction it suffices to consider the effect of the finite electron wavelength by means of the semi-classical WKB approximation [1]. However, this approach fails for the atomic fields of the specimen. In order to describe accurately the propagation of the electron wave through the object a rigorous wave-mechanical treatment is required. Owing to the extreme complexity of this problem one can tackle it only approximately by employing the Born approximation in the case of thin amorphous objects, or the multi-slice and Bloch-wave approaches for crystalline specimens, as detailed in Chap. 2.

All present procedures for determining the propagation of the electron wave within the macroscopic fields of the electron microscope assume isoplanatic conditions. This implies that the transfer properties of the system do not depend on the lateral position of the scatterers within the object. These methods thus neglect the influence of the field aberrations. These aberrations confine the isoplanatic area to a small region centred on the optic axis. The diameter of this area rapidly decreases with increasing usable aperture angle. Hence, if this angle is enlarged by correction of the spherical aberration, the number of equally-well-resolved object points will be strongly reduced. In this case the effect of the field aberrations on the transfer properties of the imaging system can no longer be neglected. To achieve isoplanatism for the imaging of extended objects in a spherically-corrected electron microscope it is, therefore, necessary to eliminate or sufficiently reduce the principal field aberrations, such as the third-order off-axial coma.

So far, a general transfer theory, which considers the effect of the field aberrations, has not been formulated. To close this gap, we describe in the following the contrast transfer in the electron microscope for the general case of non-isoplanatic conditions. In addition, we discuss means of eliminating both the spherical aberration and the off-axial coma. Isoplanatism is achieved if the system fulfils the Abbe sine condition, which will be discussed later. For reasons of simplicity we restrict our mathematical treatment to the nonrelativistic regime. Relativistic effects can be considered by taking into account a posteriori the exact relation between the wavelength λ of the electron and the accelerating voltage.

The propagation of the electron wave is governed by the Schrödinger equation

$$i\hbar\frac{\partial\Psi}{\partial t} = \hat{H}\Psi\,. \tag{5.1}$$

The Hamilton operator

$$\hat{H} = \frac{1}{2m}\left(\frac{\hbar}{i}\nabla + e\boldsymbol{A}\right)^2 - e\varphi \tag{5.2}$$

considers the influence of the magnetic field via the magnetic vector potential $\boldsymbol{A}$ and the effect of the electric potential φ on the propagation of the wave within the electron microscope. The operator ∇ denotes the gradient.

We fix the electric potential by employing the standard gauge of electron optics, which assigns the potential $\varphi = \varphi_c = 0$ to the cathode surface. For this gauge the total energy E of the electron corresponds to its initial energy of emission.

By restricting our investigation to stationary electromagnetic fields, we need to consider only solutions of the form

$$\Psi\left(\boldsymbol{r}, t\right) = \psi\left(\boldsymbol{r}\right) e^{-iEt/\hbar}\,. \tag{5.3}$$

The time-independent function $\psi = \psi\left(\boldsymbol{r}\right)$ satisfies the stationary Schrödinger equation

$$\widehat{H}\psi = E\psi\,. \tag{5.4}$$

The static electric potential

$$\varphi = \varphi_{\rm m} + \varphi_{\rm ob} \tag{5.5}$$

is composed of the potential $\varphi_{\rm m}(\boldsymbol{r})$, which describes the macroscopic electric field within the electron microscope and the atomic potential $\varphi_{\rm ob}(\boldsymbol{r})$ of the object. For simplicity we assume that the object does not contain an internal magnetic field. Accordingly, the magnetic vector potential $\boldsymbol{A}$ accounts entirely for the macroscopic magnetic field of the electron lenses and of the other elements of charged particle optics, such as deflecting magnets or magnetic multipoles.

Since the macroscopic fields do not appreciably alter the electron wavelength along a distance of several wavelengths, we can employ the WKB approximation to describe the propagation of the electron wave in the regions outside the object. However, care must be taken in the vicinity of a caustic because an infinite number of classical trajectories, emanating from a given source point, intersect one other at each point of the caustic. In the absence of a magnetic field the trajectories emanating from this source are orthogonal to the surfaces of constant phase of the corresponding elementary wave. In the case of ideal imaging, the caustic degenerates to a point, which constitutes the image of the source. The elementary wave originating from a point source located at position $\boldsymbol{r} = \boldsymbol{r}'$ is defined as the propagator $P = P(\boldsymbol{r}, \boldsymbol{r}')$. This propagator satisfies the inhomogeneous Schrödinger equation

$$\left(-i\boldsymbol{\nabla} + \frac{e}{\hbar}\boldsymbol{A}\right)^2 P - k_{\rm m}^2 P = \delta(\boldsymbol{r} - \boldsymbol{r}') , \tag{5.6}$$

where

$$k_{\rm m} = [2m(E + e\varphi_{\rm m})]^{1/2}/\hbar = 2\pi/\lambda \tag{5.7}$$

is the wavenumber of the electron. In field-free space ($\boldsymbol{A} = 0$, $\varphi_{\rm m} = \varphi_0$, $k_{\rm m} = k_0 = \text{const.}$), the solution of the propagator equation (5.6)

$$P = P_0 = \frac{1}{4\pi|\boldsymbol{r} - \boldsymbol{r}'|} e^{ik_0|\boldsymbol{r} - \boldsymbol{r}'|} , \tag{5.8}$$

is a spherical wave and represents the Green function employed in scattering and diffraction theory. To obtain a sufficiently accurate solution for the propagator in the presence of a macroscopic electromagnetic field, we employ the eikonal ansatzof light optics:

$$P = \frac{a}{4\pi} e^{iS(\boldsymbol{r}, \boldsymbol{r}')/\hbar} , \tag{5.9}$$

which is identical with the WKB approximation of wave mechanics. The point eikonal

$$S = S(\boldsymbol{r}, \boldsymbol{r}') = \int_{\boldsymbol{r}'}^{\boldsymbol{r}} \boldsymbol{p} \cdot d\boldsymbol{r} \tag{5.10}$$

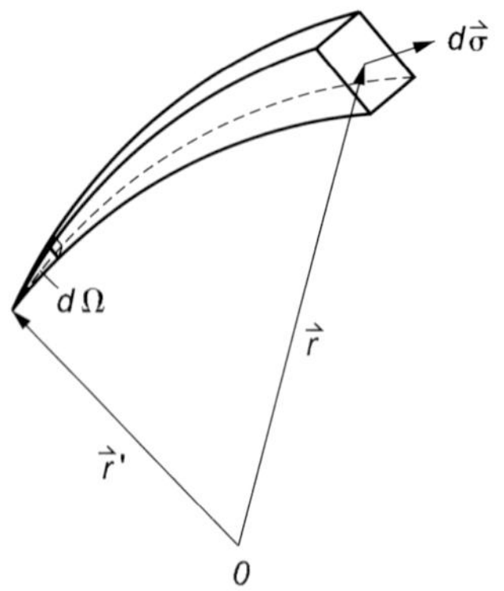

Fig. 5.1. Differential current tube for determining the amplitude of the elementary electron wave emanating from the point $\boldsymbol{r}'$

denotes the path integral taken over the canonical momentum

$$\boldsymbol{p} = m\boldsymbol{v} + e\boldsymbol{A} \tag{5.11}$$

along the classical trajectory, which connects the origin $\boldsymbol{r}'$ to the end-point $\boldsymbol{r}$.

The absolute value $|a|$ of the amplitude a is obtained most conveniently from the conservation of the current. For this purpose we consider a homocentric bundle of electron rays that originate within a differential solid angle $d\boldsymbol{\Omega}$ from the point $\boldsymbol{r}'$. The outer rays form the mantle of a conical tube, as illustrated in Fig. 5.1. The conservation of the current dJ within this tube yields the relation

$$(4\pi)^2 \frac{m}{e} dJ = |a|^2 \boldsymbol{v} \cdot d\boldsymbol{\sigma} = \boldsymbol{v}' \cdot d\boldsymbol{\Omega} \, . \tag{5.12}$$

The surface element $d\boldsymbol{\sigma}$ represents a cross-section of the tube at the end-point $\boldsymbol{r}$. Any displacement $\delta\boldsymbol{r}$ of this point is connected with a change

$$\delta\, \boldsymbol{p}' = -\delta\, \mathrm{grad}'\, S = -\left(\delta\boldsymbol{r}\, \mathrm{grad}\right) \mathrm{grad}'\, S \tag{5.13}$$

of the canonical momentum

$$\boldsymbol{p}' = m\, \boldsymbol{v}' - e\boldsymbol{A}\,(\boldsymbol{r}') = -\, \mathrm{grad}'\, S\,(\boldsymbol{r}, \boldsymbol{r}') \tag{5.14}$$

of the electron at the origin $\boldsymbol{r}'$.

The primed gradient indicates differentiation with respect to the coordinates x', y' and z' of the point source. Since the positional shifts $\delta\boldsymbol{r}$ may be performed in any direction, we can choose the orientation of the surface element $d\boldsymbol{\sigma}$ arbitrarily.

By choosing $d\boldsymbol{\sigma} = \boldsymbol{e}_z\, dx\, dy$, we find

$$v_z|a|^2 = \boldsymbol{v}'\frac{d\boldsymbol{\Omega}}{dx\,dy} = \frac{1}{v'^2}\left(v'_x\frac{dv'_y\,dv'_z}{dx\,dy} + v'_y\frac{dv'_x\,dv'_z}{dx\,dy} + v'_z\frac{dv'_x\,dv'_y}{dx\,dy}\right). \tag{5.15}$$

The velocity $\boldsymbol{v}'$ at the origin is a function of the coordinates x and y according to the relation (5.13). Using this relation, and considering that $\boldsymbol{A}(\boldsymbol{r}')$ does not depend on the coordinates of the end-point $\boldsymbol{r}$, each differential quotient becomes a two-dimensional Jacobian determinant whose terms are derivatives of S with respect to the coordinates of the origin and the end-point. Inserting these determinants into the right hand side of (5.15), we eventually obtain

$$v_z|a|^2 = \frac{1}{m^2v'^2}D_z\,. \tag{5.16}$$

The minors of the resulting determinant

$$D_z = \begin{vmatrix} v'_x & v'_y & v'_z \\ \dfrac{\partial^2 S}{\partial x\partial x'} & \dfrac{\partial^2 S}{\partial x\partial y'} & \dfrac{\partial^2 S}{\partial x\partial z'} \\ \dfrac{\partial^2 S}{\partial y\partial x'} & \dfrac{\partial^2 S}{\partial y\partial y'} & \dfrac{\partial^2 S}{\partial y\partial z'} \end{vmatrix} \tag{5.17}$$

are identical with the differential quotients in (5.15). If we place successively the surface element in the x and y directions, we find relations, which can directly be obtained from (5.16) by replacing in this expression the index z by x and y, respectively. The determinants D_x and D_y are derived by cyclic rotation of the coordinates x, y and z in the expression (5.17). Here the coordinate z appears only as an index on the left-hand side.

In order to find a representation for $|a|$ that does not depend on the chosen coordinate system, we combine the relation (5.16) with those indexed by x and y to from the vector relation

$$|a|^2\boldsymbol{v} = \frac{1}{m^2v'^2}\boldsymbol{D}\,. \tag{5.18}$$

By taking the absolute value of this equation, we obtain the symmetric expression

$$|a| = \frac{1}{mv'}\sqrt{\frac{D}{v}} \tag{5.19}$$

for the amplitude of the propagator, apart from a constant phase factor. The absolute value

$$D = \sqrt{D_x^2 + D_y^2 + D_z^2} \tag{5.20}$$

of the vector $\boldsymbol{D}$ is completely symmetric with respect to the directions x, y, and z.

It should be noted that our result differs from that derived by Glaser [2]. The validity of his approximation is restricted to the vicinity of the optic axis, while our approximation is not. Nevertheless, both approximations yield the exact result

$$a = |a| = \frac{1}{|\boldsymbol{r} - \boldsymbol{r}'|} \tag{5.21}$$

for field-free space, where $v = v'$. In this trivial case the eikonal adopts the simple form

$$S = mv\,|\boldsymbol{r} - \boldsymbol{r}'| = \hbar k\,|\boldsymbol{r} - \boldsymbol{r}'|\ . \tag{5.22}$$

For this eikonal we obtain

$$D = \frac{m^2 v^3}{|\boldsymbol{r} - \boldsymbol{r}'|^2}\ . \tag{5.23}$$

If we insert the results (5.21), (5.22) and (5.23) into (5.9) the propagator adopts the standard form (5.8) of an undisturbed spherical wave.

The semiclassical approximation of the propagator consists of a single wave (5.9), as long as the source point $\boldsymbol{r}'$ is connected to the point of observation $\boldsymbol{r}$ by a single trajectory. This is always the case for the field-free space. However, in the presence of an electromagnetic field it may occur that trajectories that start from the common source point $\boldsymbol{r}'$ intersect each other in one or more points. If such a point is connected to the point source by n trajectories, the propagator

$$P = \frac{1}{4\pi m v'} \sum_{\nu=1}^{n} \sqrt{\frac{D_\nu}{v}}\, e^{iS_\nu/\hbar} \tag{5.24}$$

consists of n terms, each of which is defined by one of the n classical paths. The determinants D_x, D_y and D_z remain finite for each eikonal S_ν. However, if the point of observation is located on the caustic, at least one of these determinants becomes infinitely large.

In this case n tends to infinity because an infinite number of trajectories, starting within a given differential solid angle, intersect each other at a conjugate point. The loci of all these points form the caustic, which generally consists of several sheets. The tip of the caustic is located at the Gaussian image point. In the case of ideal lenses the caustic shrinks to a point.

The quantity D is a measure of the trajectory density, as illustrated by the relation (5.23) for field-free space. The caustic is hence given by the condition $D = \infty$. Strong interference effects occur in the vicinity of the caustic owing to the interference of an infinite number of partial waves. The summation of theses waves can be performed by subdividing the entire space into two half-spaces, one containing the source and the other one the caustic. Generally, the diffraction plane behind the objective lens is chosen as the separating plane because each point of this plane is associated with a distinct direction of flight at the object plane. In the following, we choose the optic axis of the instrument as the z axis of the coordinate system. The plane $z = z_\mathrm{d}$ separates the object space from the image space. The latter contains the caustic formed by the trajectories emanating from the scatterers within the object. It should be noted that the tip of the caustic formed by the unscattered electrons is located at the centre of the diffraction plane.

In the case of field-free image space, the summation of the elementary waves can readily be performed by employing the Sommerfeld diffraction formula [3]:

$$\psi(\boldsymbol{r}) = \frac{1}{2\pi} \iint A(x_\mathrm{d}, y_\mathrm{d}) \frac{\partial}{\partial z_\mathrm{d}} \left(\frac{e^{ik|\boldsymbol{r}-\boldsymbol{r}_\mathrm{d}|}}{|\boldsymbol{r}-\boldsymbol{r}_\mathrm{d}|} \right) \psi_\mathrm{d} dx_\mathrm{d} dy_\mathrm{d} \,. \tag{5.25}$$

Here $\psi_\mathrm{d} = \psi(x_\mathrm{d}, y_\mathrm{d}, z_\mathrm{d})$ denotes the wavefunction at the diffraction plane. The aperture function A accounts for the absorption by a beam-limiting aperture. Since (5.25) is valid for an arbitrary incident wave ψ_d, it must also hold true for the propagator $\psi_\mathrm{d} = P(\boldsymbol{r}_\mathrm{d}, \boldsymbol{r}_0)$ emanating from the point $\boldsymbol{r}_0$ in the object space. This elementary wave is deformed by the electromagnetic field of the objective lens.

In the following we assume that the image space contains an electromagnetic field, which focuses the electrons yet does not reflect any of them back to the object space. In this case we can still apply the Sommerfeld diffraction formula (5.25) with a sufficient degree of accuracy if we substitute in this formula the propagator (5.24) with $n = 1$ for the field-free propagator (5.8). By choosing $\psi_\mathrm{d} = P(\boldsymbol{r}_\mathrm{d}, \boldsymbol{r}_0)$, we obtain the representation

$$P(\boldsymbol{r}, \boldsymbol{r}_0) = 2 \iint A(x_\mathrm{d}, y_\mathrm{d}) \frac{\partial P(\boldsymbol{r}, \boldsymbol{r}_\mathrm{d})}{\partial z_\mathrm{d}} P(\boldsymbol{r}_\mathrm{d}, \boldsymbol{r}_0) \, dx_\mathrm{d} \, dy_\mathrm{d} \tag{5.26}$$

for the propagator in the image space $z > z_\mathrm{d}$. This expression is valid regardless of whether the point of observation is located on the caustic or not.

The expression (5.26) can therefore be applied directly for calculating the electron wave in the image space $z > z_\mathrm{d}$ if this wave is known at a plane behind the object. This plane is generally the exit surface of the object because for crystalline objects both the multi-slice approach and the Bloch-wave formalism yield the electron wave at the back surface of the object. If we choose this plane as the object plane z_o, the propagation of the electron wave $\psi(\boldsymbol{r}_\mathrm{o})$ to the image space can be derived by employing the modified Sommerfeld diffraction formula with the propagator (5.26). This procedure represents a successive propagation of the wave from the object plane z_o to the diffraction plane z_d and from there to the recording plane z. Thus the wavefunction at this plane has the form

$$\psi(\boldsymbol{r}) = 4\int \left[A(\boldsymbol{\rho}_{\mathrm{d}}) \frac{\partial P(\boldsymbol{r}, \boldsymbol{r}_{\mathrm{d}})}{\partial z_{\mathrm{d}}} \int \psi(\boldsymbol{r}_{\mathrm{o}}) \frac{\partial P(\boldsymbol{r}_{\mathrm{d}}, \boldsymbol{r}_{\mathrm{o}})}{\partial z_{\mathrm{o}}} d^2\boldsymbol{\rho}_{\mathrm{o}} \right] d^2\boldsymbol{\rho}_{\mathrm{d}} . \qquad (5.27)$$

This expression does not presuppose isoplanatism and hence is valid for arbitrary systems, with the exception of mirrors.

5.1.1 Lippmann–Schwinger Equation

In the presence of an object, it is advantageous to transform the Schrödinger equation into an integral equation. This equation can be solved approximately by means of an iteration procedure. The first step yields the first-order Born approximation, which suffices to describe the image formation of thin amorphous objects. To obtain the integral equation, we write the stationary Schrödinger equation in the form

$$(-i\boldsymbol{\nabla} + e\boldsymbol{A}/\hbar)^2 \psi - k_{\mathrm{m}}^2 \psi = \frac{2me}{\hbar} \varphi_{\mathrm{ob}} \psi . \qquad (5.28)$$

The right-hand side of this equation can be considered as a source term. Since the propagator $P(\boldsymbol{r}, \boldsymbol{r}')$ is a solution for a delta-like source, the solution of (5.28) is given by the expression

$$\psi(\boldsymbol{r}) = \psi_0(\boldsymbol{r}) + \frac{2em}{\hbar^2} \int P(\boldsymbol{r}, \boldsymbol{r}') \varphi_{\mathrm{ob}}(\boldsymbol{r}') \psi(\boldsymbol{r}') d^3\boldsymbol{r}' . \qquad (5.29)$$

This integral equation is known as the Lippmann–Schwinger equation of scattering theory [4]. However, one always uses the free-wave propagator in scattering theory, whereas in our case one has to employ the propagator for a macroscopic electromagnetic field. The term $\psi_0(\boldsymbol{r})$ is a solution of the homogeneous equation obtained from (5.28) by putting $\varphi_{\mathrm{ob}} = 0$. This "undisturbed" wave describes the illuminating wave in front of the object. For a point-like source located at $\boldsymbol{r}_{\mathrm{s}}$ this wave is given by the propagator $P(\boldsymbol{r}, \boldsymbol{r}_{\mathrm{s}})$.

In the case of ideal Köhler illumination the effective source is placed at infinity by the condenser system. Accordingly, the propagator $P(\boldsymbol{r}, \boldsymbol{r}_{\mathrm{s}})$ degenerates to a plane incident wave whose wave vector $\boldsymbol{k}_0 = \boldsymbol{p}/\hbar$ defines the angular position of the effective point source. The wave originating from this source must be described by the mixed eikonal

$$W = W(\boldsymbol{r}; \boldsymbol{k}_{0\perp}, z_0) . \qquad (5.30)$$

In the case of a plane incident wave the starting plane z_0 may be chosen arbitrarily. The mixed eikonal is obtained from the point eikonal $S(\boldsymbol{r}, \boldsymbol{r}_0)$ by means of a Legendre transformation with respect to the off-axis coordinates x_0 and y_0. The vector

$$\boldsymbol{k}_{0\perp} = k_{0x}\boldsymbol{e}_x + k_{0y}\boldsymbol{e}_y \qquad (5.31)$$

is the lateral component of the wave-vector

$$\boldsymbol{k}_0 = \boldsymbol{k}_{0\perp} + \boldsymbol{e}_z k_{0z} , \quad k_{0z}^2 = (k_0^2 - \boldsymbol{k}_{0\perp}^2)^{1/2} . \tag{5.32}$$

The mixed eikonal has the property that

$$x_0 = \frac{1}{\hbar}\frac{\partial W}{\partial k_{0x}} , \quad y_0 = \frac{1}{\hbar}\frac{\partial W}{\partial k_{0y}} . \tag{5.33}$$

For field-free space, the mixed eikonal has the simple form

$$W = \hbar(\boldsymbol{k}_0 \cdot \boldsymbol{r} - k_{0z} z_0) . \tag{5.34}$$

If we place the origin of the coordinate system at the centre of the starting plane ($z_0 = 0$), the propagator for field-free space

$$P_s = P(\boldsymbol{r}, \boldsymbol{r}_s = -\infty) = e^{iW/\hbar} = e^{i\boldsymbol{k}_0 \cdot \boldsymbol{r}} \tag{5.35}$$

has the standard form of a plane wave.

In the presence of an electromagnetic field, the propagator of the point source is also defined by the mixed eikonal. Since the source is imaged into the diffraction plane by the field of the objective lens, the unscattered illumination beam forms a caustic in the region of this plane. Fortunately, the image plane is always located in a region outside of this caustic, because the image of the object is spatially well separated from that of the source. Accordingly, each point of the image plane is connected via a single trajectory to a given point of the effective source. In the case of Köhler illumination the caustic formed by the unscattered beam is centred at the back focal plane of the object plane and at consecutive images because the diffraction plane is a conjugate plane of the source. Within the frame of validity of the WKB approximation, the unscattered wave originating from a distinct point of the effective source can be written as

$$P_s = a_s e^{iW/\hbar} \tag{5.36}$$

in the presence of macroscopic fields.

To determine the amplitude a_s, we assume that the unscattered wave is neither reflected nor absorbed in the region between the object and the image. In this case the current within an infinitesimally small trajectory tube is preserved at any plane z. Choosing $a_s = 1$ at the object plane $z = z_0$, the conservation of the current yields the condition

$$v_{0z}\, dx_0\, dy_0 = |a_s|^2 v_z\, dx\, dy . \tag{5.37}$$

Employing the relations (5.33) for x_0 and y_0, the absolute value of the amplitude is given by the expression

$$|a_s|^2 = \frac{v_{0z}}{v_z}\frac{dx_0\, dy_0}{dx\, dy} = \frac{v_{0z}}{v} D_m , \tag{5.38}$$

where D_m is the second-order determinant

$$D_\mathrm{m} = \frac{\partial^2 W}{\partial p_{0x}\partial x}\frac{\partial^2 W}{\partial p_{0y}\partial y} - \frac{\partial^2 W}{\partial p_{0x}\partial y}\frac{\partial^2 W}{\partial p_{0y}\partial x} . \tag{5.39}$$

In the case of ideal imaging, the mixed eikonal at the image plane $z = z_\mathrm{i}$ is

$$W_\mathrm{i} = W(\boldsymbol{p}_{0\perp}, \boldsymbol{r}_\mathrm{i}) = W_\mathrm{i}^{(0)} + \frac{\boldsymbol{p}_0 \cdot \boldsymbol{\rho}_\mathrm{i}}{M} . \tag{5.40}$$

Here

$$\boldsymbol{\rho}_\mathrm{i} = x_\mathrm{i}\boldsymbol{e}_x + y_\mathrm{i}\boldsymbol{e}_y \tag{5.41}$$

denotes the lateral component of the position vector $\boldsymbol{r}_\mathrm{i}$, and $W_\mathrm{i}^{(0)}$ is a term that does not depend on the variables $\boldsymbol{p}_{0\perp}$ and $\boldsymbol{\rho}_\mathrm{i}$. The sign of the magnification

$$M = (-)^{\nu+1}|M| \tag{5.42}$$

is determined by the number ν of intermediate images, which are located between the object plane z_0 and the final image plane z_i. For the expression (5.40), the determinant (5.39) is $D_\mathrm{m} = M^{-2}$. This result reflects the trivial fact that in the case of perfect imaging the current density at the image plane is equal to the current density at the object plane, divided by the square of the magnification M. By inserting (5.38) into the formula (5.36), we obtain for the source propagator the expression

$$\psi_0(\boldsymbol{r}) = P_\mathrm{s}(\boldsymbol{r}, \boldsymbol{k}_0) = \sqrt{\frac{v_{0z}}{v} D_\mathrm{m}}\, e^{iW/\hbar} . \tag{5.43}$$

Here we have assumed that the condenser system transforms the elementary spherical wave into a plane wave. In the case of an extended effective source we must not superpose the individual elementary waves because each point of this source is an incoherent emitter. Accordingly, we must add up the intensities of the images formed by the individual point sources. This procedure is equivalent to the incoherent summation over the illumination angle.

To determine the intensity distribution of the unscattered beam in the diffraction plane, one must consider that this plane is conjugate to that of the source. As a consequence, the representation (5.43) for the source propagator breaks down in the region of the diffraction plane. At this plane the unscattered beam forms the central spot of the diffraction pattern. The extension of this spot depends on the size of the effective source and on the aberrations of the lenses in front of the diffraction plane. In the case of a point source, as used in the STEM, the spot size is limited by the aberrations and by the diffraction at the opening of the illumination aperture. The geometrical centre of the spot is identical with the tip of the caustic formed by the trajectories emanating from the point source. To obtain an expression for the source propagator that remains finite at the conjugate point of the source we employ the modified Sommerfeld diffraction formula at the object plane z_0. The elementary wave originating from a point $\boldsymbol{r}_0$ in this plane is described by the propagator $P(\boldsymbol{r}, \boldsymbol{r}_0)$. Since the object plane is located in front of the caustic, the source propagator is described by the elementary wave (5.43) at this plane. Considering this behaviour, the formula (5.26) yields the expression

$$\psi_o(\boldsymbol{r}) = 2 \iint A(\boldsymbol{\rho}_0) \frac{\partial P(\boldsymbol{r}, \boldsymbol{r}_0)}{\partial z_0} \psi_0(\boldsymbol{r}_0)\, d^2\boldsymbol{\rho}_0\,, \tag{5.44}$$

for the wave in the region behind the object plane where $\boldsymbol{\rho}_0$ denotes the lateral component of the vector $\boldsymbol{r}_0$, and $\psi_0(\boldsymbol{r}_0)$ is the source propagator at the object plane z_0. The expression (5.44) also describes the illuminating wave in the STEM. Here the object is located at the image plane of the point source. The integration over the object plane is equivalent to the summation of all partial waves that contribute to the scanning spot. Within the frame of wave optics each straight trajectory in front of the focus corresponds to a partial plane wave. Accordingly, a converging wave packet represents the wave optical analogue of a homocentric bundle of trajectories in geometrical optics.

5.1.2 Kinematic Approximation

The effect of the field aberrations on image formation is best illustrated for thin objects. For such objects single electron scattering prevails. This situation is generally valid for amorphous biological specimens, but not for crystalline objects, where dynamic scattering dominates. For simplicity we restrict our considerations to elastic scattering. This implies that the object potential

$$\varphi_{ob} = \sum_{\nu=1}^{N} \varphi_\nu(\boldsymbol{r} - \boldsymbol{r}_\nu) \tag{5.45}$$

can be written as the sum of the static potentials φ_ν of the constituent atoms. In a solid specimen these potentials differ somewhat from the free-atom potentials owing to bonding effects. The vector $\boldsymbol{r}_\nu$ defines the position of the ν-th atom within the specimen. The Lippmann–Schwinger equation (5.29) can be solved approximately by means of an iteration procedure. In the first step the wavefunction $\psi(\boldsymbol{r}')$ in the integrand of (5.29) is replaced by the undisturbed wave $\psi_0(\boldsymbol{r}')$. The resulting expression for the scattered wave represents the first-order Born approximation. However, there is a major departure from the standard quantum mechanical result, which is based on the free-space propagator (5.8).

The Born series is an expansion in terms of the number of scattering events. Hence the first-order Born approximation suffices if single scattering prevails. To calculate the wavefunction at the image plane we need to know the propagator from the starting point $\boldsymbol{r}_\nu$ to the point $\boldsymbol{r}_\mathrm{d}$ in the diffraction plane and the propagator $P(\boldsymbol{r}, \boldsymbol{r}_\mathrm{d})$ from this point to the point of observation $\boldsymbol{r}$, because we have to use the expression (5.26) for the propagator P in (5.29). However, the process of image formation is elucidated more clearly if we first propagate the scattered wave to the diffraction plane and from there to the image plane. To keep the mathematics as simple as possible we assume that the object, the image, and the diffraction plane are at the same electric potential. This assumption implies

$$v_\nu = v_\mathrm{d} = v_\mathrm{i} = v\,. \tag{5.46}$$

To calculate the first-order Born approximation of the scattered wave at the diffraction plane, we only need the asymptotic form of the propagator

$$P(\boldsymbol{r}_\mathrm{d}, \boldsymbol{r}') = a\, e^{iS(\boldsymbol{r}_\mathrm{d}, \boldsymbol{r}')/\hbar} \tag{5.47}$$

because the distance $|\boldsymbol{r}_\mathrm{d} - \boldsymbol{r}_\nu|$ is much larger than the extension of the atom potential $\varphi_\nu(\boldsymbol{r} - \boldsymbol{r}_\nu)$. Moreover, small variations of the amplitude a do not appreciably affect the first iteration

$$I_\nu = \frac{2em}{\hbar^2} \int P(\boldsymbol{r}_\mathrm{d}, \boldsymbol{r}')\, \varphi_\nu(\boldsymbol{r}' - \boldsymbol{r}_\nu)\, \psi_0(\boldsymbol{r}')\, d^3\boldsymbol{r}'\,. \tag{5.48}$$

of the Lippmann–Schwinger equation (5.29). In the case of ideal plane-wave illumination,

$$\psi_0 = e^{i\boldsymbol{k}_0 \cdot \boldsymbol{r}'}\,, \tag{5.49}$$

the diffraction plane is located at the back focal plane of the objective lens. If the object is placed near the front focal plane of this lens, the amplitude of the propagator (5.47) at the diffraction plane

$$a \approx \frac{1}{4\pi f_0}\,, \tag{5.50}$$

is approximately proportional to the inverse of the focal length f_0. This result can readily be verified by means of ray-optical considerations. Unfortunately, such a crude approximation does not suffice for the phase of the propagator because the phase varies rapidly over distances of a few wavelengths.

For evaluating the integral (5.48), it is advantageous to introduce new integration variables x, y, z defined by the vector

$$\boldsymbol{r} = \boldsymbol{r}' - \boldsymbol{r}_\nu\,. \tag{5.51}$$

The potential $\varphi_\nu(\boldsymbol{r}) = \varphi_\nu(r)$ of neutral atoms is roughly zero at distances $r \geq 0.1\,\mathrm{nm}$. We can therefore expand the eikonal $S = S(\boldsymbol{r}_\mathrm{d}, \boldsymbol{r}') = S(\boldsymbol{r}_\mathrm{d}, \boldsymbol{r} + \boldsymbol{r}_\nu)$ in a power series with respect to the variable $\boldsymbol{r}$:

$$S \approx S_\nu + \boldsymbol{r} \cdot \mathrm{grad}_\nu S_\nu + \ldots , \tag{5.52}$$

where

$$S_\nu = S(\boldsymbol{r}_\mathrm{d}, \boldsymbol{r}_\nu) . \tag{5.53}$$

The vector

$$-\,\mathrm{grad}_\nu S_\nu = \boldsymbol{p}_\nu = \hbar \boldsymbol{k}_\nu \tag{5.54}$$

is the canonical momentumof the scattered electron at the position $\boldsymbol{r}_\nu$ of the atom. Hence, if this momentum is given, the position $\boldsymbol{r}_\mathrm{d}$ of the electron is implicitly defined by the gradient of the eikonal at the starting point $\boldsymbol{r}_\nu$. This gradient hardly varies along distances comparable with the diameter of an atom. Accordingly, it is sufficient to consider only the first and the second terms on the right-hand side of the expansion (5.52). With this assumption and the approximation (5.50) the propagator $P(\boldsymbol{r}_\mathrm{d}, \boldsymbol{r} + \boldsymbol{r}_\nu)$ at the diffraction plane is approximately given by

$$P(\boldsymbol{r}_\mathrm{d}, \boldsymbol{r} + \boldsymbol{r}_\nu) \approx \frac{e^{iS_\nu/\hbar}}{4\pi f_0} e^{-i\boldsymbol{k}_\nu \cdot \boldsymbol{r}} . \tag{5.55}$$

By inserting this approximation together with the relation (5.49) and the new integration variables (5.51) into the integral (5.48), we eventually obtain the asymptotic expression

$$I_\nu \approx \frac{e^{iS_\nu/\hbar}}{f_0} e^{i\boldsymbol{k}_0 \cdot \boldsymbol{r}_\nu} f_\nu(\boldsymbol{q}_\nu) . \tag{5.56}$$

The function

$$f_\nu(\boldsymbol{q}_\nu) = \frac{e\,m}{2\pi\hbar^2} \int \varphi_\nu(\boldsymbol{r})\, e^{-i\boldsymbol{q}_\nu \cdot \boldsymbol{r}} d^3\boldsymbol{r} \tag{5.57}$$

is the first-order Born approximation of the elastic scattering amplitude. It should be noted that the scattering vector

$$\boldsymbol{q}_\nu = \boldsymbol{k}_\nu - \boldsymbol{k}_0 = m(\boldsymbol{v}_\nu - \boldsymbol{v}_0)/\hbar \tag{5.58}$$

does not depend on the magnetic vector potential $\boldsymbol{A}_\nu = \boldsymbol{A}(\boldsymbol{r}_\nu)$, although the wave-vectors $\hbar\boldsymbol{k}_\nu = m\boldsymbol{v}_\nu + e\boldsymbol{A}_\nu$ and $\hbar\boldsymbol{k}_0 = m\boldsymbol{v}_0 + e\boldsymbol{A}_\nu$ are both linearly related to the magnetic vector potential. This behaviour illustrates the fact that the scattering vector (5.58) is a measurable quantity, while the wave-vector is not since it depends on the gauge of the magnetic vector potential.

Using the asymptotic expression (5.56) for the integral (5.48), the first-order Born approximation of the scattered wave at the diffraction plane z_d has the asymptotic form

$$\psi_s(\boldsymbol{r}_d) = \frac{1}{f_0} \sum_\nu e^{iS_\nu/\hbar} e^{i\boldsymbol{k}_0 \cdot \boldsymbol{r}_\nu} f_\nu(\boldsymbol{q}_\nu) \,. \tag{5.59}$$

It should be noted that the eikonal $S_\nu = S(\boldsymbol{r}_d, \boldsymbol{r}_\nu)$ and the scattering vector $\boldsymbol{q}_\nu$ are both functions of $\boldsymbol{r}_d$ and $\boldsymbol{r}_\nu$. The elastic scattering amplitude of a free atom $f_\nu(\boldsymbol{q}_\nu) = f_\nu(q_\nu)$ depends only on the absolute value q_ν of the scattering vector $\boldsymbol{q}_\nu$.

In the absence of lenses, the diffraction pattern forms in the Fraunhofer (far-field) region of field-free space. This region starts at distances

$$R = |\boldsymbol{r} - \boldsymbol{r}_0| \approx z - z_0 \gg \rho_0^2/\lambda \tag{5.60}$$

from the object plane z_0; $2\rho_0$ is the lateral diameter of the object. Within the field-free Fraunhofer region, the eikonal S_ν can be replaced by its asymptotic form

$$\begin{aligned} S_\nu &= mv\,|\boldsymbol{r} - \boldsymbol{r}_\nu| \approx mv\left[R - (\boldsymbol{r}_\nu - \boldsymbol{r}_0)\,\frac{\boldsymbol{r} - \boldsymbol{r}_0}{R}\right] \\ &= \hbar\left[kR - \boldsymbol{k}\,(\boldsymbol{r}_\nu - \boldsymbol{r}_0)\right] \,. \end{aligned} \tag{5.61}$$

The vector

$$\boldsymbol{k} = k\,(\boldsymbol{r} - \boldsymbol{r}_0)\,/R \tag{5.62}$$

defines the direction of propagation. The objective lens shifts the remote diffraction pattern into the back focal plane of this lens. Substituting R for f_0 in (5.59) and considering the relation (5.62), we eventually arrive at the standard field-free asymptotic form

$$\psi_s = \frac{e^{ikR}}{R} e^{i\boldsymbol{k}_0 \cdot \boldsymbol{r}_0} \sum_\nu e^{-i\boldsymbol{q}(\boldsymbol{r}_\nu - \boldsymbol{r}_0)} f_\nu(\boldsymbol{q}) \,, \quad \boldsymbol{q} = \boldsymbol{k} - \boldsymbol{k}_0 \,, \tag{5.63}$$

of the scattered wave. The transferred momentum

$$\hbar q = 2mv \sin(\vartheta/2) \tag{5.64}$$

is entirely defined by the kinetic momentum mv and the scattering angle ϑ. Application of the formula (5.59) to thin crystalline specimens illustrates the effect of the aberrations on the intensity distribution at the diffraction plane. In the case of an ideal crystal and a perfect lens, sharp and regularly ordered diffraction spots emerge. To survey the effect of the aberrations we assume that the "field" angle $\gamma_0 = \rho_0/f_0$ is significantly smaller than the Bragg angle or, to put it another way, that the lateral diameter of the object is much smaller than the diameter of the diffraction pattern located at the back focal plane of the objective lens. With this assumption we can expand the eikonal S_ν into a power series with respect to the coordinates $x_\nu^i, i = 1, 2, 3$, of the position vector $\boldsymbol{r}_\nu$:

$$\begin{aligned} S_\nu &= S\left(\boldsymbol{r}_{\mathrm{d}}, \boldsymbol{r}_\nu\right) \\ &= S_0 + \boldsymbol{r}_\nu \operatorname{grad}_\nu S_\nu(0) + 1/2 \sum_i \sum_j x_\nu^i x_\nu^i \frac{\partial^2 S_\nu}{\partial x_\nu^i \partial x_\nu^j}(0) + \ldots \quad . \end{aligned} \tag{5.65}$$

The symbol (0) indicates that the coordinates x_ν^i must be set to zero after differentiating S_ν with respect to these coordinates.

The first term,

$$S_0 = S\left(\boldsymbol{r}_{\mathrm{d}}, 0\right) \ , \tag{5.66}$$

accounts for the aperture aberrations because it is the optical path of an axial electron emerging from the centre $\boldsymbol{r}_\nu = 0$ of the object plane. For convenience we have placed the origin of the coordinate system at this point. The vector factor

$$-\operatorname{grad}_\nu S_\nu(0) = \boldsymbol{p}\left(\boldsymbol{r}_{\mathrm{d}}, 0\right) = \hbar \boldsymbol{k}(0) \tag{5.67}$$

of the second term defines the initial canonical momentum of the axial electron that intersects the diffraction plane z_{d} at the lateral position $\boldsymbol{\rho}_{\mathrm{d}}$.

The part $S_0/\hbar$ of the phase does not affect the intensity distribution $|\psi_{\mathrm{s}}|^2$ at the diffraction plane. However, it strongly influences the resolution of the image because it represents the phase shift resulting from the axial aberrations. If we neglect the quadratic terms in the expansion (5.66) we must substitute the scattering vector

$$\boldsymbol{q}(0) = \boldsymbol{k}(\boldsymbol{r}_{\mathrm{d}}, 0) - \boldsymbol{k}_0 \tag{5.68}$$

for the scattering vector $\boldsymbol{q}_\nu$ in (5.59), in order to be consistent in the order of the expansion. Within the frame of validity of this first-order approximation the Laue condition holds true in the form

$$\boldsymbol{q}(0) = 2\pi \boldsymbol{g} \, , \tag{5.69}$$

where $\boldsymbol{g}$ is a reciprocal lattice vector. As a consequence, the sharp Bragg spots are preserved at the diffraction plane. However, the diffraction pattern will be distorted if $\boldsymbol{k}(\boldsymbol{r}_{\mathrm{d}}, 0)$ is a nonlinear function of the lateral coordinates x_{d} and y_{d}. The nonlinear terms result from off-axis coma. Therefore, an undistorted diffraction pattern can only be recorded if the coma has been eliminated or sufficiently suppressed. It should be noted that the coma affects the image of the object completely differently, because it decreases the resolution in proportion to the distance from the optic axis.

The second and higher-order terms of the expansion (5.66) affect the resolution of the diffraction image because they broaden the Bragg spots and produce a blurred background intensity. The bilinear terms in the off-axis coordinates x_ν and y_ν can be interpreted as field astigmatism and field curvature. The coefficients of these aberrations are functions of the lateral position $\boldsymbol{\rho}_\mathrm{d}$ of the electron at the diffraction plane z_d. In the paraxial region these coefficients only vanish at the centre $\boldsymbol{\rho}_\mathrm{d} = 0$ if the objective lens is perfectly aligned. A residual second-order field astigmatism has the same effect on the diffraction image as the corresponding axial astigmatism on the image of the object. The former aberration does not affect the object image, while the axial aberrations do not influence the intensity distribution of the diffraction image. This behaviour differs from that of the mixed aberrations, which diminish the resolution in the off-axis region of both images.

5.1.3 Phase Contrast

The modified Sommerfeld diffraction formula (5.26) is most appropriate for calculating the scattered wave at the final image plane if this wave is given at the diffraction plane. The corresponding propagator

$$P_\mathrm{i} = P\left(\boldsymbol{r}_\mathrm{i}, \boldsymbol{r}_\mathrm{d}\right) = a_\mathrm{i}\, e^{iS_\mathrm{i}/\hbar} \tag{5.70}$$

describes the propagation of an elementary wave, originating from the point $\boldsymbol{\rho}_\mathrm{d}$ in the diffraction plane z_d, to the point $\boldsymbol{\rho}_\mathrm{i}$ in the image plane z_i. The point eikonal $S_\mathrm{i} = S\left(\boldsymbol{\rho}_\mathrm{i}, z_\mathrm{i}; \boldsymbol{\rho}_\mathrm{d}, z_\mathrm{d}\right)$ must be used for calculating the quantity D_i (5.20), which is needed to determine the amplitude

$$a_\mathrm{i} = \frac{1}{4\pi m v_\mathrm{d}} \sqrt{\frac{D_\mathrm{i}}{v_\mathrm{i}}} \tag{5.71}$$

of the propagator at the image plane. Fortunately, this amplitude remains almost constant for all points of the aperture openings and the fields of view used in electron microscopy. The paraxial approximation

$$a_\mathrm{i} \approx \frac{1}{4\pi M f_0} \sqrt{\frac{v_\mathrm{d}}{v_\mathrm{i}}} \tag{5.72}$$

can therefore be used in all further calculations. With the additional assumption $v_\mathrm{i} = v_\mathrm{d} = v_0 = v$ we derive from the modified Sommerfeld diffraction formula

$$\psi\left(\boldsymbol{r}_\mathrm{i}\right) = 2 \iint A\left(\boldsymbol{\rho}_\mathrm{d}\right) \psi\left(\boldsymbol{\rho}_\mathrm{d}, z_\mathrm{d}\right) \frac{\partial P_\mathrm{i}}{\partial z_\mathrm{d}} d^2 \boldsymbol{\rho}_\mathrm{d} \tag{5.73}$$

the approximation

$$\psi\left(\boldsymbol{r}_\mathrm{i}\right) \approx \frac{1}{i\lambda M f_0} \iint A\left(\boldsymbol{\rho}_\mathrm{d}\right) \psi\left(\boldsymbol{\rho}_\mathrm{d}, z_\mathrm{d}\right) e^{iS_\mathrm{i}/\hbar} d^2 \boldsymbol{\rho}_\mathrm{d} \tag{5.74}$$

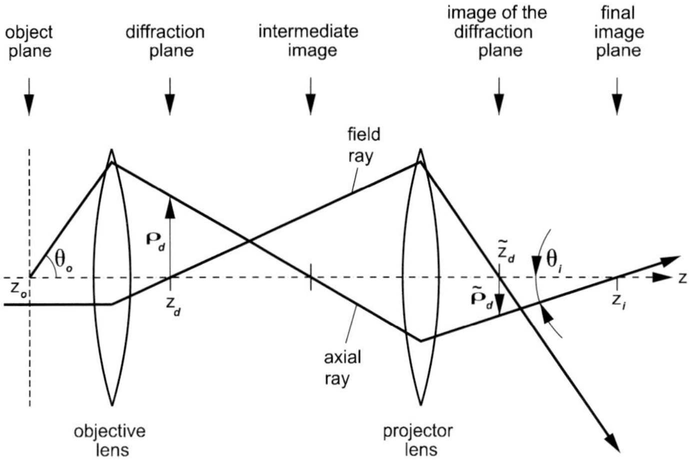

Fig. 5.2. Formation and notation of consecutive images of the object and the diffraction plane in an electron microscope

for the wave at the final image plane. This expression is obtained by assuming that $a_{\rm i}$ is constant and the propagation is confined to the region in the vicinity of the optic axis. In this case the approximation

$$\frac{\partial S_{\rm i}}{\partial z_{\rm d}} = -ip_{{\rm d}z} \approx -ip_{\rm d} = \frac{\hbar}{i}k = \frac{\hbar}{i}\frac{2\pi}{\lambda} \tag{5.75}$$

holds with a sufficient degree of accuracy. The approximation (5.74) convincingly confirms the Huygens principle of light optics, which states that the wave at a plane $z_{\rm i} > z_{\rm d}$ is formed by the coherent superposition of the elementary waves that proceed from all points of the wave from the plane $z_{\rm d}$ to the plane $z_{\rm i}$. The eikonal $S_{\rm i} = S(\boldsymbol{r}_{\rm i}, \boldsymbol{r}_{\rm d})$ considers the effect of all lenses located between the diffraction plane and the final image plane. In the absence of a beam-limiting aperture we can use any subsequent image of the diffraction plane for the summation of the elementary waves contributing to the interference pattern in the vicinity of the caustic.

The high magnification in an electron microscope is achieved by means of several projector lenses, each of which forms an intermediate image in front of the subsequent lens. Owing to the validity of the Helmholtz–Lagrange relation, which relates the positions and directions of an axial ray and a field ray, an image of the diffraction pattern must always be located between two consecutive images of the object plane. This behaviour is illustrated by the path of these rays shown in Fig. 5.2. The magnification of the intermediate images increases along the column of the microscope while that of the diffraction pattern decreases. The situation is reversed in the Boersch mode, which produces a magnified image of the diffraction pattern at the recording plane.

In the standard imaging mode a strongly demagnified image of the diffraction pattern is located close to the last projector lens. The field-free region between this image and the recording plane contains the caustic formed by the scattered electrons. We can therefore use the last diffraction image instead of the primary diffraction plane for employing the diffraction formula (5.73). However, this change of planes is only allowed in the absence of a beam-limiting aperture at the primary diffraction plane. Unfortunately, the aperture must always be placed at this plane because only there is the diffraction pattern large enough. Nevertheless, we may still use the plane $\tilde{z}_{\mathrm{d}}$ of the last diffraction image if we assume that the opening of the aperture is perfectly imaged into this plane by the intermediate lenses. In this case we may assume that a real aperture with a properly demagnified opening limits the beam at the plane $\tilde{z}_{\mathrm{d}}$. The choice of this plane has the advantage that we can employ the exact Sommerfeld diffraction formula for the field-free half-space. The distance $b = z_{\mathrm{i}} - \tilde{z}_{\mathrm{d}}$ between the final image plane and the last diffraction image is very large compared with the radius of the demagnified diffraction pattern because its magnification

$$M_{\mathrm{d}} = \frac{1}{M} \frac{b}{f_0} \tag{5.76}$$

is very much less than one for the large object magnifications

$$M = \theta_0 / \theta_{\mathrm{i}} \tag{5.77}$$

used in high-resolution electron microscopy.

The quantities θ_0 and θ_{i} denote the angles between the axial ray and the optic axis at the object plane and at the image plane, respectively. The relation (5.76) can be readily verified from Fig. 5.2 with the help of the axial ray and the relation (5.77). For very high magnifications M, the demagnified diffraction pattern at the plane $\tilde{z}_{\mathrm{d}}$ is entirely located within the first Fresnel zone for an observer at the plane z_{i}. Accordingly, we can employ the Fraunhofer approximation

$$P_i \approx \frac{1}{4\pi b} e^{ik(b - \boldsymbol{\rho}_{\mathrm{i}} \cdot \tilde{\boldsymbol{\rho}}_{\mathrm{d}} / b)} \tag{5.78}$$

for the free-space propagator. If we apply the formula (5.73) at the plane $\tilde{z}_{\mathrm{d}}$ instead at the plane z_{d} and replace the propagator by the Fraunhofer approximation (5.78), the resulting expression for the wave at the final image plane

$$\psi(\boldsymbol{r}_{\mathrm{i}}) = \frac{e^{ikb}}{i\lambda b} \iint A(\tilde{\rho}_{\mathrm{d}}) \, \psi(\tilde{\boldsymbol{\rho}}_{\mathrm{d}}, \tilde{z}_{\mathrm{d}}) \, e^{-ik\boldsymbol{\rho}_{\mathrm{i}} \tilde{\boldsymbol{\rho}}_{\mathrm{d}} / b} \, d^2 \tilde{\boldsymbol{\rho}}_{\mathrm{d}} \tag{5.79}$$

represents the Fourier transform of the wavefunction at the last image of the diffraction pattern.

To determine the wavefunction (5.79) in the case of weak-phase objects, we insert the expression (5.59) for the scattered wave into the integral (5.79).

The result,

$$\psi_s(\boldsymbol{r}_i) = \frac{e^{ikb}}{i\lambda b f_0} \sum_\nu e^{i\boldsymbol{k}_0 \cdot \boldsymbol{r}_\nu} \iint A(\tilde{\boldsymbol{\rho}}_d) e^{\tilde{S}_\nu/\hbar} e^{-ik\boldsymbol{\rho}_i \cdot \tilde{\boldsymbol{\rho}}_d/b} f_\nu(\boldsymbol{q}_\nu)\, d^2\tilde{\boldsymbol{\rho}}_d \,, \tag{5.80}$$

can be rewritten in a more familiar form by referring the plane $\tilde{z}_d$ back to the primary diffraction plane z_d in the same way as the image plane is referred back to the object plane. Nevertheless, the path integration for determining the eikonal

$$\tilde{S} = S(\tilde{\boldsymbol{r}}_d, \boldsymbol{r}_\nu) = S\left(M_d \boldsymbol{\rho}_d, \tilde{z}_d; \boldsymbol{r}_\nu\right) \tag{5.81}$$

must be performed between the starting point $\boldsymbol{\rho}_\nu, z_\nu$ and the end point $\tilde{\boldsymbol{\rho}}_d = M_d\, \boldsymbol{\rho}_d$ at the plane $\tilde{z}_d$. It should be noted however that $\boldsymbol{\rho}_d$ does not represent the real lateral position of the electron at the diffraction plane z_d. Replacing the variable distance b in the expression (5.80) by means of the relation (5.76) and referring both the image plane and the image of the diffraction plane back to their primary conjugate planes, we obtain for the wavefunction (5.80) the representation

$$\psi_s = \frac{e^{ik(z_i - \tilde{z}_d)}}{i\lambda M} \sum_\nu e^{i\boldsymbol{k}_0 \cdot \boldsymbol{r}_\nu} \iint A(\boldsymbol{\theta}) e^{i\tilde{S}_\nu/\hbar} e^{-ik\boldsymbol{\rho}\cdot\boldsymbol{\theta}} f_\nu(\boldsymbol{q}_\nu)\, d^2\boldsymbol{\theta} \,. \tag{5.82}$$

Here $\boldsymbol{\rho} = \boldsymbol{\rho}_i/M$ is the position vector of the image point referred back to the object plane. The absolute value of the two-dimensional vector

$$\boldsymbol{\theta} - \tilde{\boldsymbol{\rho}}_d/M_d f_0 \tag{5.83}$$

describes in the paraxial approximation the starting angle of the axial ray, which intersects the image plane $\tilde{z}_d$ of the diffraction pattern at the lateral position $\tilde{\boldsymbol{\rho}}_d$. Owing to the Larmor rotation introduced by the magnetic lenses, the direction of the starting angle generally differs from that of $\tilde{\boldsymbol{\rho}}_d$.

In the case of an ideal microscope the point eikonal

$$\tilde{S}_{id}(\tilde{\boldsymbol{r}}_d, \boldsymbol{r}_0) = \hbar k \left[\tilde{z}_d - z_0 - \boldsymbol{\rho}_0 \cdot \boldsymbol{\theta}\right] \tag{5.84}$$

connecting the point $\boldsymbol{r}_0$ of the ideally imaged object plane with the point $\tilde{\boldsymbol{r}}_d$ of the diffraction image describes a plane wave at the exit plane $\tilde{z}_d$. This eikonal coincides with the paraxial approximation of the corresponding eikonal for a non-perfect imaging system. The difference

$$\varDelta S_\nu = S_\nu(\tilde{r}_d, \boldsymbol{r}_\nu) - S_{id}(\tilde{\boldsymbol{r}}_d; \boldsymbol{\rho}_0 = \boldsymbol{\rho}_\nu, z_0) \tag{5.85}$$

can be considered as a perturbation eikonal, which accounts for the aberrations of the lenses and for the displacement

$$\varDelta_\nu = z_0 - z_\nu \tag{5.86}$$

of the plane z_ν with respect to the object plane z_0. The perturbation eikonal determines the deviation of the wave surface from its ideal spherical shape at the exit pupil $\tilde{z}_{\mathrm{d}}$. For Fraunhofer imaging conditions the image is located at infinity. Hence the ideal system transforms the elementary wave originating from a point at the object plane into a plane wave in the image space, as demonstrated by the relation (5.82).

The phase contrast

$$C_{\mathrm{p}}\left(\boldsymbol{\rho}_{\mathrm{i}}\right) = 2\,\mathrm{Re}\left[\psi_{\mathrm{s}}\left(\boldsymbol{\rho}_{\mathrm{i}}\right)/\psi_{\mathrm{un}}\left(\boldsymbol{\rho}_{\mathrm{i}}\right)\right] \tag{5.87}$$

in the image of a weak phase object is linearly related to the scattered wave [5]. By inserting the first-order Born approximation (5.82) for the scattered wave and the expression

$$\psi_{\mathrm{un}}\left(\boldsymbol{\rho}_{\mathrm{i}}\right) \approx \frac{1}{M} e^{iW_{\mathrm{i}}/\hbar} \tag{5.88}$$

for the unscattered wave into the right-hand side of (5.87), we obtain

$$\begin{aligned} C_{\mathrm{p}}(\boldsymbol{\rho}) = \frac{2}{\lambda}\,\mathrm{Im}\Bigg[& e^{-i\Delta W_{\mathrm{i}}/\hbar} \sum_\nu e^{i\boldsymbol{k}_0\cdot\boldsymbol{r}_\nu} \\ & \times \iint A(\boldsymbol{\theta}) e^{i\Delta S_\nu/\hbar} e^{-ik(\boldsymbol{\rho}+\boldsymbol{\rho}_\nu)\cdot\boldsymbol{\theta}} f_\nu\left(\boldsymbol{q}_\nu\right) d^2\boldsymbol{\theta}\Bigg] , \end{aligned} \tag{5.89}$$

where

$$\boldsymbol{\rho} = \boldsymbol{\rho}_{\mathrm{i}}/M \tag{5.90}$$

is the location of the image point referred back to the object plane. The quantity

$$\Delta W_{\mathrm{i}} = W(\boldsymbol{k}_{0\perp}, z_0; \boldsymbol{\rho}_{\mathrm{i}}, z_{\mathrm{i}}) - W(0, z_0; 0, z_{\mathrm{i}}) \tag{5.91}$$

with

$$W(0, z_0\,; 0, z_{\mathrm{i}}) = k\left(z_{\mathrm{i}} - z_0\right) \tag{5.92}$$

defines the difference between the optical path of an electron that travels from the object plane z_0 in the direction $\boldsymbol{k}_0$ to the point $\boldsymbol{\rho}_{\mathrm{i}}$ in the image plane and the optical path of the axial electron between these planes. The expression (5.90) represents a generalization of the standard phase-contrast formulae given in the literature since these are derived on the assumptions that the system is isoplanatic and the object infinitely thin ($z_\nu = z_0$).

In contrast to those formulae, the expression (5.90) holds for any real imaging system since it requires only the validity of the first-order Born approximation for the elastic scattering amplitude of the object.

The diameter of the field of view utilized in electron microscopes does not exceed several thousand pixels. For such image fields the curvature of the wave surface of the unscattered wave is negligibly small. Accordingly, it suffices to consider only the first two terms in the expansion of the mixed eikonal with respect to the off-axis image coordinates x_i and y_i:

$$\Delta W_\mathrm{i} = W_\mathrm{i}(\boldsymbol{k}_{0\perp},0)+\boldsymbol{\rho}_\mathrm{i}\cdot\mathrm{grad}_\mathrm{i}\, W_\mathrm{i}|_{\boldsymbol{\rho}_\mathrm{i}=0}+\dots \quad . \tag{5.93}$$

Here

$$\frac{1}{\hbar}\,\mathrm{grad}_\mathrm{i}\, W_\mathrm{i}\Big|_{\boldsymbol{\rho}_\mathrm{i}=0} = \boldsymbol{k}_\mathrm{i} = \boldsymbol{k}_\mathrm{i}(\boldsymbol{k}_0) \tag{5.94}$$

is the wave vector of the unscattered wave at the centre $\boldsymbol{\rho}_\mathrm{i} = 0$ of the image plane. The term $W_\mathrm{i}(\boldsymbol{k}_{0\perp},0)/\hbar$ is the phase shift caused by the axial aberrations of the electron optical elements. This result is readily understandable if we consider that the angular vector

$$\boldsymbol{\Theta}_0 = \boldsymbol{k}_{0\perp}/k \tag{5.95}$$

defines the angle of incidence of the illuminating plane wave. Accordingly, the vector $\boldsymbol{k}_{\mathrm{i}\perp}/k$ determines the corresponding angle of the wave at the centre of the image plane. In the paraxial approximation, the relation

$$\frac{|\boldsymbol{k}_{0\perp}|}{|\boldsymbol{k}_{\mathrm{i}\perp}|} \approx \frac{\theta_0}{\theta_\mathrm{i}} = |M| \tag{5.96}$$

holds because an axial Gaussian ray, which intersects the centre of the image plane also intersects that of the object plane. The angle θ_0 enclosed by this ray and the optic axis at the object plane coincides in our case with the illumination angle Θ_0. In the nonparaxial region the magnification $M = M(\theta_0)$ becomes a function of the angle θ_0. For this zonal magnification to be constant, the Abbe sine condition

$$\frac{|\boldsymbol{k}_{\mathrm{i}\perp}|}{|\boldsymbol{k}_{0\perp}|} = \frac{\sin\theta_\mathrm{i}}{\sin\theta_0} = \frac{1}{|M|} = \mathrm{const} \tag{5.97}$$

must be fulfilled. The relation (5.94) reveals that this is only the case if the lateral component of the gradient of the eikonal W_i satisfies the condition

$$\frac{1}{\hbar}\,\mathrm{grad}_{\mathrm{i}\perp}\, W_\mathrm{i}|_{\boldsymbol{\rho}_\mathrm{i}=0} = \boldsymbol{k}_{\mathrm{i}\perp} = \frac{1}{M_0}\boldsymbol{k}_{0\perp} \quad , \quad M_0 = M(0) \tag{5.98}$$

at the centre of the image plane. In the presence of an axial magnetic field the directions of the lateral vectors $\boldsymbol{k}_{\mathrm{i}\perp}$ and $\boldsymbol{k}_{0\perp}$ only coincide in the rotating coordinate system. The expansion of $\boldsymbol{k}_{\mathrm{i}\perp} = \boldsymbol{k}_{\mathrm{i}\perp}(\boldsymbol{\Theta}_0)$ with respect to the components Θ_{0x} and Θ_{0y} of the angular vector of incidence (5.95) yields the coefficients of the field aberrations, which are linear in the off-axial distance $\boldsymbol{\rho}_\mathrm{i}$.

In the case of rotationally-symmetric systems these aberrations are known as third and higher-order comas. Hence all comas must vanish in order to fulfil the Abbe sine condition (5.97). However, since the illumination angles employed in the TEM are generally small, the coma hardly affects the phase of the unscattered wave in the image plane.

5.1.4 Diffractograms

Diffractograms are powerful tools for determining the axial aberrationsof electron lenses, as first convincingly demonstrated by Thon [6]. Unfortunately, his procedure only allows one to determine the axial aberrations of odd Seidel order. However, in any real system aberrations of even order will be present owing to imperfect rotational symmetry or misalignment. For example, the primary second-order axial aberrations, axial coma and threefold astigmatism, must generally be compensated to attain the theoretical resolution of the electron microscope. The diffractogram is the Fourier transform of the image intensity distribution. To determine the aberrations from this diffractogram unambiguously, it is necessary to use the image of an amorphous thin object. Nevertheless, a single diffractogram is not capable of furnishing all the resolution-limiting axial aberration coefficients. To determine these aberrations, Zemlin et al. [7] introduced the diffractogram tableau, which is a set of diffractograms, each obtained from an image taken with a different direction $\boldsymbol{k}_0$ of the incident plane wave.

The present theoretical description of image formation presupposes isoplanatic imaging conditions. However, so far no experimental criterion is known that allows one to make this assumption. Fortunately, the phase-contrast formula (5.90) is valid for any imaging conditions and can hence be used to investigate the effect of the off-axial aberrations on the diffractogram.

For determining the intensity distribution of the diffractogram we must calculate the Fourier transform

$$\tilde{C}_p(\boldsymbol{\omega}) = \iint C_p(\boldsymbol{\rho})\, e^{ik\boldsymbol{\omega}\cdot\boldsymbol{\rho}} d^2\boldsymbol{\rho} \tag{5.99}$$

of the phase contrast (5.90). In order to evaluate the integral (5.99) analytically, we assume that the curvature of the wave surface of the unscattered wave is negligibly small in the region of the detector. Accordingly, it suffices to replace ΔW_i in the expression (5.90) by the approximation (5.93), which can be rewritten as

$$\Delta W_{\mathrm{i}} \approx W_0(\boldsymbol{\Theta}_0) + \hbar M \boldsymbol{\rho} \cdot \boldsymbol{k}_{\mathrm{i}\perp}(\boldsymbol{\Theta}_0) \,. \tag{5.100}$$

For clarity we introduce the angular vector

$$\boldsymbol{\Theta} = M\boldsymbol{k}_{\mathrm{i}\perp}(\boldsymbol{\Theta}_0)/k = M\boldsymbol{\Theta}_{\mathrm{i}} \,, \tag{5.101}$$

which represents the tilt angle $\boldsymbol{\Theta}_{\mathrm{i}}$ of the unscattered wave at the image space referred back to the object. For an isoplanatic system the relation

$$\Theta = M\Theta_{\mathrm{i}} = \Theta_0 \tag{5.102}$$

holds according to the Helmholtz–Lagrange relation. If we neglect the effect of the aberrations on the angular distribution of the scattering amplitudes

$f_\nu(\boldsymbol{q}_\nu)$ at the diffraction plane, the scattering vector $\boldsymbol{q}_\nu$ is linearly related to the vector $\boldsymbol{\omega}$. Since we are primarily interested in achieving an almost "white" spectrum for the spatial frequencies, we can employ this approximation without an appreciable loss of accuracy. With these assumptions the integral (5.99) can readily be evaluated, giving

$$\tilde{C}_{\mathrm{p}}(\boldsymbol{\omega}) = \sum_\nu \tilde{C}_{\mathrm{p}\nu}(\boldsymbol{\omega}) , \tag{5.103}$$

where

$$\begin{aligned}\tilde{C}_{\mathrm{p}\nu} = i\lambda f_\nu(\boldsymbol{\omega}) \Big\{ & e^{-ik\boldsymbol{\omega}\cdot\boldsymbol{\rho}_\nu} e^{-i[W_{\nu 0} - \Delta S_\nu(\boldsymbol{\omega}-\boldsymbol{\Theta})]/\hbar} A(\boldsymbol{\omega}-\boldsymbol{\Theta}) \\ & - e^{ik\boldsymbol{\omega}\cdot\boldsymbol{\rho}_\nu} e^{i[W_{\nu 0} - \Delta S_\nu(\boldsymbol{\omega}+\boldsymbol{\Theta})]/\hbar} A(\boldsymbol{\omega}+\boldsymbol{\Theta}) \Big\}\end{aligned} \tag{5.104}$$

with

$$W_{\nu 0} = W_0(\boldsymbol{\Theta}_0) - \hbar\Delta_\nu (1 - \cos\Theta_0) \tag{5.105}$$

is the contribution of the ν-th atom. In order to obtain a smooth distribution of the spatial frequencies $k\boldsymbol{\omega}$, the object must be an amorphous foil consisting of atoms of the same atomic number. In this case the intensity of the diffractogram

$$I_{\mathrm{D}}(\boldsymbol{\omega}) = \left|\tilde{C}_{\mathrm{p}}\right|^2 = \sum_{\mu,\nu} \tilde{C}_{\mathrm{p}\mu}\tilde{C}^*_{\mathrm{p}\nu} \approx \sum_\nu \left|\tilde{C}_{\mathrm{p}\nu}\right|^2 \tag{5.106}$$

is roughly equal to the sum of the single-atom intensities $\left|\tilde{C}_{\mathrm{p}\nu}\right|^2$ because the interference terms ($\mu \neq \nu$) largely cancel each other out. For simplicity we do not consider the effect of a beam-limiting aperture. By inserting the expression (5.104) in the right-hand side of the relation (5.106) and setting $A = 1$, we find

$$I_{\mathrm{D}}(\boldsymbol{w}) \sim f^2(\omega) \sum_\nu \sin^2 \gamma_\nu(\boldsymbol{\Theta}, \boldsymbol{\omega}) , \tag{5.107}$$

where the phase shifts

$$\begin{aligned}\gamma_\nu = \gamma(\boldsymbol{\omega}, \boldsymbol{\Theta}_0; \boldsymbol{\rho}_\nu) = \frac{1}{2\hbar} [& \Delta S(\boldsymbol{\Theta} - \boldsymbol{\omega}, \boldsymbol{\rho}_\nu) \\ & + \Delta S(\boldsymbol{\Theta} + \boldsymbol{\omega}, \boldsymbol{\rho}_\nu) - 2W_{\nu 0}(\boldsymbol{\Theta}_0)]\end{aligned} \tag{5.108}$$

of the elementary waves are functions of the spatial frequency $k\boldsymbol{\omega}$, the tilt angle $\boldsymbol{\Theta}_0$ and the position $\boldsymbol{r}_\nu$ of the atoms. These phase shifts result entirely from the geometrical aberrations of the image-forming system. The field aberrations produce a phase shift

$$\Delta\gamma_\nu = \gamma(\boldsymbol{\omega}, \boldsymbol{\Theta}_0; \boldsymbol{\rho}_\nu) - \gamma(\boldsymbol{\omega}, \boldsymbol{\Theta}_0; 0) , \tag{5.109}$$

which increases with increasing distance ρ_ν from the optic axis. As a result the outer rings of the diffractogram may be smeared out or even extinct, eventhough we have assumed completely coherent imaging conditions. If the off-axial aberrations are not to wash out the intensity oscillations in the high-spatial-frequency region of the diffractogram, the maximum phase shift $\Delta\gamma_{\nu\,\mathrm{max}}$ must be smaller than $\pi/4$. This phase shift belongs to the waves that emanate from atoms located at the edge of the imaged object area. The effect of the field aberrations can be assessed by subdividing the phase-contrast image into a set of subimages. If the image is recorded digitally, arbitrary configurations can be chosen for the areas of the subimages. Differences between the Fourier transforms of these partial images result from the field aberrations, which can hence be calculated from the set of diffractograms. This procedure resembles the Zemlin tableau method, which yields the axial aberrations. Combination of the two methods allows one, at least in principle, to determine the limiting axial and field aberrations. For reasons of simplicity we have assumed a monochromatic effective point source. To describe the image intensity in the case of a realistic source one must, for example, average the phase contrast (5.90) over the lateral extension and the energy distribution of the real effective source.

5.2 Properties of Aplanatic Electron Lenses

Primarily the performance of the objective lens of a microscope determines the instrumental resolution and the number of equally-well-resolved object points in the recorded image. In order to optimize these quantities the spherical aberration and the chromatic aberration of the imaging system must be eliminated or sufficiently suppressed. In addition the off-axial coma must be eliminated to provide a large enough field of view [8,30]. An optical arrangement that fulfils these requirements is called an aplanatic system. Its imaging properties do not vary for small angular tilts of the incident beam or with the position of the scatterers if they are located in the vicinity of the optic axis. For extended objects, however, the field astigmatism and the image curvature may diminish the resolution in the outer region of the image. Fortunately, these aberrations are of little concern in high-resolution electron microscopy since the high magnification under typical imaging conditions implies that only a small area of the object is imaged onto the screen or the detector.

A consequence of the eikonal formalism is that the trajectories of electrons emanating from two different source points are correlated. In the paraxial approximation this dependence is governed by the well-known Helmholtz–Lagrange relation, which relates the lateral positions and the directions of flight of any two electrons with each other.

5.2.1 Sine Condition

Imaging systems are called aplanatic if they satisfy the sine condition. As demonstrated by Abbe [11], the sine condition gives information about the quality of the image at off-axial points in terms of the properties of the pencil of axial rays. To verify that this important light-optical condition also holds in electron optics, it is advantageous to utilize the mixed eikonal

$$V = V(\boldsymbol{r}_\mathrm{o}; \boldsymbol{p}_{\mathrm{i}\perp}, z_\mathrm{i}) = S - \boldsymbol{\varrho}_\mathrm{i} \cdot \boldsymbol{p}_\mathrm{i} \,. \tag{5.110}$$

Here $\boldsymbol{p}_\perp$ denotes the lateral component of the canonical momentum (5.11). The gauge of the magnetic vector potential is chosen so that

$$\boldsymbol{A}(x=0, y=0, z) = 0 \,, \tag{5.111}$$

which guarantees that the canonical momentum of the particle coincides with its kinetic momentum at any point along the optic axis. The mixed eikonal (5.110) is obtained from the point eikonal S by means of a Legendre transformation with respect to the image coordinates x_i and y_i. Accordingly, the mixed eikonal V is a function of the four variables x_o, y_o, $p_{\mathrm{i}x}$, $p_{\mathrm{i}y}$, and of the locations z_o and z_i of the object plane and of the image plane, respectively. For mathematical simplicity it is advantageous to express the two-dimensional vectors $\boldsymbol{\varrho} = \boldsymbol{e}_x x + \boldsymbol{e}_y y$, $\boldsymbol{p}_\perp = \boldsymbol{e}_x p_x + \boldsymbol{e}_y p_y$, and $\boldsymbol{A}_\perp = \boldsymbol{e}_x A_x + \boldsymbol{e}_y A_y$ in terms of the complex quantities

$$w = x + iy \,, \quad p = p_x + ip_y \,, \quad A = A_x + iA_y ts. \tag{5.112}$$

The corresponding conjugate complex quantities are indicated by a bar. Since the variation δV vanishes for fixed w_o and p_i, the lateral component of the canonical momentum p_o at the object plane and the off-axial position of the trajectory at the image plane z_i can be obtained from the eikonal V by varying w_o and p_i, yielding

$$\delta V = -\mathrm{Re}\{p_\mathrm{o} \delta w_\mathrm{o} + w_\mathrm{i} \delta \bar{p}_\mathrm{i}\} \,. \tag{5.113}$$

Since the variations $\delta \bar{w}_\mathrm{o}$ and $\delta \bar{p}_\mathrm{i}$ can be chosen arbitrarily, we derive from the expression (5.113) the relations

$$p_\mathrm{o} = -2 \frac{\partial V}{\partial \bar{w}_\mathrm{o}} \,, \quad w_\mathrm{i} = -2 \frac{\partial V}{\partial \bar{p}_\mathrm{i}} \,. \tag{5.114}$$

To elucidate the requirements for ideal imaging, we expand the mixed eikonal in a power series with respect to w_o and $\bar{w}_\mathrm{o} = x_\mathrm{o} - iy_\mathrm{o}$:

$$V = V^{(0,0)} + \mathrm{Re}(\bar{w}_\mathrm{o} V^{(1,0)}) + V^{(1,1)} w_\mathrm{o} \bar{w}_\mathrm{o} + \mathrm{Re}(\bar{w}_\mathrm{o}^2 V^{(2,0)}) + \cdots \,. \tag{5.115}$$

The coefficients

$$V^{(\mu,\nu)} = V^{(\mu,\nu)}(z_\mathrm{o}; z_\mathrm{i}, p_\mathrm{i}, \bar{p}_\mathrm{i}) \tag{5.116}$$

are real for $\mu = \nu$. In the presence of magnetic fields the coefficients with $\mu \neq \nu$ are generally complex owing to the Larmor rotation of the electrons within these fields. Neglecting the quadratic and higher-order terms in the expansion (5.115), we obtain from (5.114) the relations

$$p_{0\mathrm{o}} = p_\mathrm{o}(\bar{w}_\mathrm{o} = 0) = -V^{(1,0)}(z_\mathrm{o}; z_\mathrm{i}, p_\mathrm{i}, \bar{p}_\mathrm{i}) , \tag{5.117}$$

$$w_\mathrm{i} = -2\frac{\partial V^{(0,0)}}{\partial \bar{p}_\mathrm{i}} - \bar{w}_\mathrm{o}\frac{\partial V^{(1,0)}}{\partial \bar{p}_\mathrm{i}} - w_\mathrm{o}\frac{\partial \bar{V}^{(1,0)}}{\partial \bar{p}_\mathrm{i}} . \tag{5.118}$$

If the system is completely corrected for spherical aberration of any order, all trajectories that originate at the centre $w_\mathrm{o} = \bar{w}_\mathrm{o} = 0$ of the object plane z_o must intersect the centre $w_\mathrm{i} = \bar{w}_\mathrm{i} = 0$ of the image plane z_i. The second expression of (5.117) shows that this is the case only if

$$\frac{\partial V^{(0,0)}}{\partial \bar{p}_\mathrm{i}} = 0 \tag{5.119}$$

at the image plane $z = z_\mathrm{i}$. In this case, the centre of the object plane is perfectly imaged into the centre of the image plane. To guarantee that all adjacent object points are also imaged ideally, the magnification

$$M = \frac{w_\mathrm{i}}{w_\mathrm{o}} \tag{5.120}$$

must be a constant $M = M_0$. It follows from the relation (5.118) that this requirement can be achieved only if

$$\frac{\partial V^{(1,0)}}{\partial \bar{p}_\mathrm{i}} = 0 , \quad \frac{\partial V^{(1,0)}}{\partial p_\mathrm{i}} = -M = -M_0 . \tag{5.121}$$

Hence the eikonal coefficient $V^{(1,0)}(\bar{p}_\mathrm{i}, p_\mathrm{i})$ of an aplanatic system must have the form

$$V^{(1,0)} = -M_0 p_\mathrm{i} . \tag{5.122}$$

The magnification

$$M_0 = |M_0| e^{i\chi_\mathrm{i}} \tag{5.123}$$

may be complex in the presence of a magnetic field, which rotates the image by the angle χ_i with respect to the object. By inserting the expression (5.122) into the equation (5.117), the condition for aplanatism is given by the simple formula

$$p_{0\mathrm{o}}/p_\mathrm{i} = M = M_0 . \tag{5.124}$$

This expression adopts a more familiar form if we take the absolute value and consider the relations

$$|p_{0\mathrm{o}}| = \sqrt{2em_0\Phi_\mathrm{o}^*}\,\sin\theta_\mathrm{o} \quad , \quad |p_i| = \sqrt{2em_0\Phi_\mathrm{i}^*}\,\sin\theta_\mathrm{i} , \tag{5.125}$$

where $\Phi^* = \Phi^*(z)$ denotes the relativistically modified electric potential along the optic axis. The lateral component of the vector potential vanishes along the axis, according to the gauge (5.111). Since the components $p_{0o} = p_o(w_o = 0)$ and $p_i = p_i(w_i = 0)$ have been taken at the centre of the object and the image plane, respectively, the expressions (5.125) do not contain the magnetic vector potential. Accordingly, the condition (5.124) may be replaced by the requirement

$$\sqrt{\frac{\Phi_o^*}{\Phi_i^*}} \; \frac{\sin\theta_o}{\sin\theta_i} = |M| = |M_o| \,, \tag{5.126}$$

which is the electron optical analogue of the Abbe sine condition in light optics. Within the frame of validity of Gaussian dioptrics the slope angles θ_o and θ_i are small. In this case $\sin\theta_o$ and $\sin\theta_i$ may be replaced by θ_o and θ_i, respectively. In this case the sine condition (5.126) reduces to the well-known Helmholtz–Lagrange relation, which is always true for any two linearly independent paraxial trajectories. To guarantee that all points of an extended object are imaged perfectly into the image plane, it does not suffice to fulfil the sine condition (5.126). In addition, the second and higher-order terms in the expansion (5.115) must also be eliminated or sufficiently well suppressed. The second-order terms account for the field astigmatism and the image curvature. It follows from these considerations that the mixed eikonal of an ideally imaging electron-optical system must have the simple form

$$V_{id} = -\mathrm{Re}(M_o \bar{w}_o p_i) + \mathrm{const}\,. \tag{5.127}$$

at the image plane z_i. To my knowledge this simple result has not yet been stated in the electron-optical literature.

5.2.2 Axial Aberrations

The term $V^{(0,0)}$ accounts for the total axial aberration of the system. Hence the absence of this term implies that the axial aberration vanishes in all orders. Owing to the Scherzer theorem this cannot be achieved with rotationally symmetric electron lenses because their third-order spherical aberration is unavoidable. Therefore, such systems can never have two axial points that are perfect stigmatic images of each other and form a pair of conjugate aplanatic points.

Fortunately, the maximum angular divergence of the trajectories in an electron microscope is much smaller than that in a light microscope. It hence suffices to eliminate or sufficiently reduce the axial aberration up to terms of fifth order. To correct the third-order spherical aberration, it is necessary to abandon rotational symmetry by incorporating multipole elements such as quadrupoles, hexapoles, and octopoles. Unfortunately, the resulting departure from rotational symmetry increases the number of aberration terms. As a result, the correction of all terms up to the fifth-order can only be achieved at the expense of a large number of additional elements. To keep the resulting increase in expenditure as small as possible, much effort has been put into finding simple quasi-aplanatic electron-optical systems that can be a part of a commercial instrument operating on a routine basis.

To obtain a familiar representation for the total axial aberration of a system, we expand the eikonal term $V^{(0,0)} = V(z_o, 0, 0; z_i, p_i, \bar{p}_i)$ in a power series with respect to the complex angular variables

$$\omega = M_0 p_i (2m_0 e \Phi_o^*)^{-1/2} \tag{5.128}$$

and $\bar{\omega}$, instead of p_i and $\bar{p}_i$. The resulting expansion may be written as

$$V^{(0,0)} = \sqrt{2em_0\Phi_o^*} \sum_{\nu=1}^{\infty} \sum_{\mu=0}^{[\nu/2]} \mathrm{Re}\left[V_{\nu\mu}^{(0)} \bar{\omega}^{\nu-\mu} \omega^{\mu}\right] . \tag{5.129}$$

The upper summation limit $[\nu/2]$ denotes the integral value of $\nu/2$. The coefficients $V_{\nu\mu}^{(0)}$ with $\nu \neq \mu$ are generally complex-valued. The spherical aberration coefficients with $\nu = 2\mu$ are always real-valued. The representation of the aberration by means of the components of the eikonal expansion (5.129) has the advantage that it immediately reveals all relations between the aberration coefficients. Within the frame of validity of the paraxial approximation the absolute value $|\omega|$ represents the starting angle θ_o of an axial ray at the centre of the object plane.

The total axial aberration at the recording plane may be written in the form of a power series

$$\Delta w_a = M_0 \sum_{n=1}^{\infty} \sum_{s=0}^{n} C_n^{(m)} \bar{\omega}^{n-s} \omega^s , \tag{5.130}$$

where

$$m = 2s - n - 1 \tag{5.131}$$

defines the multiplicity of the aberration term with Seidel order n. The multiplicity, which may be positive or negative, characterizes the azimuthal symmetry of the defocused geometrical aberration figure. For rotationally symmetric systems all coefficients with $m \neq 0$ vanish. It follows from (5.131) that in this case all aberrations with even order $n = 2s$ are zero. However, these aberrations generally do show up in a real microscope, owing to misalignment, small deviations of the lenses from rotational symmetry and inhomogeneities within the magnetic material. According to the Scherzer theorem the coefficient C_3 is always positive definite for round electron lenses [10]. In order to eliminate this coefficient, multipole elements must be introduced, which produce additional aberration terms with multiplicity $m \neq 0$. Fortunately, the accompanying increase in the number of aberrations is limited because linear relations exist between some of the aberration coefficients. These relations are obtained by inserting the expansion (5.129) of the eikonal term $V^{(0,0)}$ into

$$\Delta w_{\mathrm{a}} = -2\frac{\partial V^{(0,0)}}{\partial \bar{p}_{\mathrm{i}}} = -2\frac{M_0}{\sqrt{2m_0 e\Phi_{\mathrm{o}}^*}}\frac{\partial V^{(0,0)}}{\partial \bar{\omega}} , \tag{5.132}$$

which yields the power series

$$\Delta w_{\mathrm{a}} = -M_0 \sum_{n=0}^{\infty} \sum_{s=0}^{\left[\frac{n+1}{2}\right]} \left[(n+1-s)V_{n+1,s}^{(0)}\bar{\omega}^{n-s}\omega^s + s\bar{V}_{n+1,s}^{(0)}\omega^{n-s}\bar{\omega}^s\right] . \tag{5.133}$$

The comparison of this series with the equivalent representation (5.130) yields the following relation between the aberration coefficients and the coefficients of the eikonal (5.129):

$$C_n^{(-m)} = \frac{n+m+1}{n-m+1}\bar{C}_n^{(m)} = s(1+\delta_{0m})\bar{V}_{n+1,s}^{(0)} . \tag{5.134}$$

According to the definition (5.131) the multiplicity m depends linearly on both the order n and the summation index s. The coefficients

$$C_n^{(0)} = \bar{C}_n^{(0)} = C_n = (n+1)V_{n+1,(n+1)/2}^{(0)} \tag{5.135}$$

with multiplicity $m = 0$ and odd Seidel order $n = 2s+1$, for $s = 1, 2, \ldots$, are real quantities and determine the magnitude of the spherical component of the nth-order axial aberration.

5.2.3 Generalized Coma

If the incident illuminating beam is tilted by a relatively large angle and/or in the case of an extended field of view and wide-angle apertures, the second term of the eikonal expansion (5.115) accounts for the next important aberration after the axial aberration. This aberration, which comprises all linear terms in w_{o} and $\bar{w}_{\mathrm{o}}$, represents the generalized off-axial coma. In a system with an axis of rotational symmetry the coma of any order has a bead-like shape, whose tip is centred at the Gaussian image point. Since the coma of rotationally symmetric lenses is formed by a superposition of displaced circles, we denote this component of the generalized coma as "circular coma", in accordance with the terminology of light optics [11]. The border lines of the comet-shaped figure of the nth-order circular coma intersect each other at the Gaussian image point at an angle

$$\vartheta_n = 2\arcsin\left(\frac{n-1}{n+1}\right) . \tag{5.136}$$

For the lowest order, $n = 3$, we obtain the well-known result $\vartheta_3 = 60^0$.

Additional coma terms arise in systems with multipole elements. The most important non-circular coma components are usually characterized by particular names, according to the characteristic shape of their aberration figures. To obtain the Seidel representation of the total coma, we expand the complex eikonal term $V^{(1,0)}$ in a power series with respect to the variables ω and $\bar{\omega}$ defined in (5.128). If we insert the resulting expansion

$$V^{(1,0)} = \sqrt{2em_0\Phi_o^*} \sum_{n=1}^{\infty} \sum_{s=0}^{n} V_{ns}^{(1)} \bar{\omega}^{n-s} \omega^s \tag{5.137}$$

into (5.118), we eventually obtain for the total generalized coma the expression

$$\begin{aligned} \Delta w_{\mathrm{c}} &= w_1 - M_0 w_{\mathrm{o}} \\ &= -M_0 \sum_{n=2}^{\infty} \sum_{s=0}^{n} \left[(n-s) V_{ns}^{(1)} \bar{w}_{\mathrm{o}} \omega^{n-s-1} \omega^s + s \bar{V}_{ns}^{(1)} w_{\mathrm{o}} \bar{\omega}^{s-1} \omega^{n-s} \right] . \end{aligned} \tag{5.138}$$

To derive this expression we have assumed a distortion-free Gaussian image, which requires

$$V_{10}^{(1)} = 0 , \quad V_{11}^{(1)} = -1 . \tag{5.139}$$

The coefficient $V_{10}^{(1)}$ determines the first-order distortion, while $V_{11}^{(1)} = -1$ accounts for the paraxial magnification referred back to the object plane.

The generalized coma (5.139) can be considered as the part of the magnification that depends on the divergence of the pencils of rays forming the individual image points. If all points are imaged with the same magnification regardless of the divergence angles ω and $\bar{\omega}$, the total coma must vanish. In this case all eikonal coefficients $V_{ns}^{(1)}$ must be zero. The coma terms of even order n are caused by misalignment in systems whose fields are symmetric with respect to both the xz-section and the yz-section. Hence these fields must not contain any multipole components with odd multiplicity. For systems with an axis of rotational symmetry, only terms with $n = 2s+1, s = 0, 1, 2, \ldots$ are present in the expression (5.139) for the coma. The term with $s = 1$ represents the well-known third-order circular coma. The corresponding coefficient $V_{31}^{(1)}$ is complex in the presence of a magnetic field, the real part describes the isotropic or radial component, while the imaginary part accounts for the anisotropic or azimuthal component of the coma. This component causes the axis of the comet-shaped aberration figure to spiral about the Gaussian image point. If the azimuthal component is absent, the axis of the coma figure points in the radial direction, as known from glass lenses.

Second-order terms of the generalized coma arise in systems with a curved axis or in the presence of hexapole fields. The second-order contribution

$$\Delta w_{\mathrm{c}}^{(2)} = -2M_0 \omega \,\mathrm{Re}(V_{21}^{(1)} \bar{w}_{\mathrm{o}}) - 2M_0 \bar{\omega} (V_{20}^{(1)} \bar{w}_{\mathrm{o}} + \bar{V}_{22}^{(1)} w_{\mathrm{o}}) \tag{5.140}$$

consists of two parts, the inclination of the image field and the second-order field astigmatism. The latter aberration is determined by the two complex coefficients $V_{20}^{(1)}$, $V_{22}^{(1)}$ and by the lateral position w_o of the object point. The inclination of the image field can be considered as a defocus, which depends linearly on the off-axial coordinates of the object point. Accordingly, in the absence of the field astigmatism, a sharp image is obtained in a plane tilted with respect to the image plane by the angle

$$\tau = \arctan(2\,|V_{21}^{(1)}|)\,. \tag{5.141}$$

The inclination of the image field and the field astigmatism broaden the image points. Both aberrations form discs in the Gaussian image plane, the diameters of which increase linearly with the off-axial distance of the object points. These aberrations may strongly limit the usable field of view. Accordingly, they must be eliminated in systems containing appreciable hexapole fields such as correctors compensating for the spherical aberration of electron microscopes.

5.3 Perturbation Formalism

The components of the eikonal can be determined most appropriately by applying a perturbation technique, based on that used in celestial mechanics for calculating orbital elements. This method was first employed by Schwarzschild [12] for the calculation of the primary aberrations of glass lenses and later introduced into charged-particle optics by Glaser [13] and Sturrock [14]. Subsequently, the method was reformulated to obtain a systematic procedure for handling the higher-order aberrations of electron optical systems [15]. These aberrations must be known in order to determine the properties of lenses whose primary aberrations have been eliminated by correctors [16,17].

The eikonal approach yields great insight into the structure of the aberration coefficients because the latter are directly related to the coefficients of the eikonal regardless of their order. Different integral expressions exist for the higher-order eikonal coefficients. The number of representations increases with the order n. The integrand of the $(n+1)$th-order eikonal term generally contains path deviations up to the order $n-1$. To minimize the labour of calculation and to find the best way of eliminating the aberrations, it is, therefore, very desirable to find the representation for the eikonal terms that involves path deviations with the lowest Seidel order. Thanks to this choice of representations, the eikonal approach offers an elegant and straightforward procedure for elucidating the structure of the higher-order aberration coefficients.

Different eikonals exist, depending on the nature of their four arguments. Two of these variables belong to the object plane and two to the final plane, which need not necessarily coincide with the image plane. Two of these variables define either the lateral position coordinates or the off-axial components of the canonical momentum of the particle at one of the terminal planes. Hence the problem of finding the position and the slope of a ray at some given plane cannot be solved directly if the ray is entirely defined by the corresponding quantities at the initial plane. Owing to this difficulty it has often been argued that the eikonal method is of little use for determining the higher-order aberrations in the case of arbitrary ray-defining parameters. Fortunately, this point of view is unduly pessimistic. To prove this assertion we start from the point eikonal

$$S(\boldsymbol{r},\boldsymbol{r}_0) = \mathrm{Extr}\int_{z_0}^{z} \mu dz\,, \tag{5.142}$$

where the function

$$\mu = \boldsymbol{p}\cdot\frac{d\boldsymbol{s}}{dz} = \sqrt{2em_0\varphi*}\frac{ds}{dz} - e\boldsymbol{A}\cdot\frac{d\boldsymbol{s}}{dz} \tag{5.143}$$

represents the so-called variational function. The differential $d\boldsymbol{s} = \boldsymbol{v}dt$ represents the infinitesimal path element along the trajectory that intersects the initial point $\boldsymbol{r}_0$ and the end-point $\boldsymbol{r}$.

The factor

$$q = \sqrt{2em_0\varphi*} \tag{5.144}$$

on the right hand side of (5.143) is the kinetic momentum, where

$$\varphi* = \varphi(1+e\varphi/2m_0c^2) \tag{5.145}$$

represents the relativistically modified electrostatic potential $\varphi = \varphi(x,y,z)$. The symbol Extr in front of the integral denotes the extremum with respect to the variation of the path. It corresponds to Fermat's principle of stationary action $\delta S = 0$. Accordingly, among all paths connecting the points $\boldsymbol{r}_0$ and $\boldsymbol{r}$, the true path of the electron makes the integral an extremum.

As a consequence, the path equations can be derived from (5.142) by employing the calculus of variations resulting in the Euler–Lagrange equations

$$\frac{\partial\mu}{\partial x} - \frac{d}{dz}\left(\frac{\partial\mu}{\partial x'}\right) = 0\,, \quad \frac{\partial\mu}{\partial y} - \frac{d}{dz}\left(\frac{\partial\mu}{\partial y'}\right) = 0\,, \tag{5.146}$$

which may be combined to form a single, complex equation

$$\frac{\partial\mu}{\partial\bar{w}} - \frac{d}{dz}\left(\frac{\partial\mu}{\partial\bar{w}'}\right) = 0 \tag{5.147}$$

by employing the complex variables (5.112). If the point eikonal S is known, we can construct from it the mixed eikonal by means of a proper Legendre transformation, which replaces one set of off-axial coordinates by the corresponding canonical-momentum coordinates of the particle [18]. The eikonals can be considered as optical potentials because the conjugate variables at the terminal planes are simply obtained by partial differentiation with respect to the variables of the eikonal. Here we consider the position coordinates and the corresponding components of the canonical momentum as conjugate to each other.

5.3.1 Gaussian Optics

The paraxial trajectories represent the rays of an ideal optical system. Unfortunately, the rays never follow these paths in a real system owing to the unavoidable nonlinear terms in the differential equations (5.146). However, it may be possible to eliminate the deviations of the true path from its paraxial approximation at a distinct plane by properly adjusting the distribution of the electromagnetic field in the space between this plane and the initial plane. The problem of determining the optimum field distribution is extremely complicated and has not yet fully been solved. Without some insight into the properties of the path deviations it is almost impossible to find a suitable method of correcting the resolution-limiting aberrations.

The electromagnetic fields form an anisotropic inhomogeneous refracting medium for charged particles. Hence, in the terminology of light optics, electron lenses are gradient-index lenses, which become anisotropic in the presence of a magnetic field. In most problems of charged-particle optics one considers the propagation of beams, the constituent rays of which are confined to the neighborhood of the central trajectory, which may either be straight or curved. This trajectory is generally chosen as the optic axis of the system and forms the z axis of the coordinate system. In this case the second term $\mu^{(1)}$ vanishes in the expansion

$$\mu = \sum_{n=0}^{\infty} \mu^{(n)} \tag{5.148}$$

of the variational function (5.143) in a power series with respect to w, $\bar{w}$, w', and $\bar{w}'$. Each term $\mu^{(n)}$ represents a homogeneous polynomial of degree n in the coordinates and slopes of the ray. Here $w = w(z)$ defines the lateral position of the particle as a function of the independent variable z. Primes denote differentiations with respect to this variable. The monomial $\mu^{(0)}$ can be ignored because it does not contain any off-axial ray components and hence does not contribute to the Euler–Lagrange equations (5.146). The second-degree polynomial $\mu^{(2)}$ contains all quadratic and bilinear terms in the expansion of the variational function and completely determines the Gaussian trajectories

$$w^{(1)} = \sum_{\nu=1}^{4} a_\nu w_\nu \quad , \quad a_\nu = \bar{a}_\nu \, . \tag{5.149}$$

The four fundamental rays w_ν are particular, linearly independent solutions of the Gaussian path equation

$$2\frac{\partial\mu_1^{(2)}}{\partial\bar{w}^{(1)}} - \frac{dp^{(1)}}{dz} = 0\,. \tag{5.150}$$

The subscript 1 indicates that

$$\mu_1^{(2)} = \mu^{(2)}\left(w^{(1)}, \bar{w}^{(1)}, w^{(1)'}, \bar{w}^{(1)'}; z\right) \tag{5.151}$$

must be considered as a function of the paraxial ray. The complex-valued function

$$p^{(1)} = 2\frac{\partial\mu_1^{(2)}}{\partial\bar{w}^{(1)'}} = \frac{d}{dz}(q_0 w^{(1)'}) - \frac{i}{2}eBw^{(1)} \tag{5.152}$$

defines the paraxial lateral component of the canonical momentum, where

$$q_0 = q(x=0, y=0, z) = \sqrt{2em_0\Phi^*} \tag{5.153}$$

and

$$B = B_z(x=0, y=0, z) \tag{5.154}$$

describe the axial components of the kinetic momentum and the magnetic flux density $\boldsymbol{B}$, respectively, along the optic axis. Since $p^{(1)}$ depends linearly on $w^{(1)}$ and $w^{(1)'}$, it can also be written as the linear combination

$$p^{(1)} = \sum_{\nu=1}^{4} a_\nu p_\nu \tag{5.155}$$

of the canonical momenta p_ν of the fundamental rays w_ν. The constants $a_\nu = \bar{a}_\nu$ are real and defined by the boundary condition imposed on the ray. The fundamental rays $w_\nu = x_\nu + iy_\nu$ are complex if the two paraxial path equations cannot be decoupled by forming a linear combination of the off-axial variables x and y. In the absence of dipole and quadrupole fields, the path equation can be decoupled by introducing the coordinate system

$$u = we^{-i\chi}\,, \tag{5.156}$$

which rotates with the angle

$$\chi = \chi(z) = \frac{e}{2}\int_{z_0}^{z}\frac{B}{q_0}dz \tag{5.157}$$

about the optic axis. The angle 2χ represents the angle of the well-known Larmor rotation of the charged particles. Owing to the rotational symmetry in the paraxial region, only two of the four fundamental rays w_ν are linearly independent. The paraxial approximation of the Lagrange bracket [18], defined as

$$
\begin{aligned}
[a_\mu, a_\nu]_1 &= \operatorname{Re}\left\{ \frac{\partial \bar{w}^{(1)}}{\partial a_\mu} \frac{\partial p^{(1)}}{\partial a_\nu} - \frac{\partial \bar{w}^{(1)}}{\partial a_\nu} \frac{\partial p^{(1)}}{\partial a_\mu} \right\} \\
&= \operatorname{Re}\left(\bar{w}_\mu p_\nu - \bar{w}_\nu p_\mu\right) = q_0 \operatorname{Re}\left(\bar{u}_\mu u'_\nu - \bar{u}_\nu u'_\nu\right) \\
&= q_{0\text{o}} C_{\nu\mu} = -q_{0\text{o}} C_{\mu\nu} = -\left[a_\nu, a_\mu\right]_1 , \quad q_{0\text{o}} = q_0(z_\text{o}) ,
\end{aligned} \tag{5.158}
$$

is identical with the Helmholtz–Lagrange relation, which coincides with the Wronskian of the paraxial trajectories. The functions u_ν denote the fundamental paraxial rays in the rotating uz-coordinate system. In the case of four linearly independent fundamental rays $w_\nu = u_\nu \exp(i\chi)$, six constants of motion, $C_{\mu\nu} = -C_{\nu\mu}$, exist. The value of each constant depends on the boundary conditions imposed on the fundamental rays. It is advantageous to define these rays in such a way that the constants of motion adopt the special values

$$
C_{31} = 1 , \quad C_{42} = 1 , \tag{5.159}
$$

$$
C_{21} = C_{41} = C_{32} = C_{43} = 0 . \tag{5.160}
$$

The relations (5.160) are automatically fulfilled if the paraxial path equations are decoupled, for example in the case of quadrupole systems with plane midsection symmetry.

5.3.2 Path and Momentum Deviations

The geometrical aberrations are generally classified according to their Seidel order n. The calculation of these aberrations requires a knowledge of the lower-order path deviations

$$
w^{(m)} = w^{(m)}(z) = x^{(m)}(z) + iy^{(m)}(z) , \tag{5.161}
$$

with $m < n$ at an arbitrary current plane z between the object plane and the plane of observation. Our aim is therefore to find a systematic iteration procedure that yields successively the nth-order path deviations $w^{(n)}$ of the power series expansion

$$
w - w^{(1)} = \sum_{n=2}^{\infty} w^{(n)} , \tag{5.162}
$$

of the total deviation of the true ray w from its paraxial approximation $w^{(1)}$. Each term $w^{(n)}$ is a polynomial of order n in the expansion parameters a_ν, $\nu = 1, 2, 3, 4$, of the complex true lateral position $w = w(a_\nu; z_0, z)$ of the electron at an arbitrary plane z. The four ray parameters a_ν do not necessarily coincide with the off-axis position and/or momentum coordinates of the true ray at some given planes. In many cases it is advantageous to replace the real parameters a_ν by the complex ray parameters

$$\Omega_1 = a_1 + i a_2 \quad , \quad \Omega_2 = a_3 + i a_4 \, . \tag{5.163}$$

Using these parameters, the nth-order path deviation has the form

$$w^{(n)} = \sum_{\alpha\beta\gamma\delta} w^{(n)}_{\alpha\beta\gamma\delta}(z) \Omega_1^\alpha \bar{\Omega}_1^\beta \Omega_2^\gamma \bar{\Omega}_2^\delta \tag{5.164}$$

with the constraint

$$n = \alpha + \beta + \gamma + \delta \, . \tag{5.165}$$

The coefficient $w^{(n)}_{\alpha\beta\gamma\delta}$ of each monomial is a function of the "current" plane z and the planes z_ν at which the position and/or the slope coordinates of the ray are defined.

The coefficients $w^{(n)}_{\alpha\beta\gamma\delta}(z)$ determine the course of the nth-order path deviation along the optic axis. Their values at the observation plane, typically the final image plane z_i, determine the aberration coefficients of the nth-order aberration $w^{(n)}(z_\mathrm{i})$.

Ideal imaging is obtained if all path deviations with $n \geq 2$ vanish at the plane of observation. In this case the difference

$$\Delta S = S - S^{(2)} = \int_{z_\mathrm{o}}^{z_\mathrm{i}} \mu \, dz - \int_{z_\mathrm{o}}^{z_\mathrm{i}} \mu_1^{(2)} dz \tag{5.166}$$

between the eikonal S of the exact ray and the eikonal $S^{(2)}$ of the paraxial ray must vanish for all trajectories connecting the conjugate points $\boldsymbol{r}_\mathrm{o}$ and $\boldsymbol{r}_\mathrm{i}$. The use of the eikonal enables one to construct a systematic iteration algorithm for the calculation of the path deviations according to their order. To obtain an efficient and transparent algorithm, we introduce the perturbation eikonal

$$E_\nu = \int_{z_\nu}^{z} (\mu - \mu_1^{(2)}) dz - \mathrm{Re} \left[\left(\bar{w} - \bar{w}^{(1)} \right) p^{(1)} \right]_{z_\nu}^{z} = \int_{z_\nu}^{z} \mu_E dz \, . \tag{5.167}$$

The subtraction of the term that depends only on the ray coordinates at the terminal planes from the eikonal S does not affect the path equation because the terminal points are kept fixed when S is varied. Subtracting this term from the eikonal S is equivalent to a subtraction of the total differential

$$\frac{d}{dz} \mathrm{Re} \left[(\bar{w} - \bar{w}^{(1)}) p^{(1)} \right] = 2 \, \mathrm{Re} \left[(\bar{w} - \bar{w}^{(1)}) \frac{\partial \mu_1^{(2)}}{\partial \bar{w}^{(1)}} + (\bar{w}' - \bar{w}^{(1)\prime}) \frac{\partial \mu_1^{(2)}}{\partial \bar{w}^{(1)\prime}} \right] \tag{5.168}$$

from the integrand μ of the eikonal S. The expression on the right-hand side has been obtained by employing the relation (5.152) for the canonical momentum $p^{(1)}$ together with the paraxial path equation (5.150). The addition or subtraction of a total differential from the variational function μ corresponds to a different choice for the gauge of the eikonal. Although the eikonal is changed by this transformation, the path equation and hence the course of the electron trajectory remain unaffected. If we take into account equation (5.168) for the total differential, the integrand of the perturbation eikonal adopts the form

$$\mu_E = \mu - \mu_1^{(2)} - \mathrm{Re}\left[(\bar{w} - \bar{w}^{(1)})\frac{\partial \mu_1^{(2)}}{\partial \bar{w}^{(1)}} + (\bar{w}' - \bar{w}^{(1)\prime})\frac{\partial \mu_1^{(2)}}{\partial \bar{w}^{(1)\prime}}\right] . \tag{5.169}$$

The method of successive approximation works most effectively for integral equations such as the Lippmann–Schwinger equation for electron scattering. The iteration of this equation yields the Born series for the scattering amplitude in a straightforward way. To develop a similar approach for the calculation of the path deviations, we transform the differential path equations (5.146) into a set of integral equations. This transformation has the additional advantage that it includes the boundary conditions imposed on the ray. Hence the solution of the integral equation defines a distinct ray.

The integral equations are obtained most conveniently by varying the perturbation eikonal (5.167) with respect to the off-axis position and momentum coordinates at the terminal planes z and z_ν. The result can be written as

$$\begin{aligned} \delta E_\nu &= \mathrm{Re}\left[p\delta\bar{w} - \bar{w}\delta p^{(1)} - p^{(1)}\delta\bar{w} + \bar{w}^{(1)}\delta p^{(1)}\right]_{z_\nu}^{z} \qquad (5.170)\\ &= \mathrm{Re}\left[(p - p^{(1)})\delta\bar{w}^{(1)} - (\bar{w} - \bar{w}^{(1)})\delta p^{(1)} + (p - p^{(1)})\delta(\bar{w} - \bar{w}^{(1)})\right]_{z_\nu}^{z} . \end{aligned}$$

The perturbation eikonal E_ν, the off-axis canonical momentum p and the lateral distance w of the particle are functions of the ray-defining parameters a_ν. Therefore, we can also vary the eikonal E_ν with respect to the parameters a_ν. By considering the relations

$$w_\nu = \frac{\partial w^{(1)}}{\partial a_\nu} , \quad p_\nu = \frac{\partial p^{(1)}}{\partial a_\nu} , \tag{5.171}$$

which follow from the expressions (5.149) and (5.155), we obtain the following set of four integral equations:

$$\begin{aligned} &\mathrm{Re}\left[(w - w^{(1)})\bar{p}_\nu - (p - p^{(1)})\bar{w}_\nu\right]_{z_\nu}^{z} \\ &= -\frac{\partial E_\nu}{\partial a_\nu} + \mathrm{Re}\left[(p - p^{(1)})\frac{\partial(\bar{w} - \bar{w}^{(1)})}{\partial a_\nu}\right]_{z_\nu}^{z} , \quad \nu = 1, 2, 3, 4 . \qquad (5.172) \end{aligned}$$

The two terms in the bracket on the left-hand side are linearly related to the difference between the off-axis positions and the lateral canonical momentum, respectively, of the true ray and the paraxial ray. The second term on the right hand side is bilinear in these deviations and hence does not contribute to the primary aberrations obtained in the first iteration step. In most cases the ray is defined by its lateral position and/or momentum coordinates at distinct planes, for example the object plane $z = z_o$ and the aperture plane $z = z_a$. The number of eikonals required is identical with the number of planes at which distinct components of the ray are defined.

The most appropriate ray parameters a_ν are obtained by imposing the condition

$$\mathrm{Re}\left[(w - w^{(1)})\bar{p}_\nu - (p - p^{(1)})\bar{w}_\nu\right]_{z=z_\nu} = 0\,. \tag{5.173}$$

Inserting (5.149) and (5.155) for $w^{(1)}$ and $p^{(1)}$, respectively, and choosing the fundamental rays in such a way that the constants $C_{\mu\nu}$ of the Lagrange–Helmholtz relations (5.159) adopt the special values (5.159) and (5.160), we eventually find

$$a_\mu q_{0o} C_{\mu\nu} = \mathrm{Re}\left[w\bar{p}_\nu - p\bar{w}_\nu\right]_{z_\nu} \;;\quad \mu, \nu = 1, 2, 3, 4\,. \tag{5.174}$$

The proper fixing of the course of the paraxial fundamental ray w_ν depends on the boundary condition that is imposed on the true ray at the plane $z = z_\nu$. If this ray is defined by its lateral position at this plane, we must choose the condition

$$w_\nu(z_\nu) = 0\,. \tag{5.175}$$

On the other hand, we must impose the condition

$$p_\nu(z_\nu) = 0 \tag{5.176}$$

if the true ray is determined by the lateral component of the canonical momentum at the plane z_ν. The deviation between the true ray and its paraxial approximation strongly depends on the boundary conditions imposed on the ray, as illustrated in Fig. 5.3.

In order to solve the integral equations (5.172) by means of an iteration algorithm, we must first solve these equations with respect to the position and momentum deviations $w - w^{(1)}$ and $p - p^{(1)}$, respectively. Fortunately, the four equations are linearly related with respect to these deviations and their conjugate complex values. We can therefore solve these equations with respect to the deviations $w - w^{(1)}, \bar{w} - \bar{w}^{(1)}, p - p^{(1)}$ and $\bar{p} - \bar{p}^{(1)}$. The solution is obtained relatively easily by making use of the fundamental Lagrange brackets (5.159), which can be combined to form the expressions

$$\begin{aligned} &\sum(-)^P C_{\kappa\lambda} w_\mu p_\nu = 0\,, \sum(-)^P C_{\kappa\lambda} p_\mu p_\nu = 0\,, \\ &\sum(-)^P C_{\kappa\lambda} w_\mu \bar{w}_\nu = 0\,, \sum(-)^P C_{\kappa\lambda} p_\mu \bar{p}_\nu = 0\,, \\ &\sum(-)^P C_{\kappa\lambda} w_\mu w_\nu = 0\,, \sum(-)^P C_{\kappa\lambda} w_\mu \bar{p}_\nu = D_0\,. \end{aligned} \tag{5.177}$$

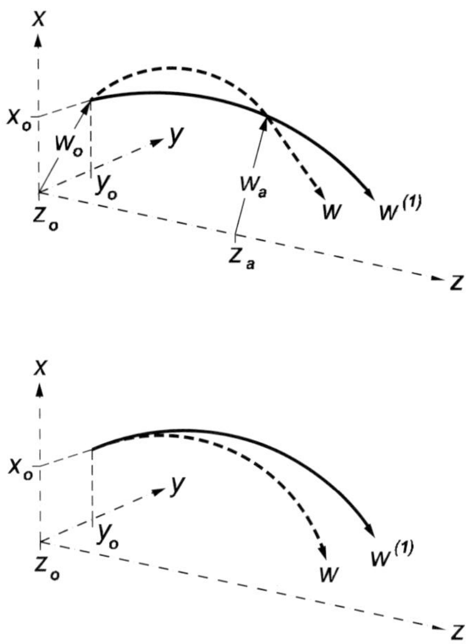

Fig. 5.3. Influence of the constraints on the deviation of the paraxial ray $w^{(1)}$ from the exact trajectory w

Here D_0 is the fourth-order Wronskian of the paraxial fundamental rays that can also be written as

$$D_0 = \frac{1}{2}\sum(-)^P C_{\kappa\lambda}C_{\mu\nu} = 4(C_{12}C_{34}-C_{13}C_{24}+C_{14}C_{23})\,. \tag{5.178}$$

The summation has to be taken over all 24 permutations of the four different indices κ, λ, μ and ν; P denotes the number of permutations. We multiply the equation (5.172) with the factors $(-)^P C_{\kappa\lambda}w_\mu$ and $(-)^P C_{\kappa\lambda}p_\mu$, respectively, and subsequently sum over all permutations. Considering further the relations (5.175) and (5.176), we obtain the deviations in the form

$$w-w^{(1)} = -\frac{2}{D_0}\sum(-)^P C_{\kappa\lambda}w_\mu G_\nu\,, \tag{5.179}$$

$$p-p^{(1)} = -\frac{2}{D_0}\sum(-)^P C_{\kappa\lambda}p_\mu G_\nu\,, \tag{5.180}$$

where

$$G_\nu q_{0o} = \frac{\partial E_\nu}{\partial a_\nu} - \mathrm{Re}\left[(\bar{p}-\bar{p}^{(1)})\frac{\partial(w-w^{(1)})}{\partial a_\nu}\right]\,. \tag{5.181}$$

The sums on the right-hand side of the expressions (5.178) reduce considerably if we fix the paraxial fundamental rays in such a way that the Lagrange–Helmholtz constants $C_{\kappa\lambda}$ adopt the values listed in (5.159) and (5.160). With these constants we obtain $D_0 = -4$, and the expressions (5.179) and (5.180) for the deviations reduce to the simple form

$$w - w^{(1)} = w_1 G_3 - w_3 G_1 + w_2 G_4 - w_4 G_3 \,, \tag{5.182}$$

$$p - p^{(1)} = p_1 G_3 - p_3 G_1 + p_2 G_4 - p_4 G_2 \,. \tag{5.183}$$

The integrand (5.169) of the eikonal E_ν is a function of $w, \bar{w}, w'$ and $\bar{w}'$. Considering the relation (5.181), it becomes obvious that the equation (5.182) represents an inhomogeneous integral equation for the lateral position w of the true ray. The paraxial approximation $w^{(1)}$ describes the inhomogeneous term. It is known from the theory of integral equations that inhomogeneous integral equations can be solved by the method of successive approximation starting with the inhomogeneous term as the zeroth-order solution. Hence the equation (5.182) represents an excellent alternative to the path equation (5.146) because the former equation can be solved approximately by means of a perturbation procedure. Moreover, this equation contains the boundary conditions and hence defines a distinct ray.

5.3.3 Iteration Algorithm

In order to solve the inhomogeneous integral equation (5.182) by iteration, we suppose that the paraxial ray $w^{(1)}$ is known. This ray represents the inhomogeneous term. Since the geometrical ray parameters a_ν are small quantities for most beam-guiding systems, the method of successive approximation yields a converging power series expansion (5.162) for the complex lateral position $w = w(z)$ of the particle. To obtain an expansion of the integrand (5.169) of the perturbation eikonal, we insert the series (5.162) for the position coordinate into the variational function (5.143) of the eikonal S and expand the resulting expression in a power series with respect to the ray parameters. It is useful to introduce a fictitious expansion parameter ε in order to separate the path deviations according to their order. Later this parameter will be set equal to unity. The ordering with respect to equal powers of this parameter yields the Seidel-order expansion

$$\mu = \sum_{k=2}^{\infty} \mu^{(k)} = \sum_{n=1}^{\infty} \varepsilon^{n+1} m^{(n+1)} \,. \tag{5.184}$$

Here $m^{(n+1)} = m^{(n+1)}(z)$ denotes the variational polynomial of degree $n+1$ in the parameters a_ν. The expansion parameter ε can be considered as a substitute for these parameters. The coefficients of each polynomial are functions of z. If we insert the expansions (5.162) and (5.184) into the expression (5.169), we find that the polynomials of the expansion

$$\mu_E = \sum_{n=1}^{\infty} m_E^{(n+1)} \tag{5.185}$$

of the integrand of the perturbation eikonal are connected with the variational polynomials $m^{(n+1)}$ via the relation

$$m_E^{(n+1)} = m^{(n+1)} - D^{(n)} \mu_1^{(2)} . \tag{5.186}$$

The operator

$$D^{(n)} = 2\,\mathrm{Re}\left[w^{(n)} \frac{\partial}{\partial w^{(1)}} + w^{(n)\prime} \frac{\partial}{\partial w^{(1)\prime}} \right] \tag{5.187}$$

replaces one of each of the four paraxial ray components $w^{(1)}$, $\bar{w}^{(1)}$, $w^{(1)\prime}$ and $\bar{w}^{(1)\prime}$ in the second-order polynomial $\mu_1^{(2)}$ by the corresponding nth-path deviation. Each polynomial $m^{(n+1)}$ is composed of all polynomials $\mu^{(k)}$ with $k \le n$. The relation between these polynomials and the variational polynomial $m^{(n+1)}$ can be expressed in operator form as

$$m^{(n+1)} = \frac{1}{(n+1)!} \sum_{k=2}^{n+1} \frac{\partial^{n+1}}{\partial \varepsilon^{n+1}} \left[\varepsilon^k \exp\left(\sum_{\ell=2}^{\infty} \varepsilon^{\ell-1} D^{(\ell)} \right) \mu_1^{(k)} \right]_{\varepsilon=0} . \tag{5.188}$$

In order to evaluate this operator expression, we must expand the exponential term in a Taylor series. Only terms with the same factor ε^{n+1} contribute to the polynomial $m^{(n+1)}$. The result of the rather lengthy calculation may be written as

$$\begin{aligned} m^{(n+1)} = {} & \mu_1^{(n+1)} + \sum_{k=0}^{n-2} D^{(k+2)} \mu_1^{(n-k)} + \frac{1}{2!} \sum_{k=1}^{r-2} \sum_{h=1}^{k} D^{(k+2-h)} D^{(h+1)} \mu^{(n-k)} \\ & + \frac{1}{3!} \sum_{k=2}^{r-3} \sum_{h=1}^{k-1} \sum_{\ell=1}^{k-h} D^{(k+2-h-\ell)} D^{(h+1)} D^{(\ell+1)} \mu_1^{(n-k)} + \cdots . \end{aligned} \tag{5.189}$$

In this formula each sum must be put zero if its upper summation index is smaller than the lower index. The formula also shows that $m^{(n+1)}$ contains the paraxial trajectory $w^{(1)}$ and all path deviations $w^{(k)}$ up to the order $k = n$. Owing to the linear relation (5.162) between w and $w^{(1)}$, we can write the complex path equation as

$$\frac{\partial \mu}{\partial \bar{w}} - \frac{d}{dz}\left(\frac{\partial \mu}{\partial \bar{w}'} \right) = \sum_{n=1}^{\infty} \varepsilon^{n+1} \left[\frac{\partial m^{(n+1)}}{\partial \bar{w}^{(1)}} - \frac{d}{dz} \frac{\partial m^{(n+1)}}{\partial \bar{w}^{(1)\prime}} \right] = 0 . \tag{5.190}$$

The last equation must be satisfied for arbitrary values of ε, which can be considered as an ordering substitute for the ray parameters. Hence the expression in the bracket must vanish identically. The result

$$\frac{\partial m^{(n+1)}}{\partial \bar{w}^{(1)}} - \frac{d}{dz} \frac{\partial m^{(n+1)}}{\partial \bar{w}^{(1)\prime}} = 0 \tag{5.191}$$

reveals that the nth-order deviation of the true lateral canonical momentum from its paraxial approximation has the form

$$p^{(n)} = 2\frac{\partial m^{(n+1)}}{\partial \bar{w}^{(1)'}} \, . \tag{5.192}$$

The two relations (5.191) and (5.192) enable one to integrate terms of the form $D^{(h)}m^{(k+1)}$ because they form total differentials. This behaviour is readily proven by partial integration of the factor $w^{(h)'}$ contained in the operator (5.187). Considering further equation (5.192), we obtain

$$\int D^{(h)}m^{(k+1)}dz = 2\,\mathrm{Re}\left[w^{(h)}\bar{p}^{(k)}\right] \, . \tag{5.193}$$

This result can be used for integrating certain terms in the expression

$$E_\nu^{(n+1)} = \int_{z_\nu}^{z} m_E^{(n+1)}dz \tag{5.194}$$

for the $(n+1)$-order component of the perturbation eikonal

$$E_\nu = \sum_{n=2}^{\infty} E_\nu^{(n+1)} \, . \tag{5.195}$$

The nth-order path deviation is obtained by substituting the power series (5.162) for the left-hand side of the integral equation (5.182) and the deviation $w - w^{(1)}$ in the expression (5.181) for G_ν. In addition we insert in this formula the corresponding expansion for the deviation $p - p^{(1)}$ of the lateral canonical momentum and the series (5.195) for the perturbation eikonal E_ν. Considering further that all terms with a fixed arbitrary order must satisfy the resulting equation separately, we eventually obtain

$$w^{(n)} = w_1 G_3^{(n)} - w_3 G_1^{(n)} + w_2 G_4^{(n)} - w_4 G_2^{(n)} \, , \tag{5.196}$$

$$p^{(n)} = p_1 G_3^{(n)} - p_3 G_1^{(n)} + p_2 G_4^{(n)} - p_4 G_2^{(n)} \, , \tag{5.197}$$

where the functions $G_\nu^{(n)}$ are defined as

$$q_{0o} G_\nu^{(n)} = \frac{\partial E_\nu^{(n+1)}}{\partial a_\nu} - \sum_{r=2}^{n-1} \mathrm{Re}\left[\bar{p}^{(r)} \frac{\partial w^{(n+1-r)}}{\partial a_\nu}\right] \tag{5.198}$$

with $n \leq 2$ and $\nu = 1, 2, 3, 4$. The sum must be set to zero for $n = 2$. In order to determine the component $E_\nu^{(n+1)}$ of the perturbation eikonal, we need to know all path deviations up to the order $n - 1$. Accordingly, considerable labour is required to calculate the higher-order deviations. Fortunately, the necessary number of path deviations can be significantly reduced by utilizing the relation (5.193) appropriately.

5.3.4 Symplectic Representation

The propagation of charged particles can be viewed as a "symplectic mapping" in the five-dimensional state space with coordinates x, y, p_x, p_y and z. The subspace x, y, p_x, p_y can be considered as the projection of the state space onto an arbitrary "plane" z. This subspace is known as the four-dimensional phase space.

Each trajectory is defined by a set of four ray parameters. If we assume that these parameters are the lateral position and momentum coordinates at a distinct plane at $z = z_0$, the initial state of the particle is given by a point in both the phase space and the state space. The solutions of the path equations (5.179) and (5.180) form a bundle of stationary trajectories in this space. Each trajectory connects a point in the initial "plane" $z = z_0$ to a distinct point in any other plane z. The trajectories do not intersect each other in the state space whereas their projections onto the "plane" $z = z_0$ may intersect. The course of the trajectories in the state space has the peculiar property that the area defined by the cross-section of a tube of such trajectories with a plane z is the same for all other planes. This behaviour is an expression of the Liouville theorem, which states that the volume of a phase-space element is conserved in time. The volume element of the phase space represents an equivalent surface element in the state space. In this space the trajectories map an arbitrary element of any initial plane onto a corresponding element of another plane. The shape of the element may change, whereas its area remains fixed. Such a projection is called symplectic.

Symplectic expressions are either symmetric or antisymmetric with respect to the position and momentum coordinates. Equations (5.196) and (5.197) for the path deviations seem to have this behaviour. However, the expression (5.198) for the functions G_ν shows that this is not the case. Fortunately, we can derive the symplectic form by means of an appropriate gauge transformation of the eikonal component $E_\nu^{(n+1)}$. To perform this transformation, we utilize the property (5.193) of the components $m^{(n+1)}$ of the variational function of the eikonal S. The appropriate gauge transforms each component $E_\nu^{(n+1)}$ of the perturbation eikonal E_ν into the corresponding component $F_\nu^{(n+1)}$ of the modified perturbation eikonal F_ν. There exist different transformations that yield a symplectic representation of the function $G_\nu^{(n)}$. We choose the transformation for which the integrands of the eikonal components

$$F_\nu^{(n+1)} = \int_{z_\nu}^{z} m_F^{(n+1)} dz \tag{5.199}$$

comprise path deviations of the lowest possible order. This can be achieved by transforming each eikonal term $E_r^{(n+1)}$ as follows:

$$E_\nu^{(n+1)} = \int_{z_\nu}^{z} \left\{ m_E^{(n+1)} - \sum_{s=\left[\frac{n+1}{2}\right]}^{n-1} \frac{1}{1+\delta_{2s,n+1}} D^{(s)} m^{(n+2-s)} \right\} dz$$
$$+ \sum_{s=\left[\frac{n+1}{2}\right]}^{n-1} \frac{1}{1+\delta_{2s,n+1}} \mathrm{Re}\left(\bar{w}^{(s)} p^{(n+1-s)}\right) . \tag{5.200}$$

If we employ the relation (5.193) and recall that either $\bar{w}^{(s)}$ or $p^{(n+1-s)}$ vanish at the ray-defining plane $z = z_\nu$, it becomes obvious that the two sums on the right-hand side cancel each other. The expression

$$m_F^{(n+1)} = m_E^{(n+1)} - \sum_{s=\left[\frac{n+1}{2}\right]}^{n-1} \frac{1}{1+\delta_{2s,n+1}} D^{(s)} m^{(n+2-s)} \tag{5.201}$$

represents the integrand of the component (5.199) of the modified perturbation eikonal F_ν. For even order $n = 2m$, $m = 1, 2, \ldots$, the Kronecker symbol $\delta_{2s,n+1}$ is always zero. The term $G_\nu^{(n)}$ is cast into the required symplectic form by inserting the relation (5.201) into the formula (5.198). By employing the expressions (5.193) and (5.199), we eventually obtain

$$G_\nu^{(n)} q_{0o} = \frac{\partial F_\nu^{(n+1)}}{\partial a_\nu} \tag{5.202}$$
$$+ \sum_{s=\left[\frac{n+1}{2}\right]}^{n-1} \frac{1}{1+\delta_{2s,n+1}} \mathrm{Re}\left[\bar{w}^{(s)} \frac{\partial p^{(n+1-s)}}{\partial a_\nu} - p^{(s)} \frac{\partial \bar{w}^{(n+1-s)}}{\partial a_\nu}\right] .$$

The integrand (5.201) of the eikonal term $F_\nu^{(n+1)}$ can be expressed in terms of (a) the polynomials $\mu^{(k)}$ of the variational function (5.148), (b) the paraxial ray $w^{(1)}$ and (c) the path deviations $w^{(\ell)}$ with $2\ell \le n+1$. Making use of (5.201) and (5.186) and changing the sequence of some summations, we eventually find the series expansion

$$m_F^{(n+1)} = m^{(n+1)} - \sum_{s=\left[\frac{n-3}{2}\right]}^{n-2} \frac{1}{1+\delta_{2s,n-3}} D^{(s+2)} \mu_1^{(n-s)} \tag{5.203}$$
$$+ \sum_{h=\left[\frac{n-1}{2}\right]}^{n-2} \sum_{s=\left[\frac{n-1}{2}\right]}^{h} \frac{1}{1+\delta_{2s,n-1}} D^{(h-s+2)} D^{(s+1)} \mu_1^{(n-h)}$$
$$+ \frac{1}{2} \sum_{k=\left[\frac{n+1}{2}\right]}^{n-3} \sum_{s=\left[\frac{n-1}{2}\right]}^{k-1} \sum_{h=1}^{k-s} \frac{1}{1+\delta_{2s,n-1}} D^{(s+1)} D^{(h+1)} D^{(k+2-s-h)} \mu_1^{(n-h)} + \ldots .$$

If the upper summation index exceeds the lower one, the sum must be put equal to zero. To determine the performance of corrected electron optical systems, we must know the lowest-order residual aberrations. Hence for systems corrected up to the fourth order inclusively, we need to know the remaining fifth-order aberrations. However, to find an optimum procedure for correcting the aberrations, we must also have some insight into the structure of the corresponding aberration coefficients. Using the relations (5.189) and (5.204), we derive the following expressions for the integrands of the first four components of the modified aberration eikonal F_ν:

$$m_F^{(3)} = m_E^{(3)} = \mu_1^{(3)} , \tag{5.204}$$

$$m_F^{(4)} = m_E^{(4)} = \mu_1^{(4)} + \frac{1}{2} D^{(2)} \mu_1^{(3)} ,$$

$$m_F^{(5)} = \mu_1^{(5)} + D^{(2)} \mu_1^{(4)} + \frac{1}{2} D^{(2)^2} \mu_1^{(3)} , \tag{5.205}$$

$$m_F^{(6)} = \mu_1^{(6)} + D^{(2)} \mu_1^{(5)} + \frac{1}{2} D^{(3)} \mu_1^{(4)} + \frac{1}{2} D^{(2)} D^{(3)} \mu_1^{(3)} \tag{5.206}$$
$$+ \frac{1}{2} D^{(2)^2} \mu_1^{(4)} + \frac{1}{3!} D^{(2)^3} \mu_1^{(3)} .$$

The corresponding polynomials (5.189) of the perturbation eikonal E_ν contain operators $D^{(\ell)}$ up to the order $\ell = n - 1$. Hence in this case we need to know the path deviations up to this order. On the other hand, the expressions (5.207) reveal that the integrands of the components $F_\nu^{(n+1)}$ of the modified perturbation eikonal F_ν contain only path deviations up to and including the order $\ell = [(n+1)/2]$. Here the bracket denotes the integer value as usual. These considerations convincingly demonstrate that the modified perturbation eikonal F_ν is optimally suited for efficient iterative calculation of the higher-order aberrations.

The path equations (5.179), (5.180) and the deviations (5.196), (5.197) can also be written in vector notation. Within the frame of this notation we introduce the four-dimensional vectors

$$\boldsymbol{W} = (w_1, w_2, w_3, w_4) , \quad \boldsymbol{P} = (p_1, p_2, p_3, p_4),$$
$$\boldsymbol{F} = (F_1, F_2, F_3, F_4) , \quad \boldsymbol{G} = (G_1, G_2, G_3, G_4),$$
$$\boldsymbol{G}^{(n)} - (G_1^{(n)}, G_2^{(n)}, G_3^{(n)}, G_4^{(n)}) \tag{5.207}$$

and the unitary anti-symmetric tensor

$$J_C = -\frac{4}{D_0} \begin{pmatrix} 0 & C_{34} & C_{42} & C_{23} \\ C_{43} & 0 & C_{14} & C_{31} \\ C_{24} & C_{41} & 0 & C_{12} \\ C_{32} & C_{13} & C_{21} & 0 \end{pmatrix} = -\frac{4}{D_0} (J_{\kappa\lambda}) . \tag{5.208}$$

Each term $J_{\kappa\lambda} = C_{\mu\nu}$ of the matrix $(J_{\kappa\lambda})$ is obtained by cyclic permutation of the indices λ, μ and ν. The index κ denotes the row and the index λ the column of the matrix element. The values of the four indices differ from each other. They adopt successively one of the numbers $1, 2, 3, 4$. The determinant of the antisymmetric matrix is zero. This determinant should not be confused with the Jacobi determinant D_0 (5.178).

The form of the matrix J_C depends on the Helmholtz–Lagrange constants $C_{\mu\nu}$, which are defined by the choice of the fundamental paraxial rays. Here we define these rays in such a way that the constants $C_{\mu\nu}$ adopt the favorable values (5.159) and (5.160). The resulting matrix

$$J = \begin{pmatrix} 0 & 0 & 1 & 0 \\ 0 & 0 & 0 & 1 \\ -1 & 0 & 0 & 0 \\ 0 & -1 & 0 & 0 \end{pmatrix} \tag{5.209}$$

is known as the fundamental symplectic matrix. Using the definitions (5.213) for the vectors $\boldsymbol{W}$ and $\boldsymbol{G}$ and the tensor (5.208) the path equation (5.179) takes the simple form

$$w = w^{(1)} + \boldsymbol{W} J_C \boldsymbol{G}^{\mathrm{t}} . \tag{5.210}$$

The superscript “t” denotes the transposed vector, which represents a column vector in the frame of our notation.

Employing this matrix and the four-dimensional vectors (5.207), we can write the nth-order path deviations in the form of symplectic vector products as

$$\begin{aligned} w^{(n)} &= \boldsymbol{W} J \boldsymbol{G}^{(n)\mathrm{t}} , \\ p^{(n)} &= \boldsymbol{P} J \boldsymbol{G}^{(n)\mathrm{t}} . \end{aligned} \tag{5.211}$$

The nth-order path deviations (5.212) are entirely determined by the paraxial fundamental rays and the vectors $\boldsymbol{G}^{(n)}$. These vectors can be obtained successively with increasing order n from a recurrence formula for the components $G_\nu^{(n)}$. This formula is derived by substituting the expressions (5.212) for the path deviations in the relation (5.203). Considering the Helmholtz–Lagrange relations (5.159) for the paraxial fundamental ray, we eventually obtain

$$G_\nu^{(n)} = \frac{1}{q_{0o}} \frac{\partial F_\nu^{(n+1)}}{\partial a_\nu} + \sum_{s=\left[\frac{n+1}{2}\right]}^{n-1} \frac{1}{1+\delta_{2s,n+1}} \frac{\partial \boldsymbol{G}^{(n+1-s)}}{\partial a_\nu} J \boldsymbol{G}^{(s)t} . \tag{5.212}$$

Owing to the nonlinearity of this equation, each component $G_\nu^{(n)}$ is composed of products of derivatives of the eikonal terms $F_\nu^{(m)}$ with $m \leq n$ apart from the linear term with order $n+1$. The number of the products and their factors grows with increasing order n. The expressions for the components $G_\nu^{(n)}$ are derived iteratively by starting from the lowest order $n = 2$. In this case the sum in the relation (5.212) does not contribute, and we find

$$G_\nu^{(2)} = \frac{1}{q_{0o}} \frac{\partial F_\nu^{(3)}}{\partial a_\nu} . \tag{5.213}$$

The third-order components $G_\nu^{(3)}$ are obtained in the first iteration step yielding

$$\begin{aligned} G_\nu^{(3)} = \frac{1}{q_{0o}} \frac{\partial F_\nu^{(4)}}{\partial a_\nu} &+ \frac{1}{2} \left[\frac{\partial^2 F_1^{(3)}}{\partial a_\nu \partial a_1} \frac{\partial F_3^{(3)}}{\partial a_3} - \frac{\partial^2 F_3^{(3)}}{\partial a_\nu \partial a_3} \frac{\partial F_1^{(3)}}{\partial a_1} \right. \\ &+ \left. \frac{\partial^2 F_2^{(3)}}{\partial a_\nu \partial a_2} \frac{\partial F_4^{(3)}}{\partial a_4} - \frac{\partial^2 F_4^{(3)}}{\partial a_\nu \partial a_4} \frac{\partial F_2^{(3)}}{\partial a_2} \right] . \end{aligned} \tag{5.214}$$

The expressions for the higher-order components become very involved. They are determined most reliably by means of computer algebra procedures.

5.3.5 Canonical Boundary Conditions

The expressions for the aberration coefficients depend on the choice of the ray parameters a_ν. Usually these parameters are the position and slope coordinates of the ray at the object plane. However, the mathematical expenditure is considerably reduced if we fix the ray in terms of the canonical momentum instead of the slope. Therefore, we denote the position and momentum boundaries as canonical boundaries. If the ray is defined by its position coordinates w_o and w_a at the object plane $z = z_o$ and the aperture plane $z = z_a$, respectively, we need the two perturbation eikonals

$$F_1 = F_2 = F_o = \int_{z_o}^{z} m_F dz , \quad F_3 = F_4 = F_a = \int_{z_a}^{z} m_F dz \tag{5.215}$$

for determining the deviation of the ray from its paraxial course. Only a single perturbation eikonal F_o is needed if the ray is fixed by its position w_o and its lateral canonical momentum p_o at the object plane $z = z_\nu = z_o$, $\nu = 1, 2, 3, 4$. In this case the nth-order path deviations can be written in a concise Lie-algebraic form by employing the familiar Poisson bracket operation defined by

$$[f, g] = \frac{\partial f}{\partial a_1}\frac{\partial g}{\partial a_3} - \frac{\partial f}{\partial a_3}\frac{\partial y}{\partial a_2} + \frac{\partial f}{\partial a_2}\frac{\partial g}{\partial a_4} - \frac{\partial f}{\partial a_4}\frac{\partial g}{\partial a_2} . \tag{5.216}$$

Here f and g are two arbitrary real or complex functions of the four ray parameters a_ν. Putting

$$f = w^{(1)} , \quad g = F_o^{(n+1)} \tag{5.217}$$

and considering the recurrence relation (5.213), the second- and third-order path deviations adopt the symplectic form

$$w^{(2)} = \left[w^{(2)}, F_{\mathrm{o}}^{(3)}\right] , \tag{5.218}$$

$$w^{(3)} = \left[w^{(1)}, F_{\mathrm{o}}^{(4)}\right] + \frac{1}{2}\left[\left[w^{(1)}, F_{\mathrm{o}}^{(3)}\right], F_{\mathrm{o}}^{(3)}\right] . \tag{5.219}$$

The path deviations at the image plane $z = z_{\mathrm{i}}$ describe the aberrations. For the primary second-order aberrations to vanish at this plane, the condition $F_{\mathrm{o}}^{(3)}(z = z_{\mathrm{i}}) = 0$ must hold. In this case the residual third-order aberrations

$$w^{(3)}(z_{\mathrm{i}}) = \left[w^{(1)}, F_{\mathrm{o}}^{(4)}\right]_{z=z_{\mathrm{i}}} \tag{5.220}$$

are entirely determined by the fourth-order perturbation eikonal

$$F_{\mathrm{oi}}^{(4)} = F_{\mathrm{o}}^{(4)}(z = z_{\mathrm{i}}) \tag{5.221}$$

at the image plane $z = z_{\mathrm{i}}$. Such a simple representation exists for the aberrations of any order if the ray is defined by its position at the object plane and its off-axis canonical momentum at the observation plane $z = z_{\mathrm{i}}$. The path deviations at an arbitrary plane $z_{\mathrm{o}} \leq z \leq z_i$ are then determined by the two perturbation eikonals

$$F_1 = F_2 = F_{\mathrm{o}} , \quad F_3 = F_4 = F_{\mathrm{i}} = \int_{z_{\mathrm{o}}}^{z} m_F dz . \tag{5.222}$$

Since the perturbation eikonal F_{i} vanishes at the image plane $z = z_{\mathrm{i}}$, the functions

$$G_3^{(n)} = G_4^{(n)} = G_{\mathrm{i}}^{(n)}(z) , \quad n = 2, 3, \dots , \tag{5.223}$$

are zero at this plane. As a consequence, the symplectic products in the expression (5.213) are also zero at the image plane. This behaviour is a direct consequence of the boundary condition

$$p^{(\ell)}(z = z_{\mathrm{i}}) = p_{\mathrm{i}}\, \delta_{1\ell} , \tag{5.224}$$

which implies that all momentum deviations $p^{(\ell)}$ with $\ell \geq 2$ are zero at the image plane. Considering further the condition

$$w_1(z_{\mathrm{i}}) = w_2(z_{\mathrm{i}}) = 0 \tag{5.225}$$

for paraxial stigmatic imaging, the nth-order aberration takes the form

$$w_{\mathrm{i}}^{(n)} = -w_{3\mathrm{i}} \frac{\partial F_{\mathrm{oi}}^{(n+1)}}{\partial a_1} - w_{4\mathrm{i}} \frac{\partial F_{\mathrm{oi}}^{(n+1)}}{\partial a_2} = -2 \frac{\partial F_{\mathrm{oi}}^{(n+1)}}{\partial \bar{p}_{\mathrm{i}}} , \tag{5.226}$$

where

$$w_{\nu \mathrm{i}} = w_\nu(z_\mathrm{i}) \,, \quad F_{\mathrm{oi}}^{(n+1)} = F_\mathrm{o}^{(n+1)}(z_\mathrm{i}) \,. \tag{5.227}$$

The second relation on the right-hand side of (5.226) is readily obtained from the expressions (5.174) for the ray parameters if we consider that we must put in these formulae $p_3(z_\mathrm{i}) = p_4(z_\mathrm{i}) = 0$ according to the boundary condition (5.176) imposed on the fundamental rays. The comparison of the formula (5.226) with the familiar relations (5.114) for the mixed eikonal $V = V(w_\mathrm{o}, z_\mathrm{o}; p_\mathrm{i}, z_\mathrm{i})$ reveals that the components $F_{\mathrm{oi}}^{(n+1)}$ of the modified perturbation eikonal at the image plane represent the Seidel components $V^{(n+1)}$ of the mixed eikonal V. Hence our perturbation procedure yields consecutively the expansion terms of the mixed eikonal with increasing order. If we fix the ray by its positions at the initial plane and the terminal plane, the expansion terms of the perturbation eikonal at this plane represent the corresponding terms of the point eikonal S. Our procedure is therefore well suited for determining the Seidel-order terms of arbitrary eikonals in an iterative, straightforward way.

5.3.6 Systems with Special Symmetry

The number of aberrations arising in an electron optical system strongly depends on the symmetry properties of the electromagnetic field. The higher the symmetry the smaller is the number of aberrations. Thus, rotationally symmetric systems produce the smallest number of aberrations. Unfortunately, such systems possess unavoidable axial aberrations, according to the Scherzer theorem. To compensate for these aberrations, one must abandon rotational symmetry, thereby increasing the total number of aberrations. To keep this number as low as possible it is advantageous to maintain a high degree of symmetry. For this reason we expand the electrostatic potential $\varphi = \varphi(\boldsymbol{r})$ and the scalar magnetic potential $\psi = \psi(\boldsymbol{r})$ in a series of multipoles about the straight symmetry axis of the system. Since this axis is a possible trajectory, an electron moving along this optic axis must not experience any lateral force. This requirement is satisfied if the dipole component of the series expansion vanishes. Each multipole component is entirely defined by its complex strength

$$\Phi_m = \Phi_m(z) = \Phi_{mc}(z) + i\Phi_{ms}(z) \,, \quad m = 0, 1, 2 \ldots \quad , \tag{5.228}$$

where the real "cosine" and "sine" parts Φ_{mc} and Φ_{ms} determine the azimuthal orientation ϑ_m of the m-fold multipole field. This angle defines the orientation of the electrodes with respect to the x-axis and is given by the relation

$$\tan(m\vartheta_m) = \Phi_{ms}/\Phi_{mc} \,. \tag{5.229}$$

If the ratio on the right-hand side depends on the z-coordinate, the electrodes are twisted about the optic axis. Assuming that the multipole strengths are known along this axis, the electrostatic potential is given by the multipole series

$$\varphi = \mathrm{Re}\sum_{m=0}^{\infty}\sum_{\ell=0}^{\infty}(-)^{\ell}\frac{m!}{\ell!(m+\ell)!}\left(\frac{w\bar{w}}{4}\right)^{\ell}\bar{w}^{m}\Phi_m^{[2\ell]}(z)\,. \qquad (5.230)$$

The upper index $[2\ell]$ indicates a 2ℓ-fold differentiation with respect to the $z-$coordinate. Since we have assumed that the straight optic axis is a possible ray, the dipole strength Φ_1 must be zero, which implies that

$$\Phi_{1c} = 0\,, \quad \Phi_{1s} = 0\,. \qquad (5.231)$$

The multipole strength of the "zero"-fold component is real, $\Phi_0 = \bar{\Phi}_0$. For rotationally symmetric systems all multipole terms with $m \geq 1$ vanish. Both the electrostatic potential and the scalar magnetic potential satisfy the Laplace equation. Accordingly, the multipole expansion of the scalar magnetic potential is readily derived by substituting ψ for φ and Ψ_m for Φ_m in the expression (5.230).

Since the variational function (5.143) is linearly related to the magnetic vector potential $\boldsymbol{A}$, we must express the components A_z and $A = A_x + iA_y$ in terms of the multipole strengths Ψ_m of the scalar magnetic potential. This connection is extremely helpful for the practical design because the surfaces of the magnetic polepieces represent surfaces of constant scalar magnetic potential in the limit of infinitely high permeability. By employing the Coulomb gauge for the magnetic vector potential, we obtain the relations [18]:

$$A = \sum_{m=0}^{\infty}\sum_{\ell=0}^{\infty}\frac{(-)^{\ell}}{2i}\frac{m!}{\ell(m+1+\ell)!}\left(\frac{w\bar{w}}{4}\right)^{\ell}w^{m+1}\bar{\Psi}_m^{[2\ell+1]}(z)\,, \qquad (5.232)$$

$$A_z = \mathrm{Re}\sum_{m=0}^{\infty}\sum_{\ell=0}^{\infty}(-)^{\ell}\frac{m!}{\ell!(m+\ell)!}\left(\frac{w\bar{w}}{4}\right)^{\ell}\bar{w}^{m}\Psi_m^{[2\ell]}(z)\,. \qquad (5.233)$$

The special case of a purely rotationally symmetric field,

$$\Psi_m = 0 \quad \text{for} \quad m \geq 1\,, \quad -\Psi_0' = -\bar{\Psi}_0' = B_z(w=0,z)\,, \qquad (5.234)$$

demonstrates that the gauge has been chosen reasonably. Since Ψ_0 is real the component (5.233) vanishes in this case. The remaining lateral component A_0 points in the azimuthal direction reflecting the fact that a rotationally symmetric magnetic field is created solely by currents that flow azimuthally around the optic axis.

Systems with two-section symmetry consist only of multipole components with even multiplicity $m = 2\mu$, $\mu = 0, 1, 2, \ldots$. Moreover, the azimuthal orientation of the multipoles must be chosen in such a way that

$$\Phi_m = \bar{\Phi}_m = \Phi_{mc}\,, \quad \Psi_m = -\bar{\Psi}_m = i\Psi_{ms}\,. \qquad (5.235)$$

In this case we have $\vartheta_m = 0$ for the electric multipole strengths and $\vartheta_m = \pi/(2m)$ for the magnetic multipoles. Hence for a system composed of electric and magnetic quadrupoles, the magnetic poles are rotated by 45° with respect to the electrodes of the electric quadrupoles. All quadrupole-octopole correctors that have been proposed so far fulfil this requirement. Owing to this double plane-section symmetry, the increase in the number of aberrations that results from abandoning rotational symmetry, is minimized. Moreover, path deviations with even Seidel order $n = 2\nu$, $\nu = 1, 2, \ldots$, do not occur. This behaviour is due to the fact that all variational polynomials (5.184) with odd order vanish. Accordingly, the perturbation eikonal consists only of components with even Seidel order. For such systems the components (5.207) of the integrand of the modified aberration eikonal significantly are reduced:

$$m_F^{(3)} = 0\,, \quad m_F^{(4)} = \mu_1^{(4)}\,, \tag{5.236}$$

$$m_F^{(5)} = 0\,, \quad m_F^{(6)} = \mu_1^{(6)} + \frac{1}{2} D^{(3)} \mu_1^{(4)}\,. \tag{5.237}$$

Systems consisting of multipoles with even multiplicity and round lenses exhibit astigmatic paraxial ray-paths. Stigmatic imaging is obtained only if the quadrupole elements are arranged and excited appropriately. Unfortunately, the quadrupole fields of correctors must be aligned and stabilized with extremely high accuracy. For this reason, a corrector that does not require quadrupole and octopole elements for eliminating the spherical aberration of round lenses has been proposed [30]. This corrector consists of a telescopic magnetic round lens doublet and two hexapoles, which are centred about the outer focal planes of the lenses. These planes coincide with the nodal planes of the doublet. The corrector produces an electromagnetic field with threefold symmetry about its straight axis. To obtain insight into the actual operation of a sextupole corrector, it is necessary to elucidate the imaging properties of electromagnetic fields with threefold symmetry up to the fifth order. Since the rotationally symmetric part satisfies arbitrary multiplicity requirements, we consider electromagnetic fields consisting of multipole components with multiplicity $3m$, where $m = 0, 1, 2, \ldots$.

5.4 Systems with Threefold Symmetry

Electromagnetic fields with threefold symmetry are produced most conveniently by means of sextupole elements. These elements are generally employed in particle optics to compensate for second-order aberrations arising in systems with a curved axis, such as spectrometers or imaging energy filters. However, sextupole elements can also be utilized for correcting the third-order spherical aberration of electron optical systems [19,20]. This behaviour is due to the fact that sextupole elements produce secondary third-order aberrations, some of which are of the same nature as those of round lenses. Moreover, the residual third-order aberrations of sextupole elements can be adjusted to compensate for the unavoidable spherical aberration and image curvature of rotationally symmetric electron lenses. However, such a correction improves the imaging properties of the overall system only if the primary second-order aberrations of the sextupoles vanish as well, and if the fourth-order aberrations can be kept sufficiently small. Hence these aberrations must also be calculated in order to determine the improvement in optical performance obtained by correcting particular third-order aberrations.

5.4.1 Paraxial Trajectories

The paraxial properties of systems with threefold symmetry are entirely determined by the round lenses because the hexapole and dodecapole fields do not affect the paraxial region. In the presence of an axial magnetic field it is advantageous to introduce the rotating u, z-coordinate system (5.156). In the frame of this system, the fundamental rays

$$
\begin{aligned}
u_\alpha &= w_1\, e^{-i\chi} = -i w_2\, e^{-i\chi}\,, \\
u_\gamma &= w_3\, e^{i-\chi} = i w_4\, e^{-i\chi}
\end{aligned}
\tag{5.238}
$$

are real and satisfy the relativistic paraxial path equation

$$
\Phi^* u'' + \frac{1}{2}\Phi^{*'} u' + \frac{\Lambda}{4}\Phi'' u + \frac{eB^2}{8m_0} u = 0\,, \tag{5.239}
$$

where

$$
\Lambda = 1 + e\Phi/(2m_0c^2) \tag{5.240}
$$

describes the relativistic factor. We select the two linearly independent solutions u_α and u_γ of the differential equation (5.239) in such a way that they are best suited for an efficient calculation of corrected aplanatic electron optical systems. This requirement is achieved by imposing the initial conditions

$$
\begin{aligned}
u_\alpha(z_o) = u_{\alpha o} &= 0\,, \quad u'_{\alpha o} = 1\,, \\
u_{\gamma o} &= 1\,, \quad u_\gamma(z_c) = u_{\gamma c} = 0
\end{aligned}
\tag{5.241}
$$

on the two fundamental rays. The axial fundamental ray u_α intersects the centre of the object plane z_o, while the field ray u_γ intersects the centre of the coma-free plane $z = z_c$. This plane is located within the field of the objective lens in front of the back focal plane. In order to eliminate the anisotropic coma, the conventional magnetic objective lens must be replaced by a compound lens producing at least two spatially separated axial fields with opposite sign. The second half of this lens can be used simultaneously as a transfer lens for imaging the coma-free plane into any given plane behind the objective lens. It should be noted that the constraint for the field ray u_γ differs from the standard constraint, which puts the zero of u_γ into the diffraction plane $z = z_d$. So that only a single eikonal needs to be considered, we fix the true ray by its lateral position w_o and its off-axial canonical momentum p_o at the object plane $z = z_o$. Moreover, we require that the paraxial ray

$$u^{(1)} = \Omega_\alpha u_\alpha + \Omega_\gamma u_\gamma \tag{5.242}$$

satisfies the same boundary conditions. Accordingly, the complex ray parameters $\Omega_\alpha = \Omega_1$ and $\Omega_\gamma = \Omega_2$ (5.163) are derived from the initial conditions

$$\begin{aligned} w_o &= w_o^{(1)} = u_o^{(1)} = \Omega_\gamma\,, \\ p_o &= p_o^{(1)} = q_{0o} u_o^{(1)'} = q_{0o}(\Omega_\alpha + \Omega_\gamma u'_{\gamma o})\,, \end{aligned} \tag{5.243}$$

yielding

$$\Omega_\alpha = \frac{p_o}{p_{0o}} - w_o u'_{\gamma o}\,, \quad \Omega_\gamma = w_o\,. \tag{5.244}$$

Since the zero of u_γ is located in front of the back focal plane of a standard objective lens, the slope u'_γ of the fundamental field ray at the object plane is always negative ($u'_{\gamma o} < 0$) in this case.

5.4.2 Second-Order Path Deviation

The course of the second-order path deviation $u^{(2)}(z)$ along the optic axis depends on the paraxial field and on the spatial distribution of the hexapole fields. In first approximation, the combined action of the electric and magnetic sextupoles is given by the hexapole function

$$H = \frac{e^{3i\chi}}{q_{0o}} \left(q_0 \frac{\Lambda}{2} \frac{\Phi_3}{\Phi^*} + ie\Psi_3 \right)\,, \tag{5.245}$$

which determines the third-order term

$$F_o^{(3)} = q_{0o} \,\mathrm{Re} \int_{z_o}^{z} H \bar{u}^{(1)^3} dz \tag{5.246}$$

of the eikonal F_o. Considering the relations (5.238) and (5.244) for the fundamental paraxial rays and the ray parameter, respectively, we derive from (5.182) the second-order path deviation in the rotating coordinate system

$$u^{(2)} = \frac{2}{q_{0o}} \left(u_\alpha \frac{\partial F_o^{(3)}}{\partial \bar{\Omega}_2} - u_\gamma \frac{\partial F_o^{(3)}}{\partial \bar{\Omega}_1} \right) = \bar{\Omega}_1^2 u_{11} + \bar{\Omega}_1 \bar{\Omega}_2 u_{12} + \bar{\Omega}_2^2 u_{22}\,. \tag{5.247}$$

The second-order fundamental rays u_{11}, u_{12} and u_{22} are given by the integral expressions

$$u_{\mu\nu} = 3\,(2 - \delta_{\mu\nu}) \left(u_\alpha \int_{z_o}^{z} H u_\alpha^{2-\tau} u_\gamma^{1+\tau} dz - u_\gamma \int_{z_o}^{z} H u_\alpha^{3-\tau} u_\gamma^{\tau} dz \right) \tag{5.248}$$

with $\tau = 0, 1, 2$ and

$$\mu = 1+[\tau/2]\,, \quad \nu = 1+[(\tau+1)/2]\,. \tag{5.249}$$

The brackets indicate the integer parts of $\tau/2$ and $(\tau + 1)/2$, respectively. In order that the three second-order fundamental rays vanish in the entire region behind a given exit plane $z = z_e$, the four conditions

$$\int_{z_o}^{z_e} H u_\alpha^{3-\tau} u_\gamma^\tau dz = 0\,, \quad \tau = 0, 1, 2, 3\,, \tag{5.250}$$

must be fulfilled. These requirements can be satisfied most easily by imposing symmetry conditions on the paraxial fundamental rays and on the total hexapole strength H. For this purpose the sextupole fields and the fundamental rays are chosen in such a way that the integrands of the integrals (5.250) are either antisymmetric with respect to the midplane of the sextupole arrangement, or with respect to the central planes of each half of the system. Since the two fundamental paraxial rays u_α and u_γ are linearly independent, they cannot posses the same symmetry about a given plane. It is therefore not possible to eliminate all second-order aberrations by a single symmetry condition. However, if we choose the paraxial path in such a way that one of the two fundamental rays is symmetric and the other antisymmetric with respect to the symmetry plane and the central planes of each half of the system within the hexapole fields, all second-order contributions can be eliminated outside the sextupole system. This is achieved if the sextupole fields are symmetric with respect to the symmetry planes. The most simple system satisfying these requirements is shown in Fig. 5.4. It consists of a telescopic round lens doublet and two identical sextupoles, which are centred on the outer focal planes of the lenses [21]. The plane midway between these lenses is the midplane of the $4f$-system, while the outer focal planes represent the centre planes of each half of the sextupole system. Accordingly, the sextupole fields and the paraxial fundamental rays fulfil the requirement for complete elimination of the second-order aberrations.

The outer focal points coincide with the nodal points N_1 and N_2 of the telescopic round lens doublet. To avoid a rotation of the image of the first sextupole, the coils of the round lenses must be connected in series opposition so that the excitations of the two lenses are equal and opposite, whatever the strength of the current. In this case the doublet images the front sextupole with magnification $M = -1$ exactly onto the second sextupole centred on the nodal point N_2 without introducing any off-axial third-order coma. Hence the front focal plane is also the coma-free plane of the $4f$-arrangement. This system can be used as a corrector for eliminating the third-order spherical aberration of an electron microscope [20].

To demonstrate this behaviour, we must also calculate the secondary aberrations of the system. In order to determine these we need to know the second-order path deviation $u^{(2)}(z)$. The constituent second-order fundamental rays (5.249) can be calculated analytically if we approximate the

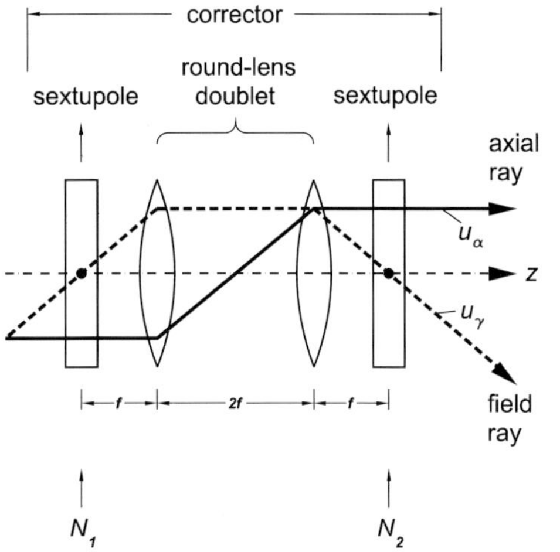

Fig. 5.4. Arrangement of the elements of a spherical-aberration corrector, which does not introduce any second-order aberrations outside of the system

sextupole strength $H = H(z)$ by two identical box-shaped distributions with axial extension 2ℓ. The resulting course of the rays u_{11}, u_{12} and u_{22} is depicted in Fig. 5.5. The course of the rays u_{11} and u_{22} is symmetric while that of the "mixed" ray u_{12} is antisymmetric with respect to the midplane z_m. Owing to this symmetry the system introduces neither off-axial coma nor third-order distortion at the image plane, as will be shown in the next section. The hexapole strength is a free parameter, which can be adjusted in such a way that the corrector compensates for the spherical aberration of the entire system.

In electron lithography the most disturbing aberrations are the image curvature and the field astigmatism because they decisively limit the usable area of the mask. Unfortunately, the third-order image curvature of rotationally symmetric systems is unavoidable and its coefficient has the same sign as that of the spherical aberration. Hence a rotationally symmetric planar system does not exist. A planar system is corrected for image curvature, field astigmatism and coma. Since sextupoles can correct the spherical aberration of round lenses, the question arises whether these elements can also be used to eliminate the unavoidable third-order field curvature of round lenses. However, to obtain a planar field of view we must also compensate for the field astigmatism. Because the corresponding aberration coefficient is complex for magnetic lenses, we need three free parameters to compensate for both image curvature and field astigmatism simultaneously.

A sextupole system that satisfies this requirement is shown in Fig. 5.6. The arrangement consists of four identical round lenses forming an 8f-system and five sextupoles, which are centred symmetrically about the midplane z_m.

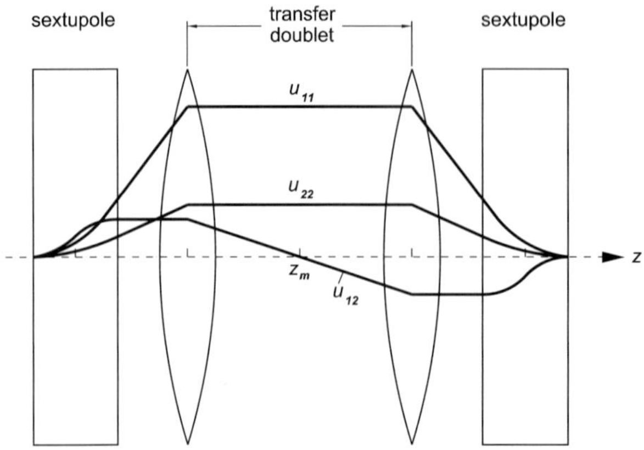

Fig. 5.5. Course of the second-order fundamental rays u_{11}, u_{12} and u_{22} within the sextupole corrector shown in Fig. 5.4

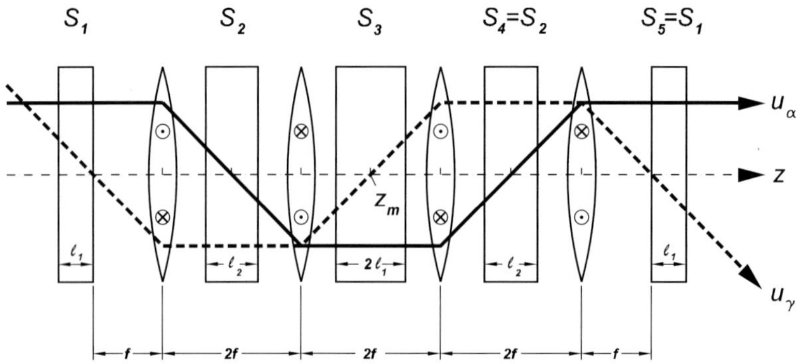

Fig. 5.6. Hexapole planator compensating for the third-order image curvature and field astigmatism. The planator also introduces a negative spherical aberration, which depends on the coefficients of the field aberrations prior to their correction

The two outer sextupoles have the same strength as the central sextupole whose thickness $2\ell_1$ is twice that of the outer sextupoles. Each half of the central sextupole is conjugate to one of the outer sextupoles because the front sextupole is imaged by the first doublet onto the first half of the central sextupole, while the second half of this sextupole is imaged by the second doublet onto the last sextupole with magnification $M = -1$. Moreover, the second sextupole is also imaged with $M = -1$ onto the fourth sextupole $S_4 = S_2$. Accordingly, the second-order path deviation vanishes in the region outside the system if the strengths of the sextupoles are chosen to be $H_1 = H_3 = H_5$, $H_4 = H_2$. Since the azimuthal orientation of the sextupoles S_2 and S_4 may differ from that of the sextupoles S_1, S_3 and S_5, we have three free parameters $|H_1|$, $|H_2|$ and $\mathrm{Re}(H_1\bar{H}_2)$. However, this does not necessarily imply that it is possible to eliminate the image curvature and the field astigmatism because non-linear relations exist between the coefficients of these aberrations and the hexapole strengths. As a result, only few systems can be found, that enable the field aberrations to be correced. The system shown in Fig. 5.6 is suitable as a planator.

5.4.3 Third-Order Aberrations

The primary aberrations of systems with threefold symmetry are of second order. Since these aberrations are large compared with the third-order aberrations produced by the rotationally symmetric fields, it is necessary to eliminate all second-order aberrations first before dealing with the third-order aberrations. In this case the relation $F_{\mathrm{o}}^{(3)}(z_{\mathrm{i}}) = 0$ holds, and the third-order aberration (5.220) at the final image plane $z = z_{\mathrm{i}}$ adopts the form

$$u_{\mathrm{i}}^{(3)} = -2u_{\gamma\mathrm{i}}\frac{\partial F_{\mathrm{oi}}^{(4)}}{q_{0\mathrm{o}}\partial\bar{\Omega}_1}\,, \tag{5.251}$$

in the case of rotationally symmetric paraxial imaging. According to the relations (5.207) the integrand

$$m_F^{(4)} = \mu_1^{(4)} + \frac{3}{2}\,\mathrm{Re}\left(H\bar{u}^{(1)^2}\bar{u}^{(2)}\right) \tag{5.252}$$

of the fourth-order term $F_{\mathrm{o}}^{(4)}$ of the modified aberration eikonal F_{o} consists of a term $\mu_1^{(4)}$, produced by the rotationally symmetric field, and a term resulting from the sextupoles. Since the conjugate complex value of the second-order path deviation (5.247) is bilinear in the ray parameters Ω_1 and Ω_2, the contribution of the hexapole fields to the fourth-order eikonal term has exactly the same structure as that resulting from the rotationally symmetric field component. Hence the hexapole fields produce exactly the same third-order aberrations as the round lenses. This surprising behaviour results from the nonlinear forces of the hexapole fields. Unfortunately, the spherical aberration produced by a sequence of sextupoles has the same sign as that of the round lenses if the second-order aberrations are eliminated. However, it is possible to reverse the sign of the spherical aberration by employing sextupoles in combination with round lenses. Systems that exhibit this property, are shown in Figs. 5.4 and 5.6.

The fourth-order term of the perturbation eikonal

$$F_{\mathrm{oi}}^{(4)} = F_{\mathrm{R}}^{(4)} + F_{\mathrm{H}}^{(4)} \tag{5.253}$$

consists of the round-lens term

$$F_{\rm R}^{(4)} = \int_{z_{\rm o}}^{z_{\rm i}} \mu_1^{(4)} dz$$

$$= -q_{0{\rm o}} \,{\rm Re} \left\{ \frac{1}{4} C_{\rm R} \Omega_1^2 \bar{\Omega}_1^2 + K_{\rm R} \Omega_1 \bar{\Omega}_1^2 \Omega_2 + \frac{1}{2} A_{\rm R} \bar{\Omega}_1^2 \Omega_2^2 \right.$$

$$\left. + \frac{1}{2} F_{\rm R} \, \Omega_1 \bar{\Omega}_1 \Omega_2 \bar{\Omega}_2 + D_{\rm R} \bar{\Omega}_1 \Omega_2 \bar{\Omega}_2^2 + \frac{1}{4} E_{\rm R} \bar{\Omega}_1^2 \bar{\Omega}_2^2 \right\} \tag{5.254}$$

and the term

$$F_{\rm H}^{(4)} = \frac{3}{2} {\rm Re} \int_{z_{\rm o}}^{z_{\rm i}} H(\bar{\Omega}_1 u_1 + \bar{\Omega}_2 u_2)^2 (\Omega_1^2 u_{11} + \Omega_1 \Omega_2 u_{12} + \Omega_2^2 u_{22}) dz \,, \tag{5.255}$$

which is produced by the combination of subsequent hexapole deflections within the corrector.

The notation for the aberration coefficients of the round lenses is in accordance to that of Hawkes and Kasper [22], which is the same as that suggested much earlier by Scherzer in his lectures on electron optics. The coefficient $C_{\rm R}$ is associated with spherical aberration, $K_{\rm R}$ with off-axial coma, $A_{\rm R}$ with field astigmatism, $F_{\rm R}$ with field curvature, $D_{\rm R}$ with distortion and $E_{\rm R}$ with spherical aberration in the diffraction plane. According to the relation (5.251), this coefficient does not affect the aberrations at the image plane. The coefficients $C_{\rm R}$, $F_{\rm R}$ and $E_{\rm R}$ are always real, while the coefficients $K_{\rm R}$ and $A_{\rm R}$ are complex in the presence of an axial magnetic field. The resulting Larmor rotation causes a rotation of the aberration figures of coma and astigmatism. The angle of rotation with respect to the line intersecting the optic axis and the Gaussian image point is proportional to the imaginary part of the corresponding aberration coefficient. It should be noted that the third-order aberration coefficients are defined as the negative values of the expansion coefficients of the eikonal. This confusing conventions goes back to the early days of electron optics and was chosen primarily to obtain a positive coefficient $C_3 = C_{\rm R}$ for the third-order spherical aberration [23].

The eikonal term (5.255) can be evaluated analytically if we employ the sharp cut-off fringing field (SCOFF) approximation and assume that the rotationally symmetric fields do not overlap the hexapole fields. In this case the paraxial fundamental trajectories form straight lines inside the field region of the sextupoles. Within the frame of the SCOFF approximation the sextupole strength is expressed as

$$H(z) = \sum_\nu H_\nu \Theta_\nu(z) \,, \tag{5.256}$$

where the step function Θ_ν is defined as

$$\Theta_\nu(z) = \begin{cases} 1 & \text{for } z_\nu \le z \le z_\nu + \ell_\nu \\ 0 & \text{otherwise} \,. \end{cases} \tag{5.257}$$

The system shown in Fig. 5.4 introduces neither coma nor distortion, owing to the symmetry of the hexapole field and of the fundamental rays $u_1 = u_\alpha$, $u_2 = u_\gamma$, u_{11}, u_{12} and u_{22} with respect to the midplane z_{m}. For this system the terms $Hu_1u_2u_{11}$, $Hu_1u_2u_{22}$, $Hu_1^2u_{12}$ and $Hu_2^2u_{11}$ in the integrand of the eikonal term (5.255) are antisymmetric functions and hence their contribution to the integral cancels out. The antisymmetry of the products can readily be verified by examining the ray-path shown in Fig. 5.4 for the fundamental paraxial rays and in Fig. 5.5 for the second-order fundamental rays.

By employing the SCOFF approximation, we derive analytical expressions for the secondary fundamental rays $u_{\mu\nu}$. In the region of the first sextupole with length $\ell_1 = \ell_2 = \ell$ the primary fundamental rays are straight lines of the form

$$u_1 = u_\alpha = f_0\,, \quad u_2 = u_\gamma = z/f_0\,, \tag{5.258}$$

where f_o denotes the focal length of the objective lens located in front of the telescopic system. By means of these relations and the expressions for the secondary fundamental rays, the fourth-order eikonal term (5.255) can be evaluated analytically. Comparison of the result,

$$\begin{aligned} F_H^{(4)} &= \frac{3}{2}q_{0o}|H|^2\ell^3 f_0^4 \operatorname{Re}\left\{\Omega_1^2\bar{\Omega}_1^2 - \frac{1}{5}\frac{\ell^2}{f_0^4}\Omega_1\bar{\Omega}_1\Omega_2\bar{\Omega}_2\right. \\ &\quad \left. + \frac{1}{5}\frac{\ell^2}{f_0^4}\bar{\Omega}_1^2 + \frac{1}{112}\frac{\ell^4}{f_0^8}\Omega_2^2\bar{\Omega}_2^2\right\}\,, \end{aligned} \tag{5.259}$$

with the representation (5.254) of the corresponding eikonal term of the rotationally symmetric field component yields the following expressions for the third-order aberration coefficients produced by the hexapole fields:

$$\begin{aligned} C_{\mathrm{H}} &= -6|H|^2\ell^3 f_0^4\,, \\ K_{\mathrm{H}} &= 0\,, \quad D_{\mathrm{H}} = 0\,, \\ F_{\mathrm{H}} &= \quad A_{\mathrm{H}} = \frac{3}{5}|H|^2\ell^5\,, \\ E_{\mathrm{H}} &= -\frac{3}{56}|H|^2\frac{\ell^7}{f_0^4}\,. \end{aligned} \tag{5.260}$$

These expressions depend quadratically on the hexapole strength H and two always have negative signs, while the coefficient F_{H} of the field curvature is always positive. Since the field curvature resulting from the round lens has the same sign, the corrector shown in Fig. 5.4 cannot compensate for this aberration. On the other hand the total spherical aberration of the system

$$C_3 = C_{\mathrm{R}} + C_{\mathrm{H}} \tag{5.261}$$

can be eliminated by choosing the hexapole strength appropriately. In order to make the off-axial coma of the entire system vanish as well, the round-lens coma must be made zero. This is the case if the coma-free plane of the

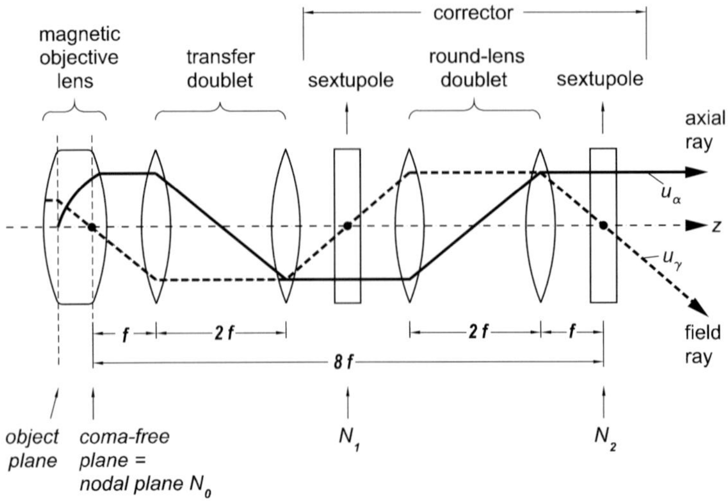

Fig. 5.7. Coma-free arrangement of the objective lens and the hexapole corrector by means of a telescopic transfer doublet

objective lens coincides with the corresponding plane of the corrector located in the centre of the first sextupole.

Since the coma-free plane of a conventional objective lens is located within its field, it is necessary to image this plane into the front focal plane of the telescopic round-lens doublet of the sextupole corrector without introducing any additional coma. This condition can, for example, be fulfilled with the aid of another telescopic transfer doublet, as shown in Fig. 5.7. However, this procedure only eliminates the radial (isotropic) component of the coma. The anisotropic coma of the objective lens can only partly be compensated by that of the weak lenses of the transfer doublet. In order to eliminate the anisotropic coma completely we must introduce a compound objective lens, consisting of two spatially separated coils with thir currents flowing in opposite directions [24]. Since the second half of the lens can simultaneously be used as the first lens of the transfer doublet, the number of coils is not increased by this concept.

The distortion does not affect the resolution of the image but does deform the geometrical structure of the imaged object. In high-resolution electron microscopy only the projector lens contributes significantly to the distortion. This aberration becomes negligibly small if the projector lens operates in such a way that an image of the effective source is located inside the field of this lens. On the other hand, the distortion is of major concern in projection electron lithography, where a large mask is imaged on the wafer with a reduction of scale of the order of 4 to 10. In this case, almost all lenses of the system contribute appreciably to the distortion. The aberration associated with the eikonal coefficient $E_3 = E_{\mathrm{R}} + E_{\mathrm{H}}$ has no effect at the Gaussian image plane. It does however cause a distortion in any defocused image. In order to avoid introducing an appreciable distortion when changing the defocus, the coefficient E_3 must be kept sufficiently small. Fortunately, this can be achieved because the sign of the coefficient $E_{\rm H}$ is opposite to that of $E_{\rm R}$. For a projection system with vertical landing angle and parallel illumination of the mask, the coefficient $E_{\rm R}$ is unavoidable and of positive sign just as the coefficient $C_{\rm R}$ of the third-order spherical aberration.

To achieve uniform imaging of all object points regardless of their lateral position, it is necessary to eliminate all off-axial aberrations. For a system consisting of round lenses this is not possible because the image curvature is unavoidable in this case. To obtain a planar system we must hence find some other means of compensating for this aberration. Unfortunately, the simple sextupole corrector shown in Fig. 5.4 cannot be used because it produces an image curvature with the same sign as that resulting from the round lenses. We must therefore look for an alternative system that produces a field curvature with negative coefficient $F_{\rm H}$; the sextupole corrector shown in Fig. 5.6 represents such a system.

In order to obtain an aplanatic system corrected for axial aberrations and off-axial coma up to the fifth-order, we do not need the sextupoles S_2 and $S_4 = S_2$ of the sextupole quintuplet. The secondary fundamental rays for this corrector are depicted in Fig. 5.8. The rays u_{11} and u_{22} are symmetric with respect to the midplane $z_{\rm m}$, while the mixed ray u_{12} is antisymmetric. If we want to adjust the image curvature and the field astigmatisms by electrical means, we must incorporate the sextupole pair S_2 and $S_4 = S_2$. These sextupoles produce additional secondary path deviations, which exhibit the same symmetry with respect to the midplane $z_{\rm m}$ as those originating from the sextupoles S_1, S_3 and $S_5 = S_1$. This behaviour can readily be verified by comparing the two sets of secondary fundamental rays shown in Figs. 5.5 and 5.8. Accordingly, the symmetry is also preserved for any linear combination formed with equivalent pairs of these rays. The aberrations introduced by the corrector depend on its location within the electron optical column because the distances of the fundamental paraxial rays u_1 and u_α and $u_2 = u_\gamma$ vary along the optic axis. The resulting aberrations will hence be affected by the telescopic intermediate magnification

$$M_{\rm c} = f_{\rm c}/f_0 \tag{5.262}$$

of the axial fundamental ray. This ray is assumed to be parallel to the optic axis at a distance $u_\alpha = f_{\rm c}$ in front of the corrector. Since the strengths of the sextupoles can be adjusted arbitrarily, it suffices to determine the aberration coefficients approximately. Employing the SCOFF approximation for the sextupole fields, we eventually obtain after a lengthy analytical calculation the following coefficients for the third-order aberrations generated by

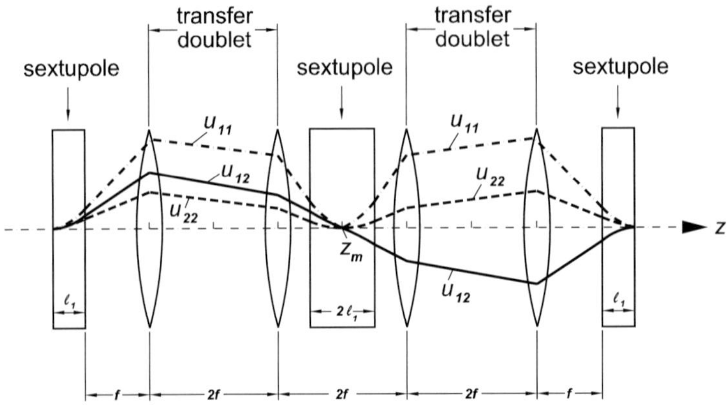

Fig. 5.8. Course of the secondary fundamental rays within the planator shown in Fig. 5.6 in the case that the sextupoles S_2 and $S_4 = S_2$ are not excited

the hexapole fields of the corrector:

$$C_{\rm H} = -3f_c^4\ell_1^3|H_1|^2 - \frac{3}{56}\frac{\ell_2^7}{f^4}f_c^4|H_2|^2 - 6\frac{\ell_1}{f}f_c^4\ell_2^3\,{\rm Re}(H_1\bar{H}_2)\,, \tag{5.263}$$

$$F_{\rm H} = -\frac{24}{5}\ell_1^5|H_1|^2 + \frac{3}{5}\ell_2^5|H_2|^2 + 4\frac{\ell_2^3}{f}\ell_1^3\,{\rm Re}(H_1\bar{H}_2)\,, \tag{5.264}$$

$$A_{\rm H} = -\frac{21}{5}\ell_1^5|H_1|^2 - \frac{3}{5}\ell_2^5|H_2|^2 - \frac{\ell_2^3}{f}\ell_1^3H_1\bar{H}_2 - 36\ell_1\ell_2f^3\bar{H}_1H_2\,, \tag{5.265}$$

$$E_{\rm H} = -\frac{6}{7}\frac{\ell_1^7}{f_c^4}|H_1|^2 - 6\frac{\ell_2^5}{f_c^2}|H_2|^2 - 24\frac{f^3}{f_c^4}\ell_2\ell_1^3\,{\rm Re}(H_1\bar{H}_2)\,, \tag{5.266}$$

$$K_{\rm H} = 0 \quad , \quad D_{\rm H} = 0\,. \tag{5.267}$$

The coefficient

$$A_{\rm H} = A_{\rm Hr} + iA_{\rm Hi} \tag{5.268}$$

of the field astigmatism is complex if the azimuthal orientation of the two sextupoles S_2 and S_4 in the rotated uz-coordinate system differs from that of the other three sextupoles, whose complex strengths $H_1 = H_3 = H_5$ are the same. The other coefficients $C_{\rm H}$, $F_{\rm H}$, and $E_{\rm H}$ are always real. The coefficients of the field curvature and astigmatism generated by the rotationally symmetric part of the electromagnetic field satisfy the Petzval relation. This relation adopts the form

$$\frac{2}{\rho_{\rm P}} = F_R - 2A_{Rr} = \frac{\Phi_o^{*\frac{1}{2}}}{16}\int_{z_o}^{z_i}\left[\frac{1+2\Lambda^2}{\Phi^{*\frac{5}{2}}}\Phi'^2 + 4\frac{e}{m_o}\frac{B^2}{\Phi^{*\frac{3}{2}}}\right]dz > 0 \tag{5.269}$$

if the electric field is zero at both the object and the image plane. Accordingly, the Petzval curvature $1/\rho_{\rm P}$ is always positive for rotationally symmetric fields if the electric field vanishes at the object and the image. In the case of short magnetic round lenses, where the focal length of each lens is large compared with the extension of its axial field, the Petzval curvature is approximately equal to the sum of the reciprocal focal lengths of all lenses located between the object and the image. The Petzval relation (5.269) also demonstrates that the coefficient F_{H} of the field curvature generated by the hexapole fields must be negative and its absolute value larger than twice that of of the field astigmatism in order that both aberrations can be eliminated simultaneously. This condition can only be fulfilled for negative values of $\mathrm{Re}(H_1 \bar{H}_2)$. Accordingly, the polarity of the sextupoles S_2 and S_4 must be chosen opposite to that of S_1, S_3 and S_5 in the rotating coordinate system.

The coefficients F_{H} and A_{H} do not depend on the distance f_{c} of the axial fundamental ray u_α in front of the corrector. Hence the action of the sextupoles on image curvature and field astigmatism is independent of the location of the corrector within the system. Since the coefficient (5.267) of the spherical aberration depends strongly on f_{c}, it should be possible to compensate for all third-order aberrations simultaneously by adjusting the free geometrical parameters $f_c/f_0, \ell_1/f_0, \ell_2/f_0$ and the hexapole strengths H_1 and H_2, appropriately.

5.4.4 Outline of a Fifth-Order Double Anastigmat

In light optics an anastigmat denotes a compound lens that is corrected for astigmatism and curvature of field. Most light-optical anastigmats are so-called double anastigmats, consisting of two symmetrical compound lenses. Owing to the symmetric arrangement of the constituent lenses with respect to the central symmetry plane of the system, the distortion and the off-axial coma cancel out. These symmetry principles have also been employed for the design of the electron-optical corrector shown in Fig. 5.6. Since this corrector can compensate for third-order field astigmatism and image curvature, we call it a "planator". Unfortunately, this planator also produces fourth-order aberrations, which prevent a substantial enlargement of the planar image area. In order to eliminate these aberrations, we again utilize symmetry considerations.

The fourth-order aberrations are derived most easily from the fifth-order term $F_{\mathrm{o}}^{(5)}$ of the perturbation eikonal F_{o}. The corresponding integrand $m_F^{(5)}$ listed in (5.207) can be written as

$$m_F^{(5)} = \mu_1^{(5)} + D^{(2)} \mu_1^{(4)} + 3 q_0 \,\mathrm{Re}\left(H \bar{u}^{(1)} \bar{u}^{(2)2} \right) . \qquad (5.270)$$

The terms on the right hand side contain only odd powers of the paraxial ray $u^{(1)}$ and of the hexapole strength H, because $u^{(2)}$ and $\mu_1^{(5)}$ depend linearly on H. The fifth-order polynomial $\mu_1^{(5)}$ consists of monomials, whose coefficients are products of powers of the fundamental paraxial rays u_α and

u_γ. These rays are either symmetric or antisymmetric with respect to the midplane $z_{\rm m}$ and the central planes of each half of the planator, as shown in Fig. 5.6. Hence all monomials are antisymmetric functions with respect to at least one of these planes. In this case the polynomial $\mu_1^{(5)}$ does not contribute to the eikonal term $F_{\rm o}^{(5)}$. The term $D^{(2)}\mu_1^{(4)}$ produces fourth-order aberrations resulting from the combination of the second-order sextupole deviations with the third-order aberrations of the round lenses located between the sextupole elements. This term consists of monomials whose coefficients are either symmetric or antisymmetric functions with respect to the midplane $z_{\rm m}$. Accordingly, only the symmetric terms produce fourth-order aberrations. Since the third term on the right-hand side of (5.270) has the same structure, the planator introduces only distinct types of fourth-order aberrations such as field astigmatism, image curvature and distortion. If we replace u_α and u_γ by their negative values the paraxial ray $u^{(1)}$ changes its sign, while the second-order path deviation $u^{(2)}$ remains unaffected.

This behaviour can be verified by means of the relation (5.249) for the second-order fundamental rays. The integrand (5.270) of the eikonal term $F_{\rm o}^{(5)}$ therefore changes sign when the sign of the paraxial fundamental rays is reversed. Hence, if we construct a system in such a way that it consists of two identical planators with opposite sign of the paraxial rays, the entire system will be free of fourth-order aberrations. Such a system can for example be obtained by connecting two identical planators of the type shown in Fig. 5.8 via a telescopic transfer doublet, which reverses the lateral positions of the paraxial rays. By incorporating such a double planator into a rotationally symmetric imaging system, we can create a system that is free of aberrations up to the fifth order. This is the case if the round lens system is free of third-order coma and distortion. A telescopic doublet consisting of two identical round lenses represents such a system in the case of unit magnification. For other magnifications that differs from unity, the proper system can be realized by four round lenses, as shown schematically in Fig. 5.9. The corrector must be incorporated in such a way that its coma-free plane, located at the front focal plane of the first corrector lens, coincides with the corresponding plane of the round-lens system. Since we can adjust two complex hexapole strengths arbitrarily, it is always possible to compensate electrically for the field astigmatism and for the image curvature or the spherical aberration. In the case of a fixed magnification it is possible to eliminate all three aberrations simultaneously by a proper choice of the free geometrical parameters $f_{\rm c}/f_0, \ell_1/f_0$ and ℓ_2/f_0. The resulting system can be considered as an electron-optical double anastigmat, free of all geometrical aberrations up to the fifth oder. The proposed corrector does not introduce a chromatic aberration of magnification. If this aberration is also eliminated for the round lens system, the double anastigmat will be achromatic, apart from the axial chromatic aberration.

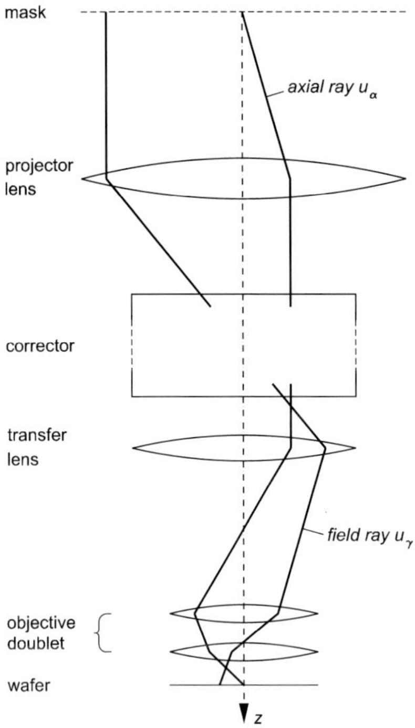

Fig. 5.9. Schematic arrangement of the double anastigmat and the round lenses forming a planar system that is free of all geometrical aberrations up to the fifth order

The proposed anastigmat is most suitable for electron projection lithography provided that the axial chromatic aberration is kept negligibly small.

5.5 W-Filter

The ultimate goal of high-resolution *analytical* electron microscopy is the acquisition of detailed information about the atomic structure, the chemical composition and the local electronic states of real objects whose structure departs from ideal crystalline periodicity. To obtain detailed information on the interatomic bonding an energy resolution of about 0.2 eV is necessary. Unfortunately, the electron microscopes currently available cannot fulfil this requirement because electron sources with a maximum energy spread of 0.2 eV at a sufficiently high current do not yet exist for conventional transmission electron microscopes. To realize such a source, we have designed an electrostatic monochromator, which reduces the energy spread of the illuminating beam [25]. The monochromator is placed behind the gun and eliminates all electrons whose energies deviate more than ±0.1 eV from the most probable energy. In the case of a Schottky field-emitter, the monochromator takes away about 70% of the emitted electrons. The incorporation of such a monochromator will make possible local electron spectroscopy with an energy resolution of about 0.2 eV, which is necessary for determining local variations of the atomic bonding near interfaces or defects. However, this information can only be obtained by means of a high-performance imaging energy filter. Such a filter must possess (a) a large dispersion to allow for sufficiently small energy windows, (b) no second-order aberrations at the image and the energy selection plane, and (c) a compact geometry to avoid undue lengthening of the microscope column. The latter requirement is especially important for corrected ultra-resolution analytical electron microscopes, because the incorporation of the monochromator and the aplanator further lengthen the column. As a result, the mechanical instabilities increase and may prevent the attainment of an information limit below 1 Å. The mechanical instabilities can be reduced by placing the heavy energy filter at the bottom of the instrument. To achieve a compact and stable microscope, it is advantageous to design it as a *twin-column* instrument, the second column containing the projector lenses and the detection system. Since the optic axis in the "image" column is parallel to that of the "object" column, the filter must also reverse the flight-direction of the electrons. Hence the total deflection of the filter must be 180°, in contrast to the straight-vision in-column filters proposed so far.

5.5.1 Geometry of the W-Filter

In accordance with the conventional nomenclature, we refer to our beam-reflecting filter as a "W-filter", owing to the pronounced W-shaped course of its optic axis. The filter can be considered as an Ω-filter placed between two bending magnets with equal deflection of the optic axis. For a total deflection of 180°, the entrance axis is parallel to the exit axis. In this case mechanical forces are avoided because the object and image columns rest on the filter, which acts as their common base. The two columns can be further stiffened by proper mechanical connections. The resulting twin column will be significantly shorter and less sensitive to mechanical instabilities than the conventional single-column electron microscopes.

The separation s of the two column axes depends on the radius of curvature R_ν and on the angular deflection Φ_ν of the optic axis in the constituent bending magnets $\nu = 1, 2, \ldots, N$ of the filter. In order to ensure that these elements focus the electrons in the two principal sections towards the optic axis, the polepieces must be conical. Such conical sector magnets can be considered as "anamorphotic" lenses with a curved axis. They provide large deflection angles yielding a high dispersion, as in the MANDOLINE filter [26]. The geometry of the tapered polepieces of an inhomogeneous sector magnet is shown schematically in Fig. 5.10.

Within the domain of validity of the SCOFF approximation, such an element produces a constant magnetic quadrupole strength Ψ_{2s} along the optic axis. In order to ensure that the filter does not introduce second-order

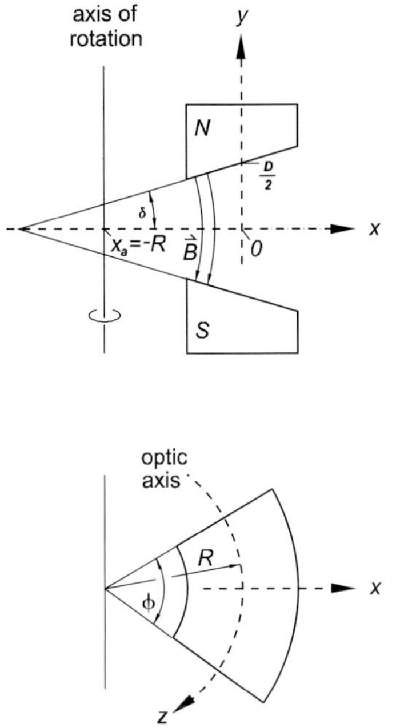

Fig. 5.10. Geometry of a conical bending magnet producing homogeneous dipole and quadrupole fields along the circular axis in the region between the tapered poles

distortions at the final image plane, pairs of identical deflection elements are arranged symmetrically about the midplane z_m of the filter, as shown in Fig. 5.11. Thanks to this symmetrical arrangement, the filter is also free of second-order axial aberrations. In order to avoid large higher-order aberrations, the fundamental paraxial rays must remain close to the optic axis throughout the filter. This condition is best achieved for a telescopic filter, where the fundamental axial rays x_α and y_β run parallel to the optic axis in front of and behind the filter.

5.5.2 Paraxial Trajectories

The properties of imaging energy filters are described most conveniently by means of a *curved xyz*-coordinate system. For conical bending magnets the curved z-axis lies in the plane midsection of the magnets. The horizontal x-axis is embedded in this section, while the vertical y-axis is perpendicular to it. Since the optic z-axis represents a possible ray, its curvature

$$\Gamma = \eta \Psi_{1s}(z) , \quad \eta = (e/2m_0\Phi^*)^{1/2} , \tag{5.271}$$

is determined by the dipole strength Ψ_{1s} of the magnets at any point along the axis. The propagation of the electron within the paraxial region of the

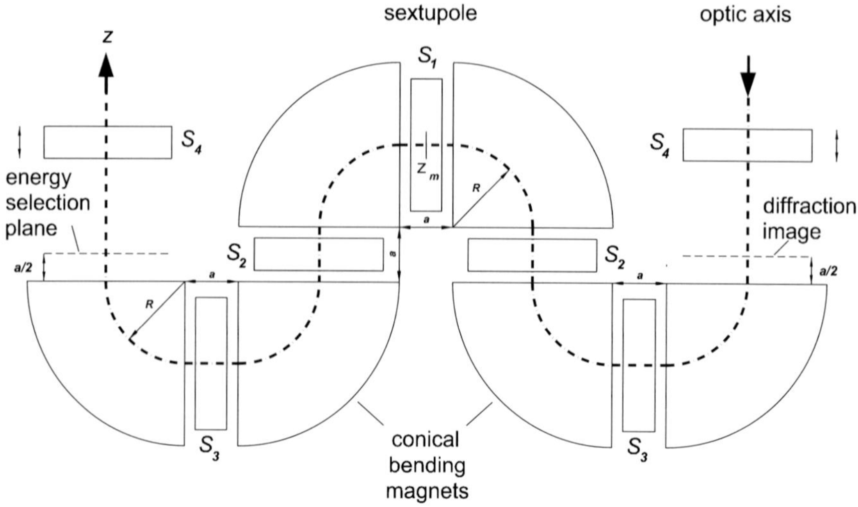

Fig. 5.11. Arrangement of the bending magnets and the sextupoles for the corrected 90° W-filter operating in the type I mode

magnets is governed by the Gaussian path equations [27,28]

$$x'' + \left(\eta^2 \Psi_{1s}^2 + 2\eta\Psi_{2s}\right) x = \kappa^* \eta \Psi_{1s}/2 \,, \tag{5.272}$$

$$y'' - 2\eta\Psi_{2s} y = 0 \,. \tag{5.273}$$

The inhomogeneous term on the right-hand side of the equation (5.272) results from the energy deviation ΔE of the electron from the nominal energy E_0. The resulting dispersion is proportional to the relativistically modified relative energy deviation

$$\kappa^* = \frac{\Delta E}{E_0} \frac{m_0 c^2 + E_0}{m_0 c^2 + E_0/2} \tag{5.274}$$

of the electron. Equation (5.272) becomes homogeneous for monochromatic electrons with nominal energy ($\Delta E = 0$). Since we restrict our considerations to the imaging energy filter, it is advantageous to define the fundamental rays at the image z_{D} of the diffraction plane located in front of the filter. The fundamental rays x_α, x_γ, y_β and y_δ are particular independent solutions of the paraxial path equations (5.272) and (5.273). In the case of a telescopic filter the axial rays x_α and y_β run parallel to the optic axis in front of and behind the filter. If these rays start with unit slope from the centre of the object plane z_{o}, they have the off-axial distance $f_0 M_{\mathrm{D}}$ in front of the filter, where f_0 and M_{D} are the focal length of the objective lens and the magnification ($M_{\mathrm{D}} \ll 1$) of the diffraction plane, respectively. Hence if we fix the axial rays by imposing the conditions

$$\begin{aligned} x_{\alpha D} &= x_\alpha(z_D) = f_0 \,, & y_{\beta D} &= f_0 \\ x'_{\alpha D} &= 0 \,, & y'_{\beta D} &= 0 \,, \end{aligned} \tag{5.275}$$

these rays will have started from the centre of the object plane with slope $1/M_D$. The two field rays x_γ and y_δ pass through the centers of the planes z_d and z_D. Here we fix these rays by imposing the conditions

$$\begin{aligned} x_\gamma(z_D) &= x_{\gamma D} = 0 \,, & y_{\delta D} &= 0 \,, \\ x'_{\gamma D} &= -1 \,, & y'_{\delta D} &= -1 \,, \end{aligned} \tag{5.276}$$

which differ from the usual conditions imposed on these rays. Owing to this choice all fundamental rays have the dimension of a length. As a result, the constants of the Helmholtz–Lagrange relations

$$x_\gamma x'_\alpha - x_\alpha x'_\gamma = f_0 \,, \quad y_\delta y'_\beta - y_\beta y'_\delta = f_0 \tag{5.277}$$

coincide with the focal length of the objective lens.

With this choice of fundamental rays, the paraxial ray is described by the linear combinations

$$\begin{aligned} x^{(1)} &= \alpha y_\alpha + \gamma x_\gamma + \kappa^* x_\kappa \,, \\ y^{(1)} &= \beta y_\beta + \delta y_\delta \,, \end{aligned} \tag{5.278}$$

where the angles α, β, γ and δ are connected with the initial parameters α_o, β_o, x_o, y_o of the ray at the object plane z_o via the relations

$$\begin{aligned} \alpha &= M_D \alpha_o & \beta &= M_D \beta_o, \\ \gamma &= \gamma_o / M_D = x_o / f_o M_D \,, & \delta &= \delta_o / M_D = y_o / f_o M_D \,. \end{aligned} \tag{5.279}$$

The dispersion ray x_κ is the inhomogeneous solution of the equation (5.272) for the special case $\kappa^* = 1$. Owing to the boundary conditions (5.276) for the field rays, the relation

$$|x_\gamma(z_i)| = |x_{\gamma i}| = |y_{\delta i}| = M_i M_D f_o \tag{5.280}$$

holds at the final image plane z_i; M_i denotes the magnification of the object at this plane. It should be remembered that the intermediate images of the diffraction plane are demagnified if those of the object are magnified.

5.5.3 SCOFF Design

Imaging energy filters have been named according to the characteristic course of their curved optic axis. For straight-vision systems the deflection angles ϕ_ν of all bending magnets add up to 0° for the "omega" filter and to 360° for the "alpha" and "gamma" filters [28]. The total deflection angle of the W-filter amounts to 180° since the optic axis reverses its direction. This angle will be in the range between 90° and 180° if the image column is inclined with respect to the object column. The curvature $\varGamma$ of the optic axis changes its sign twice within the W-filter, as it does for the $\varOmega$-filter. The W-filter must hence consist of at least two pairs of deflection magnets with opposite excitations. In order to achieve a high dispersion, conical sector magnets with large deflection angles ϕ_ν are mandatory. For simplicity, we presuppose that all N magnets have the same radius of curvature

$$R_\nu = R = 1/\varGamma\,, \quad \nu = 1, 2, \dots N\,. \tag{5.281}$$

Within the frame of validity of the SCOFF approximation, this radius is constant inside the box-shaped field of each magnet and zero outside. The quadrupole strength of the conical magnets,

$$\varPsi_{2s} = -|\varPsi_{1s}| \tan\delta/D\,, \tag{5.282}$$

depends on the inclination angle δ of the inner polefaces, the vertical distance D between these poles taken at the optic axis and on the absolute value of the dipole strength $\varPsi_{1s}$. In order to correct for the nonvanishing second-order aberrations, sextupole elements must be incorporated in pairs, arranged symmetrically about the midplane z_{m} of the system. In this case the sextupoles do not introduce distortion and axial aberrations. Owing to the symmetric arrangement of the bending magnets, these aberrations are also absent for the system as a whole. To compensate simultaneously for the axial aberration at the energy selection plane and for the aberrations (inclination of image field and field astigmatism) at the final image plane, a strongly astigmatic path of the paraxial rays within the regions between the bending magnets is mandatory [27]. The aberrations can be eliminated largely independently of each other if the sextupole elements can be placed at astigmatic images of the object plane and the diffraction plane [27,29]. For this purpose it is advantageous to insert three pairs of sector magnets, as shown in Fig. 5.11.

A sextupole centred at the midplane z_{m} of the W-filter need not be split up, since it automatically fulfils the symmetry condition. To simplify the system still further and to minimize the nonvanishing aberrations, the first and third bending magnet are made equal, but placed in opposite x-direction, in order to reverse the deflection. Owing to the required midplane symmetry, the first magnet must coincide with the sixth magnet and the third magnet with the fourth magnet. Since the deflection angles of these bending magnets cancel each other out, the second and the fifth magnet must each deflect the axis by 90° in order to achieve a total deflection of 180°. The paraxial refraction power of the entire system is symmetric with respect to the central symmetry planes at z_{s1} and z_{s2} of each half of the filter for both the xz-section and the yz-section. However, a doubly symmetric refracting power does not necessarily imply that the fundamental rays have to be either symmetric or antisymmetric with respect to these planes.

The energy selection is performed behind the filter at the energy dispersion plane z_{E}, which is conjugate to the diffraction plane z_{D} located in front of the filter. By considering the relations $x_\gamma(z_{\rm E}) = 0$ and $x_\alpha(z_{\rm E}) = f_0$, we obtain for the lateral displacement of the dispersion ray at the energy selection plane the expression

$$x_\kappa(z_{\rm E}) = C_{\gamma\kappa} = \frac{1}{2}\int_{z_{\rm E}}^{z_{\rm E}} \eta\Psi_{1s}x_\gamma dz\,, \qquad (5.283)$$

where $C_{\gamma\kappa}$ is the dispersion coefficient. In order to achieve a large dispersion, we must adjust Ψ_{1s} and x_γ in such a way that the product $\Psi_{1s}x_\gamma$ does not change its sign inside the filter. Hence if two adjacent bending magnets deflect the electrons in opposite directions, the field ray x_γ must intersect the optic axis in the field-free region between these magnets. The dispersion (displacement per eV)

$$\Delta = \kappa^* C_{\gamma\kappa}/\Delta E = \frac{m_0c^2 + E_0}{m_0c^2 + E_0/2}\frac{C_{\gamma\kappa}}{E_0} \qquad (5.284)$$

is proportional to the dispersion coefficient and influenced by relativistic effects for nominal energies E_0 larger than about 100 keV.

The field ray x_γ must run parallel to the optic axis at the midplane $z_{\rm m}$ of the filter, owing to the symmetric arrangement of the magnets about this plane. Hence the axial ray x_α must be zero at this plane to fulfill the imposed symmetry conditions. Moreover, we assume that astigmatic images of both the object plane and the diffraction plane are located at the midplane $z_{\rm m}$. This is only possible if the two line images are perpendicular to each other. To fulfill these requirements, we must impose on the fundamental field rays the conditions

$$\begin{aligned} &x'_\gamma(z_{\rm m}) = x'_{\gamma{\rm m}} = 0\,, \quad && y'_{\beta{\rm m}} = 0\,, \\ &x_{\alpha{\rm m}} = 0\,, && y_{\delta{\rm m}} = 0\,. \end{aligned} \qquad (5.285)$$

Since the fundamental rays are entirely defined by the initial constraints (5.275) and (5.276), the additional conditions (5.285) can only be fulfilled by adjusting four free parameters of the system appropriately. Such parameters are the quadrupole strengths of the conical bending magnets and the spacings between these elements. The normalized quadrupole strength,

$$\nu^2 = -2\eta\Psi_{2s}R^2 = -\frac{2}{\eta}\frac{\Psi_{2s}}{\Psi_{1s}^2}\,, \qquad (5.286)$$

defines the so-called field index of the inhomogeneous magnet. This index is zero for homogeneous bending magnets with plane-parallel inner polefaces. Since the first and third magnets are identical, we have two free field parameters $\nu_1 = \nu_3$ and ν_2. The two other parameters are the spacing a between the magnets of each half of the filter and the distance $2g$ between these halves. Hence the doubly symmetric system provides exactly the number of free parameters needed to adjust the required path of the paraxial fundamental rays. For the special system shown in Fig. 5.11, the distances a and

g are equal, as are the quadrupole strengths ($\nu_1 = \nu_2$). Although the rays x_α and x_γ do not possess any symmetry with respect to the central planes z_{s1} and z_{s2} of each half of the filter, these rays are connected by a peculiar symmetry property. Within each half of the filter the course of the axial ray x_α is mirror-symmetric with respect to that of the field ray x_γ apart from a constant factor b:

$$x_\alpha(z - z_{si}) = b\,x_\gamma(z_{si} - z)\,, \quad i = 1, 2\,. \tag{5.287}$$

In order to achieve a high dispersion, x_γ must neither be antisymmetric with respect to z_{s1} and z_{s2}, nor symmetric. However, the fundamental rays y_β and y_δ can be symmetric or antisymmetric with respect to these planes because these rays do not affect the dispersion. Since they are linearly independent of each other, one ray must be symmetric and the other one antisymmetric. Depending on the chosen symmetry for y_β and y_δ, two types of doubly-symmetric W-filters exist. We say that the filter is of type I if the course of the ray y_δ is symmetric with respect to z_{si} and of type II if it is antisymmetric. The other fundamental ray x_β exhibits the opposite symmetry. Employing the normalized axial coordinate

$$\zeta = z/R\,, \tag{5.288}$$

the SCOFF approximations of the paraxial path equations (5.272) and (5.273) adopt the form

$$\begin{aligned} &\ddot{x} + \mu_i^2 x = \operatorname{sgn}(\Psi_{1s})\,(\kappa^*/2)R\,,\\ &\ddot{y} + \nu_i^2 y = 0\,, \quad \mu_i^2 = 1 - \nu_i^2\,, \end{aligned} \tag{5.289}$$

within the region of the i-th magnet. Dots indicate differentiation with respect to ζ, and sgn denotes the sign of the argument. This sign is 1 if the magnet deflects the beam to the right-hand side and -1 otherwise. The values for the quadrupole strengths ν_i are defined by the boundary conditions (5.285). In the special case $\phi_1 = \phi_2 = \phi_3 = \phi = \pi/2$ the deflection angles ϕ_i of the magnets are equal. In this case the two resulting W-filters are entirely defined by the following SCOFF parameters:

Type I:

$$\begin{aligned} &\nu_i = \nu = 0.5573\,, \quad \mu_i = \mu = 0.8303\,,\\ &a = 2g = 0.6577R\,, \quad s = 7.93R\,,\\ &C_{\gamma\kappa} = 4.35R = 0.55s\,; \end{aligned} \tag{5.290}$$

type II:

$$\begin{aligned} &\nu_i = \nu = 0.7871\,, \quad \mu_i = \mu = 0.6168\,,\\ &a = 2g = 2.227R\,, \quad s = 12.68R\,,\\ &C_{\gamma\kappa} = 7.89R = 0.685s\,. \end{aligned} \tag{5.291}$$

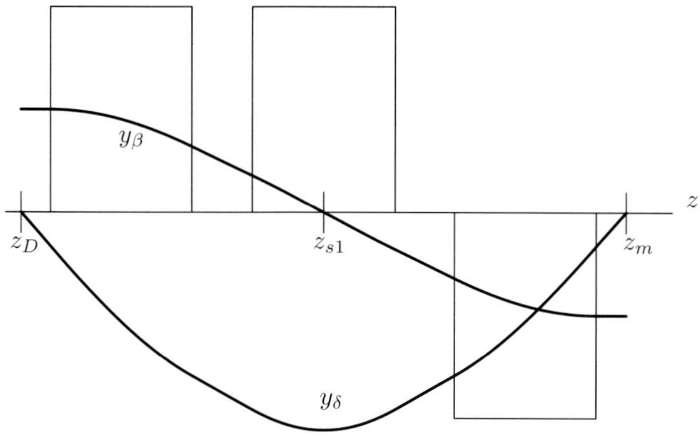

Fig. 5.12. Course of the fundamental paraxial rays along the *straightened* axis of the first half of the W-filter depicted in Fig. 5.11

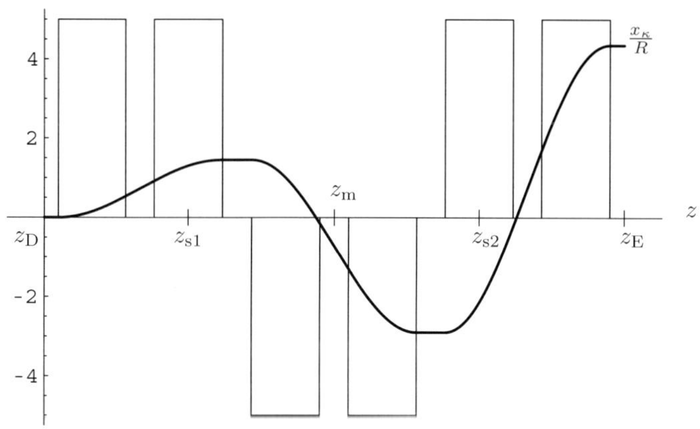

Fig. 5.13. Oscillating path of the dispersion ray x_κ along the straightened optic axis of the W-filter shown in Fig. 5.11

The schematic arrangement of the magnets for the filter of type I is shown in Fig. 5.11 together with the sextupoles for correcting the second-order aberrations. The course of the fundamental rays along the straightened optic axis within the first half of this filter is depicted in Fig. 5.12.

Owing to the relatively small spacings between the magnets, the positions of the sextupoles are largely fixed, apart from the two sextupoles that are placed in front of and behind the filter. The dispersion ray x_κ shown in Fig. 5.13 starts with zero slope in the first magnet and oscillates with increasing amplitude along the optic axis inside the filter. The ray reaches its maximum off-axial distance at the exit of the last magnet. Since x_κ leaves this magnet parallel to the optic axis, the dispersion at the energy selection plane does not depend on its distance $g = a/2$ from the filter. Comparison

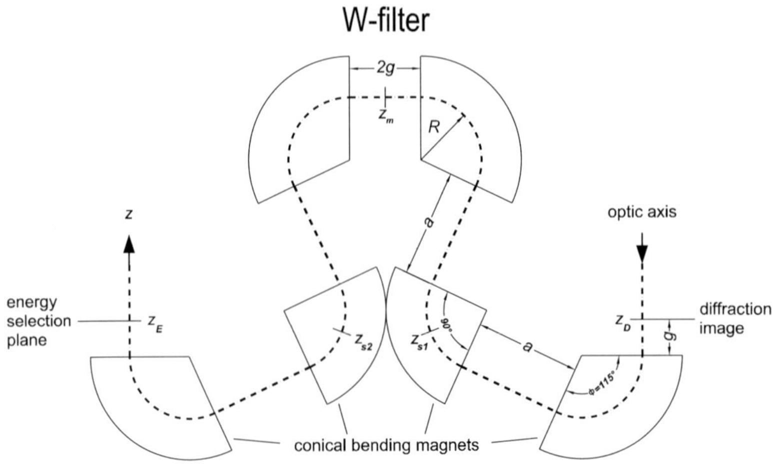

Fig. 5.14. Arrangement of the conical bending magnets for the highly dispersive 115° W-filter operating in the type II mode

of the dispersion coefficients for the two filter types reveals that the type II filter produces a significantly larger dispersion than the type I filter. Accordingly, filters of type II should be preferred in practice. Moreover, the large spacings between the magnets of this filter allow one to position more than one sextupole into each of these regions.

In order to enlarge the dispersion without increasing the distance s between the entrance axis and the exit axis, the deflection angles $\phi_1 = \phi_3$ of the first and third magnet must be larger than 90°. However, the increase in these angles cannot be made substantially larger than about 115° because the magnets must not overlap with each other or with the round lenses of the microscope. The filter depicted in Fig. 5.14 satisfies these design criteria. The SCOFF parameters of this system are:

$$\begin{aligned} \nu_1 &= 0.7906\,, \quad \nu_2 = 0.7929\,, \quad \mu_1 = 0.6123, \quad \mu_2 = 0.6094\,, \\ \phi_1 &= \phi_3 = 115°\,, \quad \phi_2 = 90°\,, \\ a &= 1.736\,R\,, \quad g = 0,581\,R\,, \quad s = 8.46\,R\,, \\ C_{\gamma\kappa} &= 7.296\,R = 0.864\,s\,. \end{aligned} \tag{5.292}$$

Comparison of the dispersion coefficient $C_{\gamma\kappa}$ with those for the systems (5.290) and (5.291) reveals that the filter (5.292) yields by far the highest dispersion for a given distance s between the axes of the two columns. The course of the paraxial rays along the *straightened* optic axis is shown in Fig. 5.15 for one half of the filter. In the vertical yz-section the axial ray y_β is symmetric and the field ray y_δ antisymmetric with respect to the central plane z_{s1}. Since such a symmetry does not exist for the corresponding rays x_α and x_γ in the horizontal xz-section, the paraxial path of the rays is largely astigmatic within the entire region of the filter.

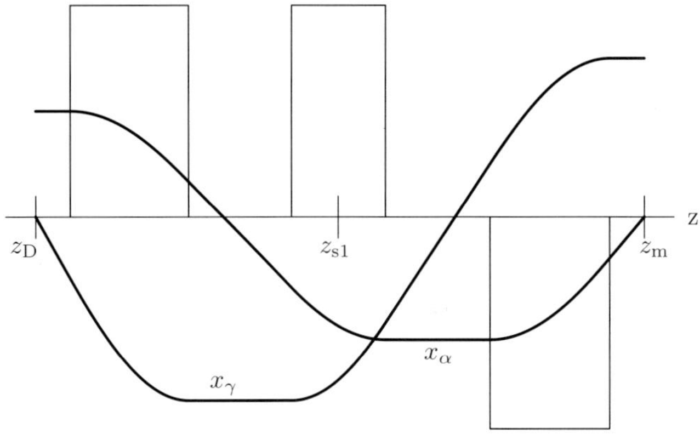

Fig. 5.15. Course of the fundamental paraxial rays along the straightened optic axis of the type II W-filter shown in Fig. 5.14

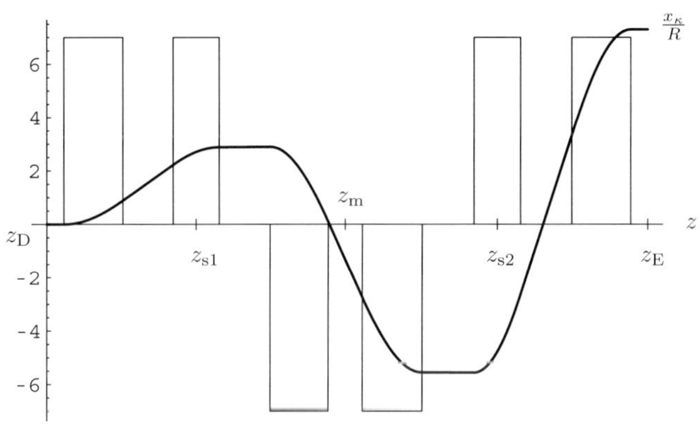

Fig. 5.16. Path of the dispersion ray x_κ within the 115° W-filter

This behaviour enables one to compensate for the nonvanishing second-rank aberrations by means of sextupoles. These correction elements should be placed at appropriate positions between the bending magnets, in order to eliminate the different aberrations largely independently. Unlike the geometrical fundamental rays, the dispersion ray x_κ does not possess any symmetry properties, as can be seen from Fig. 5.16.

The quadrupole strengths ν_1 and ν_2 of the 115° W-filter differ only slightly from each other. For practical reasons it would be very desirable to employ a system whose magnets have the same quadrupole strength. Therefore, the question arises whether such a system with a deflection angle close to 115° does indeed exist. The calculations have shown that this is indeed the case, for a deflection angle $\phi_1 = \phi_3 = 121.87°$. The parameters of the corresponding

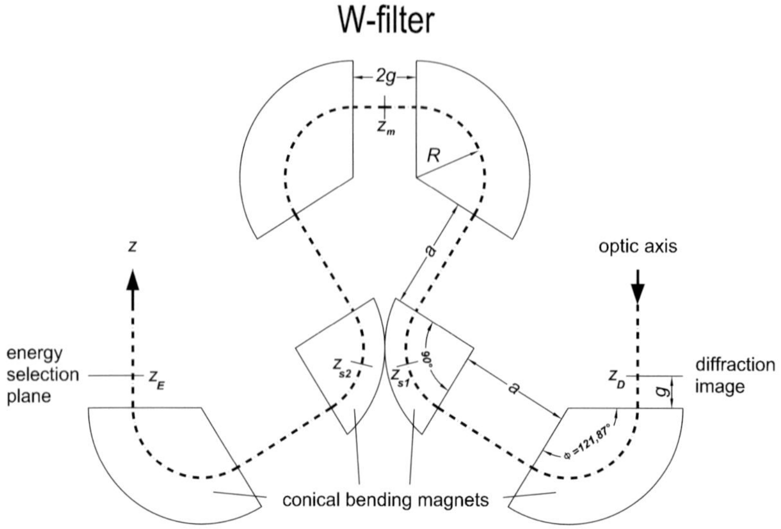

Fig. 5.17. Arrangement of the doubly symmetric W-filter in the case that the inclination angles of the tapered polepieces coincide for all magnets

filter are

$$\begin{aligned} \nu &= 0.7925\,, \quad \mu = 0.6099 \\ a &= 1.613\,R\,, \quad g = 0.46\,R\,, \quad s = 7.35\,R\,, \\ C_{\gamma\kappa} &= 7.26\,R = 0.988\,s\,. \end{aligned} \tag{5.293}$$

The courses of the fundamental paraxial rays and that of the dispersion ray within this filter hardly differ from those of the 115° W-filter. The dispersion coefficient (5.294) of the filter roughly coincides with the distance s between the axes of the two constituent columns of the microscope.

Unfortunately, the two symmetric branches of the optic axis come very close to each other within the region of the central bending magnets, as demonstrated in Fig. 5.17. Since the polefaces of the adjacent magnets are inclined in opposite directions, the quadrupole field of each magnet will be affected by the adjacent magnet. If this effect prevents a precise alignment of the filter, one can substitute a homogeneous magnet for the two central conical 90° deflection magnets. The resulting system forms a generalized MANDOLINE filter. The parameters of these filters must be determined by the calculation procedures that have been employed for the MANDOLINE filter [26].

5.5.4 Second-Rank Aberrations

The primary aberrations of imaging energy filters are of second rank in the ray-defining parameters α, β, γ, δ and κ^*. According to the nature of these parameters, the aberrations can be subdivided into two classes, one comprising the second-order geometric aberrations and the other one the second-rank chromatic aberrations. The *order* of an aberration is defined as the sum of the exponents of the geometric ray parameters, while the exponents of the ray parameter κ^* describes the *degree* of the aberration. The sum of the order and the degree is called the *rank*. It is an appropriate measure for the magnitude of an aberration [29]. The aberrations represent the deviation of the electron trajectory from its paraxial approximation at the observation plane. In an energy-filtering electron microscope this plane is either the final image plane z_{i} or the energy selection plane z_{E} where the energy-loss spectrum is formed. The second-rank aberrations are obtained most conveniently from the third-rank term of the perturbation eikonal taken at these planes. This third-rank eikonal consists of a geometric term and a chromatic term, which has not been considered in the perturbation formalism outlined in Sect. 5.3 of this chapter. The geometric term is entirely produced by the dipole and hexapole fields of the energy filter, while the chromatic term also contains contributions from the round lenses and quadrupoles of the entire electron-optical system. Here we consider only the contribution of the energy filter to the second-rank aberrations. In this case the corresponding third-rank eikonal $E_{\mathrm{F}}^{(3)}$ of the filter is constant in the region behind the filter; in particular

$$E_{\mathrm{F}}^{(3)} = E_{\mathrm{F}}^{(3)}(z_i) = E_{\mathrm{F}}^{(3)}(z_{\mathrm{E}}) . \tag{5.294}$$

To simplify the expressions for the aberrations, we introduce the modified eikonal

$$\tilde{E}_{\mathrm{F}}^{(3)} = E_{\mathrm{F}}^{(3)}/q_{0\mathrm{o}} , \tag{5.295}$$

which has the dimension of a length. It represents the third-order deviation of the optical path-length from its paraxial approximation. The eikonal term $\tilde{E}_{\mathrm{F}}^{(3)}$ is composed of at most 18 monomials in the expansion parameters. Ten of these monomials are entirely of geometric nature. In systems with midplane symmetry five of these monomials vanish and $\tilde{E}_{\mathrm{F}}^{(3)}$ adopts the reduced form

$$\begin{aligned}
\tilde{E}_{\mathrm{F}}^{(3)} &= A_{\alpha\alpha\gamma}\alpha^2\gamma + \frac{1}{3}A_{\gamma\gamma\gamma}\gamma^3 \\
&\quad + \frac{1}{2}B_{\gamma\beta\beta}\gamma\beta^2 + B_{\alpha\beta\delta}\alpha\beta\delta + \frac{1}{2}B_{\gamma\delta\delta}\gamma\delta^2 \\
&\quad + \left(C_{\alpha\alpha\kappa}\alpha^2\kappa^* + 2C_{\alpha\gamma\kappa}\alpha\gamma\kappa^* + C_{\gamma\gamma\kappa}\gamma^2\kappa^*\right)/2 \\
&\quad + \left(C_{\beta\beta\kappa}\beta^2\kappa^* + 2C_{\beta\delta\kappa}\beta\delta\kappa^* + C_{\delta\delta\kappa}\delta^2\kappa^*\right)/2 \\
&\quad + C_{\alpha\kappa\kappa}\alpha\kappa^{*2} + C_{\gamma\kappa\kappa}\gamma\kappa^{*2} .
\end{aligned} \tag{5.296}$$

At the observation planes z_{i} and z_{E} two of the four paraxial fundamental rays vanish. As a result, the second-rank aberrations at these planes are related to the third-rank eikonal via the expressions

$$x^{(2)}(z_{\rm i}) = x_{\rm i}^{(2)} = -\frac{x_{\gamma\rm i}}{f_0}\frac{\partial \tilde{E}_F^{(3)}}{\partial \alpha}, \quad y_{\rm i}^{(2)} = -\frac{y_{\delta\rm i}}{f_0}\frac{\partial \tilde{E}_F^{(3)}}{\partial \beta}, \tag{5.297}$$

$$x^{(2)}(z_{\rm E}) = x_{\rm E}^{(2)} = \frac{x_{\alpha\rm E}}{f_0}\frac{\partial \tilde{E}_F^{(3)}}{\partial \gamma}, \quad y_{\rm E}^{(2)} = \frac{y_{\beta\rm E}}{f_0}\frac{\partial \tilde{E}^{(3)}}{\partial \delta}. \tag{5.298}$$

For symmetric filters we have

$$|x_{\alpha\rm E}/f_0| = |y_{\beta\rm E}/f_0| = M_{\rm E} = 1, \tag{5.299}$$

because such filters image the plane $z_{\rm D}$, located in front of the filter, with the unit magnification $M_{\rm E} = 1$ into the energy selection plane $z_{\rm E}$ behind the filter. If we refer the image aberrations back to the object plane, the relations (5.297) must be divided by the total magnification

$$M_{\rm i} = x_{\gamma\rm i}/f_0 M_{\rm D} = y_{\delta\rm i}/f_0 M_{\rm D} \tag{5.300}$$

yielding

$$x_{\rm o}^{(2)} = M_{\rm D}\frac{\partial \tilde{E}_F^{(3)}}{\partial \alpha}, \quad y_{\rm o}^{(2)} = M_{\rm D}\frac{\partial \tilde{E}_F^{(3)}}{\partial \beta}. \tag{5.301}$$

The axial chromatic aberration of the filter is determined by the coefficients $C_{\alpha\alpha\kappa}$ and $C_{\beta\beta\kappa}$ of the eikonal term (5.296). Since these coefficients are proportional to the square of the ratio $f_0/R \ll 1$, the corresponding monomials in the expression (5.296) are negligibly small. For symmetric filters the coefficients of the chromatic aberration of magnification are related to those of the field aberrations by the relations

$$C_{\alpha\gamma\kappa} = A_{\alpha\alpha\gamma}C_{\gamma\kappa}/f_0, \quad C_{\beta\delta\kappa} = B_{\alpha\kappa\delta}C_{\gamma\kappa}/f_0. \tag{5.302}$$

The chromatic aberrations with coefficients $C_{\delta\delta\kappa}$, $C_{\alpha\kappa\kappa}$ and $C_{\gamma\kappa\kappa}$ are negligibly small at the final image plane and of no concern for the resolution of the energy-loss spectrum located at the plane $z_{\rm E}$ [28]. The remaining coefficients have been calculated for the doubly symmetric 115° W-filter by employing the SCOFF approximation. The resulting values

$$\begin{aligned} &A_{\gamma\gamma\gamma} = 17.51R, \quad B_{\gamma\delta\delta} = -6.48R, \\ &A_{\alpha\alpha\gamma} = 6.48f_0^2/R, \quad B_{\gamma\beta\beta} = 7.36f_0^2/R, \\ &B_{\alpha\beta\delta} = 0.066f_0^2/R, \quad C_{\gamma\gamma\kappa} = 34.08R, \end{aligned} \tag{5.303}$$

demonstrate that the coefficients $A_{\gamma\gamma\gamma}$ and $B_{\gamma\delta\delta}$ of the aperture aberration at the energy-selection plane and the coefficient $C_{\gamma\gamma\kappa}$ of the axial chromatic aberration at this plane are much larger than those of the field aberrations, because the focal length f_0 of the objective lens is very small compared with the radius of curvature R of the bending magnets. Although the coefficient $A_{\alpha\alpha\gamma}$ is small, the coefficient

$$C_{\alpha\gamma\kappa} = A_{\alpha\alpha\gamma} C_{\gamma\kappa} / f_0 = 47.3 f_0 \tag{5.304}$$

of the chromatic aberration of magnification is rather large. It is therefore necessary to eliminate this aberration, together with the axial aberrations at the energy-selection plane, by external sextupoles. These elements must be centred in pairs symmetrically about the midplane of the filter along the optic axis. In order to correct the individual aberrations largely independently of each other, it is advantageous to place the sextupoles at or near astigmatic images of both the object plane and the diffraction plane. The location of these images are given by the zeros of the paraxial fundamental rays x_α, y_β, x_γ and y_δ.

5.6 Conclusion

In the last decade considerable theoretical and experimental progress has been made in the field of electron optics. Thanks to the development of efficient procedures for calculating and designing complex electron-optical systems, for example, it is now possible to realize monochromators, aberration correctors, imaging energy filters, and quadrupole projector systems, to name only a few. The incorporation of these elements in an TEM will make quantitative analytical electron microscopy with sub-Ångström spatial and sub-eV energy resolution possible in the near future.

To keep the parasitic mechanical instabilities as small as possible, the instrument must be short and compact. All parts of the microscope are then easily accessible. A single-column high-performance electron microscope can hardly meet these conditions. However, by employing the W-filter the column can be split up into two parts, as illustrated in Fig. 5.18. The object column of this sub-eV sub-Ångström microscope (SESAM 2000) contains the field-emission gun, the monochromator, the condenser system, the corrected objective lens, and several intermediate lenses. The image column comprises the projector system, the CCD camera, and the viewing screen. The W-filter serves as a common solid base for both columns. Owing to this compact design, the instrument will be very stable. Assuming a separation distance of $s = 60\,\mathrm{cm}$ between the two column axes and a voltage of 200 kV, a dispersion of $3\,\mu\mathrm{m/eV}$ is obtained at the energy selection plane of the filter. This plane or the filtered achromatic image plane is imaged by the projector system onto the detector or the viewing screen, respectively.

To achieve a variable magnification in the range between 8 and 300 in the electron spectroscopic imaging mode, two quadrupole triplets must be used [21]. This system also allows one to image the stigmatic energy-loss spectrum at the dispersion plane with variable magnification into a line image at the recording plane. Hence, unduly large intensities are avoided for all pixels of the CCD camera.

The successful correction of the spherical aberration by Haider et al. [30,31] has been a decisive step on the route towards the realization of the SESAM. The basis of this success is the development of a sophisticated computer-assisted alignment procedure, which enables precise adjustment of the corrector to be completed within a short period of time.

Although computers are invaluable tools for accurate determination of the geometry of the electrodes and/or polepieces of systems consisting of numerous components, they can hardly be used as a substitute for an experienced designer! Optimum electron optical systems must satisfy many constraints, which cannot all be expressed in precise mathematical terms. Moreover, these systems consist of a large number of free parameters, which must be optimized. Without a considerable amount of preknowledge the computer is un-

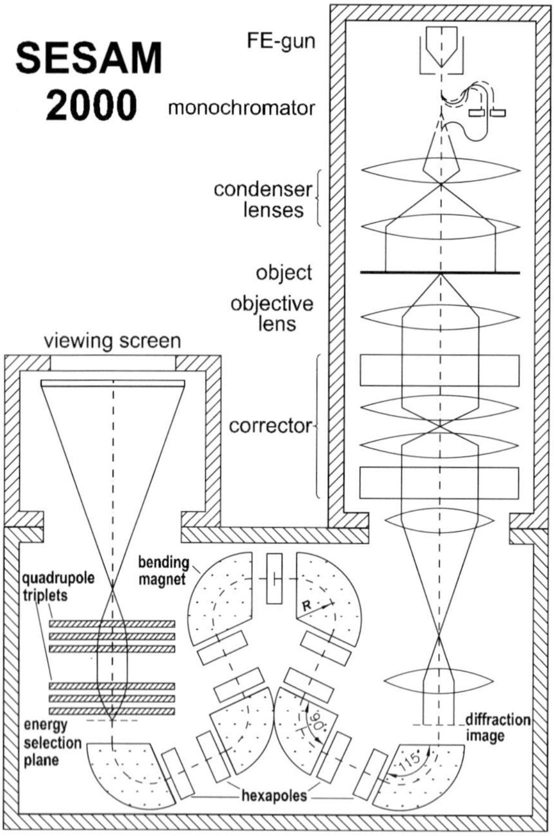

Fig. 5.18. Scheme of the SESAM 2000 illustrating the arrangement of the constituent elements

likely able to find the absolute optimum in the high-dimensional parameter space: intuition and extensive experience are necessary to find the system that meets all requirements in the best possible way. The recent success in the correction of aberrations should lead to a renaissance of electron optics. With the advancement in technology, the performance of the new instruments will no longer be limited by parasitic incoherent aberrations resulting from mechanical and electromagnetic instabilities. We can expect that future electron-optical systems will reach a degree of perfection comparable to that attained in light optics.

References

1. Sakurai J.J. (1994) Modern Quantum Mechanics. Addison-Wesley, New York
2. Glaser W. (1952) Grundlagen der Elektronenoptik. Springer, Wien
3. Sommerfeld A. (1949) Vorlesungen über Theoretische Physik, Vol. IV, Optik Dietrich'sche Verlagsbuchhandlung, Wiesbaden
4. Lippmann B. A., Schwinger J. (1950) Variational principles for scattering processes I. Phys Rev 79: 469-480
5. Scherzer O. (1949) The theoretical resolution limit of the electron microscope. J Appl Phys 20: 20-29
6. Thon F. (1966) Zur Defokussierungsabhängigkeit des Phasenkontrastes bei der elektronenmikroskopischen Abbildung. Z Naturforsch 21a: 476-478
7. Zemlin F., Weiss K., Schiske P., Herrmann K.H. (1977) Coma-free alignment of high-resolution electron microscopes with the aid of optical diffractograms. Ultramicroscopy 3: 49-60
8. Rose H. (1990) Outline of a spherically corrected semiaplanatic medium-voltage transmission electron microscope. Optik 85: 19-24
9. Haider M., Rose H., Uhlemann S., Kabius B., Urban K. (1998) Towards 0.1 nm resolution with the first spherically corrected transmission electron microscope. J Electr Micr 47: 395-405
10. Scherzer O. (1936) Über einige Fehler von Elektronenlinsen. Z Phys 101: 593-603
11. Born M., Wolf E. (1999) Principles of Optics, 7th edn. Cambridge Univ. Press, Cambridge
12. Schwarzschild K. (1905/1906) Untersuchung zur geometrischen Optik. Abhandlungen der Gesellschaft der Wissenschaften in Göttingen. Mathematisch-Physikalische Klasse. Teil 1: 3-31, Teil 2: 3-28, Teil 3: 3-54
13. Glaser W. (1933) Über geometrisch-optische Abbildung durch Elektronenstrahlen. Z Phys 80: 452-464
14. Sturrock P.A. (1952) Perturbation characteristic functions and their application to electron optics. Proc Roy Soc A 210: 269-289
15. Rose H., Petri U. (1971) Zur systematischen Berechnung elektronenoptischer Bildfehler. Optik 33: 151-165
16. Zach J., Haider M. (1995) Aberration correction in a low-voltage SEM by a multipole corrector. Nucl Instrum Meth Phys Res A 363: 316-325
17. Uhlemann S, Haider M. (1998) Residual wave aberrations in the first spherical aberration corrected transmission electron microscope. Ultramicroscopy 72: 109-119

18. Rose H. (1987) Hamiltonian magnetic optics. Nucl Instrum Meth Phys Res A 258: 374-401
19. Beck V. (1979) A hexapole spherical aberration corrector. Optik 53: 241-255
20. Rose H. (1981) Correction of aperture aberrations in magnetic systems with threefold symmetry. Nucl Instrum Meth 187: 187-199
21. Rose H. (1999) Prospects for realizing a sub-Å sub-eV resolution EFTEM. Ultramicroscopy 78:13-25
22. Hawkes P.W., Kasper E. (1996) Principles of Electron Optics, Vol. 1: Basic geometrical optics. Academic Press, London
23. Scherzer O. (1970) Communication in his lecture on Electron Optics, unpublished
24. Rose, H. (1971) Elektronenoptische Aplanate. Optik 34: 285-311
25. Kahl F. (1999) Design eines Monochromators für Elektronenquellen. PhD Thesis, Darmstadt University of Technology, Darmstadt
26. Uhlemann S., Rose H. (1994) The MANDOLINE filter – a new high-performance imaging filter for sub-eV EFTEM. Optik 96: 163-178
27. Rose H. (1978) Aberration correction of homogeneous magnetic deflection systems. Optik 51: 15-38
28. Lanio S. (1986) High-resolution imaging magnetic energy filters with simple structure. Optik 73: 99-107
29. Rose H., Krahl D. (1995) Electron optics of imaging energy filters. In: Reimer L. (Ed.) Energy-Filtering Transmission Electron Microscopy. Springer, Berlin Heidelberg, 43-149
30. Haider M., Uhlemann S., Schwan E., Rose H., Kabius B., Urban K. (1998) Electron microscopy image enhanced. Nature 392: 768-769
31. Haider M., Rose H., Uhlemann S., Schwan E., Kabius B., Urban K. (1998) A spherical-aberration corrected 200 kV transmission electron microscope. Ultramicroscopy 75: 53-60

6 Tomography by Atom Probe Field Ion Microscopy

T. Al-Kassab, H. Wollenberger, G. Schmitz, and R. Kirchheim

6.1 Introduction

The Field Ion Microscope (FIM) introduced by E.W. Müller was the first instrument capable of imaging metal surfaces with atomic resolution in real space [1–7]. The FIM was originally employed to solve problems related to surface physics and crystallography. The field of application substantially widened when the imaging mode was supplemented by the atom probe mode. This technique finally gives the chemical composition of a sample volume with atomic spatial resolution, and the sensitivity of the chemical analysis does not depend on the atom species. These advantages led to increasing and fruitful use of the method in metal physics, materials science and engineering. While the in-depth resolution corresponds to the atomic plane distance parallel to the sample axis, the lateral resolution is essentially determined by the structure of position-sensitive ion detectors that monitor the original atom positions at the sample surface before field evaporation. Such detectors, although well known for many years in numerous areas of application, have been improved substantially for use in atom probes during the past decade. In parallel, fast and extended data acquisition and handling systems have also been developed. This chapter reviews this development and presents a number of results from materials research.

6.2 Experimental Technique

Here we briefly introduce the basic design and the main features of the apparatus. First, the physical principles and the design of FIM, as well as the mode of image formation are presented. As there are various textbooks [1–7] describing in detail the different theoretical aspects of the field ionisation of imaging atoms and field evaporation of specimen surface atoms, we describe the relevant phenomena only as far as is necessary to understand the operation of the instruments and interpretation of the data. The development of the different detectors, in particular the position-sensitive detectors (PSD), is described in more detail. In addition, the determination of ion-impact position is explained and compared for the different detectors.

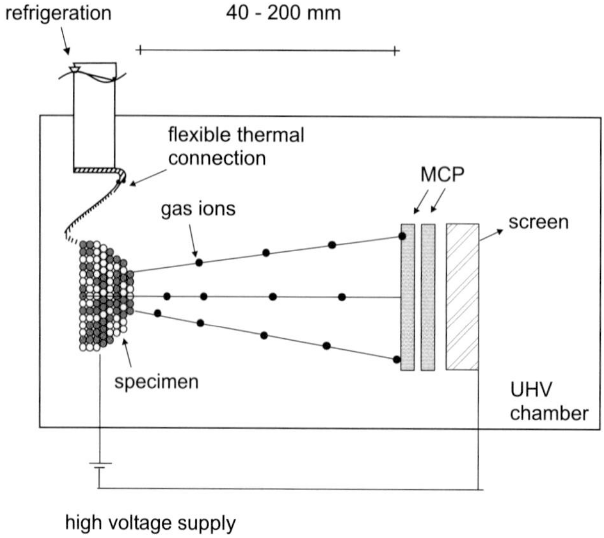

Fig. 6.1. Schematic drawing of a Field Ion Microscope (FIM)

6.2.1 The Field Ion Microscope (FIM)

The Field Ion Microscope (FIM) allows electrically conducting surfaces to be imaged with atomic resolution. The basic principle of the image formation differs considerably from those of optical and electron microscopes. As a consequence, neither diffraction theory nor Fourier analysis is needed to interpret the image. With the FIM, the image is produced by applying a high voltage to the specimen with respect to the channel plate screen in the presence of a so-called imaging gas. The field-strength required at the sample surface may reach 50 V/nm. In order to achieve this order of magnitude, needle-shaped specimens are prepared with an average radius of curvature of R = 10–100 nm at the tip. Figure 6.1 shows a schematic drawing of the FIM. The specimen (subsequently referred to as the tip) is introduced into an ultra-high vacuum (UHV) chamber with a residual pressure of 10^{-7} Pa. The tip faces the imaging system, consisting of a single or a chevron-type Micro Channel Plate (MCP) and a phosphor screen, situated at a distance of L = 4–20 cm from the specimen. During operation, the tip is kept at a temperature between 10–150 K. The magnification of the FIM is given by the relation,

$$M = \kappa L/R . \tag{6.1}$$

where κ is an image compression factor of approximately 0.7. It accounts for the fact that the tip is not ideally spherical. The average tip radius of curvature can be either exactly determined from geometrical considerations from the field ion image (see for example [2]) or estimated from the applied voltage, as discussed in Sect. 6.3.3.

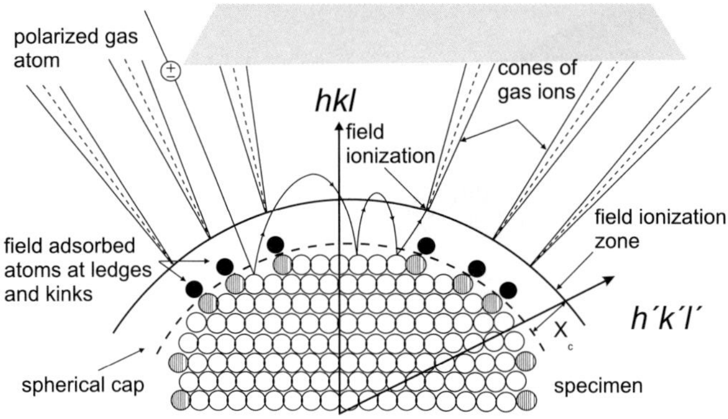

Fig. 6.2. Schematic drawing of the field ionization process of gas atoms over the high field regions of the hemi-spherical tip surface. Protruding atoms are shaded

FIM Image Formation. The so-called 'imaging gas', for example He, Ne, Ar, H_2 or mixtures of these, is leaked into the chamber with a pressure of 10^{-3} Pa. A positive high voltage (V_{DC}) is applied to the tip. The gas atoms or molecules become polarized owing to the induced electric field close to the tip surface if the applied field reaches a critical value. These atoms are attracted by the positively charged tip surface.

By successive inelastic collisions with the tip, the polarized atoms are cooled down to the tip surface temperature and are trapped in the ionisation zone, approximately 0.5 nm above the tip surface (X_c in Fig. 6.2), where electron tunneling may ionize them. After ionisation, the gas ions are accelerated along the field trajectories towards the imaging system. Since these gas atoms (or molecules) are cooled close to the tip temperature before ionization, the image quality and, hence, the lateral resolution of FIM are strongly dependent on the temperature of the tip.

The MCP acts as an electronic amplifier. After impinging on the surface of the MCP, the incoming gas ions generate electron clouds, which illuminate the phosphor screen. A few thousands of gas atoms per second must be ionized at the same site of a surface atom in order to produce an observable image on the phosphor screen. For a hemispherically shaped tip (as in the schematic drawing of Fig. 6.2), the surface atoms located on ledges or kinks at the edges of different stacks of lattice planes protrude from the surface. The tunneling probability of electrons from the gas atoms to surface atoms located at these sites, and hence the gas atom ionization probability, is very high with the result that these surface atoms image with high brightness.

The intersection of each stack of *hkl* atomic layers with the hemispherical tip, that is the plane edge, may give rise to an image. For each stack of *hkl* atomic layers the edges appear as concentric rings. This geometrical arrangement of atoms with locally high electric field strength explains the well known crystallographic pattern of FIM images of pure metals and dilute alloys as

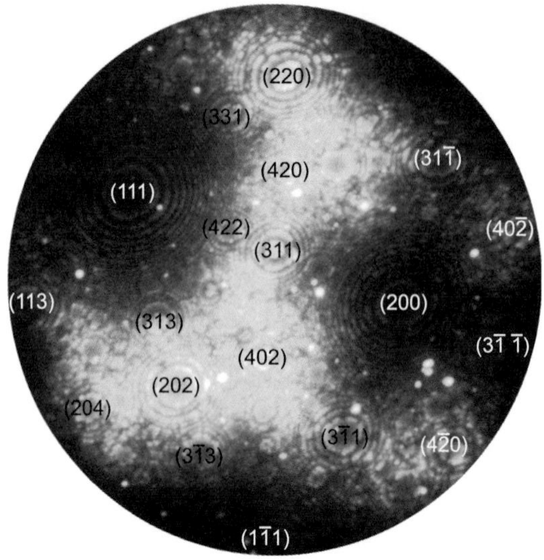

Fig. 6.3. Indexed field ion image of face-centred cubic Al-3at.% Ag

shown in Fig. 6.3. As the image is formed in real space, and predominantly originates from a single crystal (most polycrystalline specimens have a grain size of a few μm), the atomic layer edges of all lattice orientations are visible in FIM image. Several different concentric rings are therefore observed in the image, indicating the crystallographic orientation of the tip. Often, the symmetry of the image allows direct assignment of Miller indices to the different crystallographic *hkl* poles. The image resembles a stereographic projection of the tip surface (Fig. 6.3).

The ionization probability depends strongly on the chemical nature of the tip surface and, as discussed above, on the local arrangement of the different atom species on the tip surface. Mainly as a result of these dependencies, the image contrast of decomposed alloys arises from precipitates, the composition of which is different from that of the surrounding matrix. Images from disordered supersaturated alloys or even amorphous alloys are usually characterized by the absence of regular ring patterns, a consequence of the disordered distribution of the different constituents in the *hkl* atomic layers.

Field Evaporation. The influence of the electric field on the surface atoms of the tip is not limited to the lowering of the tunneling barrier for the electrons of the gas atoms and their ionization. The second important effect is the field-induced desorption or 'field evaporation', that is, removal of surface atoms allowing time-of-flight mass spectrometry. The threshold field-strength for evaporation is a function of the sublimation energy of the specific atoms and is therefore a materials parameter, known as the evaporation field-strength, F_e. The value of the F_e for different pure metals has been predicted theoretically and measured experimentally (these values are tabulated in [1,3–7]). Different models have been developed to explain the single and multicharged ionization of the surface atoms. All these models are based on the idea that one or more electrons from the surface atoms tunnel into the tip.

The surface atoms at ledges or kink-sites are less strongly bound than atoms within the planes (see Fig. 6.2) and are therefore more likley to be ionized. By increasing continuously the voltage at the tip, i.e. the applied field strength F_e at a given radius R, the surface atoms are eventually field-evaporated from the uppermost atomic layer along the edge or the corresponding first ring in the FIM image of the corresponding *hkl* plane. In the course of field evaporation, the diameter of this ring decreases until it is completely desorped, indicating the removal of exactly one plane of the *hkl*-stack. This process can be followed on the imaging screen because the field evaporation occurs in general at only slightly higher field strengths than those required for ionization of the imaging gas.

The field evaporation has many important advantages. The freshly prepared tip can be continuously field-evaporated by increasing the applied high voltage, thus cleaning oxides and other impurities from the surface as well as smoothing irregular sharp protrusions on the surface, produced during the specimen preparation, until a smooth nearly hemispherical shape of the tip is obtained. The applied field at which a fully developed field-ion image results (as in Fig. 6.3) is called the best image field (BIF) and the corresponding voltage is called the best image voltage (BIV). Field evaporation usually occurs above the BIF and controlled removal and identification of single surface atoms can hence be performed after the tip has been shaped. This is the key idea of the Atom Probe Field-Ion Microscopy (APFIM).

6.2.2 The Atom Probe (APFIM)

The combination of the FIM with a time-of-flight mass spectrometer (ToFMS) allows chemical identification of single atoms that have been field-evaporated from the tip surface. The basic design consists of a FIM, as sketched in Fig. 6.4, since an imaging device is needed to shape and image the tip. In the center of the imaging system a small 'probe hole' is situated. The field-evaporated surface ions that pass through the field-free probe hole will impinge on the detector. The detector consists of a chevron type MCP and a phosphor screen. Chemical identification of the different ions is achieved by measuring the time-of-flight from the specimen surface to the detector.

The difference between the BIF and the evaporation field strength can be created by superposing an ultrashort voltage pulse (V_P) a few ns in width and some kV in amplitude to the permanent BIV. From the experimental point of view, the amplitude of V_P is usually selected to give a ratio of 15% to 20% with respect to the BIV, and must be high enough to remove the specimen atoms by field evaporation. The ion arrival rate is usually chosen between 0.05 and

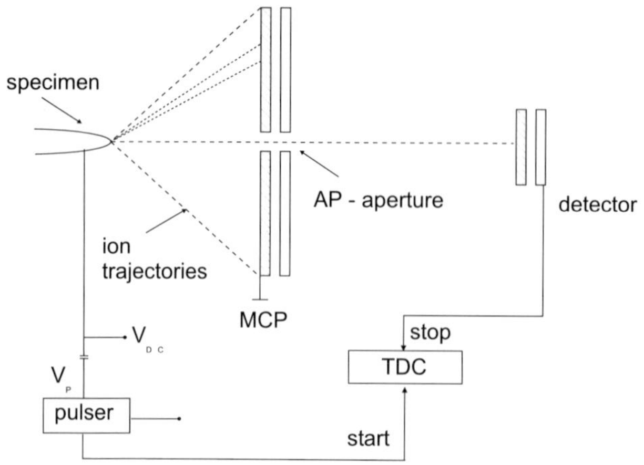

Fig. 6.4. Schematic drawing of the APFIM

0.005 detected ions per applied pulse. The pulse fraction (pulse to permanent voltage) is a very important parameter for the analysis of a multicomponent alloy, as the evaporation field-strength depends on the local chemistry of the surface. For all the different constituents, the corresponding sublimation energies must be supplied to initiate the field evaporation. By correct selection of V_{P} and setting of the analysis temperature, the preferential evaporation of any constituent with respect to the others is avoided and a reliable analysis can be achieved.

Together with the pulse applied to the tip, a synchron pulse is used to reset and start the clock (a time to digital converter TDC of the CAMAC type). The field-evaporated ion reaches its drift velocity, to a good approximation, shortly after the desorption event and drifts with this velocity v towards the detector distant L from the apex of the tip. Conservation of energy for the drifting ion implies:

$$ne(V_{\mathrm{DC}}+V_{\mathrm{P}}) = (1/2)mv^2 = (1/2)m(L/t)^2 \,. \tag{6.2}$$

After the time-of-flight t the ion hits the detector and the signal is used to stop the clock. From (6.2), the measured quantities give the mass-to-charge ratio (m/n) of the field-evaporated ion.

The total AP analysis is performed by adjusting the pulse frequency, that is the number of applied pulses per second to the tip (up to 2 kHz), and thus by gradual removal of the surface ions, layer by layer, proceeding into the volume of the tip. For each ion the ToF is measured and the corresponding value of (m/n) is evaluated. The projection of the probe hole onto the surface of the tip defines the area from which the field-desorped ions originate, and thus the diameter of the specimen cylinder analysed. For a conventional AP, this diameter is 1–5 nm. The length of this cylinder is given by the number of hkl planes evaporated. From these data the concentration of the different species of atoms are plotted versus the depth and concentration depth profiles can then be obtained [2–7]. The mass resolution of the AP, $(m/n)/\Delta(m/n)$, depends strongly on the shape of the applied pulse and the kind of ToFMS. The conventional AP with a straight flight path has usually a mass resolution of 'full width at half maximum' (FWHM) of $(m/n)/\Delta(m/n) \approx 250$. Energy-focusing ToFMS, for example those of a Poschenrieder or reflectron type, significantly improve this mass resolution to about 2000 [7].

The data obtained from such an analysis originate from the area projected by the probe hole onto the tip. The detector is used to measure the incoming atoms as a linear sequence. As a consequence, the information that can be deduced from the data is the composition of this region for each layer. No information is obtained about the distribution of the different constituents within an atomic layer. This limitation has been overcome by the development of position-sensitive detectors (PSD).

6.2.3 The Position-Sensitive Detectors (PSD)

The position-sensitive detectors supply information not only about the ToF of the impinging ions but also about their position at the specimen surface before desorption. The first proposal of attaching a position-sensitive detector to an AP was made by Walko and Müller [8] and put into practice by Panitz [9] in the Imaging Atom Probe (IAP).

This first design used a spherically curved double MCP and a screen assembly. Owing to the geometry of the microscope, the flight distance from the tip to the detector ($\approx 100\,\mathrm{mm}$) was identical for all field-evaporated ions. The acceptance angle was therefore larger than for plane detectors, and a large area of the tip surface could be mapped. The MCP was activated by a delayed synchron pulse and operated at maximum gain (10^7). Only the ions of a specific state of charge for a given element were able to trigger the MCP during the active period of time, and only these appeared as bright spots on the screen. All information on other elements or charge states of the same element were lost for the corresponding evaporation pulse. In addition, the ion trajectories in the IAP are not as simple as in the AP, owing to the rather large acceptance angle, leading to a rather poor spatial resolution of the gated images. The acquisition procedure was less refined, as the two-dimensional image was processed after each pulse before the next pulse was applied. The so-called desorption mode of the IAP was, nevertheless, used successfully to demonstrate the segregation of solute atoms in grain boundaries.

In the new age of three-dimensional data processing, many of these problems are solved when handling the PSD. Depending on the general architecture and function of the different detectors, these are classified as serial (PoSAP and OAP) or parallel (TAP, MAP, OPoSAP and OTAP). In the next section, we introduce the different types of position-sensitive detectors.

The Position-Sensitive Atom Probe (PoSAP). The PoSAP was introduced by Cerezo et al. in 1988 [10–12]. Figure 6.5 shows a schematic drawing of the main features of the instrument, in which the position of the tip for 3D-analysis and the construction of the anode are illustrated. The basic apparatus is a FIM, in which the specimen holder can be rotated around the support by 90 degrees in such a way that the tip faces either the imaging screen (mounted outside the drawing plane and not sketched in Fig. 6.5) or the detector. The PSD consists of a 35 mm chevron type MCP at a distance of 110–400 mm from the tip.

The wedge-and-strip anode (right-hand part of Fig. 6.5) is mounted behind the MCP at a distance of 10 mm and consists of about 40 wedge/strip pairs. The different electrodes for X, Y and Z-position are shown in Fig. 6.5. The X-electrode contains a series of connected strips that decrease in width from right to left. For the Y-electrode connected identical wedges are used. The remaining area between the wedges and strips forms the Z-electrode. The evaporation sites of specimen atoms are localized in the following way. The field-evaporated surface ion impinges onto the front face of the first MCP and initiates an electron cascade. The electron cloud leaving the rear face of the second MCP impinges onto the wedge-and-strip anode. Owing to the lateral extent of the electron cloud, several wedge-and-strip pairs are hit simultaneously. The partial charge given to each electrode (X, Y, Z) is measured by means of a charge sensitive preamplifier having a gain of 0.5 V/pC and a time constant of 500 ns, connected to a charge-integrating ADC (Le Croy 2249A). The timing signal is derived from the output of the MCP by means of an inverting differentiator and a simple amplifier. This signal is used after some discrimination as a stop signal for the ToF determination and to gate the ADC. The specific mass is deduced from the ToF and the applied voltage, which is required for the field evaporation of atoms. The geometry of the electrodes is such that the partial charge measured on the X-electrode is increasing when the position of the electron cloud moves from left to right, while the charge on the Y-electrode is increasing when the position moves from top to bottom of the anode. Thus, the impact position of the incident ion can be determined by calculating the charge ratios $Q_X/(Q_Y + Q_Z)$ and Q_Y/Q_Z.

The function of the wedge-and-strip anode does not allow the ion positions to be determined when several ions arrive simultaneously on the PSD (multi-impact). If several ions arrive at the same time, but at different positions, the PSD will only recognize the sum of the measured partial charges on the electrodes (X, Y, and Z), and will deduce an incorrect average position of impact. This limitation classifies the detector of the PoSAP as a serial PSD. In particular, the evaporation rate (the number of field-evaporated ions per applied pulse) must be chosen to be very small in order to impede multiple events at the PSD.

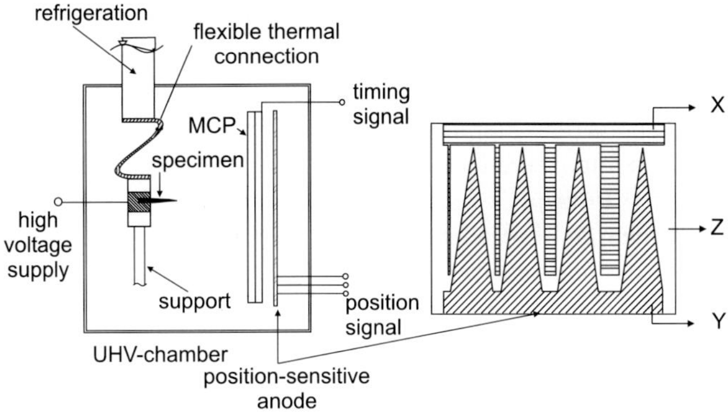

Fig. 6.5. The Position-Sensitive Atom Probe

An alternative serial form of PSD is the Optical Atom Probe (OAP), which was developed by Miller [13] and adopted by Leisch [14]. In this version, the lateral position of the impact is determined from a digitized image of the IAP detector. The time signal is supplied by the phosphor screen, and the image on the screen is read out by a fast image-intensifying CCD camera. The image processing is here much more elaborate than in the IAP. Nevertheless, when there are multiple events at the detector, the positional information arising from different ion species cannot be related to a particular ToF-signal. A detection rate as low as for the PoSAP must therefore be selected.

The Tomographic Atom Probe (TAP). A substantially different philosophy of PSD design has been developed by Blavette and co-workers [15,16]. This system provides multiple-event detection. Figure 6.6 shows a schematic drawing of the detector system. It consists of an MCP of the Chevron or *Z*-stack type MCP and a 10×10 anode array device located at about 450 mm from the tip apex. Each section of the anode consists of a thin nickel layer ($1\,\mathrm{cm} \times 1\,\mathrm{cm} \times 12\,\mu\mathrm{m}$) deposited on a glass substrate and connected to an electrical feedthrough.

The ToF is measured for every field-evaporated ion by determining the signal produced on the rear face of the MCP for every pulse, as in the PoSAP. The specific mass is deduced from the ToF and the applied voltage. As with the earlier detectors, the impinging ions generate the electron clouds, which themselves impinge onto the anode array. The charges received by each electrically isolated element of the anode array (10×10 elements) are measured separately by low-noise charge-sensitive preamplifiers connected to the elements of the anode array. For multiple events, more than one ToF must be expected and several isochronal maps of charges (8 or 5×96 ADC channels) are therefore used. These ADC cards (Le Croy 1885 F) are filled-in parallel and triggered by the time signals. The position of the impact is determined by evaluating the charge information as described in more detail in the following section. The handling and control and the interaction of the different components of the TAP, in particular that of the fast electronic parts, is significally more complex than for the types of PSD mentioned above. Nevertheless, the ability of the TAP to map a larger area of the specimen surface than the APFIM with a comparable lateral resolution, and the positioning of the different species for multiple events and simultaneous identification of the different states of charge of all elements analyzed makes this method the most powerful one. A similar approach was proposed by Miller [17] in a version called the Mapping Atom Probe (MAP), in which an IAP detector is used. The light signal produced in the desorption mode of the IAP detector is amplified and transmitted by means of fibre optics onto a photocathode. This photocathode is connected outside the UHV chamber to a pair of MCPs and a multi-anode array of 10×10 elements.

Impact Positioning with the TAP detector. The TAP was the first instrument capable of performing a truly parallel analysis. As this detector type has already been installed and successfully used at four locations throughout the world, we discuss the positioning method in detail.

In the field of electron microscopy, CCD cameras with the enormous number of $2\,\mathrm{K} \times 2\,\mathrm{K}$ pixels or more are quite common. Thus, from the view point of an electron microscopist, one might expect the spatial resolution of the tomographical atom probe to be rather limited since its detector consists of only 10×10 square anodes, each $1\,\mathrm{cm}^2$ in size (Fig. 6.6). However, what spatial resolution is physically useful? Owing to aberrations of the individual ion trajectory (Sect. 6.3.1), the determination of atom positions on the specimen surface seems to be limited to a few Å. Taking the typical magnification of

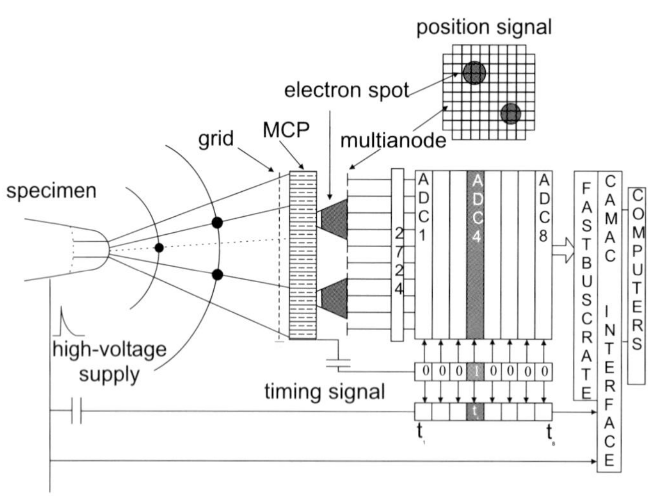

Fig. 6.6. The Tomographic Atom Probe TAP

10^6, a distance of 1 Å on the tip surface is magnified to 1 mm on the detector. Thus, improving the detector resolution much beyond the millimeter limit would not provide further useful information.

In order to obtain this resolution, increasing the number of independent anodes is not feasible because of the tremendous effort in electronics necessary for such a modification. The spatial resolution of the detector can however be sufficiently improved if the charge information stemming from the separate anodes is quantitatively evaluated.

Generalizing the ratio method used with the wedge-and-strip anode [18], D. Blavette and coworkers proposed in 1993 a charge centroiding scheme [16], which is currently employed to determine the impact positions. A single ion hitting the detector is transformed into a cloud of 10^7 to 10^8 secondary electrons via a stack of three microchannel plates. By suitable design of the detector geometry and choice of voltage supply, the diameter of these electron clouds is adjusted to an optimum size of about 2 cm. Consequently, an incoming ion will typically produce charge signals on 4 to 9 related anodes (Fig. 6.7). The charge density distribution inside the cloud may be close to a gaussian but depends in detail on the particular properties of the channel plates. However, in order to localize the center of charge, which is tacitly identified with the impact position, only two basic assumptions are necessary: (i) The shape and the width of the charge distributions are independent of the total charge of a cloud (Q_t) and remain unchanged during a measurement. (ii) The charge distributions are centrosymmetric.

For the evaluation of a separated charge cloud, a local coordinate system is chosen at appropriate anode boundaries (see Fig. 6.7) with its origin such that the two normalized partial charges above the x-axis ($q_y = Q_a/Q_t$) and

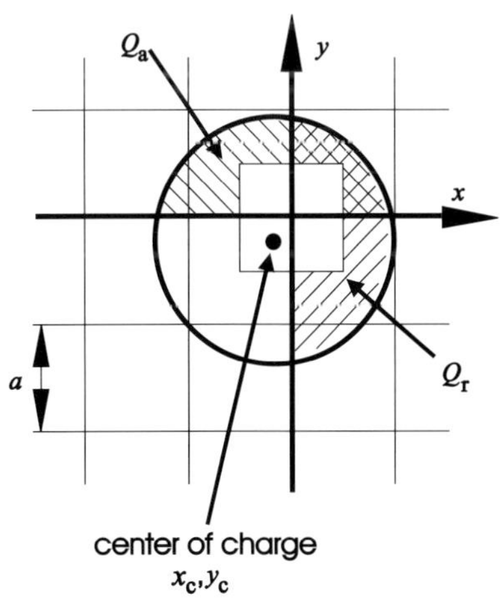

Fig. 6.7. Schematic drawing of a charge cloud hitting the square anode array. The local coordinate system (x, y) is defined in the text

on the right side of the y-axis ($q_x = Q_r/Q_t$) are close to one half of the total charge. From this definition, it is obvious that the center of charge must be located inside a squared area centered at the origin of the local system and of the same size as a single square anode. The exact position (x_c, y_c) of the charge center inside this area is then calculated from the particular values of the partial charges introduced above. As the cloud center in Fig. 6.7 is moved from $x_c = -a/2$ to $x_c = a/2$, the corresponding charge ratio q_x will monotonically increase from a minimum value $q_{\rm min}$ to a maximum value $q_{\rm max} = 1 - q_{\rm min}$. Provided that the transfer function $h := q(x)$ is known within its definition range $[-a/2, a/2]$, the impact position (x_c, y_c) may be calculated simply from the corresponding inverse function

$$x_c = h^{-1}(q_x); \quad y_c = h^{-1}(q_y). \tag{6.3}$$

In practice, the shape of the charge distribution depends on the particular instrumentation and experimental conditions and so the transfer function h is experimentally determined after analysis of a reasonably large number of detected atoms. The impact positions are then expected to be homogenously distributed across the detector area. The probability $p(x_c)dx$ of observing a specific local impact coordinate is constant and given by $1/a$, whereas the probability $p(q_x)dq$ of obtaining a specific partial charge is determined experimentally from the measurement. Since these probabilities are related by

$$p(q_x)dq = p\left(x(q)\right)\frac{dx_c}{dq}dq\,, \tag{6.4}$$

the transfer function is given by

$$x(q_x) := h^{-1}(q_x) = a\left(\int_{q_{\rm min}}^{q_x} p(q)dq + \frac{1}{2}\right). \tag{6.5}$$

This positioning procedure relies on the accurate measurement of rather small charges within very short time periods. Moreover, the two basic assumptions about the electron clouds may only be approximately valid. This might cause doubts about the feasibility of the procedure. Instead of estimating all the possible sources of error, the group around D. Blavette directly proved the accuracy of the TAP detector by experiment. A fine-structured mask was imaged onto the detector using a homogeneous beam of Ne image gas ions. The spatial resolution of the detector was shown to be better than 0.5 mm.

However, the ideal conditions of this performance test with a monoisotopic image gas are never satisfied in a real measurement. According to P. Bas [19], the total charge of an impact also varies with the kinetic energy of the arriving ion. Thus, in a real measurement, it is impossible to obtain the optimum charge figure of 4 to 9 anodes for all impacts. About 10% of all impacts cannot be localized properly either because their charge figures are too small or because of overlapping when the clouds are too large. Further difficulties may arise if the long-term stability of the charge measurement is not guaranteed. In order to overcome these problems, digital camera-based systems (OPoSAP, OTAP) are under rapid development.

The Optical Position-Sensitive Atom Probe (OPoSAP). Another PoSAP design was introduced by Cerezo et al. [20] in order to resolve the problems arising from multiple event detection. Here a conventional imaging device is used, consisting of a double MCP and a phosphor screen for the desorption mode (see IAP, Sect. 6.2.3) mounted at a distance of 280 mm from the tip. The light arising from the rear face of the screen is observed outside the vacuum chamber via a viewing port. This light signal is split by means of a partially reflecting mirror as shown in Fig. 6.8. One of the resulting beams is focused onto the CCD camera in order to monitor the position of impact. The second beam is focused onto an anode array photomultiplier tube (8 × 10 sections). The signals produced by the photomultiplier are amplified, discriminated and used for the ToF measurement. These signals are correlated with the individual light spots detected by the CCD camera. A parallel time-recording system is used in oder to evaluate multiple events.

The OPoSAP detector was tested with a 256 × 256 pixel camera yielding a spatial resolution similar to that of the TAP detector. But when compared to the charge centroiding performed with the TAP, the positioning is here rather simple and seems to be less sensitive to any kind of noise since it relies only on the position of relative light intensity maxima [21]. As no charge ratios need to be determined, the electron clouds can be made as concentrated as possible. Therefore, in the case of multiple hits, two impacts can be spatially discriminated much better than with the TAP. The minimum distance for which two impacts can be distinguished with the OPoSAP was shown to be only about 2 mm [21], whereas in the case of the TAP detector, a minimum

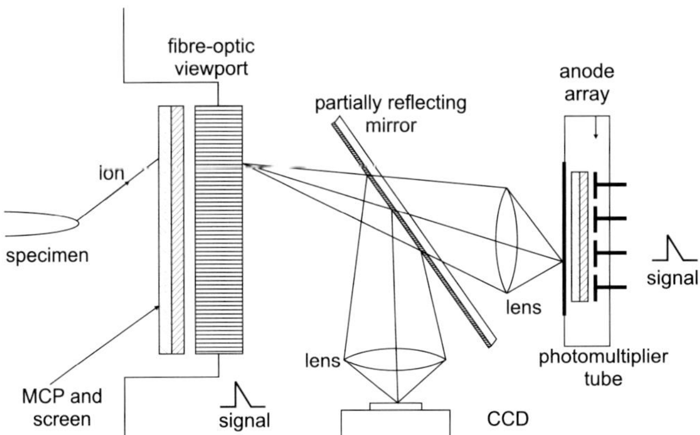

Fig. 6.8. The optical PoSAP

separation distance of 2 to 3 cm is necessary, based on the algorithm presented above. At best, this minimum distance may be decreased to 0.5 cm if much more complicated algorithms are applied [19]. However, one has to keep in mind that in the case of multi-isotopic specimens, the likelihood of multiple impacts on one charge map of the TAP detector is much reduced as several (up to 8) maps are stored one after the other subsequent to a voltage pulse.

The Optical Tomographic Atom Probe (OTAP). Blavette et al. are currently developing a new PSD characterized by a simpler configuration than that of the original TAP. The basic idea is to use a transparent anode array. The proposed detector is illustrated in Fig. 6.9. The phosphor screen is replaced by a glass substrate of optical quality, coated with an array of insulated, transparent, and conductive strips. The strips are covered with a phosphorescent material (P46). In the first test version of the detector, an array of 8 strips was used. The electron cloud generated by MCP produces both a light spot and an electrical signal at the new detector device. The ion ToF and the impact position can now be easily correlated. The position is recorded by means of a CCD-camera focused on the strip array and the ToF is measured directly on the strips. Preliminary results suggest that the data acquisition is more reliable and demonstrate the feasibility of the new design [22].

Outlook. Future development will focus on improvement of the design and operation of the PSD. In particular, some basic problems represent great challenges. One of the major disadvantages of the new detectors is the poor mass resolution compared with the conventional AP, caused by the larger acceptance angle and shorter flight distance of the ions. A high mass resolution is essential for the investigation of multi-component technical alloys like

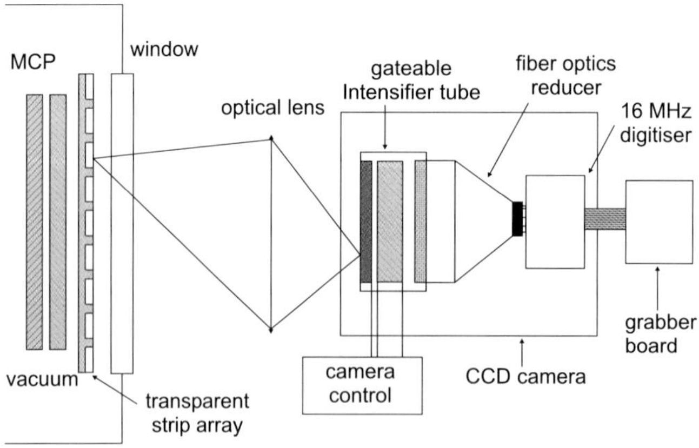

Fig. 6.9. The optical TAP

steels. Limitations of the PSD also include the accuracy of the positioning of the original location of surface atoms and the detection efficiency.

The mass resolution depends on the length of the flight distance of the ions and on the broadening of their kinetic energy distribution in the course of field evaporation. As described above, this flight distance is about 35 and 50 cm in the PoSAP and TAP, respectively. The usual way of improving mass resolution (Sect. 6.2.2) in a conventional AP is to include an energy-compensating device, such as a first-order reflectron or a Poschenrieder lens. However, including an energy-compensating system into an imaging system is a difficult task, since improvement of the mass resolution may cause degradation of the spatial resolution. This is mainly due to the fact that the position of incoming surface ions is deduced from the position of impact on the PSD after passing through the energy focusing lens. In addition, lenses usually cause imaging aberrations.

Sijbrandij et al. [23] attempted to adapt a wide-acceptance angle first-order reflectron to a 3D-AP. The design of the reflectron was optimized by using electrostatic modelling calculations. The mass resolution was improved to $(m/n)/\Delta(m/n) \approx 500$ full-width at half-maximum (FWHM) and ≈ 250 full-width at 10% maximum. However, the systematic position errors must be corrected because the reflectron lens acts only on the velocity component along its axis. Results on some multi-component materials showed that the performance of this type of instrument is promising [24].

Further improvement of the accuracy of the positioning in 3D-AP is now progressing. The first important step towards this aim is to operate the PSD in a digital fashion, by utilising ultrafast CCD devices. This has been implemented in the OPoSAP and the OTAP. The next step will certainly be the improvement of the calculation algorithm by introducing a more sophisticated physical description of the ion path.

The enhancement of the detection efficiency is a very difficult topic. As shown above, the PSD consists of an arrangement of MCP and anodes. Unfortunately, the active detection area of the MCP is physically limited to the open area ratio of the input face, which currently amounts to 60%–70%. The remaining area is blind to the incoming ions. Nevertheless, some experiments have been performed to increase this efficiency [26,25]. The idea is to accelerate secondary electrons that originate from the impact of an ion on the non-active area back to the input surface of the MCP. This can be achieved by inserting a high-transmission grid in front of the MCP and applying a small potential difference between the grid and the MCP. This method leads to an increase of the detection efficiency by about 30% [26,25]. The results obtained for the value of the applied voltage are still contradictory. In particular, the loss in spatial resolution must be further investigated.

6.3 Tomography

After the raw data have been collected, positions of the atoms inside the analysed volume are reconstructed from the sequence of impact positions, corresponding flight times, and tip voltages, without reference to specific properties of the detector. The trajectories of the ions are traced back from the detector to the tip surface to determine the original atom positions as exactly as possible. Since this atom probe tomography is currently the method with the highest spatial resolution for 3D compositional analysis, the results can hardly be tested by comparison with independent experimental methods. Instead, one has to rely on rather simple theoretical concepts to derive the final results of analysis and to test their consistency. Compared to the complex electron optical system of an electron microscope, the imaging and analysis method of modern 3D atom probes is surprisingly simple. No lenses are necessary and wave optics is not required, which makes the description of the imaging process easier. On the other hand, one always has to be aware that the specimen itself forms the essential imaging 'lens' in the field ion microscopy. The specimen not only disturbs the imaging to a certain degree, but also completely determines such essential features as the magnification and local distortions of the image. Furthermore, the specimen is continuously modified during the analysis by the field evaporation of surface atoms.

In the present section, the physics of field evaporation and imaging are discussed in so far as they are important to understand the methods of obtaining a tomographic representation of the data. First, basic properties of the ion trajectory between the tip and the detector are described, then the concept of projection microscopy is introduced, and finally a state-of-the-art algorithm for performing three-dimensional reconstruction is presented and discussed.

6.3.1 The Ion Trajectories

Both, imaging-gas ions and field-evaporated specimen ions are accelerated by the electrical field maintained between the tip and the detector entrance. As this field is determined by the particular geometry of the imaging system, consisting of the channel plates, an extraction electrode and the specimen itself, a general prediction concerning the ion trajectories would be impossible. Fortunately, it turns out that important basic features can be derived without reference to any details of the field distribution. Furthermore, the tip is developed by evaporation with a slowly increasing base voltage before starting an analysis. In this way, a universal specimen shape bounded by an equi-field surface is established.

Cylindrical symmetry around the tip axis may be assumed. The potential distribution is described by

$$\Phi(r, z) = V\phi(r, z)\,, \tag{6.6}$$

where U is the positive voltage supplied to the tip and ϕ a spatial distribution function, which remains unchanged as long as the tip shape does not change. The ion trajectories follow from the classical equations of motion

$$\frac{dr}{dt} = \frac{-neV}{m}\frac{\partial\phi}{\partial r}; \qquad \frac{dz}{dt} = \frac{-neV}{m}\frac{\partial\phi}{\partial z}. \tag{6.7}$$

where the ion mass m, the charge state n, and the tip voltage V influence the acceleration along both directions in the same way. Thus, the trajectory of an ion, originally at rest, is independent of all these variables. Different atomic species starting from the same position at the specimen surface will all hit the detector at identical positions. The relation between positions at the specimen surface and the corresponding detector positions is wholly determined by the geometry of tip and microscope.

Because of the extreme curvature of the tip, the potential drop is completely concentrated around the tip. The field strength at the apex is expressed by the equation for a field of spherical symmetry

$$F = \frac{V}{k_f R}, \tag{6.8}$$

where the additional geometrical constant $k_f \approx 5\dots7$ takes into account the influence of the specimen shaft. In Fig. 6.10, ion trajectories calculated by Smith and Wall [27] are shown together with the corresponding field lines inside a region close to the tip apex. Appreciable deviations from a straight trajectory are only expected in the high-field region shown in Fig. 6.10. Following the laws of electrostatics, the field lines leave the specimen normal to the surface but then, influenced by the specimen shaft, they are bent towards the z-axis direction. As the ions acquire most of their kinetic energy during the first few nanometers, they do not follow the field lines. Nevertheless, the trajectories are diverted towards the tip axis too. The inclination of their initially radial direction is depressed by a factor of the order of two, which must be taken into account by the image projection model. Furthermore, the ion trajectories are dominated by the tip shape. The geometry of the vacuum chamber has only a minor influence. Consequently, when the specimen is tilted, the trajectories follow this tilt as if fixed to the tip axis. This behaviour was recently confirmed by quantitative tilt experiments [28].

The atomic specimen structure is often resolved clearly in field-ion images (Fig. 6.3). Since the trajectories are independent of the ion species, one would expect atomic resolution also for field evaporation. However, it was clearly shown by field desorption images [29] that small aberrations occur in the trajectories of individual atoms stemming from the surface layer. This behaviour may be explained by the somewhat different initial position of image gas and specimen atoms. The specimen atom is influenced by local field distortions caused by the surface roughness, such as edges and kinks or adsorbed atoms close to the ionized atom, whereas the image gas, usually ionized a few Å

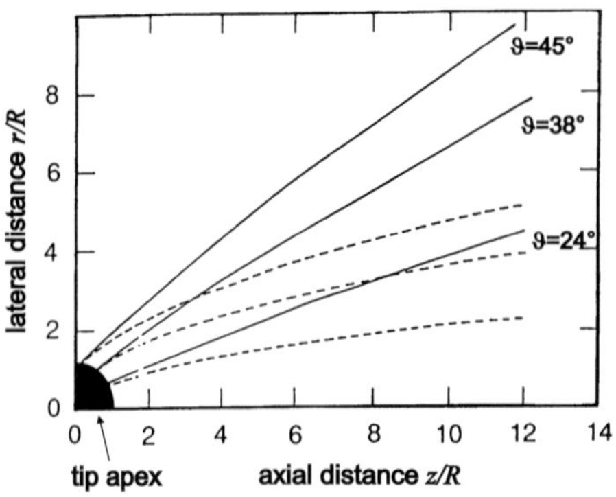

Fig. 6.10. Ion trajectories (*solid lines*) and the electrical field lines (*dashed lines*) in the vicinity of the specimen apex. ϑ describes the inclination of the corresponding tip surface normal (courtesy of R. Smith and J.M. Walls)

above the surface [30,31], 'feels' a smoothed average field. As the specific structure of the field near the surface cannot yet be taken into account in the modelling of individual trajectories, the accuracy of the lateral positioning in the tomographic reconstruction is limited to about 0.2–0.4 nm. However, theoretical work aimed to overcome this limitation is in progress and, besides the development of new detector designs, major improvements in this field may be expected in the near future. Ion trajectory calculations taking into account the atomic structure of the tip have recently been published [32].

6.3.2 Image Projections

In order to reconstruct the positions of typically one million atoms, it would be impossible to simulate all the individual ion trajectories. Instead, a simple image projection is determined based on experimental field-ion images. The important role of the tip shape for the imaging process has already been pointed out. Thus, in view of the shape variations of freshly prepared specimens, the derivation of a universal image projection seems to be a daunting task. However, two experimental facts indicate that the situation is by no means hopeless. First, it turns out that, after a careful field development, most specimens appear to have a rather simple shape. Usually, this shape can be modelled as a spherical apex combined with a truncated cone forming the specimen shaft. Second, the positions of the crystallographic poles in field ion images of well-developed specimens remain constant during continuous field evaporation [7]. When the positions of crystallographic poles in a field ion micrograph are measured, a linear relation often holds between the inclination angle ϑ of a particular zone axis and the distance r of the

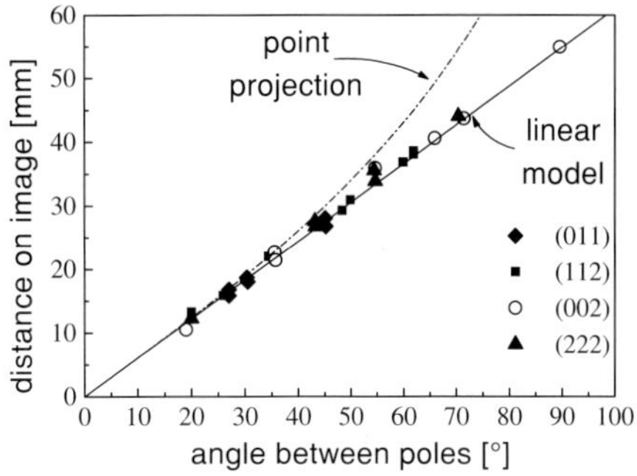

Fig. 6.11. Distances between different pole directions measured in a field ion image of tungsten. *Solid line*: linear model with $\kappa = 0.60$, *dashed*: point projection with $\mu = 0.67$ (data after A. Cerezo et al. 1998)

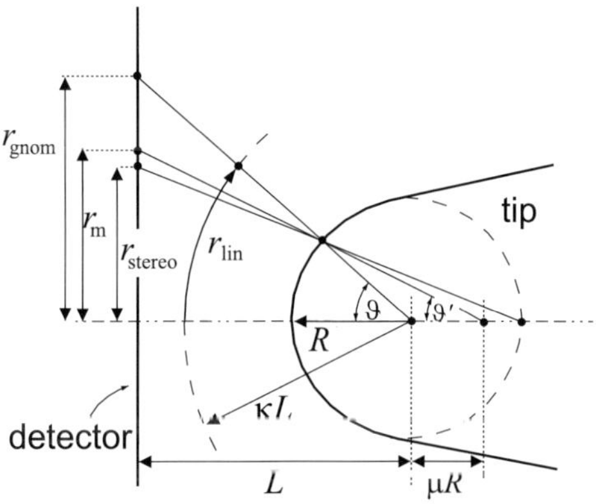

Fig. 6.12. Geometric relations between different projections connecting the surface and the impact position on the detector

corresponding pole to the center of the field-ion image [33,34],

$$r = L\kappa\vartheta\,, \tag{6.9}$$

where κ is an angular compression factor and L the specimen-detector distance. Recently, this relation was again confirmed, as shown in Fig. 6.11 [24]. A compression factor of $\kappa = 0.61$ is observed.

However, motivated by the similarity of field ion micrographs and stereographic pole figures, point projections with a center in between that of a stereographic and that of a gnomonic projection are widely used to interpret field-ion images. In Fig. 6.12 the geometric relations between different projection rules are sketched.

Compared to the simple radial (= gnomonic) projection, the distances to the poles (r_{lin}) are depressed with increasing ϑ in the linear model (6.9), which directly reflects the stronger influence of the specimen shaft on trajectories of higher inclinations. The difference between the linear relation and a point projection vanishes for small angles if the projection center, defined by the radius fraction μ, is chosen according to

$$\mu = \frac{1}{\kappa} - 1 \,. \tag{6.10}$$

In Fig. 6.11 experimental results are compared with predictions based on the point projection and the linear relation. The typical aperture angle of a PSD is only 5°. Thus, both projections can be used within the experimental accuracy as long as the specimen axis is not tilted too far away from the detector axis.

6.3.3 Tomographic Reconstruction

When data of the TAP detector are plotted versus the natural depth scale given by the data sequence, often a layered structure appears, which is due to the coordinated evaporation of the lattice planes. However, the reconstructed layers appear not to be really flat, as the surface curvature of the tip is neglected in the interpretation. To overcome this shortcoming, Blavette and coworkers established in 1995 a general data-reconstruction scheme using the point projection and taking into account the spherical shape of the tip. In this section we describe the geometrical algorithms that are needed to reconstruct the specimen volume. The formulas given here are based on published work [35]; however, this presentation is based on polar coordinates in order to treat the case of an additional tilt between detector and specimen axis within the general scheme.

The reconstruction of the data for an ion can be performed in three steps. First, the specific mass m/n is calculated from the ToF, then the lateral position at the tip surface is determined from the impact position and finally, the depth scale is established from the data sequence. In order to formulate the necessary equations, we define three geometrical parameters: impact position $(x_{\mathrm{D}}, y_{\mathrm{D}})$, inclination angle ϑ' (see Fig. 6.12) and a possible rotation of the detector relative to the tip by two angles ψ, φ along the main axis of the detector, as sketched in Fig. 6.13. The distance between the detector center and the tip is given by L.

The specific mass of an ion is calculated as for conventional AP analysis, using (6.2). However, owing to the large aperture angle, the actual length of the trajectory has to be taken into account. To sufficient approximation, the trajectory can be considered to be a straight line along which the ion travels with constant velocity, so that the specific mass is finally determined by

$$\frac{m}{n} = \frac{2t^2 e(V_{\mathrm{DC}} + V_{\mathrm{P}})}{L^2 + x_{\mathrm{D}}^2 + y_{\mathrm{D}}^2} \,. \tag{6.11}$$

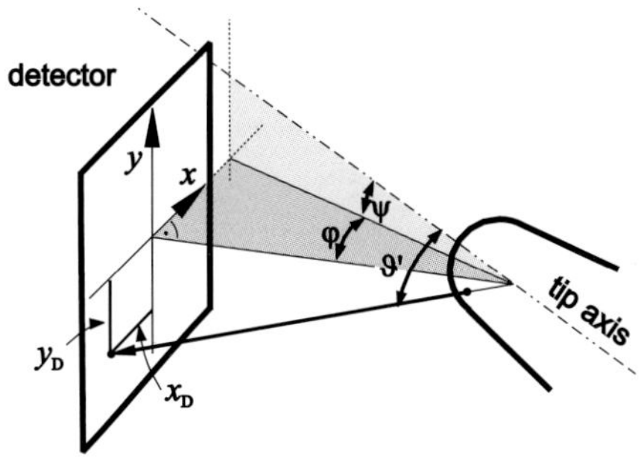

Fig. 6.13. Definition of the parameters used to describe the method of reconstruction

In order to derive the lateral position on the tip surface, the angular direction of the trajectory relative to the tip axis is determined according to

$$\Phi := \arctan\left(\frac{x_D}{L}\right) + \varphi\,; \quad \Psi := \arctan\left(\frac{y_D}{L}\right) + \psi\,. \tag{6.12}$$

The trajectory inclination relative to the specimen axis and the corresponding azimuth α follows from triangular geometry as

$$\vartheta' = \arctan\sqrt{\tan^2\Phi + \tan^2\Psi}\,; \quad \alpha = \arctan\left(\frac{\tan\Psi}{\tan\Phi}\right)\,. \tag{6.13}$$

From Fig. 6.12, it is obvious that in order to find spherical coordinates adapted to the apex sphere, we have to calculate the polar angle ϑ from the inclination ϑ'. In the case of a point projection, these angles are related by

$$\frac{\sin\vartheta}{\mu + \cos\vartheta} = \tan\vartheta'\,, \tag{6.14}$$

an implicit equation for ϑ which may be solved numerically. However, to a good approximation, a straight-line relation between ϑ and ϑ' can be used, as demonstrated in Fig. 6.14.

The result of the linear model is also shown. Again, it becomes obvious that the two models are indistinguishable, at least for the case of coincident detector and specimen axis where ϑ' is limited to 5°.

In order to transform the atom position into cartesian coordinates, the actual tip radius R is needed, which may be estimated from the tip voltage and the evaporation field of the material using (6.8), as proposed in [35]

$$R = \frac{V_{DC} + V_P}{k_f F_e}\,. \tag{6.15}$$

Finally, we obtain the coordinates relative to the tip front

$$x = R\sin\vartheta\cdot\cos\alpha\,; \quad y = R\sin\vartheta\cdot\sin\alpha\,; \quad \Delta z = R(1-\cos\vartheta)\,. \tag{6.16}$$

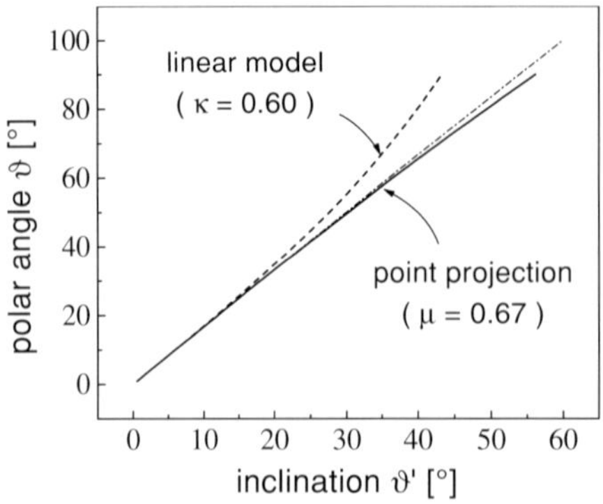

Fig. 6.14. Angular transfer function to calculate the polar angle ϑ from the inclination ϑ'. For the point projection, almost a simple proportionality between the two angles holds (dash-pointed)

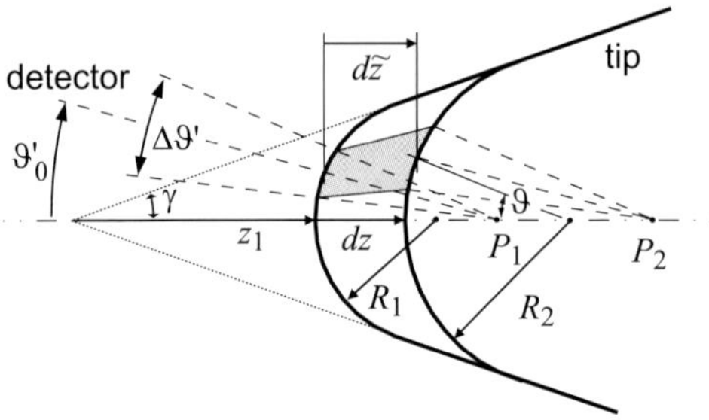

Fig. 6.15. Tip geometry during continuous field evaporation. The tip radius slowly increases since the shaft is conical with apex angle γ. The volume analysed is gray-shaded

During the analysis, the tip is continuously field-evaporated. The actual position z of the tip front is deduced from the number of the detected atoms. The geometrical situation is illustrated in Fig. 6.15 for the general case with a tilt angle ϑ'_0 between detector and tip axes.

Out of the evaporated material between the two 'hemispheres' of radius R_1 and R_2, only the shaded volume dV is registered by the detector. If the detector tilt does not exceed a reasonable limit, this volume may be estimated in linear approximation by

$$\Delta V = [\Delta\vartheta' (1+\mu)R]^2 \cos\vartheta \, dz \tag{6.17}$$

$$= \frac{A_D}{M_R^2} \cos\vartheta \, dz \,, \tag{6.18}$$

where A_D is the sensitive detector area and M_R the actual magnification depending on the tip radius. Assuming a constant atomic volume Ω and a detection probability ρ (typically $\rho > 0.5$), the shift of the tip front per detected atom is finally given by

$$\frac{dz}{dn_{\mathrm{A}}} = \frac{\Omega}{\rho} \frac{M_{\mathrm{R}}^2}{A_{\mathrm{D}} \cos\vartheta} \, . \tag{6.19}$$

Concerning the influence of a possible detector tilt, this depth calculation could be performed more exactly, but given the uncertainty in quantifying the actual tip radius and shaft angle, such refinements are not reasonable.

6.3.4 Data Analysis

A successful measurement with a three–dimensional atom probe will contain up to one million atoms, distributed inside a volume of about $15 \times 15 \times 100\,\mathrm{nm}^3$. Thus, the experimentalist is confronted with an enormous table containing the quantitative data for the atomic species and the positional coordinates. In order to extract useful information about the morphology of the microstructure, commercially available 3D visualization tools are widely used.

The plot of dots representing single atoms inside the analysed volume is helpful to represent the spatial atomic arrangement. When the measurement is performed close to a low-indexed pole, the lattice plane structure usually becomes visible, provided that the viewing angle is suitably chosen (Fig. 6.16) [36]. The occurrence of lattice planes definitely confirms the atomic resolution of the tomography along the analysis direction.

The correctness of the choice of the reconstruction parameters k_{f}, F_{e} and ρ may be tested by the flatness and the correct spacing of these planes. In order to also detect such planar arrangements in the case of nanocrystalline

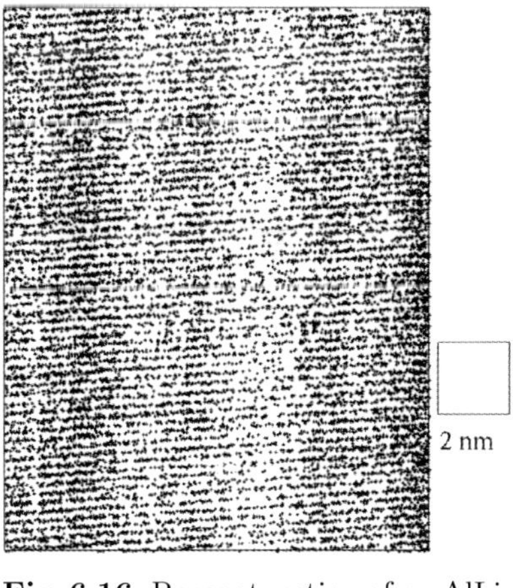

Fig. 6.16. Reconstruction of an AlLi alloy. Detected atoms are represented by small dots. The (111) lattice planes are clearly resolved with the proper spacing of 2.3 Å. For comparison, a small square of 2 nm edge size is plotted at the right side

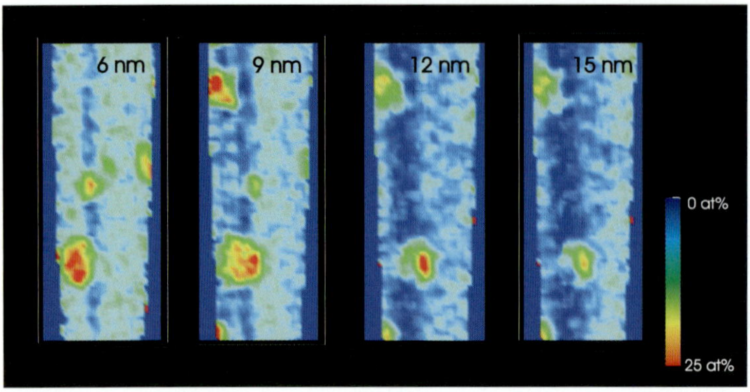

Fig. 6.17. Two-dimensional intersections through the Li–composition field of a decomposed AlLi alloy

samples, a Fourier analysis similar to a diffractogram has recently been successfully employed [37].

Beside the direct plot of single atoms, the calculation of local compositions proves to be very fruitful. For this, local evaluation volumes are defined around each point and the atomic fraction inside these volumes is determined. Their size must be determined by a suitable compromise between optimum spatial and optimum compositional resolution, as the compositional accuracy is limited by the number N of atoms considered according to

$$\Delta c \approx \sqrt{\frac{c(1-c)}{N}}, \tag{6.20}$$

in the case of a binary solution. In Fig. 6.17, two-dimensional intersections of the local composition are shown; here, Al_3Li particles are embedded into a matrix of 6 at.% Li [36].

Unlike electron microscopy, where thin foils are always projected along the optic axis, atom-probe tomography yields a real three–dimensional analysis. By forming a stack of intersections 3 nm apart, the arrangement of the particles is resolved in the third dimension as well. After determining the local compositions, it is an easy task to calculate iso-concentration surfaces. Often, these surfaces are highly suitable for clarifying the morphology of a heterogeneous microstructure, as shown in Fig. 6.18 for the arrangement of Co precipitates inside a Cu matrix [38].

6.3.5 Artefacts of the Reconstruction

Atom-probe tomography using the reconstruction algorithm introduced above has yielded convincing experimental results in the case of decomposed materials with finely dispersed particles. Outstanding insight into the chemistry

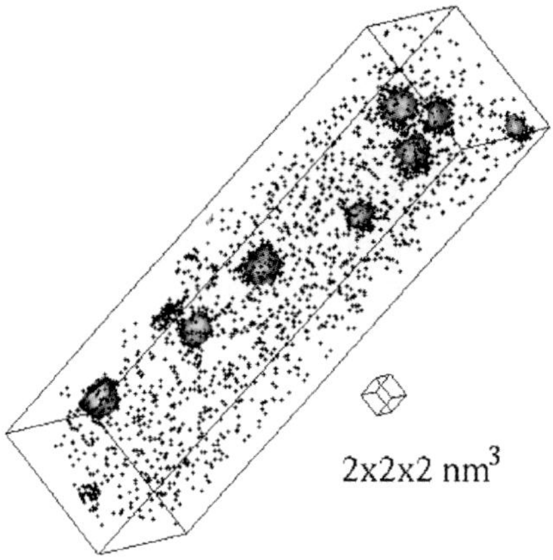

Fig. 6.18. Co precipitates inside a CuCo alloy emphazised by iso-concentration surfaces at 50 at.% Co

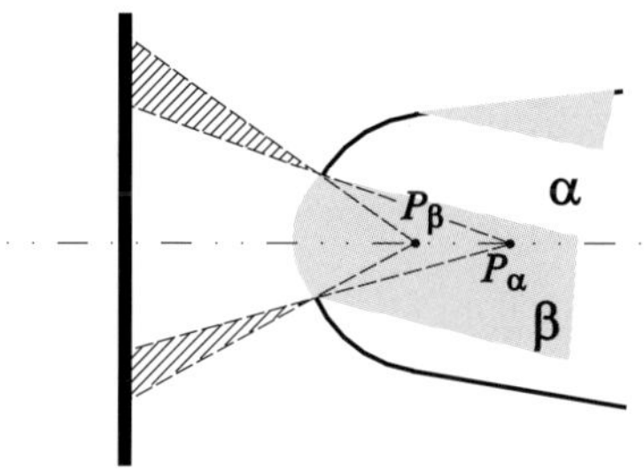

Fig. 6.19. Projective geometry of a tip containing two phases of different evaporation fields. The atom trajectories of the two phases overlap inside the hatched regions

and the morphology of the microstructure is provided in this way. However, during a careful analysis one has to keep in mind that artefacts in the reconstruction may arise if the specimen analysed contains heterogeneities on a length-scale comparable to the tip radius. When the phases investigated are characterized by different evaporation fields, the ideal hemispherical specimen shape is not attained even after a proper field development. Instead, the local curvature will vary inversely with the evaporation field. Evaporation fields of pure metals vary up to a factor of four; those of intermetallic phases are usually unknown, but they may be estimated from the sublimation energy of the material [7].

A typical situation in which two phases are located at the tip surface is sketched in Fig. 6.19.

Neglecting aberrations introduced by the complicated field distribution close to the interface, the ion trajectories may be estimated from the projective geometry. Owing to the varying curvature radius however, different projection centers must be assumed for the phases. Consequently, the protruding β phase is imaged with a higher magnification than the α phase, resulting in a decreased atomic density well inside the β region. At the interface, the atom trajectories of the two phases do overlap. Thus, an increased atomic density is observed in this region. These local magnification effects are rather well known in the case of conventional field-ion microscopy, where one has to be very careful in determining the size of a precipitate from its lateral extent. During the analysis with a conventional 1D atom probe the depth scale varies locally. But only tomography offers – for the first time – an opportunity to test directly whether this effect strongly influences the analysis, as fluctuations of the atomic density become obvious in the reconstructed volume.

However, since the specimen shape, the field distribution, and the resulting ion trajectories would have to be calculated in detail, a quantitative prediction of the geometric distortions in the reconstruction is rather difficult. Instead, a recent investigation of grain boundary segregation gives an estimate of the influence of the local magnification effect. Blavette and coworkers investigated the boron segregation at incoherent γ/γ' interphase boundaries in a Ni-based superalloy [39]. The width of the segregation zone, determined from 3D reconstructions, strongly depends on the boundary orientation relative to the tip axis (see the data in Fig. 6.20).

As the grain boundaries are imaged brightly in field ion micrographs, the authors assume a protruding boundary owing to the very high evaporation field of boron. Thus, a segregation zone orientated along the tip axis ($\vartheta = 0$) appears overmagnified. Defining a local magnification factor λ, the apparent

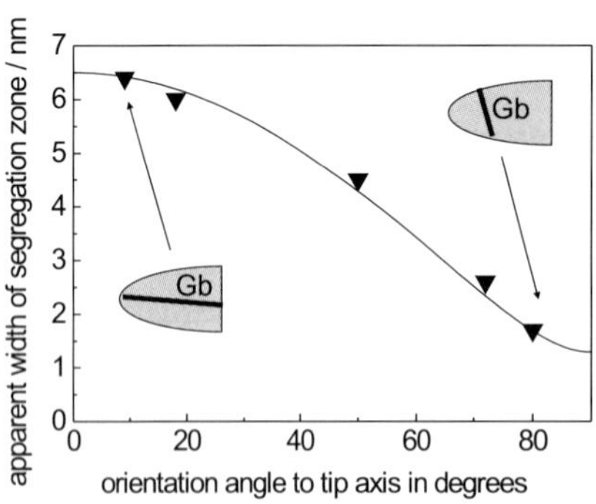

Fig. 6.20. Width of segregation zones measured at γ/γ' interfaces of a Ni base superalloy. Data after Blavette et al. [39]. The solid line is calculated according to (6.21)

width is expected to vary with the orientation according to

$$w_{\mathrm{a}} = w_{\mathrm{r}} \sqrt{\lambda^2 \cos^2 \vartheta + \sin^2 \vartheta}\,, \tag{6.21}$$

where w_{r} represents the real width. The solid line shown Fig. 6.20 is calculated from this relation with $\lambda = 5$ and $w_{\mathrm{r}} = 1.3\,\mathrm{nm}$. It describes the experimental data quite well, indicating a correct interpretation. Thus, local magnification effects may have drastic consequences when the evaporation fields are strongly inhomogeneous. In order to determine the real width of the segregation zone, measurements of grain boundaries oriented normal to the tip axis should be preferred.

Interfaces or layered structures oriented normal to the tip axis do not lead to local magnification effects, as is demonstrated in the example above. In the case of thicker layers however, the tip voltage must be adjusted according to the varying evaporation field during the analysis. If the tip radius is derived from the actual voltage as described in Sect. 6.3.3, the lateral dimensions of the analysed volume and, as a consequence, the z direction is scaled incorrectly between the different materials. This effect may be demonstrated by considering the example of the interface region between an Al top-layer and a Ni substrate tip [40]. In Fig. 6.21a the atom distribution at the interface is reconstructed using the evaporation field-strength of pure Ni.

As Al evaporates at a much lower field-strength, i.e. at lower tip voltage, the actual tip radius is greatly underestimated. Thus, the volume containing the Al appears to be compressed in the lateral direction whereas, in order to keep the atomic density constant, its dimension along the z-axis is drastically overestimated. In this specific case, it is obviously advantageous to estimate the tip radius from the actual z coordinate using the geometrical relation

$$\frac{dR}{dz} = \frac{\sin\gamma}{1 - \sin\gamma}\,, \tag{6.22}$$

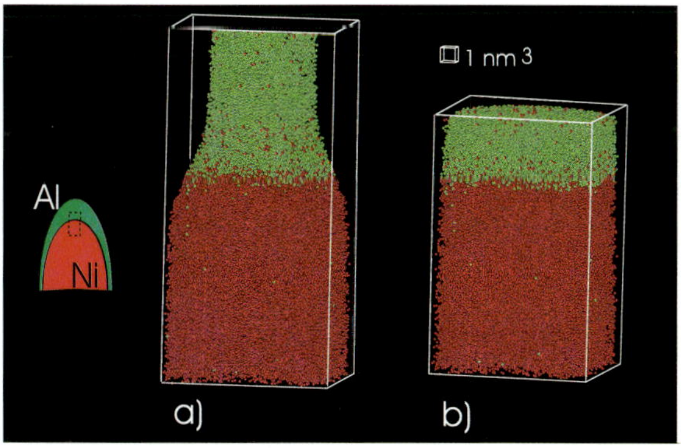

Fig. 6.21. Tomographical reconstructions of the data obtained at the interface between an Al surlayer and a Ni tip. (**a**) Tip radius inferred from tip voltage, (**b**) tip radius estimated geometrically

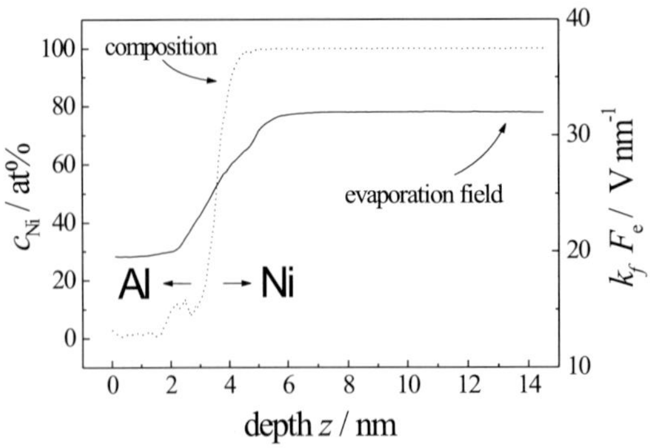

Fig. 6.22. Composition profile normal to the interface and evaporation field as calculated from the geometrical reconstruction shown in Fig. 6.21. Inside the homogeneous layers, the evaporation field remains constant but it increases drastically at the interface

where γ is the shaft angle of the specimen (see Fig. 6.15). The reconstruction shown in Fig. 6.21b was calculated in this way from the same raw data as that of the reconstruction a), yielding a rectangular specimen volume and a reliable depth scale even in the case of the heterogenous layered specimen. To perform this kind of reconstruction, the composition-dependent evaporation field is not needed a-priori. Instead, it follows from the calculation, as shown in Fig. 6.22. The initial tip radius and the shaft angle, required geometrical parameters, may be determined by electron microscopy. However, it is also feasible to consider them as adaptable parameters which are adjusted to obtain $k_f F_e$ of Al at the beginning of the analysis and constant evaporation fields inside the homogeneous regions of the measurement.

6.4 Atom Probe Tomography in Materials Studies

The present section gives examples of fruitful application of tomographic atom probes. They represent our own work as well as work of other authors and will show the wide rang of problems that can be investigated by means of this method. Performance and limitations will become obvious.

6.4.1 Distribution of Solutes

Cu/Pd Multilayer Preparation and Analysis. Cu and Pd layers have been alternately sputter-deposited onto a tungsten tip for the purpose of diffusion studies [41]. Figure 6.23a shows the 3D reconstruction of the as-prepared sample. Individual atomic planes arranged in parallel through the entire sample can be seen and indicate epitaxial growth at each interface.

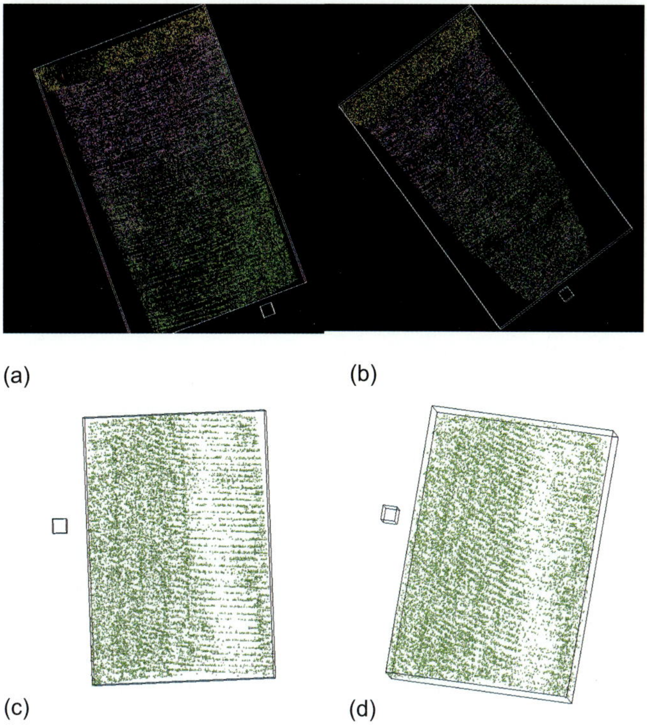

Fig. 6.23. 3D reconstruction of atom distribution for a Pd/Cu/Pd layer sample sputter-deposited onto a W tip (**a**) as-prepared, (**b**) annealed at 673 K for 100 min, (**c**) enlarged section of (**a**), (**d**) same data set as in (**c**) but slightly tilted in space. Only Cu atoms are shown in (**c**) and (**d**) for clarity. Small cubes give $1 \times 1 \times 1\,\mathrm{nm}^3$

In the lower left-hand corner, the structure appears to be less regular. Figures 6.23c and d show that the lattice planes can be visualized everywhere by appropriately tilting the whole data set in the computer.

The difference in orientation of the two sets of lattice planes indicates the presence of a grain boundary. By calculating depth profiles parallel to the sample axis, the interface profile is obtained and is shown in Fig. 6.24. For the as-prepared sample, we have a transition range of about 500 ions from one element to the other, which means a spatial width of 1 to 2 nm. The size of this region is predominantly caused by atomic mixing during the sputter deposition (see Sect. 6.4.2). After annealing for 100 min at 673 K, the Pd atoms that have diffused into the Cu layer are clearly visible in Fig. 6.23b. The ladder diagrams in Fig. 6.24 show the effect quantitatively.

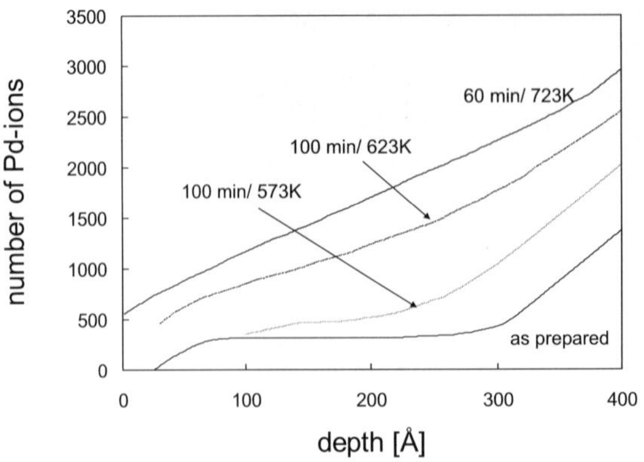

Fig. 6.24. Pd ladder diagrams for Pd/Cu/Pd layer samples measured at 50 K by means of a conventional APFIM

Distribution of Fe in Decomposed Cu-Ni-Fe. The alloy system CuNiFe exhibits a miscibility gap extending from the binary CuNi into the ternary field. The decomposition of supersaturated $Cu_{44.7}Ni_{47.3}Fe_{8.0}$, for example, has been studied extensively in the past (see references in [42]). The alloy decomposes into two ternary solid solutions with considerable enrichment of Fe in the Ni-rich phase. From measurements of the anomalous small angle x-ray scattering, Lyon and Simon [43] concluded that Fe might segregate near the interface of the two phases. The TAP allows a reliable check of this suggestion. Figure 6.25 shows a 3D reconstruction derived from a TAP data set taken around a Ni-rich precipitate. The atom positions for Cu (a) and Fe (b) are shown. No indication of an Fe-enrichment in the interfacial region can be detected.

A concentration depth profile taken from the same data set as in Fig. 6.25 is shown in Fig. 6.26. Again, no Fe-enrichment can be detected near the interfaces. The homogeneous distribution in the Ni-rich phase is obvious.The conflicting result obtained from anomalous x-ray scattering remains to be re-examined.

6.4.2 Early Stages of Phase Formation

Cobalt Precipitation in Copper. The work reported here [44] was desigend to check the fundamental Gibbs assumption of a sharp interface between precipitate and matrix. For Cu–Co, extensive studies on the early stage of the Co precipitation were reported previously [44]. The size and shape of nuclei, in particular, were subjects of controversy. Figure 6.27 shows the reconstruction of a sample volume for Cu-1 at.% Co annealed at 723 K for 24 h.

The precipitates are large enough to allow a reliable analysis of the interface structure. Examples of interface composition profiles are given in

Fig. 6.25. 3D reconstruction of atom positions for a $Cu_{44.7}Ni_{47.3}Fe_{8.0}$ alloy sample aged at 773 K for 100 h. Cu: green, Fe: white. Probe volume: $10 \times 10 \times 30\,nm^3$

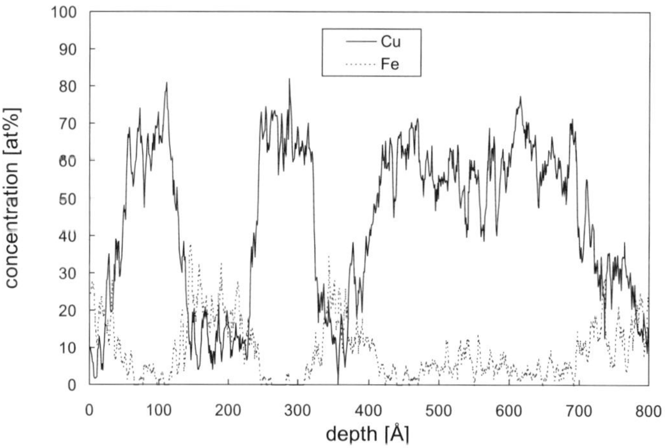

Fig. 6.26. Concentration depth profile derived from the data set of Fig. 6.25

Fig. 6.28a. The agreement of horizontal and vertical profiles indicates that the inherent scatter of the atom positioning in the horizontal direction (the lattice planes perpendicular to the sample axis) does not influence the profile analysis. By data evaluation with respect to the moment of the precipitate on all axes in three dimensions, it was ensured that a spherical shell will correctly sample the interface. A profile (spherical-shell sampling) for

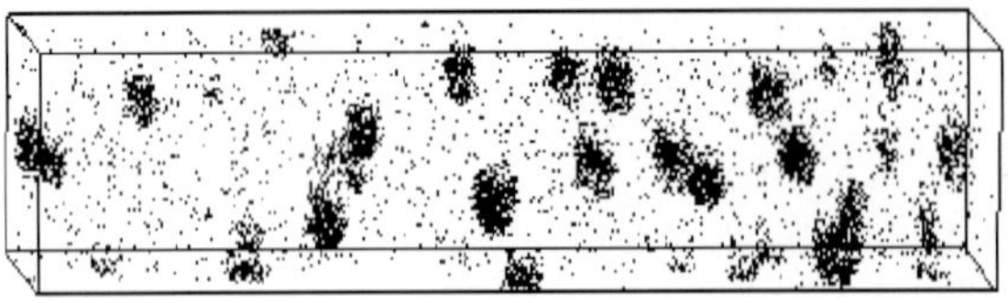

Fig. 6.27. 3D reconstruction of the Co atom distribution for Cu-1 at.%Co aged at 723 K for 24 h. Box size: $15 \times 15 \times 70\,\mathrm{nm}^3$. By courtesy of A. Cerezo

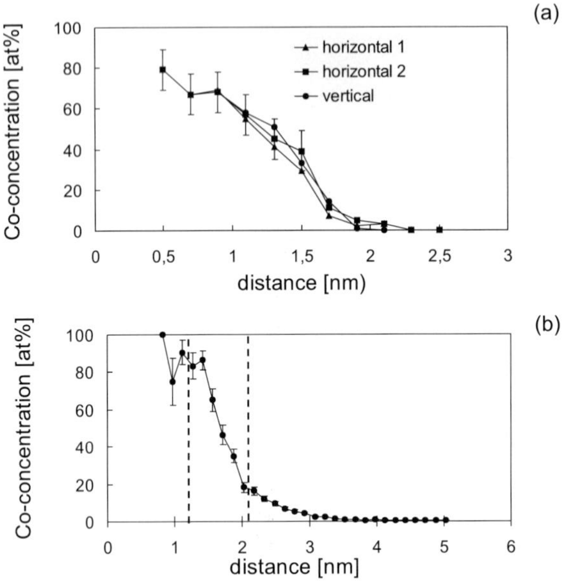

Fig. 6.28. Concentration profiles across precipitate interfaces for Cu-1 at.%Co. (**a**) Aged at 723 K for 24 h, rectangular probe volume taken parallel (vertical) or perpendicular (horizontal) to the tip axis. (**b**) Aged at 823 K for 10 min, dashed lines represent inner and outer bounds defined for the interface. By courtesy of A. Cerezo

the late nucleation or beginning growth state established by annealing at 823 K for 10 minutes is given in Fig. 6.28b. This heat treatment produces 10^{23} precipitates/m^3 which implies that only 1 to 2% of the total sample volume investigated contained entire precipitates. The precipitate composition was determined for the spherical volume in which the composition gradient did not exceed 10% of the maximum value found for the interface. Table 6.1 gives the results for composition, radius and interface width. For comparison the results of Monte Carlo simulations are also given.

These simulations were based upon a dynamical Ising model with and without lattice strain and nearest neighbour interaction to reproduce the solubility at 723 K and 823 K. Larger precipitates than 4 nm radius began

Table 6.1. Composition and size of precipitates in Cu–Co (experimental and calculated) according to [44]

Treatment	Co content (at.%)	Radius (nm)	Interface width (nm)
823 K, 10 min	85	2.0 ± 0.2	0.7 ± 0.2
823 K, 10 min	89	1.8	1.1
723 K, 24 h	58	1.4	0.7
723 K, 24 h	70	1.3	0.7
723 K, 24 h	70	1.8	0.9
723 K, 24 h	100	0.9	0.9
MC simulation	100	2.0	0.4
MC simulation	100	4.2	0.5

to facet on ⟨100⟩ and ⟨111⟩ planes. The interface width found along these facets also amounted to 0.5 ± 0.5 nm. The resulting widths were equal within the error limits for the models with and without lattice strain consideration. The reason for the difference between experimental and calculated widths are unknown. The authors note, however, that long-range energy terms may induce more diffuse interfaces in the calculations. Obviously, for precipitates in the nanometer range the experimental interface width is not much smaller than the particle radius and, hence, not sharp in contrast to the postulate of the classical nucleation theory.

γ' Precipitation in Superalloys. Formation of the γ' phase in superalloys depends on unmixing and ordering. How do the early stages evolve? Figure 6.29 shows the spatial distribution of Al, Cr and Ni for a Ni–22.5 at.% Cr–3.1 at.% Al alloy aged at 923 K [45]. The structure consists of a fine dispersion of Al-enriched particles of about 7 nm diameter. Within these particles Al-rich lattice planes alternate with Ni-rich ones. The periodicity equals half of the γ' lattice parameter. In agreement with the equilibrium phase diagram, the Al concentration in the precipitates is close to 18 at.% while 1.5 at.% is found in the matrix. The isosurfaces shown for a 10 % threshold give the interface positions approximately. The envelopes are somewhat irregularly shaped, which is due to the statistical error of the measurement rather than to the real particle shape. The interface width can be derived from depth profiles as in Fig. 6.30. It was found by several investigators that this width is not larger than 0.2 nm. As the depth profile is measured along ⟨100⟩ the plane by plane modulations of the Al concentration are clearly resolved. For the left-hand precipitate in Fig. 6.30 the cross-section analyzed remained almost totally within the precipitate while for the second and third one, precipitate and matrix were only partially involved by this cross-section. The Al concen-

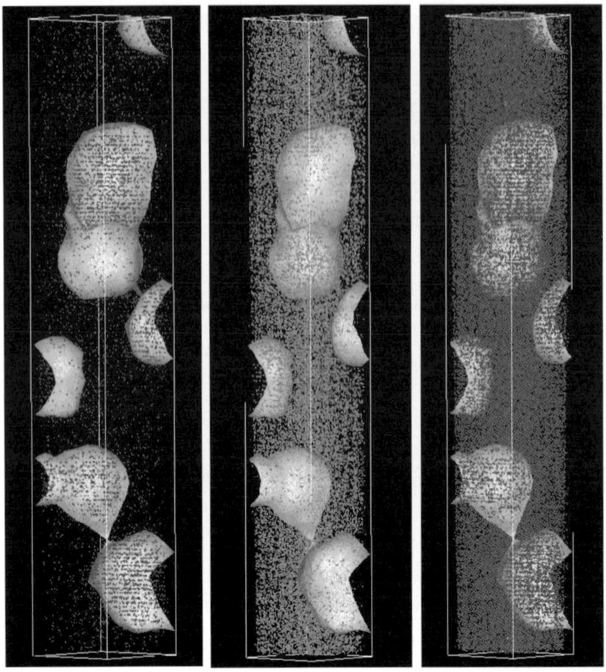

Fig. 6.29. 3D–reconstruction of the Al and Cr atom distribution for the superalloy Ni–22.5 at.% Cr–3.1% Al aged at 923 K. Al: red, Cr: yellow, Ni: pink, isosurfaces shown for 10 at.% Al, box size $8 \times 8 \times 45\,\mathrm{nm}^3$. By courtesy of D. Blavette

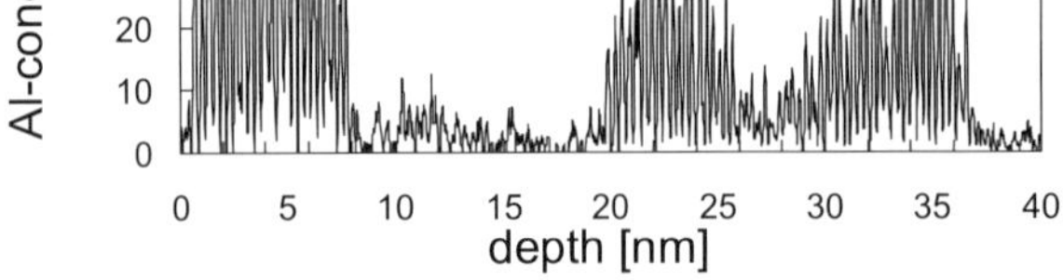

Fig. 6.30. Depth profile of Al related to the 3D reconstruction shown in Fig. 6.29. The probe volume cross-section was approximately 1/3 of that shown in Fig. 6.29 and positioned such that mixed analysis of precipitate and matrix was minimized. The concentration data were derived from slices of 0.1 nm thickness. By courtesy of D. Blavette

tration amplitude is thereforere lower than in the pure γ' case. From pure γ' profiles, the mean compositions of both types of planes within the ordered structure may be derived. From this, the degree of long-range order follows. Defferences between the evaporation field-strengths of the different elements may however affect the output.

Ω Precipitation in an Al-Cu-Mg-Ag Alloy. Aluminium alloys are considered for high temperature applications in the aerospace industry. The required thermal stability of the mechanical strength calls for more stable hardening precipitates. This challenge fuels many studies on the evolution of such precipitates in alloys. For the Ω phase in Al-Cu-Mg, detailed atom-probe studies [46] have answered a number of questions. In particular, the role of Ag addition, which was known to prevent precipitate coarsening, has been investigated. Figure 6.31 demonstrates that Cu, Mg, and Ag are essentially uniformly distributed in the as-quenched state. After annealing for 15 s at 453 K, co-clustering of Mg and Ag atoms is observed as shown by Fig. 6.32. The ratio of the species in the clusters analyzed is close to unity,

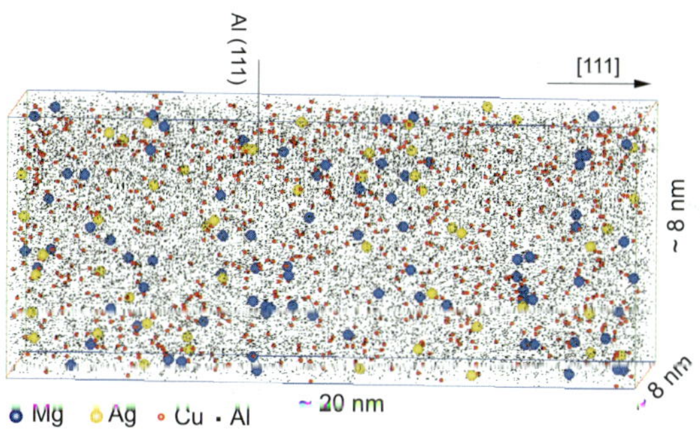

Fig. 6.31. 3D reconstruction of the atom distribution for an as-quenched sample of Al-1.9 at. % Cu-0.3 at.% Mg-0.2 at.% Ag alloy. By courtesy of K. Hono

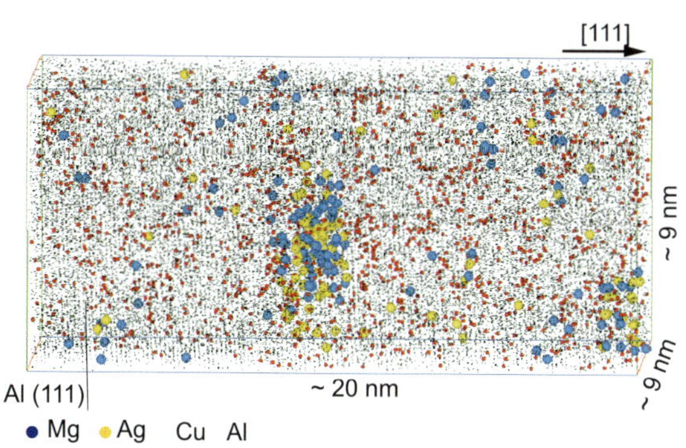

Fig. 6.32. 3D reconstruction of the atom distribution for a sample of the same alloy as in Fig. 6.31 annealed at 453 K for 15 s. By courtesy of K. Hono

the number of atoms per species ranging from 40 to 80. The cluster shape is not yet well defined. The statistical assessment of the atomic distribution in Fig. 6.32 yielded no correlation for Cu atoms with Mg and Ag atoms. Further annealing causes the cluster evolution shown in Fig. 6.33.

This series of pictures reveals the process in greater detail than had ever been seen. With increasing Cu accumulation at the cluster, its shape becomes more and more plate-like. After 300 s annealing time the thickness of the cluster was estimated to be about four matrix layers, which would match to the unit cell size of Al_2Cu. The depth profiles show the depletion of Mg and Ag in the cluster center after 1800 s and 7200 s. The final state may have been reached after 10 h annealing time (Fig. 6.33e). The precipitate has developed a ledge at which its thickness abruptly changes. The thickness in the thinner area amounts to about 10 to 13 $\{111\}$ layers, corresponding to the thickness of three Ω unit cells. Mg and Ag atoms are absent from the interior of the precipitate and are segregated on the flat Ω/α interface. The segregated layers are monolayers. The missing segregation at the ledge suggests that Mg and Ag do not segregate on the narrow side of the platelet. The concentration of Cu appears to be higher near the precipitate than in the bulk of the α matrix. The effect is clearly seen near the ledge. Such an atom distribution is not expected for a growing precipitate but indicates that dissolution occurs during coarsening. Figure 6.34 shows a sample volume containing one Ω precipitate and two Θ'' precipitates. Clearly, the latter are not decorated by Mg and Ag atoms. The depth profile gives the same Cu content as for the Ω particle after the same aging time (Fig. 6.33c). The thickness of the Θ'' particle must not be derived from the depth profile as the cut was not performed perpendicular to the particle. The difference between the phases Θ'' and Ω was a subject of controversy for a long time in the literature (references are given in [46]). The findings of this work also falsify earlier postulations concerning a precursor phase Ω'.

α-Fe Precipitation in Fe-Zr-B-Au. The favourable magnetic properties of soft magnetic alloys of the types Fe-Si-B-Nb-Cu (FINEMET) and Fe-M-B (M = Zr, Hf, or Nb) are mainly due to the presence of α-Fe(Si) crystallites in the nm grain-size range embedded in an amorphous matrix. For FINEMET, the addition of Cu is essential to reach a grain size around 10 nm, for Fe-M-B it substantially enhances the permeability. In both FINEMET and Fe-Zr-B-Cu alloys, Cu was found to form clusters prior to primary crystallization. During primary crystallization, α-Fe nucleates at the Cu clusters, leading to a significantly enhanced density of nuclei and a corresponding decrease of particle size. The clustering of Cu is believed to be due to the positive mixing enthalpy of Cu and Fe. As it is also positive for Au and Fe, it is worthwhile to know whether Au addition causes α-Fe refinement as well. This question was answered for a Fe-7Zr-5B-1Au amorphous alloy by means of TAP analysis [47]. First, it was found that Au does not cluster prior to primary crystal-

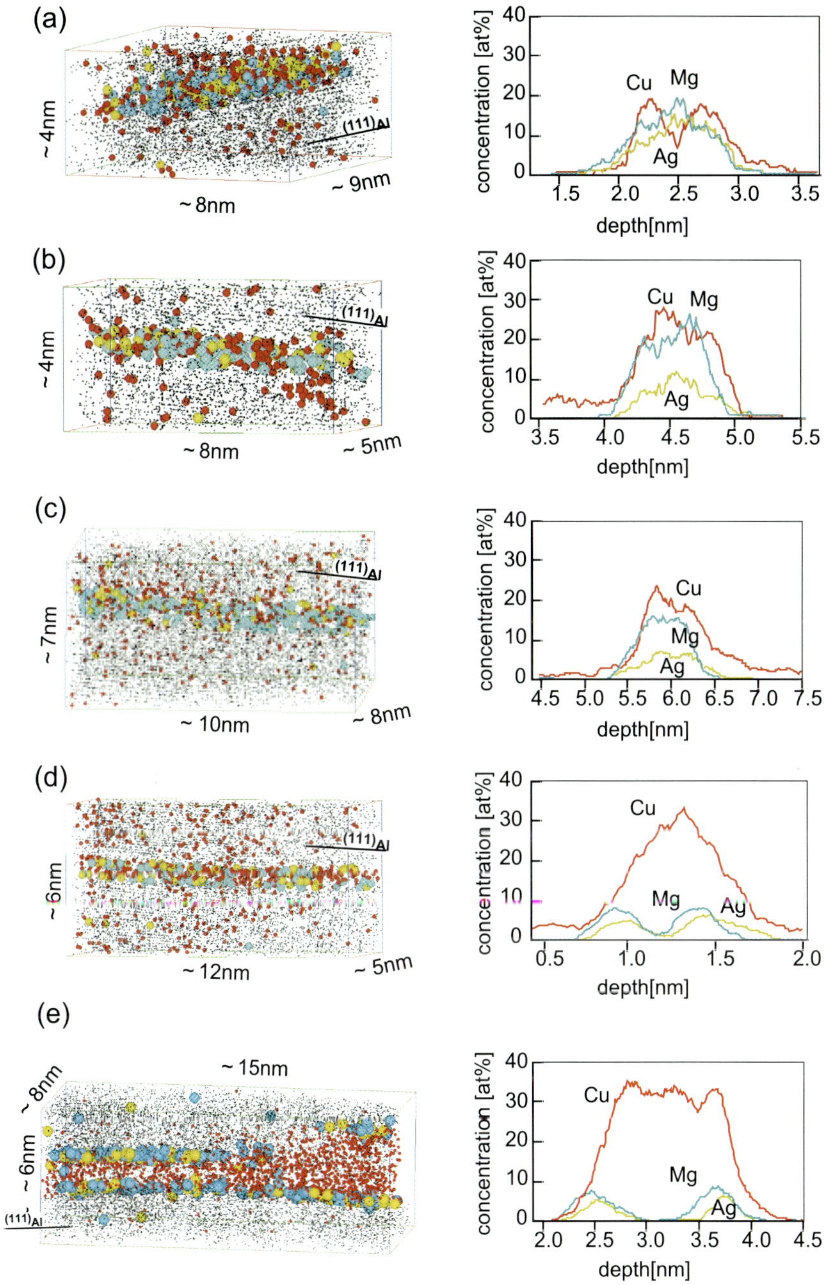

Fig. 6.33. 3D reconstruction of the atom distribution and corresponding concentration depth profiles for a sample of the same alloy as in Fig. 6.31 annealed at 453 K for (**a**) 2 min, (**b**) 5 min, (**c**) 30 min, (**d**) 2 h and (**e**) 10 h. By courtesy of K. Hono

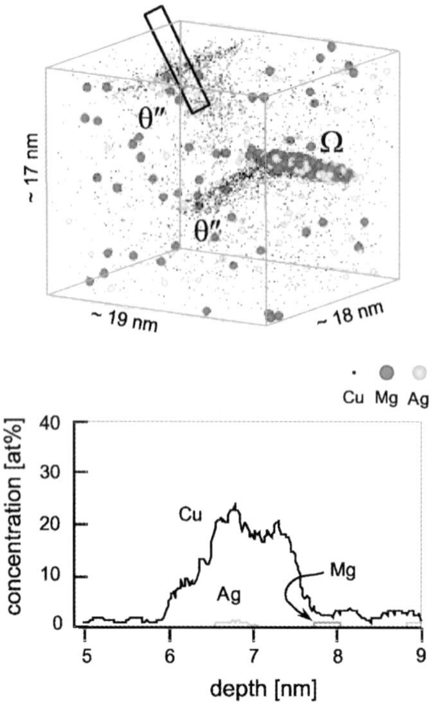

Fig. 6.34. 3D reconstruction of the atom distribution for a sample of the same alloy as in Fig. 6.31 annealed at 453 K for 30 min showing three different precipitates and concentration depth profile for the upper precipitates cut as shown. By courtesy of K. Hono

lization of α-Fe. After annealing of the samples at 723 K for 10 h, both α-Fe particles and Au clusters were observed, but rather well separated from each other as demonstrated in Fig. 6.35a. The detailed composition analysis in Figs. 6.35b and c also does not show any significant Au-Fe anticorrelation, as is oberserved in the case of Cu-Fe. Such anticorrelation will enrich Fe around the nascent Cu cluster and thus favour heterogenous α–Fe nucleation there.

Reference [47] gives more details about the atomic distribution of all alloy constituents in the amorphous matrix before and after primary crystallization. The passive role of Au is explained by its weak repulsive interaction with Fe and the strong attractive interaction with Zr. In addition, the slow diffusivity of Au retards its clustering. The actual growth restriction for α-Fe must have the same origin as in the alloy without Au addition. As Zr is rejected from α-Fe and presents the largest atom size in the amorphous alloy, its diffusion rate is probably the slowest and might control the α-Fe growth rate.

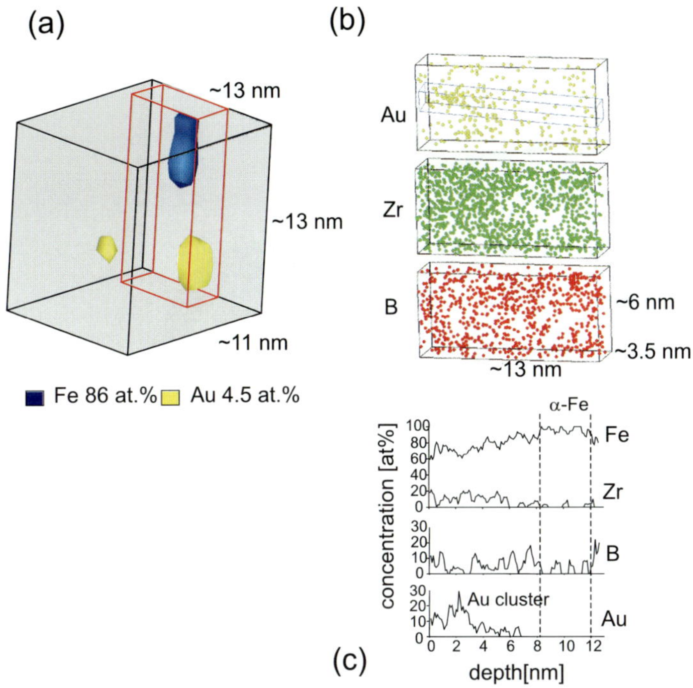

Fig. 6.35. (**a**) 3D isoconcentration surfaces for Fe_{87} Zr_7 B_5 Au_1 after annealing at 723 K for 600 min, (**b**) 3D elemental maps for the volume drawn in red in part (**a**), (**c**) concentration depth profiles for the blue box in part (**b**)

Amorphization of NiZr Multilayers. Phase formation may also be studied in multilayer samples. For this purpose the multilayers have to be prepared on a substrate tip. In the present case this is done by sputter deposition on a tungsten FIM tip [48]. In Fig. 6.36, such a tip is sketched together with the FIM images observed at different depths reached by field evaporation of the material. The uppermost Ni layer consists of a large number of crystal grains with an average size of about 10 nm. The sample cross-section shown in (c) shows the nanocrystalline Ni layer as a brightly imaged ring while the dark center must arise from the Zr layer. The tiny bright spots do not show any crystal symmetry.

The third cross-section (d) shows the bottom Ni layer in the center as an almost perfect single crystal. Around this, concentric rings corresponding to the structureless Zr layer and the nanocrystalline Ni are visible. The 3D reconstruction of the field-evaporation data is shown in Fig. 6.37. The bottom Ni layer clearly shows the lattice plane arrangement known for a monocrystalline tip. The upper layer consists of both elements, Ni and Zr and does not show any planes. Nor did rotation of the data set uncover planes in this layer for any orientation. The authors concluded that this Ni/Zr layer has

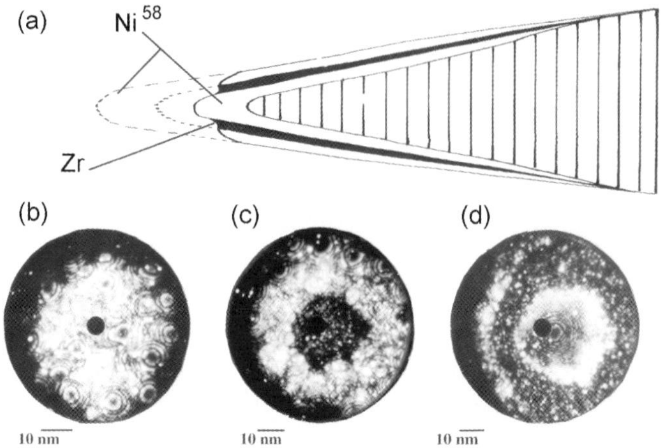

Fig. 6.36. Sketch of a multilayer FIM tip (**a**), and FIM images after field evaporation to the states shown in the sketch, (**b**) uppermost Ni^{58} film, (**c**) uppermost film partly removed, Zr film cut, (**d**) cut through uppermost Ni^{58} film, Zr film and bottom Ni^{58} film

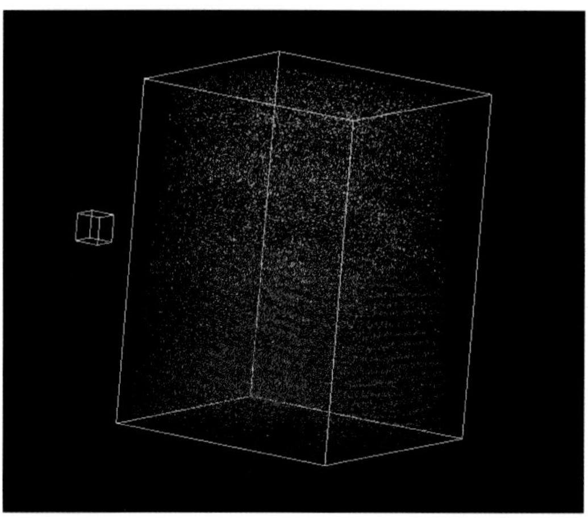

Fig. 6.37. 3D reconstruction of the atom distribution for the as-prepared Ni/Zr/Ni layer on a W-tip, red: Ni, green: Zr, probe volume $1 \times 1 \times 1\,\mathrm{nm}^3$

an amorphous structure. The missing crystal structure in Fig. 6.36c supports this conclusion. The concentration depth profile in Fig. 6.38 shows that the Zr concentration steeply rises to about 35 at.% at the interface with the bottom Ni layer. It then increases gradually to about 65 at.% before again steeply increasing to 100 at.%. The composition range between 35 and 65 at.% Zr in Ni is known as the existence range for an amorphous phase [49]. The rapid

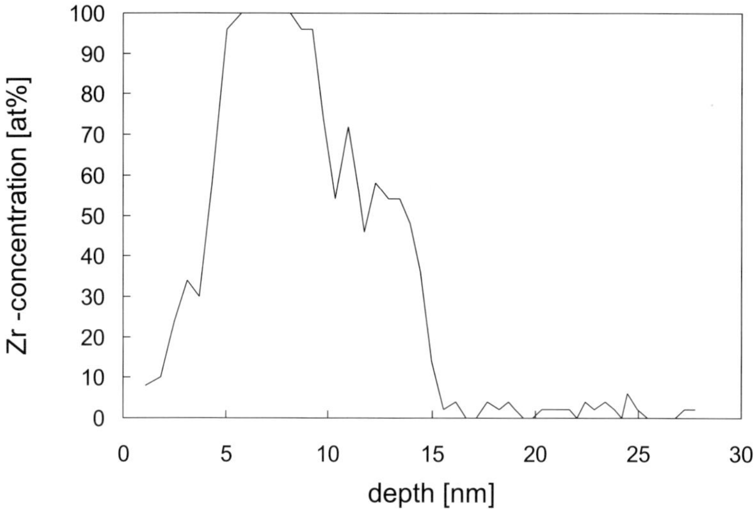

Fig. 6.38. Depth concentration profile from the data set shown in Fig. 6.37

concentration changes adjacent to the intermediate range clearly indicate interphase interfaces. The TAP analysis reveals the presence of an amorphous Ni-Zr layer of less than 5 nm thickness. It must have been produced by the sputter preparation of the sample. The kinetic energy of the sputtered ions is believed to range up to about 100 eV, which corresponds to a penetration range for Zr in Ni of a few nanometers, accompanied by significant atomic mixing and point-defect production. In addition, sputtering Ar ions of 10 keV energy may be reflected by the target towards the sample and enhance the radiation effect during the sputter deposition [50].

6.4.3 Segregation Phenomena

Chromium Distribution in Co-Cr-Ta Layers. Co-Cr thin films are used as recording media. The addition of Ta significantly increases the coercivity and reduces the magnetic noise. The underlying mechanism of this effect is not really understood, but tomographic atom-probe analysis reveals the role of Cr in this process [51]. Figure 6.39 shows digital FIM images taken near a grain boundary. In the alloy with added Ta, Co is depleted and hence Cr enriched along the grain boundary. Nothing like this is seen for the binary alloy. The 3D concentration mapping in Fig. 6.40 clearly shows the Cr-enriched phase near the grain boundary, but Ta is apparently not involved in this phase. Hence, the role of Ta is restricted to a destabilization of the Co-12 Cr binary. The good magnetic properties are due to the nanostructure composed of ferromagnetic and paramagnetic regions [52]. With sufficient Cr enrichment, the Curie temperature falls below room temperature.

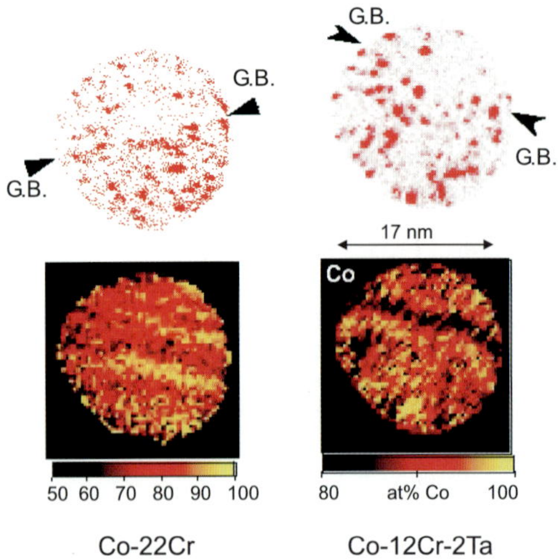

Fig. 6.39. Digital FIM images of Co-22Cr and Co-12Cr-2Ta films containing a grain boundary (*top*) and two-dimensional concentration maps (*bottom*) taken from the same area. By courtesy of K. Hono

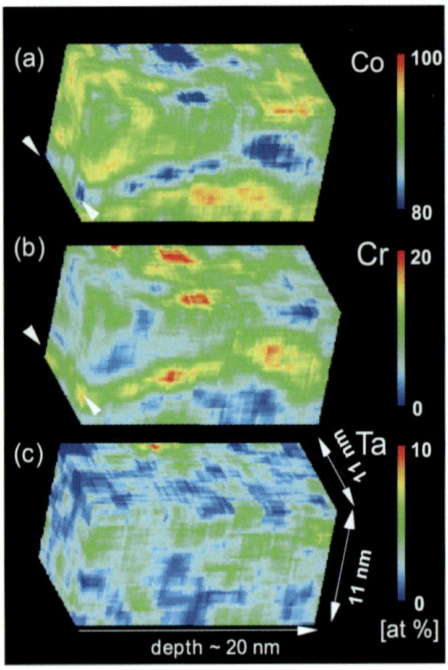

Fig. 6.40. 3D reconstruction of the atom distribution for a Co-12Cr-2Ta thin film. By courtesy of K. Hono

Phosphorus in Nanocrystalline Nickel Layers. Nickel-phosphorus layers have high resistance against corrosion and wear. For P contents around 10 at.% and less, the material is nanocrystalline and exhibits even higher wear resistance than the amorphous material found for higher P contents.

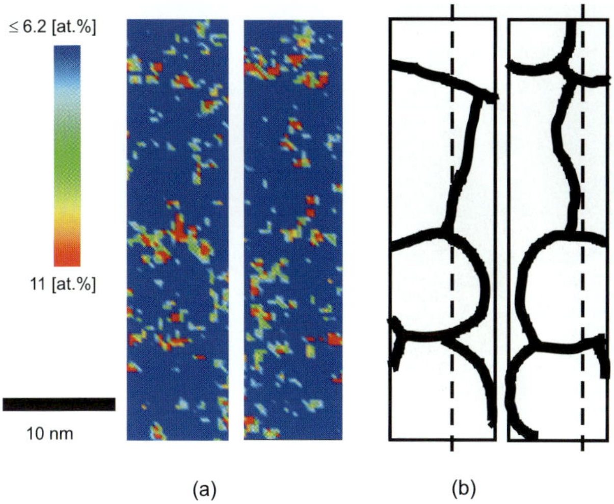

Fig. 6.41. (**a**) P–concentrations of a specimen aged at 523 K for 60 min. The two planes are cross-sections through the analysed volume, which are aligned perpendicular to each other, (**b**) schematic redrawing of a possible grain boundary structure. The intersecting line of the two cross-section planes is marked by a dashed line

A distinct step forward in the quantitative understanding of the chemical nanostructure was achieved by means of TAP analysis [53]. The sample material consisted of Ni-3.6 at.% P layers electroless plated to polished titanium substrates. After the layers had been peeld off the substrates, they had a thickness of 40 μm and were polycrystalline with a grain size of about 9 nm. Annealing at 523 K for 1 h produces a grain size of about 11 nm. Such a sample was probed with the TAP within a probe volume of $10 \times 10 \times 40\,\mathrm{nm}^3$. Local concentration data were evaluated for a probe unit volume of $1 \times 1 \times 1\,\mathrm{nm}^3$.

The obtained P distribution is shown in Fig. 6.41a for two cross-sections aligned perpendicular to each other. The blue colour indicates regions with P concentrations below the 1σ limit. The regions of higher P concentrations seem to be aligned as sketched in Fig. 6.41b. This line arrangement very much resembles the cross-section of a grain boundary system. The size of the enveloped areas accords well with the grain size of 11 nm, as obtained from TEM analysis [54], suggesting the segregation of P to the grain boundaries.

More details on the P enrichment have been obtained for a sample heated up isochronally to 673 K at 5 K/min. In this case, considerable grain coarsening was observed with a mean final grain-size of about 25 nm. The P distribution found within a $10 \times 10 \times 60\,\mathrm{nm}^3$ probe volume is shown in Fig. 6.42a. The P concentration was calculated here by using a probe unit volume of $2 \times 2 \times 2\,\mathrm{nm}^3$. Again the inhomogeneous P distribution is obvious. The topology of the P-enriched zones is illustrated by the isoconcentration surface plot in Fig. 6.42b. The enriched zones form shells, which envelop P-depleted

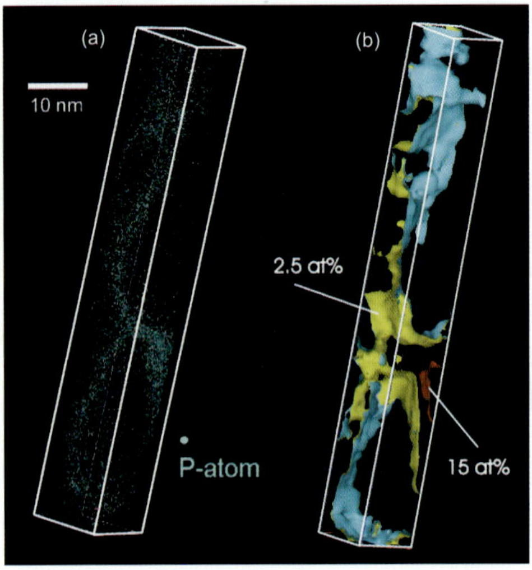

Fig. 6.42. (**a**) 3D reconstruction of the P–distribution after heating up to 675 K at 5 K/min., (**b**) isoconcentration surfaces of the same volume showing the P–segregation in grain boundaries. Red represents a level of 15 at.% P, whereas yellow and blue assign 2.5 at.% P. The direction from lower P–content in the grains (*blue*) to higher P–concentration in the grain boundaries (*yellow*) is obvious

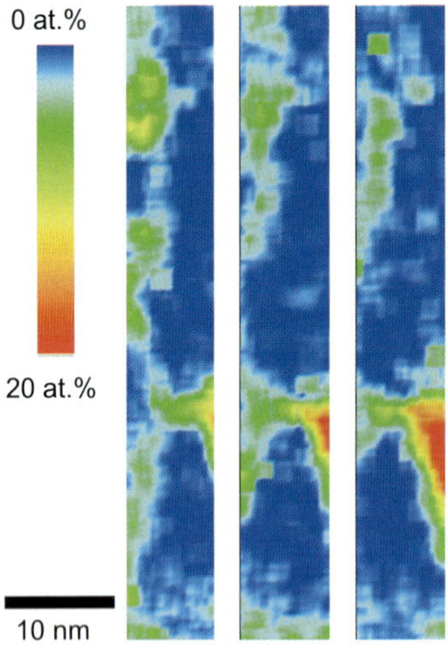

Fig. 6.43. P–concentration cross-sections through the volume of Fig. 6.42. The distance between each section is 2 nm

volumes with sizes corresponding to the observed grain size. Three parallel cross-sections of the analyzed volume are shown in Fig. 6.43. The blue area indicates P contents less than 1 at.%. A concentration profile taken through the red area in Fig. 6.43 shows a maximum of 25 at.% P (see Fig. 6.44) and

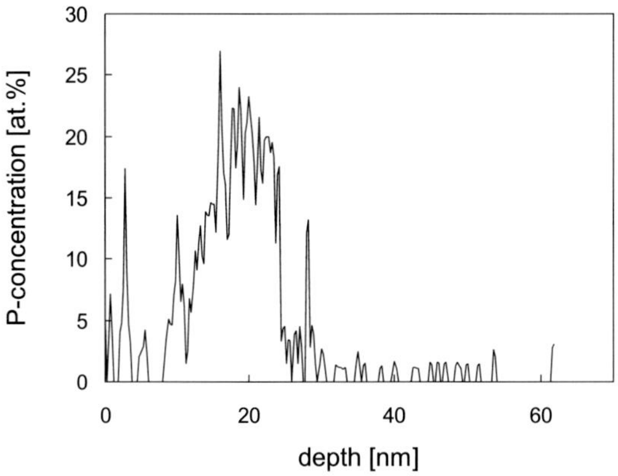

Fig. 6.44. Concentration depth profile through the triple junction of Fig. 6.42a. The P–concentration indicates the formation of a Ni_3P–precipitate

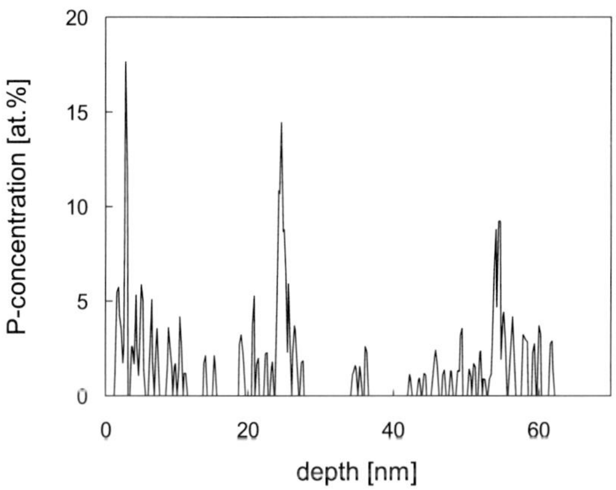

Fig. 6.45. Concentration depth profile through the volume of Fig. 6.42a along the grain boundaries. The P concentration indicates the segregation of P at the grain boundaries and a clear P-depletion in the grains

is to be interpreted as a Ni_3P particle formed at a grain boundary triple junction.

By using a probe unit volume of $2.0 \times 2.0 \times 0.4\,nm^3$ the width of the P-enriched zone was obtained as shown in Fig. 6.45. The three enriched zones are certainly broader than a monolayer and extend to a thickness of about 1 nm. The amplitudes of the remaining peaks correspond to the statistical fluctuation, which is large, owing to the small probe unit volume. The authors [53] conclude from the data that in the Ni-3.6 at.% P layers a metastable grain boundary phase is formed, which predominantly controls the grain growth up to 573 K annealing temperature.

By a simple conservation-of-mass calculation, they obtain for this alloy composition a segregation amplitude of 15 at.% (Fig. 6.45), a width of 0.8 nm (Fig. 6.45), and a P-content of 1 at.% (Fig. 6.43) within the grains. The grain size of 12 nm agrees with the measured data after 573 K annealing. As Ni_3 P formation is not observed here, the critical temperature for its formation must lie between 573 K and 673 K. The grain growth observed above this critical temperature is governed by P-accumulation in Ni_3 P precipitates. The study convincingly explains the grain-growth kinetics in NiP layers as reported in the literature. It disproves an earlier model based upon the solute-drag effect [55].

References

1. Müller E.W., Tsong T.T. (1969) Field Ion Microscopy-Principles and Applications. American Elsevier Publ Comp, New York
2. Bowkett K.M., Smith D.A. (1970) Field–Ion Microscopy. In: Amelinckx S., Gevers R., Nihoul J (Eds.) Defects in crystalline Solids Vol 2, North–Holland Publishing Company-Amsterdam. London
3. Wagner R. (1982) Field Ion Microscopy in Materials Science. Crystals, Vol. 6, ed. Freyhardt H.C., Springer–Verlag, Berlin
4. Haasen P., Wagner R. (1985) Application of analytical field–ion microscopy to the decomposition of alloys, Ann Rev Mater Sci, **15**:43
5. Sakurai T., Sakai A., Pickering H.W. (1988) Atom–Probe Field Ion Microscopy and its Application (Review), Part I and II. Technical Report of ISSP; Ser A No 1904 and 1905, The University of Tokyo, Tokyo
6. Miller M.K., Smith G.D.W. (1989,1992) Atom Probe Microanalysis, 1st and 2nd edn. Materials Research Society, Pittsburgh, PA
7. Miller M.K., Cerezo A., Hetherington M.G., Smith G.D.W. (1996) Atom Probe Microanalysis. Monographs on the Physics and Chemistry of Materials, 52. Clarendon Press, Oxford
8. Walko R.J., Müller E.W. (1972) Self–imaging of a surface by field desorption. Physica Stat Sol (a) 9:K9
9. Panitz J.A. (1973) The 10 cm atom probe. Rev Sci Instrum 44:1034
10. Cerezo A., Godfrey T.J., Smith G.D.W. (1988) Development and initial applications of a position–sensitive atom probe. J de Phys C66, 49:25
11. Cerezo A., Godfrey T.J., Smith G.D.W. (1988) Application of a position–sensitive detector to atom probe microanalysis. Rev Sci Instrum 59(6):862
12. Cerezo A., Godfrey T.J., Grovner C.R.M., Hetherington M.G., Hoyle R.M., Jakubovics J.P., Liddle J.A., Smith G.D.W., Worrall G.M. (1989) Materials analysis with a position–sensitive atom probe. Journal of Microscopy, Vol 154:215
13. Miller M.K. (1992) Implementation of the optical atom probe. Surf Sci 266:494
14. Leisch M. (1994) Three–dimensional field ion mass spectrometry. Fresenius J Anal Chem, 349:102
15. Bostel A., Blavette D., Menand A., Sarrau J.M. (1989) Toward a tomographic atom probe. J de Phys C8, 50:501

16. Blavette D., Deconihout B., Bostel A., Sarrau J.M., Bouet M., Menand A. (1993) The tomographic atom probe: A quantitative three–dimensional nanoanalytical instrument on an atomic scale. Rev Sci Instrum 64 (10):2911
17. Miller M.K. (1991) Concepts in atom probe designs. Surf Sci, 246:428-33
18. Martin C., Jelinsky P., Lampton M., Malina R.F., Anger H.O. (1981) Wedge–and–strip anodes for centroid–finding position–sensitive photon and particle detectors. Rev Sci Instr 52:1067
19. Bas–Chambreland P. (1997) Reconstruction Tridimensionelle des Images de Tomographie Atomique. PhD Thesis, University of Rouen, France
20. Cerezo A., Godfrey T.J., Hyde J.M., Sijbrandij S.J., Smith G.D.W. (1994) Improvements in three–dimensional atom probe design. Applied Surface Science, 76/77:374
21. Cerezo A., J.M. Hyde J.M., Sijbrandij S.J., Smith G.D.W. (1996) Data analysis in the optical PoSAP. Appl Surf Sci 94/95:457
22. Deconihout B., Renaud L., Da Costa G., Bouet M., Bostel A., Blavette D. (1998) Implementation of an optical TAP: preliminary results. Ultramicroscopy vol 1–4:253
23. Sijbrandij S.J., Cerezo A., Godfrey T.J., Smith G.D.W. (1996) Improvements in the mass resolution of the three–dimensional atom probe. Applied Surface Science, 94/95:428
24. Cerezo A., Godfrey T.J., Sijbrandij S.J., Smith G.D.W., Warren P.J. (1998) Performance of an energy–compensated three–dimensional atom probe. Rev Sci Instrum 69:49
25. Sijbrandij S.J., Cerezo A., Deconihout B., Godfrey T.J., Smith G.D.W. (1996) Characterization of efficiency enhancement in microchannel plate detectors. J de Physique IV C5:297
26. Deconihout B., Gerard P., Bouet M., Bostel A. (1996) Improvement of the detection efficieny of channel plate electron multiplier for atom probe application. Appl Sur Sci 94/95:422
27. Smith R., Walls J.M. (1978) Ion trajectories in the field ion microscope. J Phys D: Appl Phys, 11:409
28. Cerezo A., P.J. Warren P.J., Smith G.D.W. (1999) Some aspects of image projection in the field–ion microscope. Ultramicroscopy 79:251
29. Waugh A.R., Boyes E.D., Southon M.J. (1976) Investigations of field evaporation with a field–desorption microscope. Surf Sci 61:109
30. Southworth H.N., Ralph B. (1970) Image formation from ordered alloys in the field ion microscope. Phil Mag 21:23
31. Berg H, Tsong T.T., Cohen J.B. (1973) Local atomic arrangements in partially ordered CoPt II Field–ion microscopy. Acta Metall 21:1589
32. Vurpillot F., Bostel A., Menand A., Blavette D. (1999) Trajectories of field emitted ions in 3D atom–probe. Eur Phys J AP 6:217
33. Newman R.W., Sonwald R.C., Hren J.J. (1967) A method of indexing field ion micrographs. J Sci Instr 44:127
34. Wilkes T.J., Smith G.D.W., Smith D.A. (1974) Quantitative analysis of field–ion micrographs. Metallography 7:403
35. Bas P., Bostel A., Deconihout B., Blavette D. (1995) A general protocol for the reconstruction of 3D atom probe data. Appl Surf Sci 87/88:298
36. Al-Kassab T., R. Kirchheim R., Investigation of phase decomposition in alloys with the tomographic atom probe. Materials Sci and Eng A, in press

37. Warren P.J., Cerezo A., Smith G.D.W. (1998) Observation of atomic planes in 3DAP analysis. Ultramicroscopy 73:261
38. Al-Kassab T., Kirchheim R., to be published
39. Blavette D., Duval P., Letellier L., Guttmann M. (1996) Atomic–scale APFIM and TEM investigation of grain boundary microchemistry in astroloy nickel base superalloys. Acta Mater 44:4995
40. Jeske T., Schmitz G., Influence of the microstructure on the interreaction of Al/Ni investigated by tomographic atom probe. Mater Sci and Eng A, in press
41. Al-Kassab T., Macht M.-P., Wollenberger H. (1995) FIM/AP analysis of Cu-Pd multilayers. Applied Surface Science 87/88:329
42. Al-Kassab T., Wollenberger H., Blavette D. (1997) Application of the tomographic atom probe to selected problems in materials science. Z Metallkd 88:2
43. Lyon O., Simon J.P. (1988) An analysis of the partial structure function in unmixed Cu-Ni-Fe alloys studied by anomalous small–angle X–ray scattering. J Phys F: Met Phys 18:1787
44. Rozdilkski I., Cerezo A., Smith G.D.W., Watson A. (1998) Atomic scale study of precipitate/matrix interfaces in a metallic alloy. In: Ma E., Bellon P., Atzmon M., Trivedi R. (Eds.) Phase Transformation and System Driven far from Equilibrium, Boston, December 1-5, 1997. Symposium Mater Res Soc, Warrendale, PA, 481:521
45. Blavette D., Deconihout B., Chambreland S., Bostel A. (1998) Three–dimensional imaging of chemical order with the tomographic atom probe. Ultramicroscopy 70:113
46. Reich L., Murayama M., Hono K. (1998) Evolution of Ω–phase in an Al-Cu-Mg-Ag alloy - a three–dimensional atom probe study. Acta Materialia 46:6053
47. Zhang Y., Czubayko U., Wanderka N., Zhu F., Wollenberger H. (2000) Effect of gold addition on the nanostructure of amorphous Fe-Zr-B alloy. J Mater Res 15:1271
48. Al-Kassab T., Macht M.-P., Naundorf V., Wollenberger H., Chambreland S., Danoix F., Blavette D. (1996) Characterization of sputter–deposited multilayers of Ni and Zr with APFIM/TAP. Applied Surface Science 94/95:306
49. Samwer K., Fecht H.J., Johnson W.L. (1994) Amorphization in metallic systems. In: Beck A., Güntherodt H.J. (Eds.) Glassy Metals III, Vol 72 of Topics in Applied Physics, Springer, Berlin:5
50. Fu–rong–Ding, Averback R.S., Hahn H. (1988) Radiation–enhanced diffusion in Ni/Zr diffusion couples. J Appl Phys 64: 1785
51. Nishimaki J., Hono K., Hasegawa N., Sakurai T. and M. (1996) Three–dimensional atom probe analysis of Co-Cr-Ta thin film. Appl Phys Lett 69:3095
52. Ouchi K., Iwasaki S. (1982) Perpendicular magnetization structure of Co-Cr films. IEEE Trans Mag MAG-18:1110
53. Färber B., Cadel E., Menand A., Schmitz G., Kirchheim R. (2000) Phosphorus segregation in nanocrystalline Ni–3.6 at.% P alloy investigated with the tomographic atom probe (TAP). Acta Materiallia 48:789
54. Müller F. (1996) Herstellung, Mikrostuktur und Eigenschaften von außenstromlos abgeschiedenen Nickel–Phosphor–Schichten mit nanokristallinem Aufbau. PhD Thesis. Fortshr.–Ber. VDI–Reihe 5 Nr. 427, VDI–Verlag, Düsseldorf
55. Mehta S.C., Smith D.A., Erb U. (1995) Study of grain growth in electrodeposited nanocrystalline nickel 1.2 wt% phosphorus alloy. Materials Sci and Eng A 204:227

List of Standard Abbreviations

Acronyms	
ADC	Analogue-to-Digital Converter
AP	Atom-Probe
APFIM	Atom-Probe Field-Ion Microscopy
BIF	Best Image Field
BIV	Best Image Voltage
CAMAC	Computer Automated Measurement and Control
CCD	Charge-Coupled Devices
d.c.	direct current
DAC	Digital-to-Analogue Converter
FEM	Field-Emission Microscope
FIM	Field-Ion Microscope
HV	High Voltage
IAP	Imaging Atom Probe
MAP	Mapping Atom Probe
MCP	Micro-Channel Plate
OAP	Optical Atom Probe
OPoSAP	Optical Position-Sensitive Atom Probe
OTAP	Optical Tomographic Atom Probe
P46	Technical classification of phosphorous material
PLAP	Pulsed Laser Atom Probe
PoSAP	Position-Sensitive Atom Probe
PSD	Position-Sensitive Detector
TAP	Tomographic Atom Probe
TDC	Time to Digital Converter
ToF	Time of Flight
ToFMS	Time-of-Flight Mass Spectrometry
UHV	Ultra-High Vacuum

List of Standard Notations

Symbols	
A_{D}	detector area
a	size of square anode
α	polar co-ordinate of trajectory
$h(x)$	transfer function
L	flight path distance for ions
e	elementary charge
F_0	electric field on the surface of the tip
F_{e}	evaporation field
Φ, ϕ	angular directions of ion trajectory
$\Phi(r, z)$	electrical potential
γ	shaft angle of conical tip
k_{f}	geometrical constant
κ	image compression factor
M, M_R	magnification
m	atomic mass in amu
μ	relative position of projection center
n	charge state of ions
n_a	number of detected atoms
Ω	atomic volume
P_α, P_β	centers of projection
p	probability
Ψ, ψ	angular directions of ion trajectory
Q_X, Q_Y, Q_Z, Q_r, Q_t	fractions of detected charge
q_x, q_y	normalised charge fractions
R	average radius of the FIM-tip
r	co-ordinate of ion trajectory
ρ	detector efficiency
t	time of flight
ϑ	polar co-ordinate of atom position
ϑ'	inclination of trajectory
ϑ'_0	angle between tip and detector axis
v	velocity of ions
V	applied total high voltage
V_{DC}	applied base voltage
V_{P}	applied high voltage pulse
x, y, z	co-ordinates of atom position
$x_{\mathrm{c}}, y_{\mathrm{c}}$	co-ordinates of charge center
$x_{\mathrm{D}}, y_{\mathrm{D}}$	position of impact on detector

7 Scanning Tunneling Microscopy (STM) and Spectroscopy (STS), Atomic Force Microscopy (AFM)

H. Neddermeyer and M. Hanbücken

7.1 Introduction

The breakthrough of scanning tunneling microscopy (STM) [1] in science and technical applications results from its potential in real space imaging of surfaces. As shown by early work [2,3], a lateral resolution can be achieved down to the nanometre or even atomic scale. Moreover, the idea of STM is conceptually simple. A strongly localized tunneling current between a sharp metal tip and the sample and scanning are used to image the surface. The localization of the current determines the lateral resolution. Since the instrumental requirements and the electronic circuitry of an STM are neither extremely sophisticated nor expensive, numerous of groups embarked on STM work right after the first publications had appeared and contributed to the development of this technique. This was certainly very different from electron microscopy which, for decades remained a subject for specialists and relied upon demanding equipment.

Even in the early years of STM, many modifications of the original concept had been proposed and new branches in this kind of microscopy were developed, which are now often summarized by the term scanning probe microscopy (SPM) or sometimes by SXM [4]. Among the proposals of the first years are the spectroscopic use of STM, which was correspondingly called scanning tunneling spectroscopy (STS), and atomic force microscopy (AFM) where, in the first experimental realizations, the strong dependency of the tunneling current on the sample–tip separation was utilized to detect the small forces acting on a tip while scanning the surface [5]. Another proposal of the first years was the measurement of losses in the tunneling current, but this has only been accomplished in a convincing way very recently [6]. The imaging of biological objects by STM was also attempted very early but was successfully established as a routine method only much later [7]. Owing to the high sensitivity of biomaterials to the tunneling current (and sometimes to the electric field at the tip apex) and to difficulties in the interpretation of the experimental STM data of such materials, the measurement of biosamples has now essentially become a field of AFM.

The fact that many possible modifications of the STMwere proposed long ago have not so far been realized, or only for selected systems, may be explained by various reasons. It is quite clear that some of the proposed modifications only concerned experiments on very special samples and a more general use was not planned. Another limiting point is that, in most cases, tunneling requires the presence of a well-defined surface not covered with contamination or other unwanted adsorbates. Otherwise details of the images cannot be assigned to real structures of the sample surface but are more probably caused by unknown artifacts. Moreover, measurements of ill-defined or of contaminated surfaces mostly give rise to instabilities in the tunneling current and do not allow imaging of the sample surface with satisfactory resolution. Owing to the sample–tip interaction and diffusion of adsorbates, the measured surface and the tip do not remain in a stable state and the sample cannot be reproducibly studied by SPM methods. It has been demonstrated in recent years that very low temperatures facilitate the access to local properties on an atomic scale. Most STM experiments on single atoms or on individual adsorbed molecules are therefore carried out at low temperature, by cooling with liquid He.

We conclude these introductory remarks with a brief description of the concept of this chapter. It is mainly based on individual projects of the Schwerpunktprogramm which means that not all the activities of groups in Germany and in other parts of the world can be considered in any detail. However, since the individual projects cover a wide field of SPM work and deal with subjects of broad interest, we nevertheless provide a general survey on recent developments in the field of SPM. In addition, some highlights that have not been treated in the Schwerpunktprogramm will also be mentioned in the text. The majority of the individual projects were focused on the development of specific STM- or AFM-related techniques or on their refinements. At a later stage, the combination of SPM with complementary techniques became important, mostly techniques from surface science. The application of the newly developed techniques to resolve problems of fundamental or more technical relevance in materials science was the main aim of the investigations in the last stage of the Schwerpunktprogramm. This chapter is intended to provide a self-contained review of SPM without going into too many details. More detailed information can be found in a number of excellent monographs [8–11].

7.2 Scanning Tunneling Microscopy (STM)

The basis of STM is transport of electrons through a potential barrier by tunneling. A sharp metal tip is brought into close proximity d (typically 1 nm) of the surface of a conducting sample (see Fig. 7.1).

If a voltage U is applied between tip and sample, a tunneling current I is generated, which depends exponentially on the sample–tip distance and which can be used for adjusting the sample–tip distance to a constant value. The sample is then scanned under the conditions of a constant tunneling current and an image can be recorded that contains information on structural and the

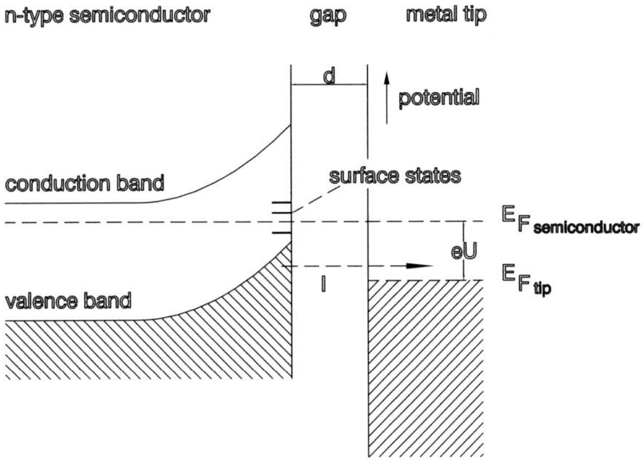

Fig. 7.1. Schematic energy diagram for tunneling between a metallic tip and a semiconductor surface which are separated by a small distance d (typically 1 nm or less). The polarity of the sample–tip bias U determines the direction of the tunneling current I. Here, a negative sample-bias voltage U (throughout this chapter, U is referred to the sample) gives rise to tunneling from the occupied states of the sample into the empty states of the tip. Consequently, the occupied (unoccupied) states of the sample are imaged by negative (positive) sample–tip bias

electronic properties of the surface. The spatial resolution in the STM images depends on the degree of localization of the electrical tunneling current and may reach the atomic scale if the tunneling current flows mainly through one atom of the tip apex. For planar systems the tunneling effect has been treated in textbooks and is used, for example to explain the characteristics of an electrical current through a metal–insulator–metal (MIM) structure [12], which is the first approximation of STM. To understand atomically resolved STM images the atom structure of the surface has to be included in the theoretical description of the tunneling process. Early theoretical models to describe atomically resolved STM images, have been reported by Garcia et al. [13], Stoll et al. [14], and Tersoff and Hamann [15]. The latter work has attracted particular attention, since it relates the observed corrugation to a simple quantity of the surface, namely, the local density of states of valence or conduction electrons. Although a number of approximations have to be made in the Tersoff–Hamann picture of STM, it will be summarized briefly, since it can be used as a first guideline for the interpretation of the experimental results. Moreover, exact theories have so far only been applied to a small number of systems and cannot be generalized in a simple way.

Tersoff and Hamann based their work on the concept of a transfer Hamiltonian, as introduced by Bardeen for the electrical current through a MIM structure [16]. According to this, the tunneling current I can be expressed as

$$I = \frac{2\pi e}{\hbar} \sum_{\mu\nu} f(E_\mu)[1 - f(E_\nu + eU)]|M_{\mu\nu}|^2 \delta(E_\mu - E_\nu) \, ,$$

where $f(E)$ is the Fermi function, E the energy, $M_{\mu\nu}$ the tunneling matrix element between states Ψ_μ of the metal probe tip and Ψ_ν of the sample surface, E_μ and E_ν the energy of the unperturbed states in the absence of tunneling, and U the sample bias voltage. The contribution of reverse tunneling has been neglected here. The tunneling matrix element $M_{\mu\nu}$ may be computed from the quantum mechanical expression for the current between states Ψ_μ and Ψ_ν, irrespective of the height and form of the potential barrier between the sample and the tip. If we model the tunneling tip by a point probe and assume a small sample bias voltage, and low temperature, the equation for the local tunneling current may be reduced to

$$I \propto \sum_\nu |\Psi(\boldsymbol{r}_0)|^2 \delta(E_\nu - E_\mathrm{F}) ,$$

which is clearly a localized property and corresponds to the charge density of states at the Fermi level E_F and the position $\boldsymbol{r}_0$ of the point probe. For a tip with spherical radius corresponding to an atomic s wavefunction, the lateral resolution of STM was found to be consistent with the experimental data on a reconstructed metal surface. However, it was shown later that the measured corrugation (and consequently also the lateral resolution) of a dense-packed metal surface cannot be explained by tunneling into or from spherical tip states [17]. Tip states with non-zero angular momentum provide a better explanation for the experimental resolution.

A quantitative analysis of the corrugation measured on Al(111) as a function of tunneling resistance showed that the interaction between tip and sample states and the dynamical effects of the tunneling electrons have to be considered to obtain agreement between experimental and theoretical corrugation on a dense-packed metal surface [18]. On the basis of the latter work it appears rather difficult to draw conclusions, in which case the Tersoff–Hamann picture provides a correct approximation for the description of experimental STM data. Although a large number of theoretical STM images calculated on the basis of the transfer Hamiltonian model agree with the experiment [19], one always has to ask whether the approximations apply to the specific sample–tip system and the measurement conditions, and in particular to the sample–tip distance. It is reasonable to assume that at larger sample–tip distances the sample–tip interaction becomes smaller. On the other hand, the resolution is also reduced. More work is needed to elucidate the physical basis for interpretation of STM images. It should be mentioned in this context that the structure and the electronic nature of the tip apex are mostly unknown, at present, which is another complication for a satisfactory modeling of experimental STM data.

According to the pioneering work of Binnig et al. [1], an STM construction consists of a fine positioning piezoelectric xyz system and a coarse approach between sample and tip. In the literature many STM devices have been described that allow high-quality imaging of surfaces. If a number of rules are respected, the construction of an STM is not too critical. (i) The resonance frequencies of the STM should be high, if possible in the 10 kHz range or higher. High resonance frequencies are obtained by using rigid and light-weight construction. Difficulties mostly arise from the tip holding device and the sample carrier including mounting of the sample. (ii) The electronic equipment consists of a feed-back loop to regulate the tunneling current to a constant value, a scanning unit and a data acquisition system. The precision of the tip movement should be controllable within a few 10 pm laterally and a few picometres normal to the sample surface, which requires electronic equipment with correspondingly small noise. (iii) Normally, the STM is supported by a damped spring or a viton stack system with a low resonance frequency, below 10 Hz. (iv) To achieve high resolution in STM images it is advisable to work in UHV, since the sample surface and the tip can then be prepared and kept in a well defined clean state. In principle, imaging of surfaces is also possible in environmental conditions.

The emphasis in the design of STM equipment during the first period of the Schwerpunktprogramm was concentrated on finding ways of measuring samples above room temperature [20,21] and increasing the speed of data acquisition [22]. Instrumentation for the low-temperature range was also developed in some of the groups (see, for example [23]). The progress in this field allowed in situ studies of adsorption and reaction on surfaces [24] as well as deposition experiments, in which the deposition and the growth of the condensing species could be observed simultaneously [21]. Examples of such measurements are described below. An STM for high-temperature experiments and fast measurements is shown schematically in Fig. 7.2 [24]. On small sample areas (for example $25 \times 25\,\mathrm{nm}^2$) the image repetition rate can be chosen up to nearly $10\,\mathrm{s}^{-1}$. However, the noise is then fairly high and data acquisition at 1 image per 10 s is usually sufficient to follow changes on the surface with atomic resolution. In adsorption experiments, for example, one has then to use a sufficiently low partial pressure (typically 10^{-6} Pa) in order to minimize rearrangements of the surface between successive measurements. For the high-temperature measurements, the high measurement speed is also advantageous to minimize drifts in the area selected. In general, when using an STM with a special self-compensating design, high-temperature drifts are not a serious problem.

A second direction in instrumental development was the combination of STM with other surface sensitive techniques, such as photoelectron and Auger electron spectroscopy [25], in order to characterize the elemental and chemical nature of surfaces. Other work was concerned with the combination of STM with methods for structural characterization, such as scanning electron microscopy [26], low-energy electron [27] and surface sensitive X-ray diffraction [28]. A few representative examples of such combined studies are given below. It should be noted that the availability of such supplementary methods proved to be very important since a typical difficulty of STM lies in the fact that in normal measurements the various species on a surface cannot simply

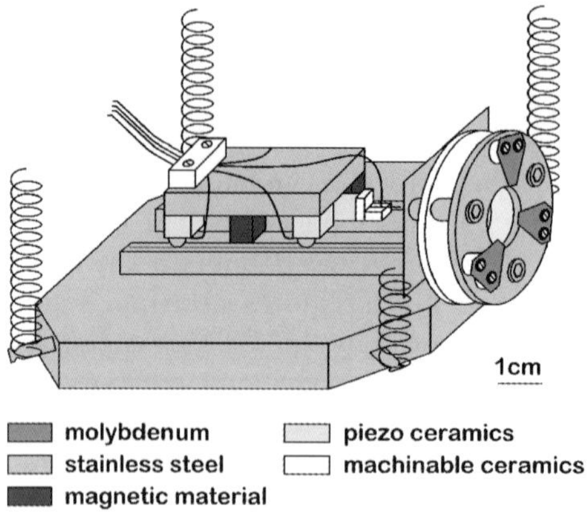

Fig. 7.2. Principle of a high-temperature STM [24]

be distinguished from each other in a simple way. Many attempts have been made to use STM directly as an analytical tool, but it was mostly found to be applicable only for specific systems.

7.3 Scanning Tunneling Spectroscopy (STS)

The possibility of local spectroscopy of electronic states is a very attractive issue in STM work and many groups have concentrated on this technique, which is called scanning tunneling spectroscopy (STS). It has been shown that STS works very satisfactory in the vicinity of the Fermi level of the sample and measurements of surface states and their local behaviour are unique possibilities of STS [29,30]. A satisfactory theory of STS measurements has not yet been developed (the problems are the same as those encountered in calculating STM images) and the spectroscopic information is usually derived in a simplified way. In general, to obtain information on the local electronic structure, the tunneling current I is measured as a function of sample bias voltage U, while keeping the sample–tip separation constant. In a first approximation it is assumed that the derivative dI/dU is proportional to the local density of states of the sample. The density of states of the tip does indeed have to be fairly constant in order to obtain reasonable results. A problem is that the slope of the I–U curves always drastically increases with U. This effect is related to the increase of the tunneling probability with U. To consider the tunneling probability in an approximate way, Stroscio et al. have divided their experimental dI/dU values by I/U which in many cases leads to better agreement between the experimental results and other measurements or calculations of the density of states [31].

7.4 Atomic Force Microscopy (AFM)

Atomic force microscopy (AFM) [5] has now found wide acceptance as a routine microscopic technique to explore the topography of surfaces (an account on the basic concepts and a number of applications have been given in [9]). In addition, it has been established as a method of studying the interactions between a sample surface and a probe tip. It was originally developed for the imaging of insulating samples and for direct probing of the atomic surface potential [5]. In general, AFM utilizes the atomic forces which exist in the small distances (typically in the nm region) between a sample surface and a sharp probe tip to provide a microscopic image of the surface. For detection of the forces the tip is mounted on a cantilever, sensitive enough to be deflected by forces in the 10^{-8} to 10^{-9} N range or even smaller. In the original and simplest concept of AFM, the sample surface is scanned by the probe tip under the condition of a constant force, giving rise to an image that represents the contours of constant force between sample and tip. Such images are usually obtained in the so-called "contact mode" of the AFM, where the tip is pushed towards the surface under a constant load and repulsive interaction between sample and tip apex has been established. In a first approximation, the constant force (repulsive) mode provides a topographical image of the surface, but may also lead to severe damage. Other (non-contact) modes of AFM have been developed, which utilize the attractive part of the sample–tip potential. Here the interactions between sample and tip are much weaker and the influences on the surfaces are correspondingly smaller. The non-contact methods have been found very suitable for measurements on sensitive samples, for example biological material or polymers. In addition, true atomic resolution, that is, the possibility of imaging individual atomic features, such as defects or other irregularities is usually only possible in non-contact operation; this was first described by Giessibl [32] and is discussed in more detail below. The presence of a regular atom pattern in an image acquired in the contact mode does not in fact mean that true atomic resolution has been achieved. In the early days of AFM, a large number of images (measured, for example on graphite) showing atom structures were misinterpreted until it was realized that only general properties of the surface potential and its symmetry are probed in this case [33,34].

The attractive mode is usually implemented by making dynamical measurements where the cantilever is vibrating with its resonance frequency. During tip approach, the sample–tip interaction leads to changes of the resonance frequency of the cantilever, which control the sample–tip distance [35]. In earlier work this change was associated with the force gradient of the potential [36] (see, however, discussion below). It should be mentioned that the response of a sample on a vibrating cantilever is also used in the so-called "tapping" mode, where the amplitude (up to 100 nm) of the vibration (which is chosen near the resonance frequency of the cantilever) is held constant [37]. In the tapping mode and in dynamical methods, in general, the sample–tip interaction is weaker by several orders of magnitude than in the contact mode. Another important issue of the attractive mode operation is the possibility to detect long-range interactions (on the length scale of up to a few 10 nm) resulting from van der Waals, magnetic or electrostatic forces. By using such forces, magnetic [38] or ferroelectric [39] domains may be imaged. It should also be mentioned that AFM provides experimental access to mechanical properties of surfaces on the atomic scale, for example to friction [40], which can hardly be measured by any other methods.

The efforts of AFM projects in the Schwerpunktprogramm have been focused on two major problems, namely on the construction of an UHV-compatible low-temperature AFM and on the development and analysis of measurement techniques that allow true atomic resolution to be attained and provide direct access to the sample–tip interaction. Both problems have been solved convincingly and constitute remarkable steps in the progress of AFM for further applications in surface science and material science. The construction of an AFM for low-temperature applications in an UHV environment is very demanding. The reduction of the mechanical vibrations, which result from external sources, the development of electronic equipment with sufficient sensitivity, resolution and low noise level, the mechanical design, and the compatibility with UHV and low-temperatures are the main problems to be solved. Figure 7.3 depicts the scheme of the entire setup [41]. Figure 7.4 shows a sketch of the AFM design. For measurements at low temperature, the AFM is transferred into a bath cryostat. The forces acting on the tip are detected either by deflection of the cantilever, which is determined by optical interferometry, or dynamically by excitation of vibrations of the cantilever, as explained below.

The requirements that concern the imaging of surfaces with true atomic resolution have been explained and discussed in detail in [42–44] but will also be described here since they are of general importance for our understanding of how a surface is probed by AFM. In the early days of AFM and in most applications nowadays, AFM is used in the static deflection mode of the cantilever. This means that the feedback circuit regulates the deflection of the cantilever from its zero position to a preset value, which controls the force the tip exerts on the sample. In practice, this mode can only be used while a contact between tip and sample surface has been established. The exact position of the tip in the contact mode with respect to the sample surface is then defined by the balance of repulsive and attractive interactions between sample and tip and is a stable quantity if the tip is pushed towards the sample by the cantilever. True atomic resolution cannot be obtained in this mode of AFM operation since the actual forces on the tip apex are so large that they give rise to inelastic deformations with the result that the contact area comprises hundred of atoms (see, for example [44] for more details). To understand the physical situation from another viewpoint one has to consider the well known "jump to contact" effect [45] of the tip during the coarse

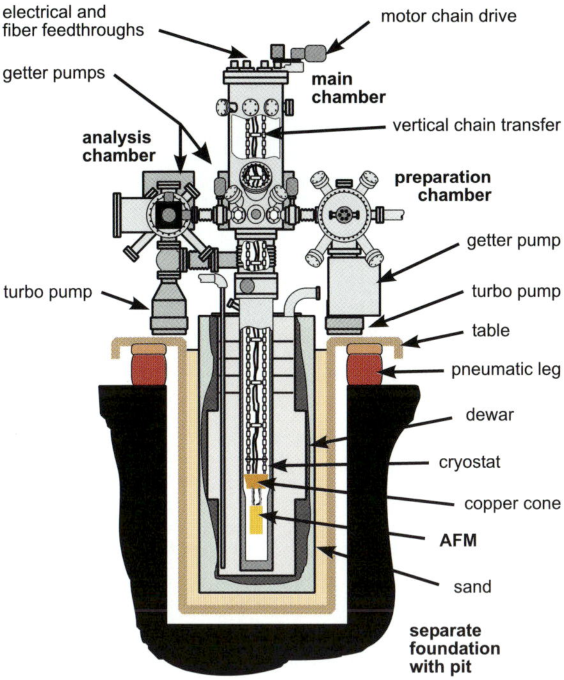

Fig. 7.3. Design of an apparatus for UHV low-temperature AFM [41]. The AFM (denoted SFM in the figure) can be lowered to the cooling position by a chain-driven lift system. Removal of the AFM out of the cryostat is necessary for sample and tip exchange

approach to the sample. Below a distance somewhat smaller than 100 nm, an attractive force begins to develop between sample and tip. On further approaching the sample surface the absolute value of the force gradient of the sample–tip interaction may exceed the spring constant of the cantilever (this will always happen with the commonly used cantilevers) and as a consequence the tip jumps towards the sample surface. After contact between tip and surface has been established, the aforementioned force balance is given by the repulsive interaction of the front part of the tip apex with the sample, the attractive interaction of the entire tip with the sample surface and the actual deflection of the cantilever (which is bent towards the sample surface during the tip approach).

In the case of a static (deflection sensitive) control of the sample–tip distance the instability, caused by the "jump to contact" effect, prevents true atomic resolution from being attained in the attractive part of the sample–tip potential. As has already been mentioned above, true atomic resolution will generally only be possible when dynamical modes of AFM control are employed. In this case the cantilever is vibrating at its resonance frequency (typically 300 kHz). Changes of the resonance frequency caused by the ad-

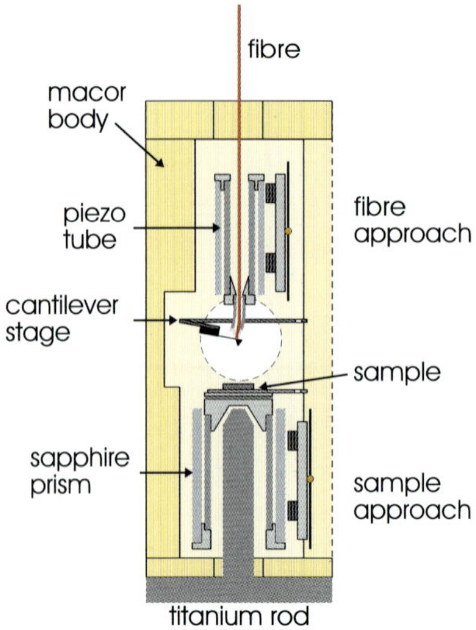

Fig. 7.4. Schematic drawing of the low-temperature AFM in a section through the symmetry plane [41]

ditional sample–tip interactions are utilized to probe the surface locally. For small vibration amplitudes, the changes are proportional to the force gradient at the sample. In actual measurements, this condition will normally not be fulfilled since the amplitude of the vibration normally exceeds this distance range [42].

By not only measuring the changes of the resonance frequency but also analysing their dependency on the vibration amplitude and on the (relative) sample–tip distance Hölscher et al. have demonstrated that quantitative information on the sample–tip interaction potential may indeed be derived [43]. In particular, the sample–tip interaction potential has been quantitatively obtained for a graphite sample probed by a clean Si tip. Under the experimental conditions, the potential minimum has been found to be 40 eV, which seems to be a very reasonable result; it shows that although the sample–tip interaction was provided by more than one atom, the number of atoms was still rather small (less than ten in a very crude estimation). To clarify the vibration behaviour of the cantilever-tip system, when it is already in the vicinity of the surface and is experiencing the additional sample–tip interaction (Fig. 7.5a), the potential of the tip is schematically shown in Fig. 7.5b. The dashed line corresponds to the sample–tip interaction potential, and the dotted line to the cantilever potential (referred to the position of the tip) without sample–tip interaction, i.e., when the tip is far from the sample surface. The cantilever potential follows the quadratic dependency of E on x known for a harmonic oscillator. If the tip approaches the surface more closely, the effective total potential (full line) obtained by adding the two contributions,

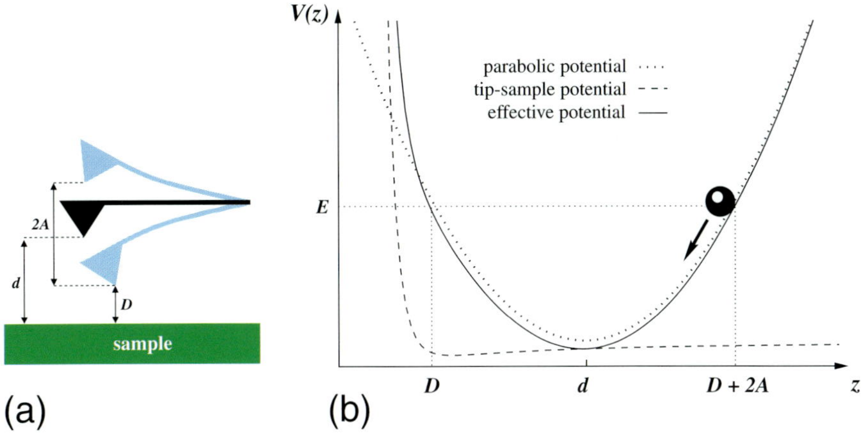

Fig. 7.5. Schematic behaviour of a cantilever-tip system in dynamic measurements. (**a**) The position of the tip without (*in black*) and at the inflection points (*shaded*) during the oscillations. (**b**) Harmonic potential of cantilever (*dotted line*), sample–tip interaction potential (*dashed line*) and effective potential (*full line*) (see [42–44])

becomes inharmonic. As a consequence, both, the resonance amplitude and the resonance frequency, are changed. It is evident from the potential diagram that the effective interaction potential depends on the sample–tip distance particularly on the left hand side of the minimum, while on the right-hand side the differences may be neglected.

The basis for true atomic resolution in the dynamical mode is that the tip apex must not touch the sample surface during the vibrations, particularly at the inflection point, when the distance between tip and sample is at its minimum. At the same time the vibration amplitude has to be so large that the elastic forces of the cantilever at the inflection points are larger than the force resulting from the sample–tip interaction. This ensures the absence of the "jump to contact" instability and possible damage of the tip in the contact regime. Wiesendanger and coworkers have demonstrated by a number of examples that the conditions for true atomic resolution can indeed be realized [46–48]. Figure 7.6 shows an atomically resolved AFM measurement for an InAs(110) surface. Individual vacancies are seen in the upper part of the figure, demonstrating true atomic resolution. Previously, the detection of such vacancies and other point defects was the domain of STM work (see, for example [49]) and the progress made in non-contact AFM is immediately evident.

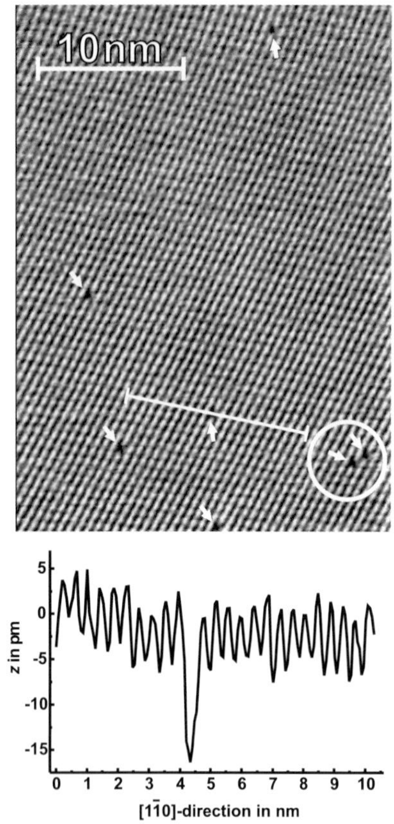

Fig. 7.6. AFM image obtained on InAs(110) at 14 K showing true atomic resolution [46–48]. Seven missing protrusions (*white arrows*) corresponding to As vacancies are identified. The line section (*lower part of the figure*) shows a depression of about 15 pm at the position of the As vacancy

7.5 Special Techniques

7.5.1 Generalities

The specific and controllable interaction between a sharp tip and a surface can be employed as a general basis for microscopic measurements of surfaces. The interaction can be modified in many ways to achieve sensitivity for particular surface properties. For example, magnetic domains can be imaged by using magnetized tips in an AFM [38], or by employing tips with spin-polarized surface states in an STM [50]. The domain structure of ferroelectric material becomes accessible by application of additional potentials to an AFM tip [39], including the use of modulation techniques [51]. It is also possible to modify an STM experiment to obtain sensitivity for quantities that are not accessible by a conventional STM setup. For example, a small temperature gradient between sample and tip, leading to a thermovoltage, can be used to obtain particular sensitivity to states in the vicinity of the Fermi level E_F. A material contrast can then be seen in STM images [52] or measurement allows surface states close to E_F to be imaged [53]. Another experiment is devoted to the recording of photons that have been generated by the tunneling process. In the first application of this technique to metal surfaces the decay of plasmons gave rise to photon emission [54]. In the Schwerpunktprogramm, a photon-detecting STM is utilized for measurements of the time-dependence of the luminescence intensity on GaAs-based homostructures and heterostructures [55]. A very broad range of applications has been found for the use of the STM in electrochemistry, which will be described in greater detail in the following subsection.

7.5.2 STM in Electrochemistry

Sonnenfeld and Hansma were the first to show that imaging of surfaces with high resolution is possible in an electrolytic environment [56]. This observation opened the wide field of applications of STM in electrochemistry, where not only can structural changes of the surface during or after the electrochemical reactions be visualized but the electrochemical processes themselves also become accessible on a local scale, for example by potentiometric measurements with the STM tip. When surfaces are imaged in solutions, conventional STM instrumentation can be adapted directly. If one is interested in electrochemical reactions, however, it is necessary to introduce a new type of STM control, in which the sample–tip distance and the electrochemical processes are determined by different electronic circuits. Without going into details (more information can be found in a review article by Siegenthaler in [9]), a typical scheme of an STM for electrochemical applications is reproduced in Fig. 7.7. The usual two-electrode (sample–tip) configuration is replaced by one with four electrodes, which allow independent control of sample (E_s) and tip potential (E_t). Both potentials are referred to a currentless reference electrode (RE,) and the actual sample–tip voltage is U_t. In order to obtain structural information about the sample surface, the sample–tip voltage U_t should ideallylead only to an electronic tunneling current between sample and tip (and not to unwanted Faraday current contributions, which implies that the tip should not affect the electrolytic reaction on the sample surface). The potential-barrier for the tunneling current is then determined by the thin layer of the electrolytic solution between sample and tip. The presence of a Faraday current between them could be avoided by applying a suitable potential E_t. Since this is not always possible, the tip shaft is usually covered by an insulating layer, except for the front part of the tip apex. Unwanted currents and effects may therefore be minimized and can be neglected in most cases.

Sample surfaces in electrolytic solutions may be imaged at atomic resolution [57] (which allows an interesting comparison of the effect of the different

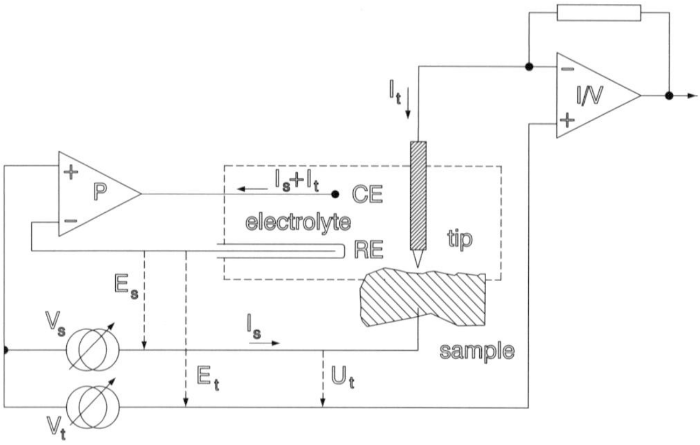

Fig. 7.7. Principle of the design and the electronic circuit of an STM used in electrochemistry (according to Siegenthaler in [9]). In the electrolytic solution four electrodes are immersed: sample, tip, reference electrode (RE) and counterelectrode (CE). The potentials of the sample (E_s) and tip (E_t) are referenced relative to the currentless reference electrode RE and are set by the voltage sources V_s and V_t. Potentiostatic control against the reference electrode is provided by the operational amplifier denoted P. Since two potentials are concerned, the circuitry is called bipotentiostat. The difference between sample and tip potential corresponds to the tip voltage U_t responsible for tunneling

kind of interfaces on the stability of the surface) and furhtermore the influence of the sample potential E_s during measurement or preparation of the surface may be examined in situ. Electrolytic deposition or dissolution of materials has therefore been studied, for example, of metal layers on metal substrates [58]. Nucleation and the observed growth modes of metals from electrolytic solutions are very similar to those observed in vacuum for deposition from the gas phase. For example, for the electrolytic deposition of Cu onto Au(111) one observes the Stranski–Krastanow growth mode (growth of a Cu interface layer with subsequent growth of three-dimensional Cu islands) [59]. The first layer already begins to form at underpotential of the sample, that is, when the potential of the sample relative to the reference electrode is not negative enough for subsequent growth of thicker layers. For continuous deposition, one needs overpotentials, which, in the case of Cu/Au(111), lead to the three-dimensional growth mode of Cu islands. Further experiments showed that not only can the influence of the sample potential and of the nature of the electrolytic solution on the growth of the metal deposit be directly visualized, but so too can the effect of additives [60]. The formation of a patterned surface by means of an electrochemical STM will be described in a later section.

For the study of reactions of electrode surfaces and of adsorption, electrochemical STM has also become a unique tool. The measurement of surface

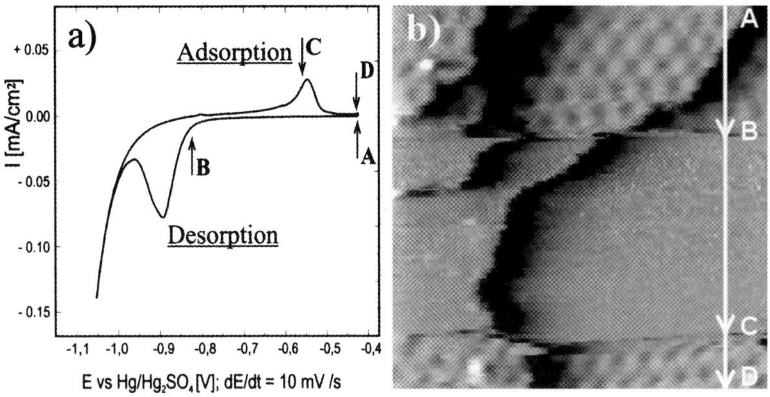

Fig. 7.8. STM on Cu(111) immersed in 5 nm H_2SO_4 measured with a speed of $10\,mV{\cdot}s^{-1}$. In (**a**) the cyclic voltammogram is displayed and in (**b**) the STM image acquired simultaneously. Image size: $39 \times 39\,nm^2$ [62]

structures does not depend on the potential that controls them. A a typical example in this field, we cite the anodic oxidation of Au(111) in acid solution, studied by Schneeweiss et al. [61]. In contrast to the generally accepted picture that the oxidation of Au(111) proceeds via two oxidation stages it was demonstrated that these stages are in fact related to oxide formation and growth at step edges and on terrace sites.

In another example, the formation and disappearance of ordered adsorbate structures simultaneously acquired during measurement of a cyclic voltammogram could be observed [62]. The period of the cyclic voltammogram was synchronized with the data acquisition speed for one image, and peaks in the voltammogram could hence be directly related to the appearance and disappearance of structures in the image. In Fig. 7.8, the results for the adsorption of sulfate (coadsorbed with water) onto Cu(111) are reproduced. In (a) the cyclic voltammogram is shown and in (b) the STM image obtained in one cycle. Scanning started in the region A of the voltammogram, where an ordered adsorbate layer is present on Cu(111). It can be recognized by the hexagonal moiré structure in the upper part of the image. If the electrode potential (relative to a Hg/Hg_2SO_4 reference electrode) is decreased the adsorbate starts to desorb at B (the desorption is correlated with a local minimum in the voltammogram) and the pattern of the adsorbate disappears in the image (centre part of (b)). On further increasing the voltage, the adsorption gives rise to a maximum in the voltammogram at C and the moiré pattern of the adsorbate is recovered (bottom part of (b)).

7.6 Combination of STM with Other Techniques

The combination of STM with other surface-sensitive measurement techniques is desirable in order to obtain a more quantitative description of the sample surface. This is particularly important for chemical characterization of the surface, which is not yet possible by routine STM techniques, although in certain well-defined cases different species (for example on surfaces of binary alloys) may exhibit an element-specific "chemical" contrast [63]. In such this cases, local order [64] or atomic details during growth processes [65] could be determined. For more general use and for chemical analysis, X-ray photoelectron and Auger electron spectroscopy may be combined with STM [25]. Ion scattering spectroscopy is more sensitive for examination of the surface layer but may damage the surface by sputtering. For the identification of reaction products, high-resolution electron energy-loss spectroscopy would be very well suited but this is not yet a very common combination. It has already been mentioned that the measurement of characteristic energy-losses of individual adsorbed species by analysis of the inelastic part of the tunneling current is now possible [6] and the combination of STM with energy loss spectroscopy is of great interest. Regarding the structural analysis, the determination of interatomic distances or directions on the measured surfaces with the STM or AFM alone is not very precise, in general. Owing to possible drifts or creep of the piezoelectric scanning device, STM measurements are often not accurate enough, to assign an observed long-range periodicity of structures (for example a moiré pattern) to exact values of the lattice parameter. The technique is useful only for single-crystal surfaces and epitaxial layers where atomic resolution can be obtained and self-calibration is possible. Another problem concerns the structural analysis of large unit cell systems where the STM images very often only show featureless protrusions. For the large scale analysis of a sample surface, STM has already been combined with scanning electron microscopy [26], which is particularly useful to select relevant parts of the surface for the STM measurements. In the following sections a number of typical combinations of STM with structure-sensitive methods are described.

7.6.1 STM and Low-Energy Electron Diffraction

For structural studies on ordered surfaces the use of low-energy electron diffraction (LEED) to complement STM represents the most common combination of methods. From an analysis of the kinematic diffraction spots in LEED, accurate values of the lattice parameters and the azimuthal orientation of the sample surface can be obtained. Moreover, by using dynamic LEED (in particular the I–V analysis), atom positions in layers below the surface can be obtained, which are not accessible by STM. In the surface layer, too, the atom positions may be determined from dynamical LEED with a precision that cannot be obtained with STM or AFM (this is quite independent of the vertical or lateral sensitivity of the relative positioning of an STM or AFM tip, which can certainly be controlled with sufficient accuracy). The dynamical LEED analysis becomes increasingly difficult for large unit cells with a correspondingly large number of atoms. Here, STM may facilitate the LEED analysis by providing information on subunits and on the possible symmetry of the unit cell.

As a typical study in this field, we mention the structural analysis of the stepped Si/SiO_2 interface after thermal oxidation [66]. In this case also the profile of the diffraction spots has been analysed (spot-profile analysis of low-energy electron diffraction, SPALEED) and compared with the STM images. It was found that the STM results confirmed the statistical values obtained by SPALEED. Moreover, the STM results provided additional information on local deviations from the mean surface level, on singular defects and grooves on the surface and on the detailed shape of islands formed after different oxidation processes. The combination of the spot analysis in reciprocal space with real space measurements constitutes the value of this work.

In another study, the determination of a detailed atomic structure was the main reason for combining STM and LEED [67]. In particular, the development of structural phases of iron silicide films on Si(111) has been studied by LEED, STM and Auger electron spectroscopy (AES). In LEED, two types of surface structures are observed, namely a (1×1) and a (2×2) phase, depending on the annealing conditions. It was hoped that the STM measurements would provide additional information on the surface structure of these phases in order to reduce the number of models to be tested in the full LEED analysis, involving dynamical calculations. As a general result of this work, a phase diagram for the existence of these two phases for various coverages with Fe and as a function of annealing temperature could be derived. In the atomically resolved STM images, (2×2) reconstructed iron silicide islands were identified. It was not possible to deduce a detailed structural model of the (2×2) iron silicide.

7.6.2 STM and Surface X-ray Diffraction

Surface X-ray diffraction (SXRD) has become a very valuable tool for the structural analysis of large unit cell systems, in particular, for surface reconstructions that involve different constituents [68]. As in the case of LEED, the rearrangement of atoms in layers below the surface can be evaluated once the approximate atomic model is known. The advantage of the structural analysis compared to LEED lies in the fact that the data evaluation is based on the kinematic description, where the incoming X-ray beam is only scattered once and does not interact with the backscattered beam (in LEED, a full dynamic calculation has to be performed for data analysis). On the other hand, SXRD is a difficult technique, unsuitable for laboratory use, since it needs the high beam intensity of synchrotron-radiation sources. The procedure for determination of surface structures is rather complicated (for details see, for example, [68]), diffraction intensities from the surface structure must be obtained under grazing incidence of the incoming X-ray beam. The so called "in-plane" measurements obtained under grazing exit angle can be used to calculate a Patterson function, which is projected on the surface plane. This modified Patterson function (in contrast to the Patterson function used in crystallography for three-dimensional structures) provides information on the relative coordinates of the atomic scatterers projected onto the surface layer. It is particularly suitable for deriving atom models for adsorbate-induced reconstructions. The atom model cannot be obtained directly from the analysis of the SXRD data. However, by considering the relative coordinates and additional information provided by experiments such as STM, a reasonable atom model can be chosen as a starting set. The diffraction intensities are then computed for the starting set and compared with the experimental data. In an optimizing process the atom positions are systematically changed (including the number of atoms present in the unit cell) until a model is reached for which the deviations between the theoretical and experimental diffraction intensities are minimized.

As an illustrative example, the structure determination of the In-induced Si(001)-(4×3) reconstruction by SXRD and STM is explained and discussed [28]. It will be demonstrated that only by the combination of both SXRD and STM results obtained in reciprocal and real space, respectively, could a detailed atom model of this complicated surface structure be derived.

In the first step of such an analysis, a flat sample with sufficient long-range order and perfection has to be prepared. For this step, STM plays a crucial role. The surface morphology of the clean substrate has to be measured, the general cleanliness (as deduced from irregularities of the (2×1) reconstruction of clean Si(001)) and the domain formation of the (4×3) reconstruction, which covers the surface completely at a coverage of 0.5 monolayers of In, must be confirmed. As a consequence, in the atom model of the (4×3) reconstruction, six In atoms per unit cell have to be considered. The number of Si atoms in the surface layer of the (4×3) unit cells can also be determined. The characteristic fact that, during formation of the (4×3) reconstruction from the metastable (2×2) geometry (which is observed at lower temperatures for the same In coverage), the (4×3) reconstruction is not only developing in the initial Si surface layer but also in the form of additional islands is utilized. The Si atoms in the (4×3) islands obviously result from mass transport from the initial surface layer during (4×3) island growth. By mass balance the number of Si atoms in the top layer of the (4×3) unit cells is found to be seven. The STM measurements have, therefore, provided the number of In and Si atoms in the unit cells, information that can hardly be derived to this accuracy by other methods.

The exact atom positions of the Si and In atoms in the unit cell can then be determined from the SXRD data. On the basis of the STM measurements alone this would be impossible, as demonstrated in Fig. 7.9, where STM im-

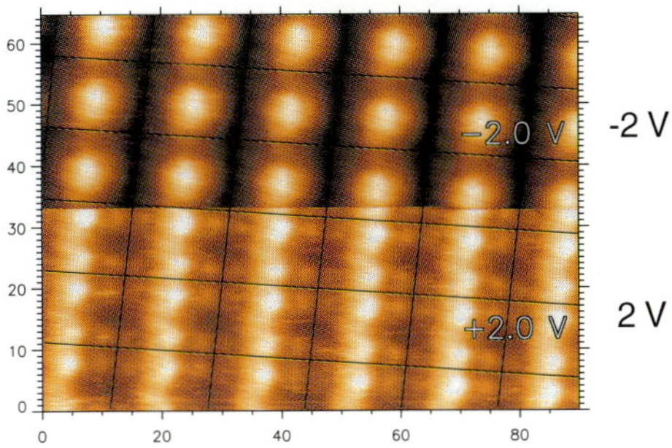

Fig. 7.9. Si(001)-(4 × 3)-In reconstruction. The top part was measured with $U = -2\,\mathrm{V}$ (occupied states) and the lower part with $U = 2\,\mathrm{V}$ (empty states). The black lines correspond to a (4 × 3) lattice [28]

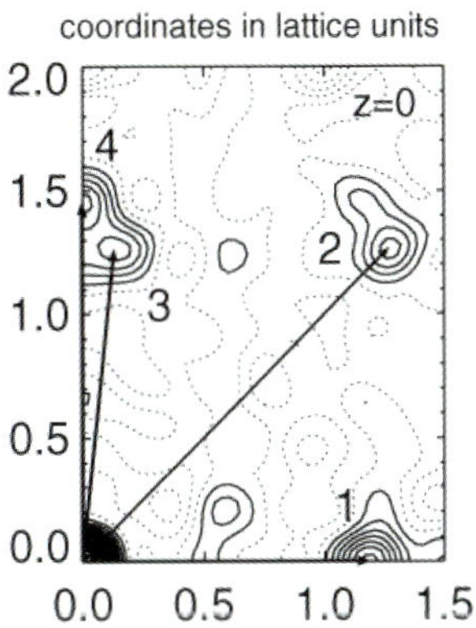

Fig. 7.10. Contour lines of the Patterson function computed for Si(001-(4 × 3)-In on the basis of experimental in-plane diffraction data. The coordinates are given in units of the surface lattice [28]

ages of the In-induced Si(001)-(4 × 3) reconstruction are reproduced. For imaging with the filled states, only one big protrusion per unit cell is visible, and for imaging with the empty states two more distinct and four weaker features appear. In view of the large number of atoms in the (4 × 3) unit cell, the lack of features in the STM images is surprising (but is often found on similar samples) and is probably related to the formation of extended electronic states.

The analysis of the Patterson function of this surface as shown in Fig. 7.10 provides characteristic details on the relative position of the atomic scatterers in the unit cell. In this particular case, only scattering of the In atoms have to be considered since scattering at the Si atoms can be neglected in a first approximation owing to the large differences in the number of core electrons (only these contribute essentially to scattering of the X-rays). By considering the known experimental information on this system, a first atom model could

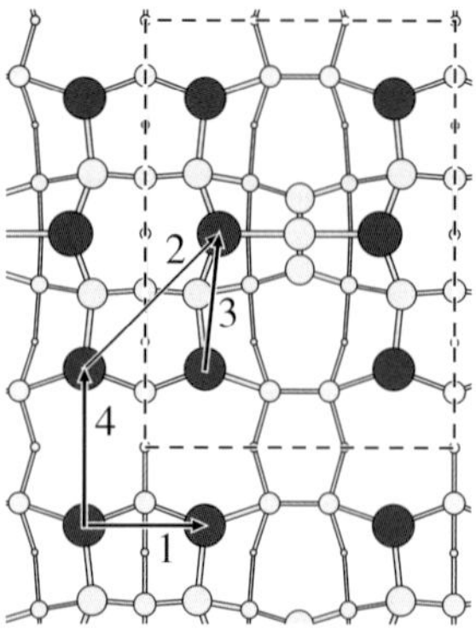

Fig. 7.11. Final atom model of the Si(001-(4×3)-In unit cells. The vectors indicated by the arrows correspond to those shown in Fig. 7.10 [28]

be derived as a starting point. This model is then optimized by a fitting procedure which represents one of the major tasks in the structural analysis. Figure 7.11 shows the final model of this reconstruction, as obtained by this fitting procedure. The complexity of the structure is striking. It has to be emphasized, however, that the experimental details presently available for this system are now understood – the relative position of the In atoms as depicted in the Patterson function, for example.

7.7 In situ Studies of Adsorption, Reaction and Growth

The examples selected in the previous section already indicate that the study of semiconductor surfaces was a main issue of STM work in the Schwerpunktprogramm. This does not mean that these cannot be applied successfully to other systems. For example, other problems of fundamental interest in surface science have been investigated, for example the reconstruction [69] and adsorption [70] behaviour of clean metal surfaces, metal film deposition in the nucleation and growth stages [70], the growth of oxide films [71] and the preparation of metal/oxide model catalysts [72]. However, from a more practical point of view, the new techniques in STM were very often first applied to semiconductor samples. The reason lies in the technical importance of semiconductors, the interesting behaviour of their surfaces, and in their ready availability in the form of well-oriented polished samples, which are not to difficult to prepare for STM. In the following subsections we discuss representative examples mainly in the field of semiconductor surfaces. The results concern vicinal surfaces of Si, the high-temperature behaviour of clean Si(111), initial stages of O_2 adsorption and oxidation at high temperatures, and the CVD growth of Si on Si(111).

7.7.1 Vicinal Surfaces of Silicon

Depending on their initial crystallographic orientation (*hkl*) and the individual misorientation, the vicinals of Si form stepped surfaces of very different morphologies [73–75]. Unfortunately, owing to the lack of a comprehensive surface energydiagram, it is still not possible to predict the exact morphology of an arbitrary vicinal surface of Si. The structural characterization of stepped, vicinal surfaces has attracted the attention of many research groups during the last decade – in particular those groups employing STM. A surprising variety of ordered surfaces have thus been reported. Si surfaces reconstruct in very large unit cells and as a consequence, atom models which have been confirmed by a detailed structural analysis only exist for the (100) [76], (111) [77], and (113) orientations [78]. Most work on vicinals has been devoted to surfaces near these orientations, and a few representative results will be presented here. Some remarks on the more open Si(110) surface will also be made.

It is well known that Si(100) samples, slightly misoriented towards (111), rearrange into terraces with (2×1) and (1×2) reconstruction, rotated by 90° with respect to each other and separated by single-layer steps. The structural features are orthogonal dimers, which combine across them on atomic step edges, giving rise to two energetically different steps [79]. With increasing polar misorientation relative to the [100] direction, double-layer steps are progressively created and finally become dominant [80]. The presence of double-layer steps is a necessary pre-requisite for creating a one-domain Si(100) surface, an important substrate for heteroepitaxial growth experiments. For higher polar misorientation, moving along the $[0\bar{1}\bar{1}]$ azimuthal direction, several stable facets are formed on Si(100). The orientations of the facets are summarized in a stereographic projection (see Fig. 7.12) which

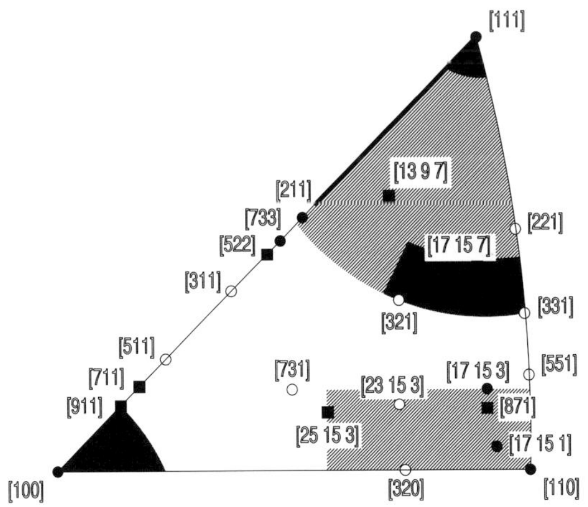

Fig. 7.12. Stereographic projection of measured faces on vicinal Si

helps us to derive more general trends for facet formation [81]. The data represented by the open circles have been taken from the literature (see, for example [81]), the data shown by the filled circles have been obtained during the work in the Schwerpunktprogramm,and the data represented by the filled squares have been identified for the first time. Most of the stable facets are formed when vicinals are prepared in low-index azimuthal directions. A possible explanation is the presence of dense-packed atomic rows of bulk silicon, which can then form low-energy surface structures in a simple way. The observed structural subunits at the origin of these rearrangements are mainly dimers and adatoms.

On Si(111), stable step structures are created on vicinals misoriented exactly towards[100]. They form a combination of single- and triple-layer steps [82], separating small, (7×7)-reconstructed terraces, which are stable under equilibrium up to nearly the (211) orientation (indicated in Fig. 7.12 by a solid line). The morphological features in the opposite direction are very different. The vicinal surface decomposes here into large (7×7)-reconstructed terraces, separated by step bunches [82].

As can be seen in Fig. 7.12, many stable facets are found near the (110) orientation. This observation correlates well with the fact that the (110) surface reconstructs with less long-range order than the low-index (100) and (111) surfaces. Figure 7.13 presents an STM image of a slightly misoriented (110) surface. Two different surface structures can be clearly distinguished. The zoom in the inset gives more structural details. The lines marked "16" correspond to the "16×2" reconstruction, the 1×5 reconstruction is indicated by "5". The zig-zag pattern follows the monatomic steps and reflects

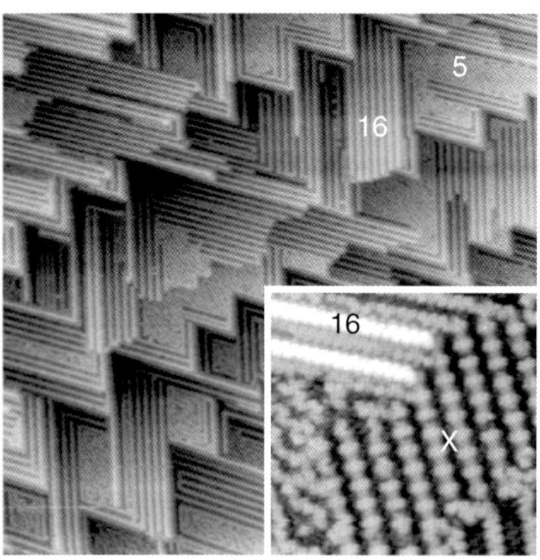

Fig. 7.13. STMimage from Si(110) ($300 \times 290\,\mathrm{nm}^2$). The inset showsan area of $31 \times 31\,\mathrm{nm}^2$ on the same sample

the misorientation of the surface. Very small parts, marked "X", show a $\begin{pmatrix} 3 & 4 \\ 10 & \bar{1} \end{pmatrix}$ reconstruction. Taking into account the complexity of the observed reconstructions on Si(110), it seems plausible that vicinals slightly misoriented relative to (110) are able to form new low-energy facets.

The different morphologies observed on vicinal Si surfaces are all composed of only a few structural subunits, namely dimers, adatoms, re-bonded- and non-re-bonded steps and π-bonded chains [74]. All these reduce the density of dangling bonds. Surface stress is created through unusual bond angles and the lengths of these structures. The local minimum in the surface free-energyis obtained by balancing these two parameters. Depending on the initial orientation, stable planar reconstructions or stable facets are formed on vicinal surfaces.

7.7.2 Silicon Surfaces at High Temperatures

STM measurements at high temperatures represent a fascinating fieldsince thermally-induced changes of surfaces, including thermal excitations or phase transitions and kinetically hindered reactions or growth processes, can be followed in situ. Commercial equipment is now available and the number of studies in this field is very large. Only one representative example is considered here. It concerns the investigation of hole and island decayon Si(111)–(7×7) in the temperature range from 720 K to 830 K.

The problem of fabrication and thermal stability of nanoscale structures on silicon surfaces at elevated temperatures is of current technological interest. In situ experiments in real time of the decay of holes and islands artificially created in a high-temperature STM are able to yield quantitative information on attachment/detachment or transport-related kinetics while the surface relaxes to thermodynamic equilibrium. In addition, elementary annealing mechanisms may be identified on the near-atomicscale.

For the Si(111) surface, the important question of whether high-temperature fluctuations of step edges and the decomposition of nanostructures are governed by attachment or by diffusion-limited kinetics has previously been studied by several authors. In particular, from investigations of the thermal relaxation of step bunches at 1200 K, Liu et al. [83] concluded that mass is conserved locally on a single terrace between neighbouring step edges. However, from their data they could not distinguish between the two limiting cases. In general, experiments involving the decay of well-defined non-equilibrium structures should be favoured when attempting to determine kinetic parameters. This is because they include a net mass transport rather than just thermal fluctuations around equilibrium as observedon the relaxed surface. Other authors have interpreted both kinds of experimental data, assuming a constant-rate, attachment/detachment-controlled hole or island decay [84]. Obvious deviations from the linear behaviour were attributed to the influence of the STM tip. In more recent results [85], which will be shown and discussed here briefly, the decay curves show distinct non-linearities in the temporal behaviour in contrast to previous experiments. The correct quantitative analysisis compatible with diffusion-controlled decay kinetics where the dependence of hole or island size on time is approximately described by a $(t_0 - t)^{2/3}$ law with t_0 being the time of complete annealing of the structure. Furthermore, similar investigations on Ag(111) [86] revealed significant differences between the decay behaviour of islandsand holes owing to the Schwoebel effect , the presence of a potential barrier on the edge of a terrace which prevents downward diffusion. Hence, the comparison of island and hole decay kinetics should allow an estimate of the possible existence of a Schwoebel barrier on the Si(111) surface.

For the experiment, hole and island nanostructures with a vertical extent of one to three Si bilayers were produced on the Si(111)-(7 × 7) surface by abruptly switching the tunneling current from typically 1 nA to 10–40 nA. The annealing of the structure is then observed by dynamic imaging with frame intervals varying between 7 s and 200 s. Figure 7.14 shows the decay kinetics of a hole at 830 K. This curve is well described by a power law function $S = A(t_0 - t)^b$ with $b = 0.688 \pm 0.019$. The results are clearly incompatible with linear decay. This is particularly evident from the double-logarithmic plot in the inset of Fig. 7.14, where the quasilinear fit yields $b = 0.645 \pm 0.011$. From the analysis of a large number of other hole and island decay experiments using the power law, values of b in the range from 0.62 to 0.75 have been obtained. It can therefore be concluded that the annealing behaviour of both holes and islands on Si(111)-(7×7) follows diffusion-limited kinetics. Regarding the Schwoebel barrier, no distinct differences were observed between the kinetic decay behaviour of both holes and islands. The Schwoebel barrier must therefore be much smaller than the activation barrier for diffusion.

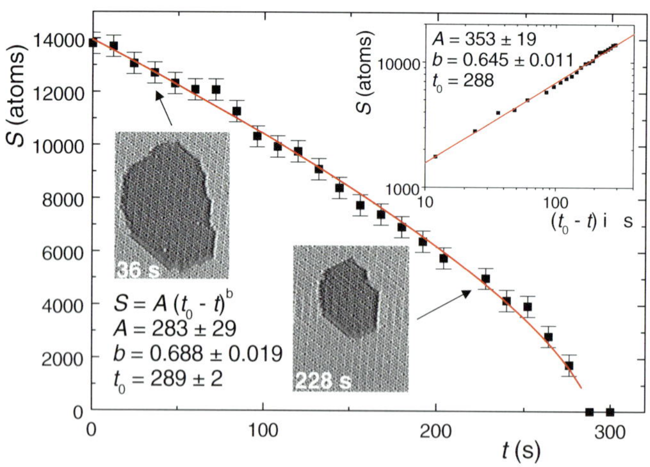

Fig. 7.14. Decay kinetics of a single bilayer hole at 830 K [85]

7.7.3 Initial Stages of Oxygen Interaction and Oxidation of Silicon Surfaces

Oxygen on Si is perhaps one of the most studied systems in surface physics. Apart from the technological importance of the SiO_2/Si interface, an essential element in semiconductor electronics, the achievement of a fundamental understanding of the interaction mechanisms of oxygen with the Si surface is of great interest.

With respect to the oxidation stage following the initial adsorption, it has been known for a long time that two regimes can occur, depending on the silicon substrate temperature T and oxygen partial pressure [87]. These are the passive oxidation resulting in oxide (ideally SiO_2) growth on the surface and the "active" oxidation, where the formation of volatile SiO leads to the reactive etching of the Si surface in step flow mode in such a way that a clean surface is preserved during the oxygen exposure. The latter regime dominates at high temperatures and below a critical partial pressure. The established phase diagram reveals that no sharp boundary can be drawn between the two modes, but a rather broad transition region appears where both mechanisms compete with each other. The study of the oxidation kinetics and the surface and oxide morphology arising from this interaction in the transition region is, therefore, especially important. Despite general agreement on the oxidation behaviour, different opinions about details of the reactions can be found, which concern the oxide nucleation and film or island growth mode, the temperature dependence of the etch rate in the active regime and the step flow mechanism (see, for example [88]).

Real-time experiments of the in situ oxidation in a high-temperature STM are ideally suited to study the temporal evolution of the interaction of oxygen with the (7×7) reconstructed Si(111) substrate, starting from the first adsorption events up to the oxidation of the entire surface. It appears that stable imaging conditions with satisfactory resolution of the surface covered with a thin oxide layer can be obtained if the tunneling voltage is increased to overcome the higher potential barrier although full atomic resolution may be lost after oxygen exposure. In [24] the adsorption in the initial phase of the reaction and the evolution of the surface oxide film has been investigated, especially in connection with Si etching in the transition region between the two regimes. In the transition region between the passive and active oxidation regimes (775 K–875 K, 10^{-8}–10^{-7} mbar O_2 partial pressure), heterogeneous oxide nucleation at surface defects and at the same time etching of step edges was mostly observed. This competition leads to a rather rough surface morphology where the step edges are locally pinned by the oxide.

Figure 7.15 reproduces a sequence of images. Figure 7.15a shows the initial state of the sample before O_2 exposure. If we compare this with the next image in Fig. 7.15b, which corresponds to an exposure of 10 L (1 L corresponds to an exposure of 1.3×10^{-4} Pa · s) a fast step-flow-like retraction of the step edge with formation of the (7×7) structure on the newly exposed terrace

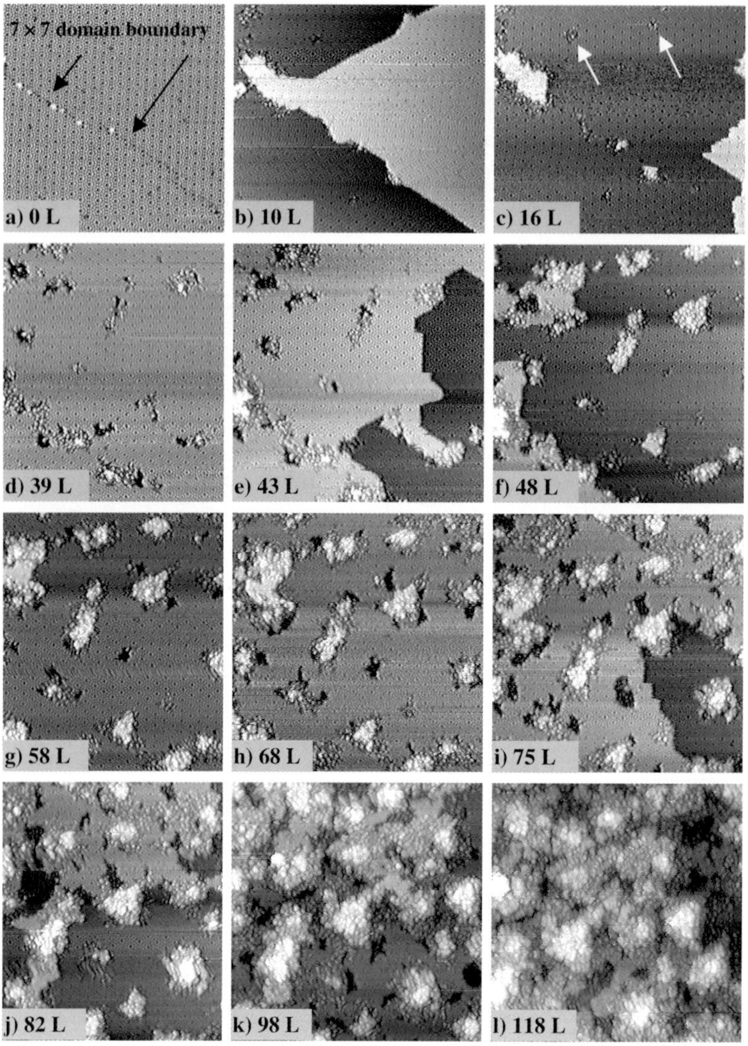

Fig. 7.15a–l. Sequence of STM images of Si(111)-(7 × 7) during O_2 exposure at 845 K and a partial pressure of 10^{-8} mbar. U was changed while imaging from 2.6 V to 4.0 V. A (7 × 7) domain boundary is indicated by black arrows, oxide nuclei by white arrows [24]

is seen. During the removal of the upper terrace, small disordered structures form on the Si surface and on some intermediate step edge positions. Especially near domain boundaries, the step retraction is already delayed. These regions do not exhibit any reconstruction and are therefore identified as oxide nucleation sites. Owing to their higher tunneling resistance, oxide clusters are expected to be imaged as apparent depressions. However, since the upper terrace has already been removed, they appear here with the same grey level or even slightly brighter than the surrounding lower-lying bare Si surface. The oxide areas slowly grow further laterally. At the edges of the oxidized regions, the formation of depressed areas is now indeed observed (see, for example Figs. 7.15d and h). These areas of dark contrast clearly indicate the beginning of oxidation of the next Si layer. During the following stage of the surface reaction, the oxidized regions growing in number and size hinder the etching process more and more and cause the step edges to be locally pinned (Fig. 7.15e). The step retraction is effectively stopped after the removal of the third Si double layer, see Figs. 7.15j and k. The step flow in the transition region is thus able to proceed over several layers and a longer period of time than reported previously [88], where the etching was stopped after removal of the first layer. The simultaneous processes of slow lateral oxidation and rapid etch removal in the remaining oxide-free areas leads to an apparent vertical island growth. After an exposure of about 100–110 L, a very rough surface morphology with a maximum corrugation of 1.5 nm has been established. For more details see [24].

7.7.4 Growth of Silicon by Chemical Vapour Deposition

Investigations of the growth of metals or semiconductors on semiconductor substrates probably form one of the most fruitful applications of STM [89]. The real-space imaging capability and the high resolution obtained on such systems are advantages that can hardly be achieved by other methods. Regarding the resolution, however, it must be remembered that the resolved features can normally not be directly assigned to individual atoms. The expression "atomic resolution" often encountered in the literature does not necessarily mean that the observed features correspond to atoms (although for some important surfaces, for example for Si(111)-(7×7), this is actually the case). To understand the atomically resolved structures in detail one needs theoretical information on the local electronic structure of the surface, as discussed above. Nevertheless, information about the growth processes derived from STM images remains unique and a knowledge of the detailed atom structure of the deposit is not always of crucial importance.

We have chosen the chemical vapour deposition (CVD) of Si on Si(111) [90,91] as an illustrative example. Two groups of the the Schwerpunktprogramm have been concentrating on this subject by STM techniques. While the work of Behm et al. was more devoted to the reaction mechanism and kinetics, deduced from the structural information in the initial stage (see, for example a survey by Rauscher and Behm [92]), Köhler et al. concentrated on in situ STM measurements of the growth process (see, for example [21]). By admitting a chemical precursor gas into the STM chamber and heating the substrate to a preselected temperature, the growth behaviour of the Si deposit is immediately apparent in the STM images. It should be mentioned that molecular beam epitaxy (MBE) can also be followed by in situ STM [93]. However, shadowing of the atomic or molecular beam by the tunneling tip may sometimes be a problem, if the tip is not regularly retracted during scanning of the sample surface [21]. Usually, the front part of a tunneling tip is fairly broad, except for a sharp microtip, which has been developed during tip formation. In the case of CVD, the precursor molecule can reach the tunneling area by continued collisions with the tip and substrate surface, while for MBE, shadowing of the substrate surface may only be prevented by grazing incidence of the molecular beam and by retracting the tip between scanning of lines or frames.

Figure 7.16 shows a series of images obtained at a partial pressure of 8×10^{-3} Pa of disilan (Si_2H_6) as a precursor gas, as reproduced from [91]. The sample temperature was 783 K. Initially, the main part of the surface corresponds to one terrace and only one double-layer step is visible at the edges of Fig. 7.16a. During the exposure (Figs. 7.16b–e) one observes the development of triangular "pyramids" until they completely cover the surface, a few monolayers of Si having then been condensed on the surface. The surface roughness $D^2 = \langle (h - \langle h \rangle)^2 \rangle$ systematically increases with the exposure. More specifically, D^2 shows an initial small jump to a finite value after the first Si islands have been formed and then increases linearly with time and coverage. After the deposition has been stopped (by closing the gas supply) the surface smoothens out asymptotically until again only Si(111) terraces may be identified (Fig. 7.16f) as in the case of a more two-dimensional growth mode.

The value of such measurements lies in the fact that the growing features can be observed continuously, the experimental conditions for deposition (for example partial pressure of the precursor gas and substrate temperature) can be systematically changed, their influence on the growth behaviour is immediately recognized, and growth parameters can be optimized. In this particular case it was found that the presence of adsorbed H on the surface is of particular importance for the observed growth mode. For lower deposition rates of Si, H species from dissociating disilan molecules preferentially desorb and the Si growth is then similar to the MBE case. With increasing deposition rate (this means, increasing partial pressure of the precursor gas) the desorption of H is not fast enough and the surface eventually becomes saturated with H. The growth of Si on a H-saturated surface is completely different from that observed in MBE experiments. The differences arise from changes of the diffusion coefficient of the Si species on the H-saturated surface and depend on the development of H-saturated facets that have a smaller surface energy than that of Si(111).

As already been mentioned, the more chemical aspects of precursor gas adsorption and reaction in relation to the observed structural details of Si homoepitaxy have been studied in detail by Behm and coworkers (see overview article by Rauscher and Behm on growth structure of Si homoepitaxy by

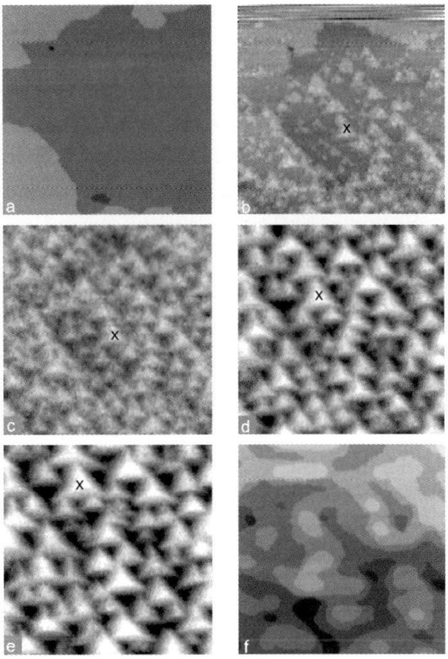

Fig. 7.16. Development and decay of three-dimensional islands during CVD growth of Si on Si(111) at a sample temperature of 783 K [91]. The partial pressure of disilane was 8×10^{-5} mbar. The imaged area ($250 \times 250\,\mathrm{nm}^2$) on the sample drifted only slightly during the measurements (one of the islands is marked by an "x"). (**a**) Clean surface, (**b–d**) continuous growth of the islands in the form of triangular "pyramids", (**e**) smoothing of the surface after stopping the gas exposure [91]. The images (**a–b**) were measured at 0 s, 90 s, 450 s, 1530 s, 2300 s, $t_{\mathrm{stop}} + 500$ s

chemical vapour deposition [92]). Many results have been obtained for the first time, thus establishing the importance of the STM techniques for the understanding and optimization of the growth mechanisms. For example, the interaction of SiH_4 [94] and Si_2H_6 [95] with Si(111)-(7 × 7) for Si epitaxy has been carefully studied. The STM data are correlated with kinetic model calculations, from which one obtains the time dependence of the development of dissociation and reaction products [96]. One particular example will be explained here. It concerns the observation of specific subunits at a certain stage of SiH_4 exposure on Si(100)-(2 × 1) [97], which had previously not been seen by other authors. The STM results on this stage of the Si growth show, in addition to small anisotropic Si islands and patches of the hydrogen-terminated substrate, metastable cross-shaped structural units. In Fig. 7.17, an image of such a surface is reproduced, where two kind protrusions may be recognized. One kind does not show much fine structure, whereas the other kind clearly has developed an atom pattern with a 4-fold symmetry. Both structures are shown more clearly in the inset. To understand the nature of the cross-shaped features, the reaction scheme of SiH_4 decomposition on Si(100)-(2 × 1) was examined. In the first step, SiH_4 is dissociatively adsorbed into H and SiH_3, which further decomposes into SiH_2 and desorbing H_2 and/or adsorbed H. When calculating the thermal evolution of the SiH_4 dissociation products, the increase of the coverage with SiH_2 species nicely correlates with the observation of the cross-shaped features. It was, therefore, plausible to explain these structures as due to SiH_2 tetramers that occupy bridge positions between Si atoms on the locally rebonded Si surface. The proposed structure needs further confirmation by theory or other methods.

That other methods provide more details on the dissociation process of the incoming molecule has also been demonstrated by Behm and coworkers [25]. By using XPS they were able to show how the dissociative adsorption of SiH_2Cl_2 proceeds. In general, the combination of STM results with data obtained by spectroscopic techniques such as XPS and AES allows adsorption and reaction mechanisms in CVD growth to be eludicated in great detail.

In this context it is interesting to note that the possibility of X-ray excitation of the sample made it possible to investigate the effect of X-ray

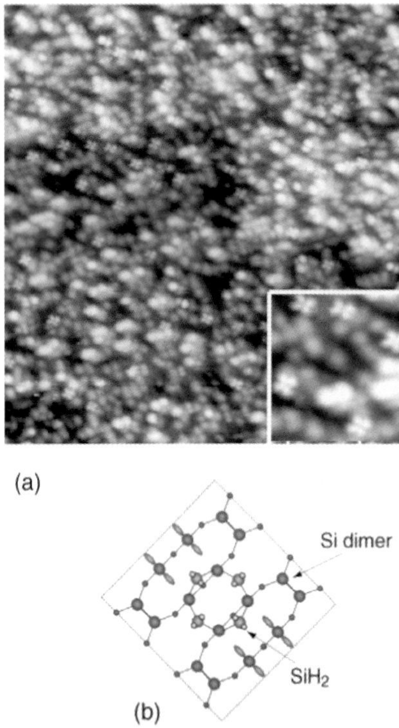

Fig. 7.17. (**a**) STM image ($42 \times 42\,nm^2$) of the Si(100) surface after exposure of 340 L SiH_4 at 570 K substrate temperature. The inset shows an enlarged detail of the image. In (**b**) the proposed tetramer structure with SiH_2 groups is reproduced [97]

irradiation of a substrate on the growth process. In the case of Ge growth on Si(111)-(7 × 7), by using the precursor gas GeH_4 it was found that the irradiation drastically changed the growth mode [98]. On the initial substrate surface, at most 2.5 bilayers of Ge could be grown under the chosen experimental conditions, whereas for the irradiated surface, the growth of Ge continued in the form of three-dimensional islands.

7.7.5 Lithography: Fabrication of Nanostructures

The idea of using one of the SPM techniques to generate nanometer or even atom scale structures was suggested immediately after these methods were developed and found its first spectacular success in the manipulation of individual Xe atoms on a metal single-crystal surface at low temperature (4 K) by an STM tip [99] (see also [100]). At the nm scale, several techniques have been developed for the modification of surfaces, including for example, simple mechanical scratching with an AFM tip [101] and the application of voltage pulses in the tunneling regime of an STM (for example on Si(111)-(7 × 7) [102]). In the latter case, the voltage pulses probably give rise to a local desorption or to more complicated effects owing to the sudden increase of the tunneling current. The decay of structures produced in this way at higher temperature (typically 800 K) has already been discussed above. It has also been found that in both, STM and AFM, voltage pulses on the tips can initiate local chemical reactions, for example an oxidation, which can be used for patterning of surfaces [103]. Some particularly relevant work will be summarized here. It concerns the fabrication of nanostructures by STM-induced decomposition of SiH_4 and SiH_2Cl_2 [104], the refinement of AFM tips for mechanical patterning of surfaces, and the application of these tips to generate a regular pattern of holes [105], also known as antidots in semiconductor heterostructures. In another experiment of Irmer et al. such tips have been used for structuring a surface layer, for example, to fabricate a thin-film Josephson device [106]. Finally, the concept of local oxidation by an AFM tip to prepare a metal-insulator-metal tunneling barrier [107] will be explained.

AFM super tips composed of electron beam deposited material have first been described by Akama et al. [108]. They mainly consist of C and hydrocarbons cracked in the high vacuum (in the range of 10^{-4} to 10^{-6} Pa) by the electron beam of a scanning electron microscope and have a typical radius of only a few 10 nm. In earliest experiments, such tips were used for imaging rough surfaces. Later, Wendel et al. [105] utilized these tips for lithographic purposes. After this group had introduced a mechanical patterning technique for the photoresist on a GaAs surface by using an ordinary AFM tip in the tapping mode and pushing the tip towards the surface by a force of approximately 1 μN [109], they applied the same technique with a super tip and wrote a regular arrangement of holes in the resist. By etching the resist/GaAs sample and removing the rest of the resist, they obtained a hole pattern in GaAs, as shown in Fig. 7.18. The size of the holes and their distance could be further

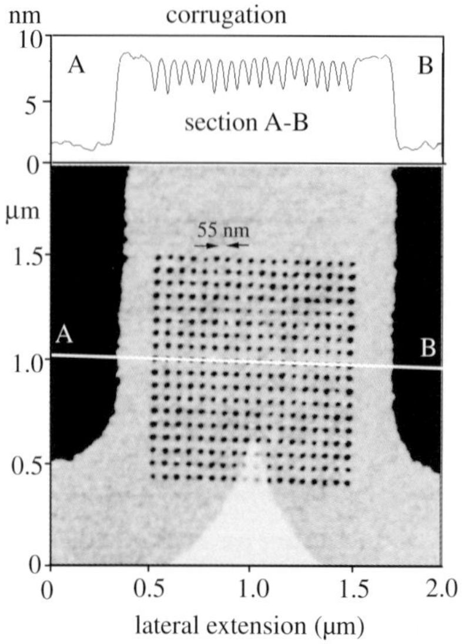

Fig. 7.18. Antidot array in a GaAs mesa [105]. Depth and period of the holes are 3 and 55 nm, respectively. The holes have been written in a thin resist layer by using an AFM electron beam deposited tip, by subsequent wet chemical etching and final removal of the residual resist with acetone. In the upper part of the figure the cross section along line AB is shown [105]

reduced by sharpening the electron-beam-deposited tip in an oxygen plasma. A minimum period in the resist of smaller than 10 nm could be obtained in this way. Details of the growth conditions of the super-tips and their etching behaviour in the oxygen plasma are described more explicitly in [105].

In subsequent work, electron beam-deposited tips have been used to prepare Josephson junctions in a thin Al layer by ploughing away part of the film [106]. By optical lithography, 2 µm wide Al wires and loops with a thickness of 100 nm were first prepared and then cut by the "nanoplough", leaving behind a small "bridge" with a size of $100 \times 100\,\mathrm{nm}^2$. This bridge forms a Josephson junction if the sample is cooled down below the critical temperature T_c for superconductivity. The measured $I-V$ curve clearly demonstrated the behaviour typical of a single Josephson junction. If the critical current was exceeded, the $I-V$ dependency followed an ohmic behaviour, owing to the conductivity in the normal state. Below the critical current density, the bridge did not show any voltage drop at all. In addition, by preparation of a second Josepson junction, which was connected to the first one by a ring-like Al film structure, the entire device acted as a SQUID, as was demonstrated by the expected modulation of the critical current density as a function of applied external magnetic field.

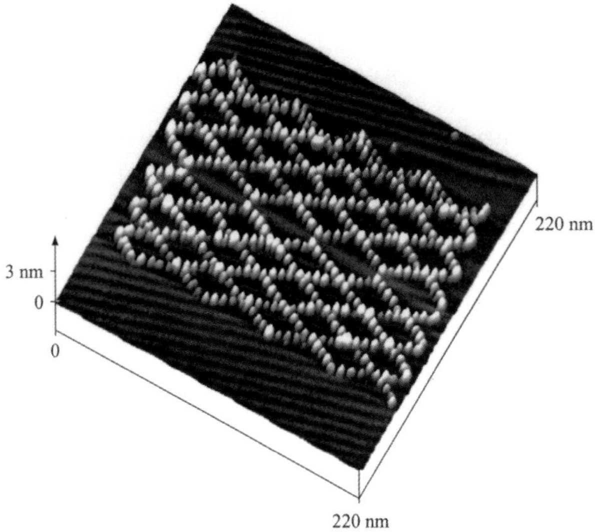

Fig. 7.19. STM image of an array of Cu clusters on Au(111) in 0.05 M H_2SO_4 + 1 mM $CuSO_4$, covering an area of $0.22 \times 0.14\,\mu m^2$ [114]

This group has also studied the structuring of materials by anodic oxidation while scanning the surface with a negatively biased AFM tip in order to fabricate a metal-insulator-metal contact [107]. For these experiments the sample (a thin Ti layer with a thickness of 5 nm) was structured by local oxidation in such a way that a narrow Ti layer (with a width of around 70 nm) was interrupted by a small (15 nm) line of TiO_x, which was prepared by applying −10 V to the tip, while using the AFM in the tapping mode. By plotting the conductance of a number of metal-insulator-metal structures against the width of the TiO_x lines (which had been prepared individually) it was found that the conductance depends exponentially on the barrier thickness. This clearly indicates that tunneling through the barrier is the dominant conduction mechanism.

It could be shown that STM can also be used for the patterning of surfaces in an electrolytic environment. Nanostructuring of electrode surfaces has been reported by a number of groups [110,111]. On the basis of the results described in [111], Kolb et al. have developed this technique still further and are now able to deposit Cu clusters on a Au(111) surface with remarkably uniform size and in a very regular pattern [112,113] (see Fig. 7.19).

The principle of their experiment is as follows. In the first stage the tip (in this case etched from a PtIr wire) is covered by a Cu layer. This is achieved by using a copper sulfate electrolytic solution and by applying a slight negative potential on the tip (−30 mV relative to the reference electrode). If the tip is then approached to the surface by approximately 1 nm, a metallic Cu cluster is deposited onto the Au(111) surface (which is already covered by one monolayer of Cu by underpotential deposition). The tip approach is effected by application of an external voltage pulse to the z-piezoelectric system of the scanner. The control unit still attempts to regulate the tunneling current to the preset value but initially shows a strong increase owing the differences in the time constants of the regulating circuit and of the z voltage pulse. After the end of this pulse the tip is withdrawn to the initial position, and subsequent measurements demonstrate that a Cu cluster with a height of 0.6 nm and a width of 3.5 nm has formed on the surface. A 20×20 pattern of Cu clusters on a surface area of $400 \times 400\,\mathrm{nm}^2$ can be written in 90 s, without apparent changes of the Cu cluster size [112]. This means that the Cu coverage of the tip must remain constant owing to a permanent redeposition of Cu from the Cu electrolyte onto the tip surface in the normal tunneling mode.

7.7.6 Biological Material and Polymers

The use of SXM methods for the investigation of biological material including large molecules and polymers has become a very important issue and the published work is overwhelming. Developments in this field have also been initiated by projects of the Schwerpunktprogramm, and these will be summarized here to some extent. It has to be recalled, however, that a broad range of materials is covered in these investigations and that a detailed explanation of the image formation has not yet been obtained in most cases. This fact may not have been important for the specific work but makes it unsatisfactory to include descriptions of results in as much detail as for the solid samples discussed above. Moreover, these systems have become a major subject in material science only more recently. Their importance can be expected to increase in the future.

The measurements of elastic properties of biological samples [115] represent milestones in this kind of experiments. Subsequent progress in such experiments to access individual molecules with the AFM for individual force spectroscopy [116] and to change the degree of folding of individual domains of large molecules mechanically is likewise impressive [117]. The manipulation and measurement of individual atoms (see [99] and [100]) and of large molecules belong to the ultimate experiments that have become possible by using SXM techniques. In more conventional applications of AFM, thin polymer films are investigated. As a typical problem in this research field the de-wetting dynamics at a polymer-polymer interface [118], the structure formation after demixing of a polymer blend [119], or the microdomain morphology of a thin ABC triblock copolymer film [120] may be mentioned.

7.8 Prospects for the Future

An attempt has been made to summarize recent progress in SPM techniques, in projects of the Schwerpunktprogramm, and to demonstrate current achievements by illustrative examples. The developments concerned the refinement of techniques in the direction of new measuring conditions, for example for in situ experiments at high temperatures, and in real time in the field of adsorption, reaction and growth, the possibility of reaching true atomic resolution in AFM in non-contact mode and at low-temperatures, the development of STM equipment for electrochemistry, and methods to of producing nano-scale structures. Many studies have been devoted to semiconductor surfaces; other classes of materials belong to the biology and polymer sciences. It was not possible to consider all results in the same depth, and the selection inevitably reflects the special interests of the authors. A similar comment may be made concerning the selection of referenced publications. Future applications of SPM techniques will not only be devoted to routine structural characterization of surfaces, but particularly to the possibility of controlling the position of the SPM tip on the atomic scale. This not only provides local access to properties of individual atoms and molecules but also allows them to be manipulated. The sample–tip interaction is still a fairly unknown quantity. To some extent, lack of more precise knowledge of the sample–tip interaction results from the unwanted changes of the tip apex during the measurements. These changes are more frequently found for measurements in environmental conditions than in UHV. One task of future work will be, therefore, to prepare well-defined tips, which can also be used for the measurement of systems under non-UHV conditions.

Acknowledgments

R.J. Behm, O. Bunk, H.L. Hartnagel, R.L. Johnson, U. Köhler, D.M. Kolb, J.P. Kotthaus, G. Krausch, H. Rauscher, U.D. Schwarz, K. Wandelt, R. Wiesendanger.

References

1. Binnig G., Rohrer H., Gerber Ch., Weibel E. (1982) Surface studies by scanning tunneling microscopy. Phys Rev Lett 49:57–61
2. Binnig G., Rohrer H., Gerber Ch., Weibel E. (1983) 7 x 7 reconstruction on Si(111) resolved in real space. Phys Rev Lett 50:120–123
3. Binnig G., Fuchs H., Gerber Ch., Rohrer H., Stoll E., Tosatti E. (1986) Energy-dependent state-density corrugation of a graphite surface as seen by scanning tunneling microscopy. Europhys Lett 1:31–36
4. References from the literature to these more general remarks will be given in the sections below
5. Binnig G., Quate C.F., Gerber Ch. (1986) Atomic force microscope. Phys Rev Lett 56:930–933
6. Stipe B.C., Rezaei M.A., Ho W. (1999) Localization of inelastic tunneling and the determination of atomic-scale structure with chemical specificity. Phys Rev Lett 82:1724–1727

7. Guckenberger R., Wiegräbe, W., Baumeister W. (1988) Scanning tunelling microscopy of biomacromolecules. J Microscopy 152:795–802
8. Güntherodt H.-J., Wiesendanger R. (1994) Scanning Tunneling Microscopy I: General Principles and Applications to Clean and Adsorbate-Covered Surfaces. Springer Series in Surface Sciences 20, Springer, Berlin
9. Wiesendanger R., Güntherodt H.-J. (1993) Scanning Tunneling Microscopy II: Further Applications and Related Scanning Techniques. Springer Series in Surface Sciences 28, Springer, Berlin
10. Wiesendanger R., Güntherodt H.-J. (1995) Scanning Tunneling Microscopy III: Theory of STM and Related Scanning Probe Methods. Springer Series in Surface Sciences 29, Springer, Berlin
11. Chen C.J. (1993) Introduction to Scanning Tunneling Microscopy. Oxford University Press, New York, Oxford
12. Simmons J.G. (1963) Electric tunnel effect between dissimilar electrodes separated by a thin insulating film. J Appl Phys 34:2481–2590
13. Garcia N., Ocal C., Flores F. (1983) Model theory for scanning tunneling microscopy: Application to Au(110) (1 x 2). Phys Rev Lett 50:2002–2005
14. Stoll E., Baratoff A., Selloni A., Carnevali P. (1984) Current distribution in the scanning vacuum tunneling microscope. J Phys C:Solid State Phys 17:3073–3086
15. Tersoff J., Hamann D.R. (1985) Theory of the scanning tunneling microscope. Phys Rev B 31:805–813
16. Bardeen J. (1961) Tunneling from a many-particle point of view. Phys Rev Lett 6:57–59
17. Chen C.J. (1990) Origin of atomic resolution on metal surfaces in scanning tunneling microscopy. Phys Rev Lett 65:448–451
18. Drakova D., Doyen G. (1997) Local charge injection in STM as a mechanism for imaging with anomalously high corrugation. Phys Rev B 56:R15577–R15580
19. Tsukada M., Kobayashi K., Isshiki N., Kageshima H. (1991) First-principles theory of scanning tunneling microscopy. Surf Sci Rep 13:265–304
20. Wilhelmi G., Kampschulte T., Neddermeyer H. (1995) Step dynamics on highly oriented and vicinal surfaces of Si(111). Surf Sci 331–333:1408–1413
21. Köhler U. (1995) Kristallwachstm unter dem Rastertunnelmikroskop. Phys Bl 51:843–945
22. Kliese R., Röttger B., Badt D., Neddermeyer H. (1992) Real-time STM investigation of the initital stages of oxygen interaction with Si(100)2x1. Ultramicroscopy 42–44:824–831
23. Wengelnik H., Badt D., Neddermeyer H. (1994) Scanning tunneling microscopy on W(100) at 80 K and room temperature. Surf Sci 307–309:619–624
24. Hildebrandt S., Kraus A., Kulla R., Neddermeyer H. (1999) From active to passive oxidation: O_2 on Si(111)7x7. Appl Surf Sci 141:294–304
25. Fehrenbacher M., Rauscher H., Behm R.J. (1997) Dissociative adsorption of SiH_2Cl_2. Z Physik Chemie 198:205–220
26. Fuchs H., Laschinski R. (1990) Surface investigations with a combined scanning electron-scanning tunneling microscope. Scanning 12:126–132
27. Pietsch G.J., Köhler U., Jusko O., Henzler M., Hahn P.O. (1992) Structure of the stepped Si/SiO_2 interface after thermal oxidation: Investigations with scanning tunneling microscopy and spot-profile analysis of low-energy electron diffraction. Appl Phys Lett 60:1321–1323

28. Bunk O., Falkenberg G., Seehofer L., Zeysing J.H., Johnson R.L., Nielsen M., Feidenhans'l R., Landemark E. (1998) Structure determination of the indium induced Si(001)-(4x3) reconstruction by surface X-ray diffraction and scanning tunneling microscopy. Appl Surf Sci 123/124:104–110
29. Hasegawa Y., Avouris Ph. (1993) Direct observation of standing wave formation at surface steps using scanning tunneling spectroscopy. Phys Rev Lett 71:1071–1074
30. Li J., Schneider W.-D., Berndt R., Bryant O.R., Crampin S. (1998) Surface-state lifetime measured by scanning tunneling spectroscopy. Phys Rev Lett 81:4464–4467
31. Feenstra R.M., Mårtensson P. (1988) Fermi-level pinning at the Sb/GaAs(110) surface studied by scanning tunneling spectroscopy. Phys Rev Lett 61:447–450
32. Giessibl F.J. (1995) Atomic resolution of the silicon (111)-(7x7) surface by atomic force microscopy. Science 267:68–71
33. Ohnesorge F., Binnig G. (1993) The atomic resolution by atomic force microscopy through repulsive and attractive forces. Science 260:1451–1456
34. Hölscher H., Schwarz U.D., Zwörner O., Wiesendanger R. (1998) Consequences of the stick-slip movement for the scanning force microscopy imaging of graphite. Phys Rev B 57:2477–2481
35. Albrecht T.R., Grütter P., Horne D., Rugar D. (1991) Frequency modulation detection using high-Q cantilevers for enhanced force microscope sensitivity. J Appl Phys 69:668–673
36. Dürig U., Gimzewski J.K., Pohl D.W. (1986) Experimental observation of forces acting during scanning tunneling microscopy. Phys Rev Lett 57:2403–2406
37. Zhong Q., Inniss D., Kjoller K., Ellings V.B. (1993) Fracture polymer/silica fiber surface studied by tapping mode atomic force microscopy. Surf Sci 290:L688–L692
38. Grütter P., Mamin H.J., Rugar D. (1993) Magnetic force microscopy (MFM). In: [9], 151–207
39. Lüthi R., Haefke H., Meyer K.-P., Meyer E., Howald L., Güntherodt H.J. (1993) Surface and domain structures of ferroelectric crystals studied with scanning force microscopy. J Appl Phys 74:7461–7471
40. Mate C.M., McClelland G.M., Erlandson R., Chiang S. (1987) Atomic-scale friction of a tungsten tip on a graphite surface. Phys Rev Lett 59:1942–1945
41. Allers W., Schwarz A., Schwarz U.D., Wiesendanger R. (1998) A scanning force microscope with atomic resolution in ultrahigh vacuum and at low temperatures. Rev Sci Instrum 69:221–225
42. Hölscher H., Schwarz U.D., Wiesendanger R. (1999) Calculation of the frequency shift in dynamic force microscopy. Appl Surf Sci 140:344–351
43. Hölscher H., Allers W., Schwarz U.D., Schwarz A., Wiesendanger R. (1999) Determination of tip-sample interaction potentials by dynamic force spectroscopy. Phys Rev Lett 83:4780–4783
44. Schwarz U.D. (1999) Nanomechanics–Nanomechanical Investigations with the Scanning Force Microscope. Habilitationsschrift, Universität Hamburg
45. Burnham N., Colton R.J. (1989) Measuring the nanomechanical properties and surface forces of materials using an atomic force microscope. J Vac Sci Technol A 7:2906–2913

46. Schwarz A., Allers W., Schwarz U.D., Wiesendanger R. (1999) Simultaneous imaging of the In and As sublattices on InAs(110)-(1x1) with dynamic scanning force microscopy. Appl Surf Sci 140:293–297; Schwarz A., Allers W., Schwarz U.D., Wiesendanger R. (2000) Dynamic-mode scanning force microscopy study of n-InAs(110)-(1×1) at low temperatures. Phys Rev B 61:2837–2845
47. Allers W., Schwarz A., Schwarz U.D., Wiesendanger R. (1999) Dynamic scanning force microscopy at low temperatures on a van der Waals surface: graphite(0001). Appl Surf Sci 140:247–252
48. Allers W., Schwarz A., Schwarz U.D., Wiesendanger R. (1999) Dynamic scanning force microscopy at low temperatures on a noble-gas crystal: Atomic resolution on the xenon(111) surface. Europhys Lett 48:276–279
49. Ebert Ph., Urban K., Lagally M.G. (1994) Charge state dependent structural relaxation around anion vacancies on InP(110) and GaP(110) surfaces. Phys Rev Lett 72:840–843
50. Bode M., Getzlaff M., Wiesendanger R. (1999) Spin-polarized vacuum tunneling into the exchange-split surface state of Gd(0001). In: Prelim. Proceedings STM '99, ed. by Y. Kuk *et al.*, Seoul, 5–6
51. Franke K., Besold J., Haessler W., Seegebarth C. (1994) Modification and detection of domains on ferroelectric PZT films by scanning force microscopy. Surf Sci 302:L283–L288
52. Hoffmann D., Rettenberger A., Baur C., Läuger K., Grand J.Y., Möller R. (1995) Variation of the thermovoltage across a vacuum tunneling barrier: Copper islands on Ag(111). Appl Phys Lett 67:1217–1219
53. Hoffmann D., Grand J.Y., Möller R., Rettenberger A., Läuger K. (1995) Thermovoltage across a vacuum barrier investigated by scanning tunneling microscopy: Imaging of standing electron waves. Phys Rev B 52:13796–13798
54. Berndt R., Gimzewski J.K. (1993) Photon emission in scanning tunneling microscopy: Interpretation of photon maps of metallic systems. Phys Rev B 48:4746–4754
55. Horn J., Vogt A., Aller I., Hartnagel H.L.(1996) Heterostructure interface characterization using scanning tunneling microscope excited time-resolved luminescence. J Vac Sci Technol B 14:820–823
56. Sonnenfeld R., Hansma P.K. (1986) Atomic-resolution microscopy in water. Science 232:211–213
57. Manne S., Butt H.J., Gould S.A.C., Hansma P.K. (1990) Imaging metal atoms in air and water using the atomic force microscope. Appl Phys Lett 56:1758–1759
58. Siegenthaler H. (1995) STM in electrochemistry. In: [9], 7–49
59. Magnussen O.M., Hotlos J., Nichlos R.J., Kolb D.M., Behm R.J. (1990) Atomic structure of Cu adlayers on Au(100) and Au(111) electrodes observed by in situ scanning tunneling microscopy. Phys Rev Lett 64:2929–2932
60. Hölzle M.H., Apsel C.W., Will T., Kolb D.M. (1995) Copper deposition onto Au(111) in the presence of thiourea. J Electrochem Soc 142:3741–3749
61. Schneeweiss M.A., Kolb D.M., Liu D.Z., Mandler D. (1997)Anodic oxidation of Au(111). Can J Chem 75:1703–1709
62. Wilms M., Broekmann P., Kruft M., Park Z., Stuhlmann C., Wandelt K. (1998) STM investigation of specific anion adsorption on Cu(111) in sulfuric acid electrolyte. Surf Sci 402–404:83–86; Wilms M., Broekmann P.,

Stuhlmann C., Wandelt K. (1998) In-situ STM investigation of adsorbate structures on Cu(111) in sulfuric acid electrolyte. Surf Sci 416:121–140

63. Ritz G., Schmid M., Biedermann, A., Varga, P. (1996) Strain-induced local surface chemical ordering observed by STM. Phys Rev B 53:16019–16026
64. Wouda P.T., Schmid M., Nieuwenhuys B.E, Varga P. (1998) STM study of the (111) and (100) surfaces of PdAg. Surf Sci 417:292–300
65. Lundgren E., Stanka B., Leonardelli G., Schmid M., Varga P. (1999) Interlayer diffusion of adatoms: A scanning-tunneling microscopy study. Phys Rev Lett 82:5068–5071
66. Pietsch G.J., Köhler U., Jusko O., Henzler M., Hahn P.O. (1992) Structure of the stepped Si/SiO_2 interface after thermal oxidation: Investigations with scanning tunneling microscopy and spot-profile analysis of low-energy electron diffraction. Appl Phys Lett 60:1321–1323
67. Weiß W., Kutschera M., Starke U., Mozaffari M., Reshoeft K., Köhler U., Heinz K. (1997) Development of structural phases of iron silicide films on Si(111) studied by LEED, AES and STM. Surf Sci 377–379:861–865
68. Robinson I.K., Tweet D.J. (1992) Surface X-ray diffraction. Rep Prog Phys 55:599–651
69. Wöll Ch., Chiang S., Wilson R.J., Lippel P.H. (1989) Determination of atom positions at stacking-fault dislocations on Au(111) by scanning tunneling microscopy. Phys Rev B 39:7988–7991
70. Wintterlin J., Behm R.J. (1994) Adsorbate covered metal surfaces and reactions on metal surfaces. In [8], 39–82
71. Sebastian I., Bertrams Th., Meinel K., Neddermeyer H. (1999) Scanning tunneling microscopy on the growth and structure of NiO(100) and CoO(100) thin films. Faraday Discussions 114:129–140
72. Bäumer M., Freund H.-J. (1999) Metal deposits on well-ordered oxide films. Prog Surf Sci 61:127–198
73. Ranke W. (1990) Atomic structure of Si and Ge surfaces: Models for (113), (115), and stepped (001) vicinal surfaces. Phys Rev B 41:5243–5250
74. Baski A.A., Erwin S.C., Whitman L.J. (1995) A stable high-index surface of silicon: Si(5 5 12). Surf Sci 392:69–85
75. Hanbücken M., Röttger B., Kliese R., Vianey I., Neddermeyer H. (1993) An STM study of mechanochemically prepared Si(111) substrates: An extended set of vicinal surfaces. Europhys Lett 23:573–578
76. Haneman D. (1997) Surfaces of silicon. Rep Prog Phys 50:1045–1086
77. Brommer K.D., Needels B.E., Larson B.E., Joanopoulos J.D. (1992) Ab initio theory of the Si(111)-(7 x 7) surface reconstruction: A challenge for massively parallel computation. Phys Rev. Lett 68:1355–1358
78. Knall J., Pethica J.B., Todd J.D., Wilson J.H. (1991) Structure of Si(113) determined by scanning tunneling microscopy. Phys Rev Lett 66:1733–1736
79. Chadi D.C. (1987) Stabilities of single-layer and bilayer steps on Si(001) surfaces. Phys Rev Lett 59:1691–1694
80. Poon T.W., Yip S., Hon P.S., Abraham F.F. (1990) Equilibrium structures of Si(100) stepped surfaces. Phys Rev Lett 65:2161–2164
81. Röttger B., Hanbücken M., Neddermeyer H. (2000) Nanostructures obtained by self-organization of silicon surfaces. Appl Surf Sci 162–163:595–598
82. Williams E.D., Phaneuf R.J., Wei J., Bartelt N.C., Einstein T.L. (1993) Thermodynamics and statistical mechanics of the faceting of stepped Si(111). Surf Sci 294:219–242

83. Liu D.-J., Fu E.S., Johnson M.D., Weeks J.D., Williams E.D. (1996) Relaxation of the step profile for different microscopic mechanisms. J Vac Sci Technol. B 14:2799–2808
84. Ichimiya A., Tanaka Y., Ishiyama K. (1996) Quantitative measurements of thermal relaxation of isolated silicon hillocks and craters on the Si(111)-(7x7) Surface by Scanning Tunneling Microscopy. Phys Rev Lett 76:4721–4724
85. Hildebrandt S., Kraus A., Kulla R., Neddermeyer H. (2001) STM investigations of the diffusion-controlled decay of hole and island structures on Si(111)7x7. Surf Sci: submitted
86. Morgenstern K., Rosenfeld G., Lægsgard E., Comsa G. (1998) Measurement of energies controlling ripening and annealing on metal surfaces. Phys Rev Lett 80:556–559
87. Smith F.W., Ghidini G. (1982) Reaction of oxygen with Si(111) and (100): Critical conditions for growth of SiO_2. J Electrochem Soc 129:1300–1306
88. Feltz A., Memmert U., Behm R.J. (1994) High-temperature scanning tunneling microscopy studies on the interaction of O_2 with Si(111)-(7x7). Surf Sci 314:34–56
89. Kubby J.A., Boland J.J. (1996) Scanning tunneling microscopy of semiconductor surfaces. Surf Sci Rep 26:61–204; Neddermeyer H. (1996) Scanning tunneling microscopy of semiconductor surfaces. Rep Prog Phys 59:701–769
90. Köhler U., Andersohn L., Dahlheimer B. (1993) Time-resolved observation of CVD-growth of silicon on Si(111) with STM. Appl Phys A 57:491–497
91. Köhler U., Andersohn L., Bethge H. (1997) The influence of hydrogen on CVD-growth on Si(111) surfaces. phys stat sol (a) 159:39–51
92. Rauscher H., Behm R.J. (1998) Growth structures of silicon homoepitaxy by chemical vapour deposition. In: "Morphological organizations in epitaxial growth and removal", Series on Directions in Condensed Matter Physics, Vol. 14., Z. Zhang and M. Lagally, Eds., World Scientific, Singapore, 242–261
93. Andersohn L., Berke Th., Köhler U., Voigtlaender B. (1996) Nucleation behaviour in molecular beam and chemical vapor deposition of silicon on Si(111)-(7x7). J Vac Sci Technol A 14:312–318
94. Fehrenbacher M., Rauscher H., Memmert U., Behm R.J. (1997) SiH_4 chemical vapour deposition on Si(111)-(7x7) studied by scanning tunneling microscopy. Surf Sci 385:123–145
95. Braun J., Rauscher H., Behm R.J. (1998) Adsorption of disilane on Si(111)-(7x7) and initial stages of CVD growth. Surf Sci 416:226–239
96. Fehrenbacher M., Spitzmüller J., Pitter M., Rauscher H., Behm R.J. (1997)Structures on Si(100)2x1 at the initial stages of homoepitaxy by SiH_4 decomposition. Jpn J Appl Phys 36:3804–3809
97. Spitzmüller J., Fehrenbacher M., Pitter M., Rauscher H., Behm R.J. (1997) Local arrangement of silylene groups on Si(100)2x1 after SiH_4 decomposition. Phys Rev B 55:4659–4664
98. Braun J., Haupt M., Thonke K., Sauer R., Rauscher H., Behm R.J. (2000) Irradiation-induced Ge multilayer growth from GeH_4 on Si(111). Surf Sci 454–456:811–817
99. Eigler D.M., Schweizer E.K. (1990) Positioning single atoms with a scanning tunneling microscope. Nature 344:524–526
100. Meyer G., Zöphel S., Rieder K.-H. (1996) Scanning tunneling microscopy manipulation of native substrate atoms: A new way to obtain registry information on foreign adsorbates. Phys Rev Lett 77:2113–2116

101. Allers W., Hahn C., Löhndorf M., Lukas A., Pan S., Schwarz U.D., Wiesendanger R. (1996) Nanomechanical investigations and modifications of thin films based on scanning force methods. Nanotechnology 7:346–350
102. Uchida H., Huang D., Grey F., Aono M. (1993) Site-specific measurement of adatom binding energy differences by atom extraction with the STM. Phys Rev Lett 70:2040–2043
103. Snow E.S., Campbell P.M., McMarr P.J. (1993) Fabrication of silicon nanostructures with a scanning tunneling microscope. Appl Phys Lett 63:749–751
104. Rauscher H., Behrendt F., Behm R.J. (1997) Fabrication of surface nanostructures by scanning tunneling microscope induced decomposition of SiH_4 and SiH_2Cl_2. J Vac Sci Technol B 15:1373-1377
105. Wendel M., Lorenz H., Kotthaus J.P. (1995) Sharpened electron beam deposited tips for high resolution atomic force microscope lithography and imaging. Appl Phys Lett 67:3732–3734
106. Irmer B., Blick R.H., Simmel F., Gödel W., Lorenz H., Kotthaus J.P. (1998) Fabrication of Ti/TiO_x tunneling barriers by tapping mode atomic force microscopy induced local oxidation. Appl Phys Lett 73:2051–2053
107. Irmer B., Kehrle M., Lorenz H., Kotthaus J.P. (1997) Josephson junctions defined by a nanoplough. Appl Phys Lett 71:1733–1735
108. Akama Y., Nishimura E., Sakai A., Murakami H. (1990) New scanning tunneling microscopy tip for measuring surface topography. J Vac Sci Technol A 8:429–433
109. Wendel M., Kühn S., Lorenz H., Kotthaus J.P., Holland M. (1994) Nanolithography with an atomic force microscope for integrated fabrication of quantum electronic devices. Appl Phys Lett 6:1775–1777
110. Li W., Virtanen J.A., Penner R.M. (1992) Nanometer-scale electrochemical deposition of silver on graphite using a scanning tunneling microscope. Appl Phys Lett 60:1181–1183
111. Ullmann R., Will T., Kolb D.M. (1995) Nanostructuring of electrode surfaces by tip-induced metal deposition. Ber Bunsenges Phys Chem 99:1414–1420
112. Kolb D.M., Ullmann R., Will T. (1997) Nanofabrication of small Cu-clusters on Au(111) by an STM. Science 275:1097–1099
113. Engelmann G.E., Ziegler J.C., Kolb D.M. (1998) Electrochemical fabrication of large arrays of metal nanoclusters. Surf Sci 401:L420–L424
114. Kolb D.M. (2000) private communication
115. Radmacher M. (1997) Measuring the elastic properties of biological samples with the atomic force microscope. IEEE Eng Med Biol 16:47–57
116. Rief M., Oesterhelt F., Heymann B., Gaub H.E. (1997) Single molecule force spectroscopy on polysaccharides by atomic force microscopy. Science 275:1295–1297
117. Rief M., Gautel M., Oesterhelt F., Fernandez J.M., Gaub H.E. (1997) Reversible unfolding of individual titin immunoglobulin domains by AFM. Science 276:1109–1112
118. Qu S., Clarke C.J., Liu Y., Rafailovich M.H., Sokolov J., Phelan K.C., Krausch G. (1997) Dewetting dynamics at a polymer-polymer interface. Macromolecules 30:3640–3645
119. Walheim S., Böltau M., Mlynek J., Krausch G., Steiner U. (1997) Structure formation via polymer demixing in spin-cast films. Macromolecules 30:4995–5003

120. Elbs H., Fukunaga K., Stadler R., Sauer G., Magerle R., Krausch G. (1999) Microdomain morphology of thin ABC triblock copolymer films. Macromolecules 32:1204–1211

8 Multi-Method High-Resolution Surface Analysis with Slow Electrons

E. Bauer and T. Schmidt

8.1 Introduction

Electron microscopy with fast electrons has reached a high state of sophistication allowing not only structural imaging with atomic resolution but also chemical analysis in the sub-nanometre range and magnetic imaging with Lorentz microscopy and holography below ten nonometres. These possibilities are due to the strong elastic forward scattering, to the signal accumulation along the path of the electron and to the small aberrations of the employed electron optical elements. In surface imaging, the signal must come from a small depth and fast electrons are useful in only a few imaging modes that emphasize the surface features. Examples are weak-beam imaging, secondary electron emission microscopy and reflection electron microscopy, the last being the most versatile.

Instead of trying to overcome the low surface sensitivity of fast electrons by employing modes of operation that emphasize the signal contributions from the surface, it appears logical to seek conditions under which electrons interact only with the surface. This is the case at very low electron energies, which led in 1962 to the idea of low-energy electron microscopy (LEEM). The history of this field up to 1993 has been reviewed [1] and will not be repeated here. Only the principles of the electron-surface interaction, the basic electron optical problems, the extension of LEEM to Spin-Polarized LEEM (SPLEEM) and, foremost of all, the combination of LEEM with Spectroscopic Photo Emission Electron Microscopy (SPELEEM), the development of which was supported by the Volkswagen Foundation, will be discussed. Imaging can always be combined with local spectroscopy and diffraction so that a very comprehensive surface characterization is possible.

8.2 Interaction of Slow Electrons with Condensed Matter

Slow electrons interact very strongly with matter so that their effective penetration depth, which determines the signal, is only a few atomic layers. Calculations made in the early sixties, which have been reviewed recently [2,3], show very clearly that the elastic backscattering cross-sections of atoms in

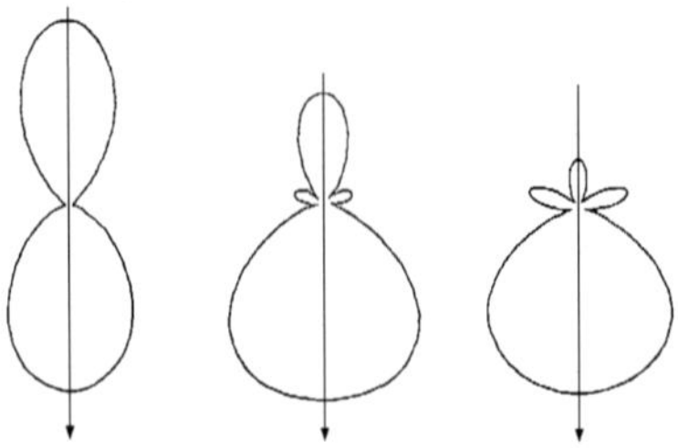

Fig. 8.1. Angular distribution of electrons scattered from solid-state Cu atom potentials (polar diagrams). (**a**) 20 eV, (**b**) 40 eV and (**c**) 60 eV

solids are so high at very low energies that more than 50% of the electrons are reflected. Figure 8.1 illustrates this for the scattering by solid-state Cu atom potentials. At 20 eV backscattering is as strong as forward scattering. If the specimen is a single-crystal, most of the electrons will be confined by diffraction into the specular beam. By using this beam and its environment, very high intensities can be expected, therefore. Not only is the angular distribution dominated by diffraction, but so too is the energy dependence of the reflectivity as seen in Fig. 8.2 [4]: while the reflectivity of the W(110) surface is high at very low energies, that of the W(100) surface is very small. This is a band structure effect. Along the [110] direction, there is a band gap from about 0 eV to 5 eV so that there are no allowed states for the electrons in the crystal and the incident electrons are reflected. In the [100] direction the density of unoccupied states is large below 3 eV so that the electrons can easily penetrate the crystal and thermalize in it. A band gap approximately 2 eV wide centred at about 4 eV causes the narrow reflectivity peak at 4 eV.

The phase relations that cause the high backscattering cross-sections of atoms are thus overridden by those caused by the periodicity of the atomic distribution. The former determine only the general trend of the backscattering as a function of atomic number Z and energy while the details are controlled by diffraction. This dependence of the reflectivity upon the band structure is of fundamental importance for the formation of the magnetic contrast in SPLEEM. In ferromagnetic materials, the energy bands of minority (spin down) electrons are displaced on the energy axis by the exchange splitting to higher values as indicated in Fig. 8.3 for cobalt along the [0001] axis; the energy gaps are hence shifted relative to each other. As a consequence, at the lower energy, electrons whose spin polarization $\boldsymbol{P}$ is parallel to the magnetization $\boldsymbol{M}$ can penetrate into the crystal while those with $\boldsymbol{P}$ antiparallel to $\boldsymbol{M}$ are reflected from the (0001) surface. This reflectivity difference is the basis of the contrast between ferromagnetic domains.

While Z-contrast plays an important role in addition to phase contrast in conventional high-energy electron microscopy, contrast in LEEM is much

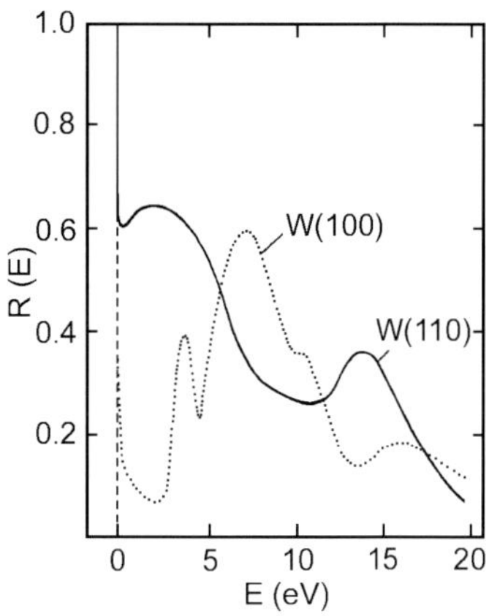

Fig. 8.2. Normal incidence reflectivity of two tungsten single-crystal surfaces at very low energy

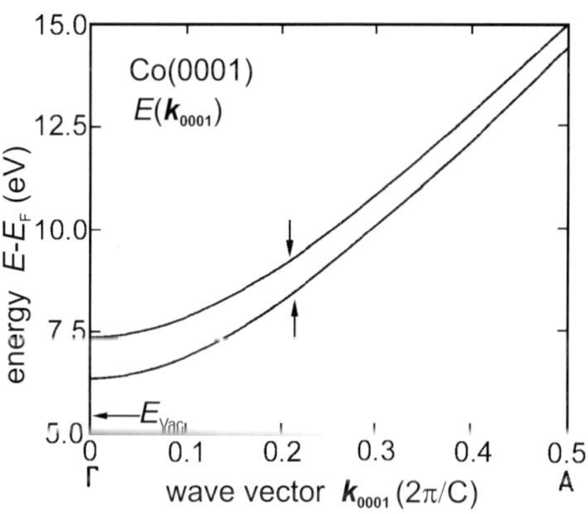

Fig. 8.3. Spin-up (↑) and spin-down (↓) band structure of cobalt along the [0001] direction above the vacuum level E_{Vac}, measured from the Fermi energy E_{F}

more dominated by phase contrast. There is no simple relationship between atomic number and backscattering cross-section as can be seen, for example, in Fig. 8.4 for 100 eV electrons and free atom potentials. Another example is Si which has a very high reflectivity around 40 eV [2] while that of W is high at much lower energies (Fig. 8.2). A more detailed discussion of backscattering and diffraction contrast may be found in [2,3] and the references cited there.

Phase contrast is particularly striking in step contrast and in quantum size contrast. A clean, completely defect-free single crystal surface acts like a structure-free mirror. There are always steps on the surface, which separate terraces of different height. The waves reflected from the two sides of the step have a phase difference, which can be converted to an amplitude

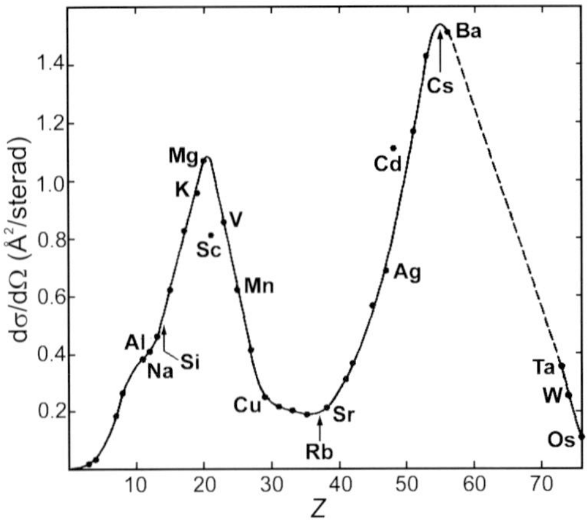

Fig. 8.4. Differential scattering cross-sections of free atoms for 100 eV electrons close to 180° as a function of atomic number

difference by slight defocusing. In general, the spherical aberration of the objective lens is sufficient for this conversion. In surface layers that are so thin that the wave still has a notable amplitude at its backside, destructive and constructive interference in the layer can occur just as in reflection-reducing or -increasing films in light optics. In summary, phase contrast, whether it arises from the periodic crystal lattice (diffraction contrast) or from the topography of the surface region (interference contrast), is the most important contrast mechanisms in LEEM.

The next question which needs to be addressed is what is the information depth of LEEM and SPLEEM? Two processes limit the information depth: elastic and inelastic scattering. Elastic scattering determines the penetration depth at energies that fall in band gaps at low energies, inelastic scattering dominates otherwise. As seen in Fig. 8.2, the reflectivity is not 100% in the band gaps. This is due to the fact that the incident wave continues into the crystal as an evanescent wave. The latter can transfer energy and momentum into surface states, surface phonons and other surface waves and thus reduce the reflectivity. The penetration depth λ_e (1/e value of the intensity) of the evanescent wave is smallest at the centre of the gap and increases to the value determined by the inelastic mean free path λ_i at the band edges. At the centre of the gap it scales approximately inversely as the band gap [5]. In the case of Fig. 8.2, λ_e is 0.23 nm and 0.56 nm in the gap centre for the (110) and (100) surface, respectively.

Band gaps apart(Bragg conditions), the information depth is determined by inelastic scattering. The most important energy-loss processes are one-electron excitations (electron–hole pair creation) and collective electron excitations (plasmon creation). In free-electron metals, that is in materials with nearly parabolic band structure, the joint density of states, which is important

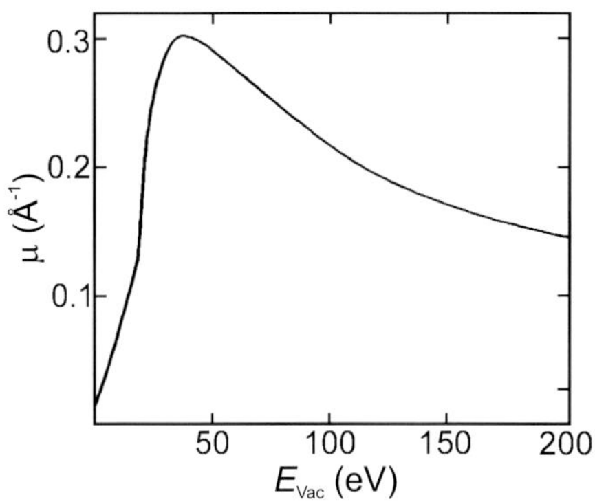

Fig. 8.5. Absorption coefficient of Al in the jellium approximation

for single electron excitations [6], is small, so that the inelastic damping of the incident wave rises with the number of accessible states until the threshold for plasmon creation is reached. Above this energy it increases rapidly to a maximum at a few tens of eV, beyond which it decreases slowly. At higher energies, inner shell excitations, which will be discussed in connection with spectroscopic imaging, play an increasing role. As an example, Fig. 8.5 shows the absorption coefficient $\mu = 1/\lambda_i$ of Al derived from the jellium model of Al [7], which determines the inelastic mean free path λ_i. The minimum penetration depth is here about 0.35 nm. For non-free-electron-like materials λ_i can be calculated from the imaginary part of the dielectric constant [8] but the accuracy of the data below 50 eV is questionable because of the approximations involved in the calculations.

In non-free electron-like materials, in particular in d and f electron metals, the joint density of states is frequently very large and damping due to single-electron excitations is correspondingly strong. Because of the exchange splitting of the bands the spin-up bands are much less occupied than the spin-down bands, which leads to a spin-dependent excitation probability. The inelastic mean free paths of slow electrons are therefore not only small but also spin-dependent, which is important for the contrast formation in SPLEEM. Figure 8.6 shows the spin averaged $1/\lambda_i$ values of some metals for electron energies within 5–10 eV from the Fermi level as a function of the number of unoccupied d-orbitals [9]. The spin dependence is known only for a few materials and energies. For example for Co and 1.5–3.5 eV electrons, $\lambda_{i\uparrow} = 1.47$ nm, $\lambda_{i\downarrow} = 0.54$ nm [10], while for Fe and 9.4 eV electrons $\lambda_{i\uparrow} = 1.1$ nm, $\lambda_{i\downarrow} = 0.67$ nm [11]. These examples demonstrate that, at least in transition metals, the information depth at the low energies frequently used in LEEM and SPLEEM can be much smaller than the value predicted by the widely used "universal curve" and that it is spin-dependent.

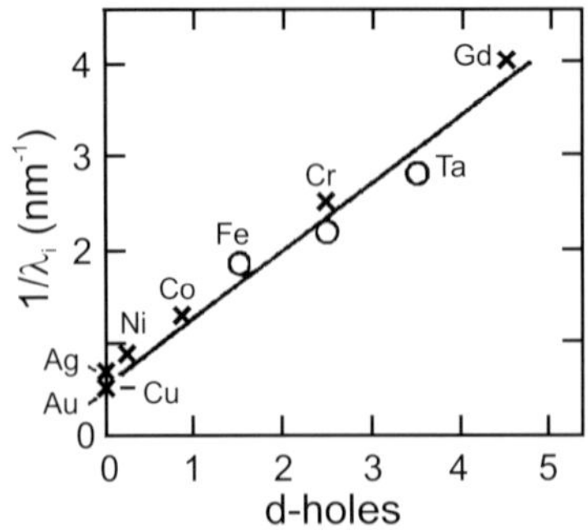

Fig. 8.6. Inverse inelastic mean free path of slow electrons in some non-free-electron-like metals as a function of number of d(f)-holes

Summarizing these considerations, the information depth of slow electrons, that is the depth at which the intensity has decreased to $1/e$ of the incident intensity, can be varied over a wide range by proper choice of the energy. In the centre of wide band gaps, it can be made as small as one monolayer or at the maximum of the plasmon damping as small as two monolayers; outside band gaps it can be many monolayers as long as no strong single-electron or plasmon excitations occur (see for example Cu, Ag and Au in Fig. 8.6).

All these phenomena make the theory of electron–specimen interactions much more complicated for slow electrons than for fast electrons. Not only does it require a dynamical theory of diffraction but the surface barrier, the deviations of the positions of the atoms in and near the surface from those in the bulk, the energy-dependent exchange-correlation interaction of the electron with the specimen, and the energy-dependent inelastic attenuation of the incident wave must all be taken into account. Only highly simplified models have been calculated to date but at the present state of image interpretation this state of affairs is not a major hurdle. With increasing development of LEEM and SPLEEM more sophisticated theoretical work in this field will certainly become necessary.

8.3 Electron-Optical Considerations

The wave field emanating from the specimen is transferred by the electron optical system with some degree of modification into the image plane of the microscope. The phase changes upon reflection are very large at low energies. Therefore, this modification cannot be described by the contrast-transfer function used at high energies, which is based on the approximation that the phase shift in the specimen is small. Instead a more general treatment in terms of a reflection cross-coefficient – analogous to the transmission cross-coefficient [12] – is needed [13], which makes the comparison of object

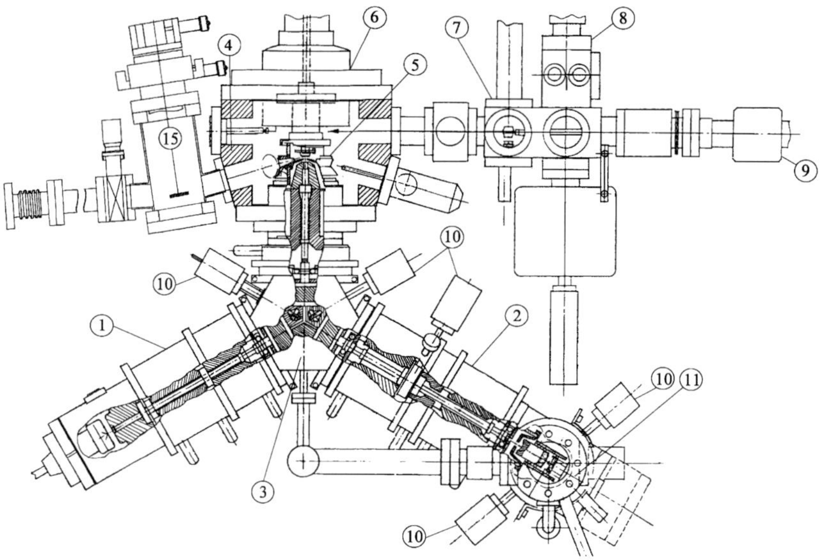

Fig. 8.7. Schematic of the Volkswagen Foundation spectroscopic low energy electron microscope

and image more difficult. This is not the only difference between imaging with fast and slow electrons. A common feature of imaging with emitted and reflected electrons is the acceleration of the slow electrons to high energies in the objective lens, which thus becomes a cathode lens. Such lenses have much larger aberrations than the objective lenses used with fast electrons, mainly because of the initial accelerating field. A second difference is the need for a beam separator in the case of imaging with reflected electrons, which introduces additional aberrations. These limitations can be largely overcome by using correcting elements, as discussed below.

The instrument for spectroscopic imaging with low-energy electrons, which was developed with support of the Volkswagen Foundation, will now be described briefly. The present version, which differs somewhat from the original design of Veneklasen [14], is shown in Fig. 8.7. The illumination and imaging columns (1 and 2, respectively) use conventional electron optics but with bakeable lenses outside the vacuum. The major differences are the beam separator (3), the specimen chamber (4), the cathode lens (5) and the specimen manipulator (6). The remaining components are the specimen-preparation chamber (7), air-lock (8) and the transfer rod (9), the illumination, the selected-area and the contrast aperture and the energy slits(10) behind the energy filter (11), and in addition a refocusing mirror (15) for the soft X-rays from the synchrotron used in XPEEM. The energy filter as shown here is an electrostatic 90° analyser but has been replaced in the meantime by an electrostatic 180° analyser.

Various versions of the instrument and its components have been described repeatedly [3,14–18] so that only a few data will be given, which are important for the application of the instrument. The cathode lens is a magnetic triode, which has the best resolution of all lenses considered [16]. For a final electron energy of 20 kV, a field strength of 10 kV/mm, an initial energy of 10 eV and an energy-width of 0.5 (0.1) eV the theoretical resolution is about 5 (3) nm in the Gaussian image. The electrostatic tetrode used in SPLEEM in order to be able to work in zero magnetic field has a slightly worse resolution (about 7 and 5 nm, respectively). The beam separator is a close-packed prism array, similar to those described in [17,19] but with 60° deflection instead of the 90° deflection discussed there. The image astigmatism of the 60° deflector is smaller than that of the 90° deflector and does not limit the practical resolution (8 nm) of the instrument, which is determined by operational problems such as specimen, lens-current and high-voltage stability as well as alignment. Better resolution requires solution of these problems, a more complicated beam separator and correction of the chromatic and spherical aberration of the objective lens. Some improvement has been achieved with an improved beam separator [20]. Further progress should be made with the beam separator and mirror-corrector developed by Rose's group, initially with support of the Volkswagen Foundation [21], subsequently improved [22] within the SMART (spectromicroscope for all relevant techniques) project [23]. The SMART has a theoretical resolution of better than 1 nm, which is made possible by aberration correction with an electron mirror [24–26]. In addition, it will be equipped with an omega filter [27] for energy-selected imaging with a resolution down to 0.1 eV. Once in full operation it will have all the operational modes of the SPELEEM but with better resolution and significantly higher transmission. The operational experience gained with the SPELEEM should help considerably in achieving the goals of the SMART.

In LEEM, SPLEEM and MEM (mirror electron microscopy) the transmission of the system is not very important because most of the intensity is concentrated in a small angular range and the energy width of the imaging electrons is small. Therefore, the small contrast aperture, that is needed to obtain optimum resolution as determined by the chromatic and spherical aberrations and by diffraction, does not limit the intensity. In emission microscopy, however, electrons are emitted over a wide angular range and have a wide energy distribution. Unless the aberrations are corrected, the aperture needed to achieve an acceptable resolution δ severely limits the transmission T. For reasonable image acquisition rates, one then has to make a compromise between resolution and transmission, which can be characterized by a quality factor $Q_n = T^n/\delta^2$ [28]. For LEEM, SPLEEM, and MEM, $n = 0$. While for emission microscopies, $n = 1$ or 2, depending upon angular and energy distribution and available total intensity. Q_1 or Q_2 may be optimized by increasing the size of the contrast aperture. This is indicated in Fig. 8.8 for the magnetic triode and some electrostatic tetrodes for the case $n = 1$. At

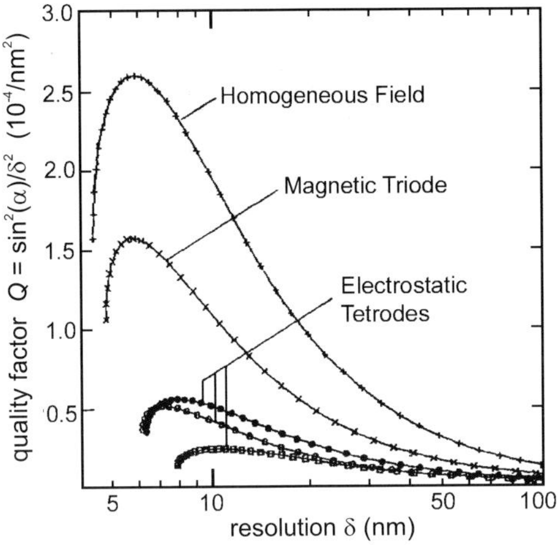

Fig. 8.8. Quality factor $Q_1 = T/\delta^2$ of various cathode lenses as a function of the resolution, which is determined by the maximum transmitted angle α. For reference the curve for the homogeneous field is shown too

the optimum Q the resolution limit is in the 10 nm range. An increase of the aperture beyond the optimum Q increases the transmission at the expense of the resolution. This is the area in which aberration correction will have its main impact.

8.4 Analytic Methods in the SPELEEM

The SPELEEM combines microscopy, spectroscopy and diffraction or – more generally – imaging of the angular distribution of the electrons from the selected specimen region. Although all imaging modes with reflected and emitted electrons are in principle possible, it is not efficient to try to do everything in one instrument. This is true of SPLEEM in particular, which requires a spin-polarized electron source. This source is a GaAs or more complex photocathode, which is activated to have negative electron affinity by deposition of Cs together with O_2. A very low residual gas pressure is needed in the gun region in order to maintain the activation over a long time. This makes a SPLEEM electron source much less suited for routine operation than the sturdy LaB_6 cathode. For this reason SPLEEM has not been incorporated yet into the SPELEEM but is used as a separate instrument [3,29]. However for completeness of the subject of multi-method analysis, a brief description and an example of SPLEEM will be given.

8.4.1 SPLEEM

The magnetic contrast in SPLEEM is a consequence of the exchange interaction $J \sum \boldsymbol{s}_i \cdot \boldsymbol{s}_j$ between the spins $\boldsymbol{s}_i$ and $\boldsymbol{s}_j$ of the incident and specimen electrons, respectively, J being the exchange constant. For a spin-polarized beam with polarization $\boldsymbol{P}$ and a ferromagnetic specimen with magnetization $\boldsymbol{M}$ this interaction – which determines the magnetic contrast – is proportional to $\boldsymbol{P} \cdot \boldsymbol{M}$. In general the magnetic contrast is much weaker than the $\boldsymbol{P}$-independent contrast but can be obtained in pure form by subtracting images taken with opposite $\boldsymbol{P}$. The vector $\boldsymbol{P}$ can be aligned in any direction in space so that it can be made (anti)parallel to $\boldsymbol{M}$ for maximum contrast. Alternatively, three mutually orthogonal directions may be chosen to determine the three components of $\boldsymbol{M}$: one is usually normal to the specimen surface, the other two parallel to the surface of the specimen along the "easy" and "hard" axes of magnetization, when they exist. The sensitivity and resolution of SPLEEM depend upon the magnitudes of $\boldsymbol{P}$ and $\boldsymbol{M}$, the reflected intensity and the image acquisition time. They are somewhat poorer than those of LEEM since image subtraction is neccessary and leads to a decrease of the signal-to-noise ratio. Magnetic imaging with SPLEEM is illustrated in

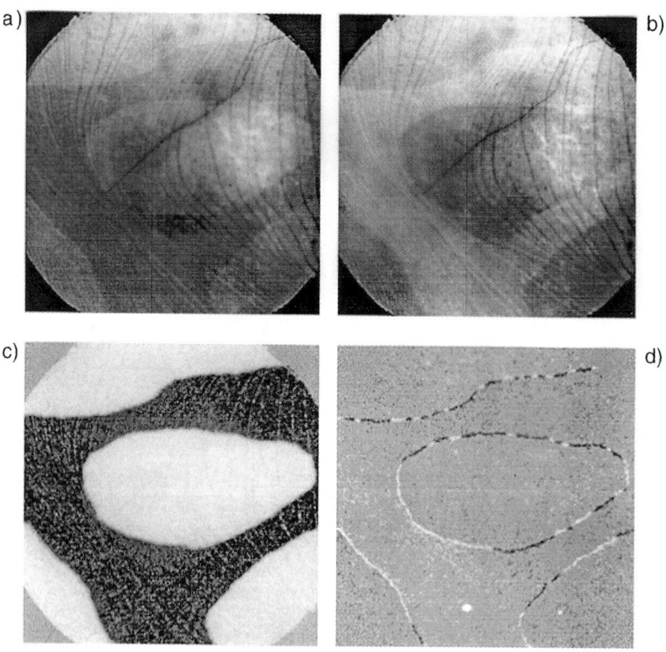

Fig. 8.9. SPLEEM image acquisition and processing. The dark line in (**a**) and (**b**) is a crack in the microchannel plate which is eliminated in the image subtraction which gives the magnetic images (**c**) and (**d**). The specimen is a 6 monolayer thick Co film on a W(110) surface. Field of view 13 μm, electron energy 2 eV

Fig. 8.9 [30]. Subtraction of the upper two images taken with opposite $\boldsymbol{P}$ (parallel to $\pm\boldsymbol{M}$), which contain both magnetic and structural contrast, gives the lower-left image with (nearly) pure magnetic contrast. With $\boldsymbol{P}$ perpendicular to $\pm\boldsymbol{M}$, the domain-wall image on the lower-right is obtained. The main application of SPLEEM is in the study of the connection between the magnetic properties and the microstructure and topography of thin ferromagnetic film systems. Some recent examples are given in [31].

8.4.2 LEEM

LEEM, for which ten years ago only one instrument existed [19], has in the meantime spread slowly around the world but the number of instruments is still small compared to (S)TEMs. The method has been reviewed repeatedly, for example in [3] so that only a few highlights and its limitations will be discussed here. The most important aspects of LEEM are its high surface sensitivity and depth resolution, its lateral resolution of about 10 nm and its high intensity, which allows surface processes on bulk materials to be studied. This last aspect explains why a LEEM instrument should not be considered primarily as an electron microscope – with poor resolution compared to a (S)TEM – but rather as a multi-method UHV system for in-situ studies with high resolution, especially when equipped with an energy filter. This is illustrated in Fig. 8.10, which shows the combination of reflection and synchrotron radiation-excited photo-emission methods in the SPELEEM. It is relatively straightforward to switch between microscopy, diffraction and spectroscopy as will be shown below. Of course, all other modes possible with reflected or emitted electrons can be used too: mirror electron microscopy (MEM) and electron-excited secondary electron emission microscopy (SEEM), Auger electron emission microscopy (AEEM), thermionic (TEEM), UV light-excited photo electron emission microscopy (PEEM), and other emission electron microscopies. The modes MEM and SEEM are particularly useful for specimens without diffraction or interference contrast and MEM also allows surface potential variations and electric and magnetic field distributions in front of the specimen to be imaged. AEEM was originally believed to be the major chemical imaging mode [32,33] before high-brightness synchrotron radiation sources became available in the nineties. Now, that these sources are increasingly accessible, soft x-ray excited PEEM (XPEEM) has largely replaced AEEM, for reasons given below. The other emission modes mentioned above are mainly of use for specimens with work-function variations.

Before discussing the relative merits of AEEM and XPEEM three of the major limitations of cathode lens systems – in addition to UHV compatibility and sufficiently low vapour pressure of the specimen, obstacles that in principle can be overcome by cooling – must be mentioned. First of all, the surface must be reasonably smooth. The rougher the surface, the more difficult the image interpretation because of the field distortion caused by the roughness, in particular when the roughness features are closely spaced. The

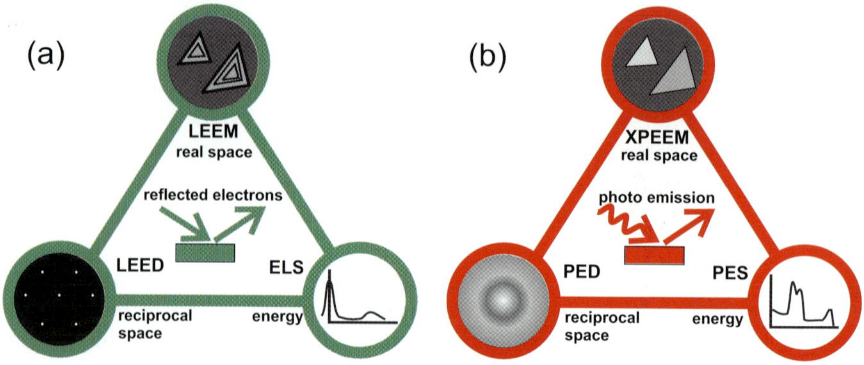

Fig. 8.10. The basic surface analysis combinations in the SPELEEM: (**a**) reflected and (**b**) photo-emitted electrons

lower the energy, for example in MEM, the greater the difficulty. In extreme cases, such as the surfaces used in field-emission displays, field emission sets in long before the field is high enough for one of the other imaging methods. Of course, if one wants to study field emission, the cathode-lens system is a good FEEM. The second limitation is charging. Insulating specimens tend to charge more or less strongly at the high current densities needed at high magnifications. This is true not only for methods based on incident electrons but also for emission methods. Overall specimen charging can be reduced or even eliminated by a variety of techniques but frequently local charge fluctuations (in space and time) at the submicroscopic level remain which can preclude, for example, MEM imaging. The third limitation, radiation damage, is common to all electron microscopies. In LEEM it can be reduced or even completely avoided by using electron energies below the radiation damage threshold, in general below 10 eV. On the other hand, variation of the electron energy allows the study of radiation damage processes at the 10 nm level with LEEM.

8.4.3 Comparison of AEEM and XPEEM

Both AEEM and XPEEM make use of the core levels of the atoms in the specimen to characterize their chemical nature and bonding. Both methods can be used in two modes: in the secondary electron emission mode and in the characteristic-electron emission mode. In the first mode, the secondary electrons produced by the primary photo electron provide the level-characteristic signal. Chemical contrast is obtained from difference images below and above the core level ionization threshold or, in the case of magnetic materials, from difference images obtained with different excitation conditions. No energy filter is necessary in this mode but the resolution is limited by the chromatic aberration of the objective lens. At the high core level binding energies used in element-specific imaging with magnetic dichroism XPEEM

[34], this is possibly the best mode because of the built-in secondary electron multiplication of the weak primary signal (see below). In the second mode, the characteristic Auger or photo electrons are selected for imaging with an energy filter. The SPELEEM is operated mainly in this mode, and sometimes in the secondary electron mode but with an energy filter to improve the resolution. AEEM has been demonstrated with the SPELEEM [15] but was largely abandoned once the instrument had good access to the synchrotron radiation source ELETTRA, for reasons that will now be discussed.

The first point to notice is that core level ionization is less efficient with electrons than with photons. This is evident from Table 8.1, which compares the maximum ionization cross-sections of two core levels of various atoms for electrons and photons. Secondly, Auger electron "lines" are much wider than photo electron "lines". The "line width", that is the energy width, is determined in photo electron emission by the lifetime of the core hole and is typically of the order 0.5–1 eV. The most intense Auger electron transitions involve two electrons in the topmost occupied band, which is usually several eV to 15 eV wide. To a rough first approximation, the Auger electron line width is twice as large as the bandwidth (self-convolution of the band), so that the characteristic signal is smeared out over an energy range between about 5 eV and 30 eV. For good spatial resolution an energy window of less than 1 eV has to be selected, which allows the use of most of the photo electrons but only a fraction of the Auger electrons with the concomitant loss of intensity. Thirdly, Auger electron signals sit generally on a background that is much larger than the signal itself while the background in photo-emission is much smaller. The fourth reason is an operational one: in AEEM the energy of the incident beam is usually much higher than that of the emitted Auger electrons. The sector of the beam separator for the incident beam therefore has to be set at an energy different from that of the exit sector. This makes the combination of AEEM and LEEM, where input and output energy are equal, inconvenient. In addition, the incident beam should be focused onto the specimen in order to maximize the Auger electron signal,

Table 8.1. Maximum ionization cross-sections of several atoms for electrons (*left column*) and photons (*right two columns*) in 10^{-22} m^2. Data from various sources

K shell	Calculated		Experimental
C	0.31	0.9	0.7
N	0.18	0.7	0.66
O	0.10	0.5	0.57
L_{23} shell			
P	2.25	5	4
S	1.5	5	4.1
Cl	1.2	5	3.6

whereas parallel illumination is optimal in the LEEM mode. These problems do not arise in XPEEM. The arguments in favor of AEEM made more than twenty years ago [32] and still valid ten years later [33] were based on the much higher primary current available at that time with electrons than with photons. With the development of undulators with high brightness, the arguments had to be revised at the beginning of this decade [28] and, with the construction of spectromicroscopy-dedicated undulator beamlines, they have become reversed. The rest of this contribution will, therefore, be limited to the combination XPEEM–LEEM, as indicated in Fig. 8.10.

One of the additional advantages of XPEEM over AEEM is the free choice of the electron energy. Three considerations determine this choice: (i) a high photo electron yield, (ii) a high transmission of the objective lens and (iii) a low background signal. Consideration (i) is optimally fulfilled at the maximum of the ionization cross-section as a function of photon energy and for outer core levels (see Fig. 8.11). Some materials also have a high density of states in the valence/conduction band, which produces a useful sharp peak in the photo-emission spectrum. Figure 8.11 [28] clearly shows that the core levels with the smallest ionization energies E_i are the most desirable ones because of the $1/E_\mathrm{i}^2$ dependence of the maximum cross-section. For light atoms such as C, N or O, only the K levels with their low cross-sections are available. The choice dictated by consideration (i) fulfills to a certain extent the other two considerations as well. In many cases the energy at which the cross-section reaches its maximum is 10–100 eV above E_i, so that the photo electrons have the same energy (minus the work function). If this energy is in the range with high secondary electron emission, the energy can be increased without much loss of intensity by increasing the photon energy in order to improve the contrast. On examining the problem in more detail, considerations (i) and (ii) are found to be conflicting. The transmission increases strongly with decreasing energy [28] but so does the secondary electron background. The photo electron energy range between about 20 eV and 50 eV has been found to be a good compromise. Thus, for ordinary spectromicroscopy, work at low photon energy is advantageous.

The situation is quite different in magnetic imaging with magnetic dichroism XPEEM where E_i is about 800 ± 100 eV as opposed to say 80 eV in ordinary XPEEM, resulting in a photo electron current that is about 100 times weaker. As already mentioned, imaging with the secondary electrons generated by the primary photo electrons is here the best choice. The SPELEEM had access to photons with sufficiently high energy for only a few days at BESSY I so that only very limited data could be obtained [15]. For this reason, this subject will not be discussed here. A detailed up-to-date description of this mode of operation and its possibilities has been published recently [34]. The intermediate ionization energy range (e.g. C, N,O) should become accessible to direct imaging with photo electrons with the dedicated beamlines that are now in construction.

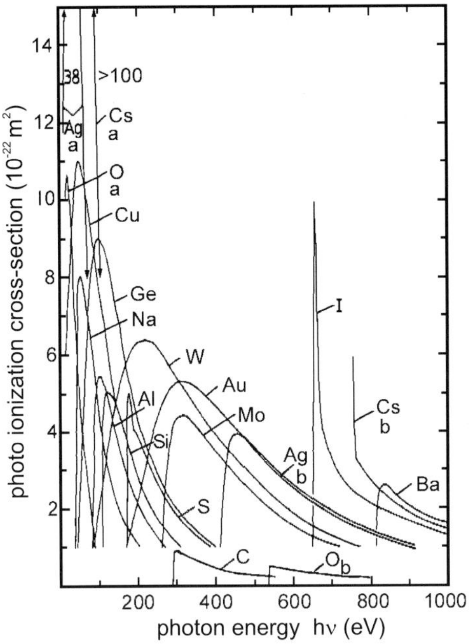

Fig. 8.11. Selected core level ionization cross-sections of free atoms as a function of photon energy

8.4.4 XPEEM with the SPELEEM

How the basic operation modes of the SPELEEM indicated in Fig. 8.10 are realized in practice is shown in Fig. 8.12 [35]. For easier comparison of the modes the optical system is shown straight. As a consequence the objective lens has to be shown twice and the input and output part of the beam separator are separated in the drawing, which represents three electron reflection modes: imaging (a), diffraction (b) and spectroscopy (c). For the corresponding photo electron emission modes, the components to the left of the surface are not used, e.g. by closing the Wehnelt cylinder, but replaced by the photon beam. The excitations of all lenses up to and including the field lens (FL) are identical for all modes. Switching between the modes is achieved by changing the excitations of the subsequent lenses and inserting apertures (square symbols) in the proper positions. In energy-filtered imaging, an aperture, the contrast aperture, is inserted in the first image of the diffraction pattern in the centre of the field lens (FL). The subsequent lenses are excited in such a way that the image plane is transferred to the centre of the energy filter (HP). Simultaneously the diffraction pattern is imaged into the dispersive plane of the filter where the energy selection slit selects the desired energy for final imaging with the projector lenses. In order to go from imaging to diffraction, the contrast aperture is removed from the diffraction pattern, an

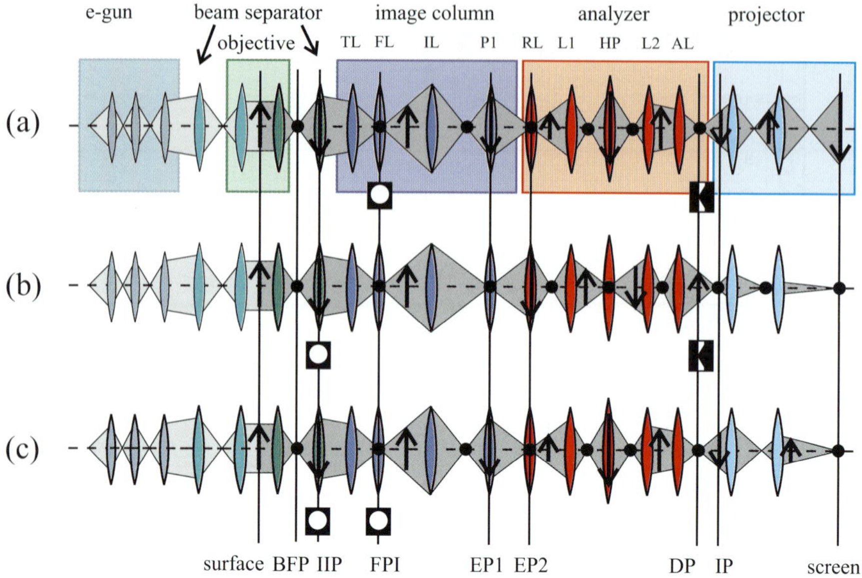

Fig. 8.12. Basic operation modes of the SPELEEM. (**a**) imaging, (**b**) diffraction and (**c**) spectroscopy. At the top the various sections and optical elements are listed, at the bottom the various planes

aperture, the selected area aperture, is introduced into the intermediate image plane (IIP) and the excitation of the intermediate lens (IL) is changed so that the diffraction pattern is imaged into the centre of the filter. The specimen is now imaged into the dispersive plane where energy selection again takes place, followed by imaging of the diffraction pattern with the projector system on the final screen.

While in imaging only angular acceptance and in diffraction only area acceptance have to be limited by apertures for good results, spectroscopy requires both limitations as shown in (c). Again, as in (a), the diffraction pattern is imaged into the dispersive plane but now no energy is selected; unlike (a) it is not the image plane of the filter but the dispersive plane that is imaged by the projector onto the final screen. As seen in the figure, the contrast and selective-area aperture act simultaneously as virtual entrance apertures of the energy filter. They can be chosen to be so small that diffraction patterns and spectra of areas less than 1 μm in diameter can be obtained. The instrument is operated at 18 kV with a 10:1 retardation in the energy filter, which allows an energy resolution of better than 0.5 eV. The spatial resolution of this instrument is not worse than that of a pure LEEM instrument in spite of its greater complexity ($\delta = 10\,\text{nm}$). Unlike LEEM, where the signal-to-noise ratio is usually no problem and images can be acquired in real time on video tape, imaging, diffraction and spectroscopy with photo electrons in general require signal integration for several to many seconds,

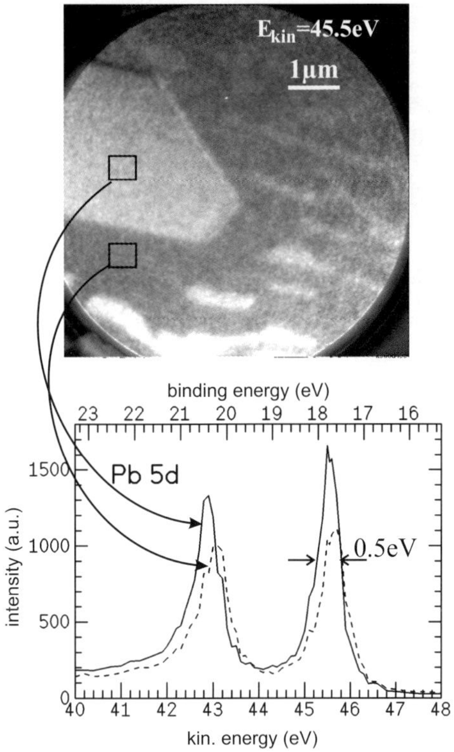

Fig. 8.13. XPEEM image taken with 45.5 eV photo electrons (*top*) and selected-area microspectra of a Pb layer on a W(110) surface (*bottom*). Photon energy 65 eV

depending upon the signal available. This is done with a Peltier-cooled CCD camera and a frame grabber, which also allows convenient image processing.

The following three figures illustrate the three operation modes using thin Pb layers as examples. Figure 8.13 [35] shows a XPEEM image taken with 45.5 eV photo electrons excited with 65 eV photons (above). This image is one of about 40 images which were taken in electron energy steps of 0.1–0.2 eV. The bright regions are thick Pb crystals; in the darker regions only one monolayer of Pb is present on the W(110) surface on which the Pb film was grown. In all the images the intensity in the 0.25 μm^2 areas was integrated and plotted in the lower part of the figure for local spectroscopy. In spite of the 0.5 eV half-width of the Pb 5d lines, the chemical shift of about 0.15 eV between thick Pb and the Pb monolayer is clearly measurable. The spatial resolution in this particular image is only 70 nm but in other cases a 15%/85% edge resolution of better than 25 nm has been measured. When a precise 1:1 correlation of spectrum and image is not needed, then it is much faster to use the spectroscopy mode. This has been chosen in Fig. 8.14,

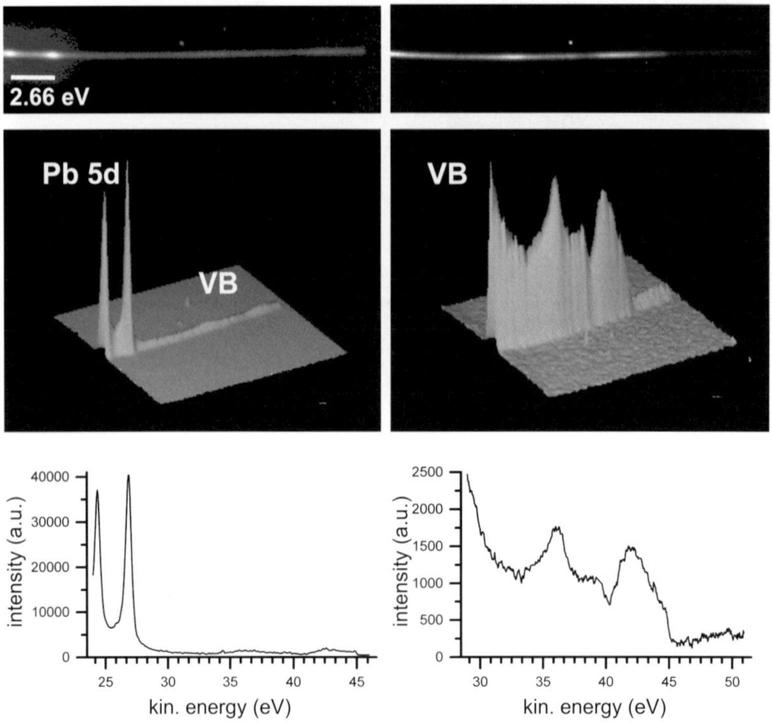

Fig. 8.14. Photo electron spectrum of an 8 monolayer thick Pb film on a Si(111) surface pre-covered with a Ag submonolayer. Photon energy 48 eV

which shows the spectrum of a Pb layer on a Si(111) surface covered with less than one monolayer of Ag. The two spectra from the same specimen were obtained with somewhat different instrument settings in order to bring out the desired features: the overall spectrum from 24 eV to 46 eV with the Pb 5d peaks (left) and the valence band region (right). The top shows the spectra as acquired with the CCD camera, the centre the 3d intensity plots and the bottom the averaged intensity obtained by integrating the intensity perpendicular to the energy dispersion. The photon energy was 48 eV. In this "microspectroscopy" mode 500 data from an area of 0.8 μm^2 were taken in only 60 sec. In the "spectromicroscopy" mode used in Fig. 8.13 the acquisition of a spectrum with comparable quality would have taken at least one hour. This shows the importance of combining the various modes.

Finally, Fig. 8.15 illustrates the diffraction mode. Both patterns are from an area of 1 μm diameter of the same Pb layer from which the spectra of Fig. 8.14 were taken. The picture on the left is the photo electron diffraction (PED) pattern of the 5d electrons with 25 eV kinetic energy, but the picture on the right is actually not a diffraction pattern. It is rather a map of the $\boldsymbol{k}_{\|}$ dependence of the transition probabilities of the valence electrons with

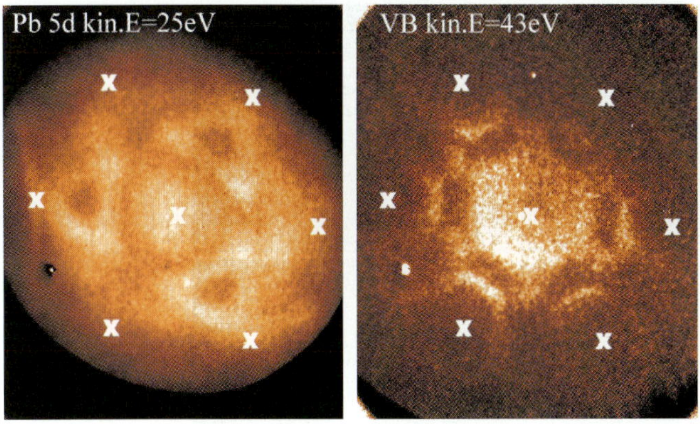

Fig. 8.15. Photo electron diffraction pattern (*left*) and "valence band map" of the same Pb layer as in Fig. 8.14, obtained with the same photon energy

about 2 eV binding energy (Fig. 8.14, right) to the final states accessible with the photons used (energy, angle of incidence, polarization). These valence electron patterns vary rapidly with energy and should allow an efficient local electronic structure analysis to be made, while the PED pattern gives information about the neighbourhood of the emitting atom. The white crosses mark the positions of the LEED spots and provide an accurate angular coordinate system. The PED pattern is also important for the understanding of the XPEEM contrast because the contrast aperture selects only a small fraction of the pattern. Photo electron diffraction can actually produce more intensity along the optic axis from a monolayer than from a thick crystal and hence cause contrast inversion. Thus the diffraction mode opens a wide field of microscopic analysis. The combination of all these emission modes with the reflection modes makes SPELEEM a unique method for the study of surfaces and thin films.

8.5 Some Applications of SPELEEM

There are many problems in the field of microcharacterization of materials that do not require chemical characterization. An example is nucleation and growth of homogeneous materials ("homo-epitaxy"). Only the topography is needed here and this is best obtained via interference contrast as illustrated in Fig. 8.16, which captures the spiral growth and two-dimensional nucleation and growth of Pb on Pb(111) at various times during the deposition process via step contrast [36]. This is also true in heteroepitaxy after it has been established by spectroscopy that no material exchange between film and substrate occurs [37].

Another example is surface diffusion. Surface diffusion is always connected with concentration gradients, which in turn are connected in general with

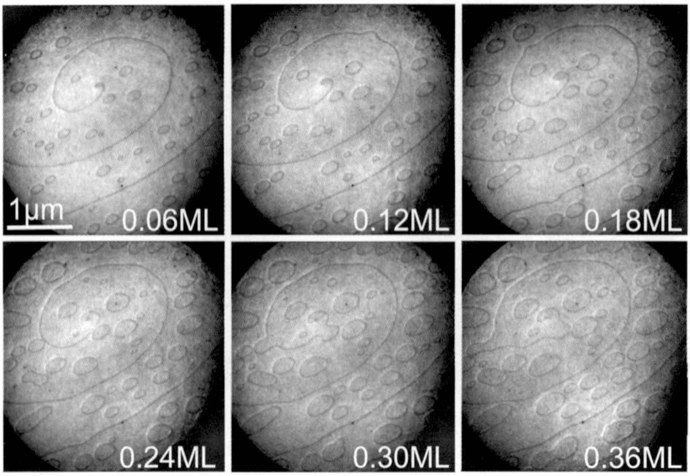

Fig. 8.16. Nucleation and growth of Pb on a Pb(111) surface at room temperature. LEEM images taken with 8.6 eV electrons during growth

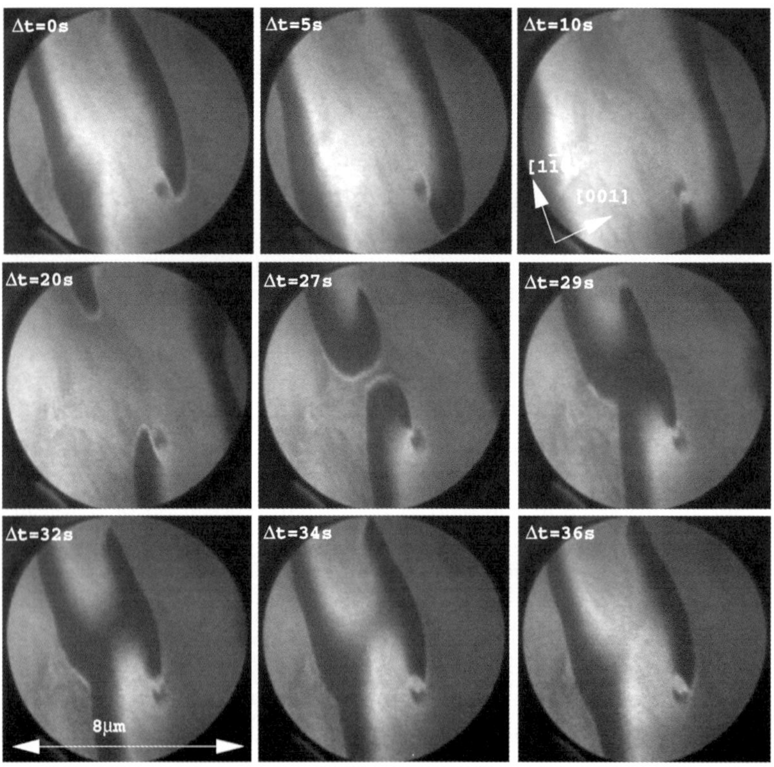

Fig. 8.17. LEEM images taken with 61.2 eV electrons during the reaction between NO and H_2 on a Rh(110) surface. p(NO) $= 4 \times 10^{-5}$ Pa, p(H_2) $= 2.5 \times 10^{-5}$ Pa

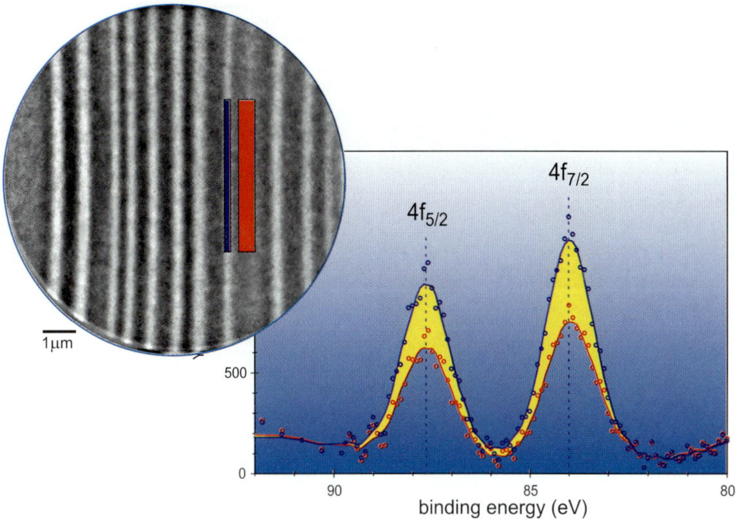

Fig. 8.18. XPEEM image and local PES spectra of a 0.6 ML Au-covered Si(100) vicinal surface. The spectra were obtained by integration over the areas marked in the image from images taken at many energies

work function changes and in many cases – when ordered structures are formed at certain coverages – with diffraction contrast as well. Surface diffusion can hence be studied with LEEM when high resolution is needed or with UV-excited PEEM making use of work-function contrast. This was demonstrated in the early LEEM/PEEM work on the surface diffusion of Au on Si(111) [19]. Similarly, the study of chemical reactions on surfaces does not necessarily require spectroscopy because the local chemical composition of the surface is frequently connected with a well-defined order giving rise to diffraction contrast. An example is provided by the reaction fronts in the reaction between NO and H_2 on the Rh(110) surface shown in Fig. 8.17 [38], which can move very rapidly, depending upon the pressure of the gases and the surface temperature. The long image acquisition times needed for O and N photo electrons will probably preclude such studies with XPEEM even with the next generation of optimized undulators. LEEM with its high intensity is here the ideal imaging method.

There are, however, many characterization problems in which the chemical composition or concentration cannot be deduced from the LEED pattern or the work function. One phenomenon that illustrates this is the Au-induced faceting of vicinal Si(100) surfaces [39]. If Au is deposited at about 850°C onto such a surface with a miscut of about 4° along the [110] direction, then the surface breaks up into very long reconstructed (100) terraces separated by (119) facets. The kinetics of faceting can be studied very well with LEEM while XPEEM allows the determination of the local Au coverage on the ter-

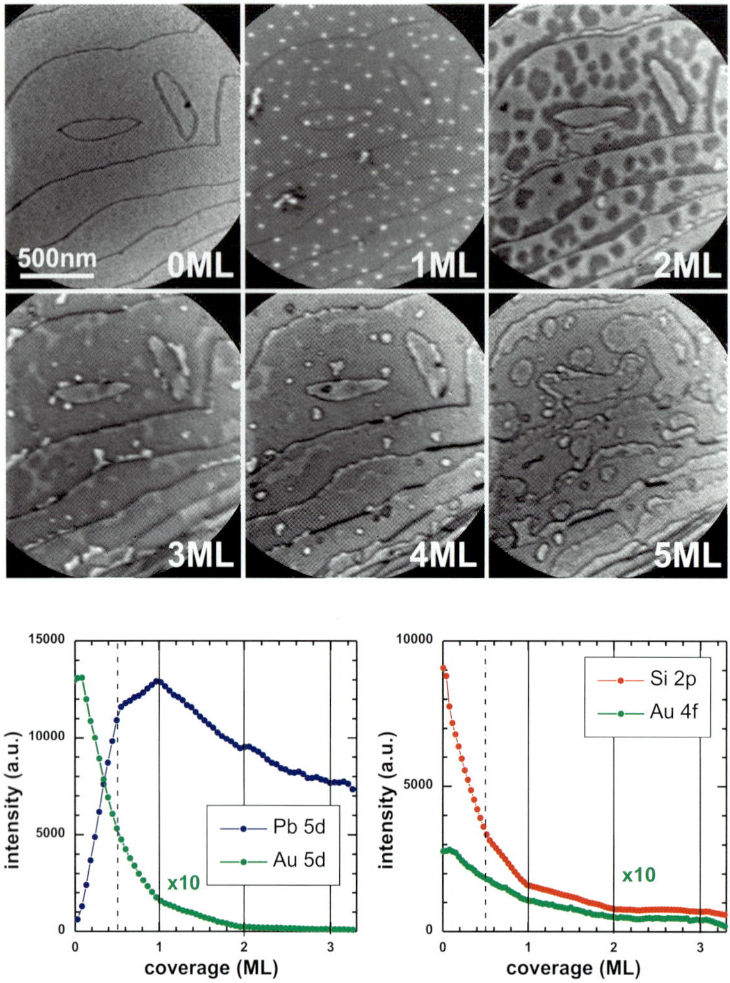

Fig. 8.19. Au-interfactant-mediated growth of Pb on a Si(111) surface. *Top*: LEEM images taken with 8 eV electrons, *bottom*: Pb 5d and Au 5d (*left*) and the Au 4f and Si 2p (*right*) photo electron signal as function of Pb coverage, $h\nu = 72\,\text{eV}$ (Pb 5d, Au 5d) and $h\nu = 133\,\text{eV}$ (Si 2p, Au 4f)

races and facets. The results shown in Fig. 8.18 were obtained after terminating the Au deposition at 0.6 monolayers and cooling immediately [40]. The Au 4f photo electron signal difference between the terraces (bright, blue line) and the facets (dark, red line) is about 40%. It appears unlikely that emission angle and photo electron diffraction differences can cause such a large difference so that the signal difference is attributed to coverage differences.

In order to control the morphology of heteroepitaxial films, so-called surfactants and interfactants are being used increasingly. Ideal surfactants "swim" completely on the surface of the growing film, ideal interfactants remain completely at the interface. Whether or not this is the case can be decided by XPEEM. A well-studied model system for an interfactant is Pb on a Si(111) surface with Au and Ag as interfactant. In Fig. 8.19 [37] a submonolayer of Au deposited at 850 K on a Si(111)-7 × 7 surface to produce a $(\sqrt{3} \times \sqrt{3})$-R30° reconstruction was used as an interfactant for the growth of a Pb layer. The interfactant transforms the three-dimensional growth into a layer-by-layer growth in which the two-dimensional periodic lattice modulation (lattice constant $\lambda = 34$ Å) is clearly visible; it appears as dark areas in the 2–4 ML images of Fig. 8.19. The substrate step structure is reproduced as a result of the perfection of the growth. The Au 4f and Si 2p photo electron signals decrease as expected but the Pb 5d signal decreases after the initial increase up to one monolayer despite the monolayer by monolayer growth. The photo electron diffraction pattern shows an overall decrease of the intensity, and not only in the forward direction, so that the decrease of the Pb 5d signal cannot be attributed to diffraction changes with increasing thickness. This shows that the physics of the processes involved remains imperfectly understood.

A similar phenomenon has been observed in the study of surfactants. The system in this case was Fe on a W(100) surface with Pb and Ag as interfactants. Understanding the place exchanges occurring in ferromagnetic films in the presence of materials such as Pb, Cu, Ag or Au is important for the understanding of their magnetic properties. In particular Pb is a popular surfactant. Iron grows on W(100) initially in two pseudomorphic monolay-

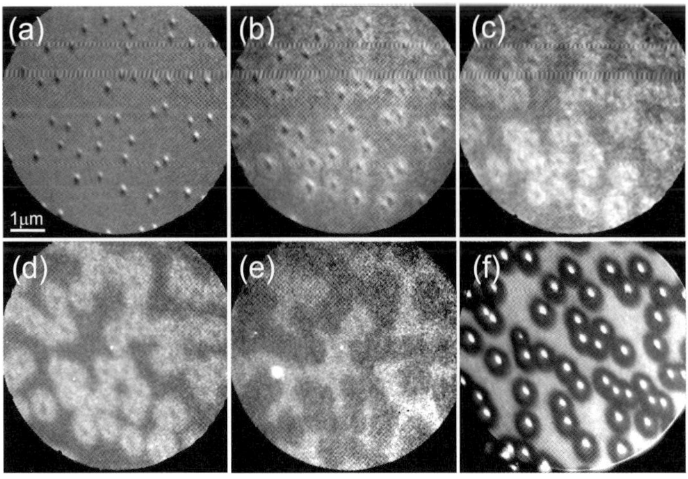

Fig. 8.20. Spreading of Pb during the growth of Fe on a Pb-covered W(100) surface. XPEEM images taken with the Pb 5d electrons during growth (**a**) 0 ML, (**b**) 5 ML, (**c**) 10 ML, (**d**) 13 ML), with Fe 3d electrons at 1 eV binding energy (**e**) and with LEEM after the deposition (**f**). The surface region in the LEEM image is shifted

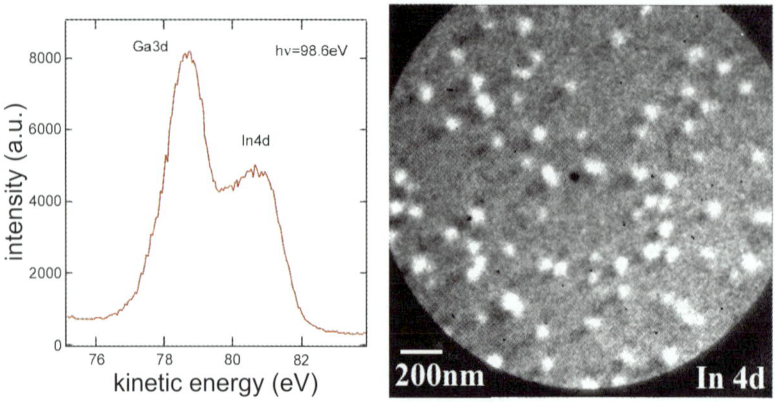

Fig. 8.21. InAs nanocrystals on a GaAs(100) surface. *Right*: XPEEM image taken with In 4d electrons; *left*: photo electron spectrum of this region

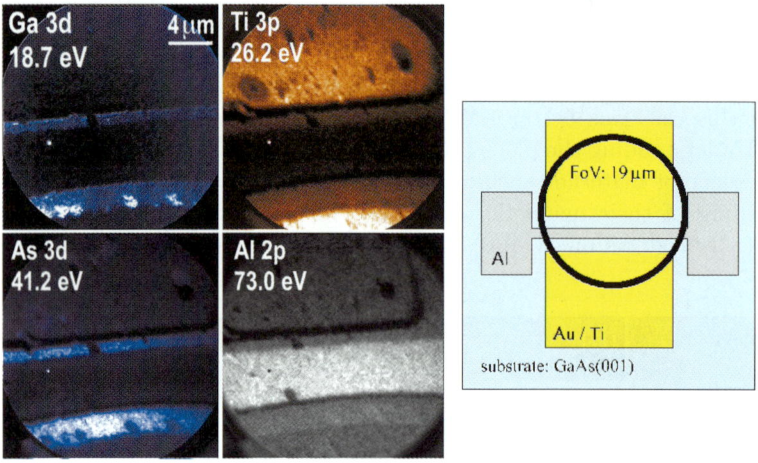

Fig. 8.22. XPEEM images of a field-effect transistor taken with Ga 3d, Ti 3p, As 3d and Al 2p photo electrons. The diagram on the right shows the lay-out and the field of view of the images

ers followed by the growth of three-dimensional crystals. With the transition from two-dimensional to three-dimensional growth, the Fe 3d signal decreases and this can be used as a fingerprint of the growth mode. At room temperature, Ag does not act as a surfactant as can be seen in XPEEM from the disappearance of the Ag 4d signal of the Ag decoration of the substrate steps. When the Fe film is grown at 480 K, however, the decoration pattern "swims" on the surface and spreads somewhat laterally, and the Fe signal continues to increase after deposition of the second monolayer. Lead already acts as a surfactant at room temperature. The lateral spreading of the surfactant can be seen very nicely when Pb in excess of one monolayer is deposited. The excess forms three-dimensional crystals, which act as Pb sources during the growth of the Fe film (Fig. 8.20 [41]). The lower part of this figure shows not only the Pb 5d image but also the complementary Fe 3d image and the LEEM image at the end of the Fe deposition.

The last two examples of the application of SPELEEM give a hint of what the method can contribute to the semiconductor industry. Figure 8.21 [42] shows on the right the image of InAs nanocrystals on a GaAs(100) surface taken with the In 4d part of the photo electron spectrum shown on the left with an energy resolution of 0.5 eV. SPELEEM is also useful for defect analysis as is seen from the images of a field effect transistor of which Fig. 8.22 [43] is a general view. The defects seen in the images taken with the various characteristic photo electrons can be analysed with selected area microspectroscopy.

8.6 Concluding Remarks and Outlook

The goal of this contribution was to give a general picture of a the microanalysis field that the support of the Volkswagen Foundation has opened up. Whenever the possibilities of SPELEEM were presented in public, whether it was the unequalled combination of spatial and energy resolution of submicron areas, the possibility of performing photo electron diffraction and electronic structure analysis of submicron areas or the high image acquisition rate relative to other methods, they excited not just great interest but frequently enthusiasm. With undulator beamlines optimized specifically for spectromicroscopy, exciting possibilities for surface and thin-film analysis in the 10 nm range and below will be opened up.

What is the future of SPELEEM? The SPELEEM studies with the Volkswagen Foundation instrument at ELETTRA terminated in summer 1999 because the instrument had to be returned to the TU Clausthal where it was developed. However, for the beginning of the new millenium the future for SPELEEM looks promising. ELETTRA is replacing the first SPELEEM instrument with a second generation commercial instrument on a new very bright undulator beamline dedicated to spectromicroscopy. The SMART instrument, which grew out of the SPELEEM instrument in a large collaboration, should slowly become operational. It has much better starting conditions than the SPELEEM instrument: immediate permanent access to a high-brightness undulator source, much larger human resources available and, most important of all, all the experience gained with the SPELEEM instrument. Similar instruments are planned at other synchrotron radiation facilities so that the seed sown by the Volkswagen Foundation will certainly bear many fruits.

Acknowledgements

The contribution should end with thanks to the Volkswagen Foundation. Its generous support of the project has not only opened up a new era in surface science but has given those involved in it great scientific and technical challenges, excitement and satisfaction, in particular to the designer of the basic instrument, Lee Veneklasen. Only two more names out of many who have contributed to the success should be mentioned: Wolfgang Telieps and Herbert Steinhardt. Wolfgang Telieps started the project with great zeal together with the first author, Herbert Steinhardt accompanied it for the Volkswagen Foundation with great understanding and patience. Both of them died an untimely death. This contribution is dedicated to their memory.

Finally, the first author thanks once more the Volkswagen Foundation for giving him the opportunity to realize his dreams and the second author thanks the Bundesministerium für Bildung und Forschung (BMBF, project no. 05 SL8 WW8-1) and the Training and Mobility of Researcher (TMR) program of the European Community for the grant no. ERB FMBI-CT96-1749, which enabled him to bring SPELEEM to its present state after the end of the Volkswagen Foundation project. Both authors thanks the staff at BESSY I and, in particular, that of ELETTRA where most of the synchrotron radiation work was done, for technical support.

References

1. Bauer E. (1994) Surface electron microscopy: The first thirty years. Surf Sci 299/300:102–115
2. Bauer E. (1998) LEEM basics. Surf Rev Lett 5:1275–1286
3. Bauer E. (1994) Low energy electron microscopy. Rep Prog Phys 57:895–938
4. Herlt H.-J. (1982) Elastische Rückstreuung sehr langsamer Elektronen an reinen und an gasbedeckten Wolfram-Einkristalloberflächen. Ph.D. Thesis, Technical University Clausthal, Clausthal, Germany
5. Feibelman P.J., Eastman D.E. (1974) Photoemission spectroscopy – correspondence between quantum theory and experimental phenomenology. Phys Rev B 10:4932–4947
6. Bauer E. (1969) Inelastic scattering of slow electrons in solids. Z Physik 224:19–44
7. Bauer E. (1970) Interaction of slow electrons with surfaces. J Vac Sci Technol 7: 3–12
8. Tanuma S., Powell C.J., Penn D.R. (1990) Material dependence of electron inelastic mean free paths at low energies. J Vac Sci Technol A 8: 2213–2216; (1990) Electron inelastic mean free paths in solids at low energies, part I. J Electr Spectr Rel Phen 52:285–291; (1991) Part II. Surf Interface Anal 17:911–926; (1991) Part III. Surf Interface Anal 17:927–939; (1994) Part V. Surf Interface Anal 21:165–176
9. Siegmann H.C. (1994) Surface and 2D magnetism with spin polarized cascade electrons. Surf Sci 307–309: 1076–1086

10. Gröbli J.C., Oberli D., Meier F. (1995) Crucial tests of spin filtering. Phys Rev B 52:R13095–R13097
11. Passek F., Donath M., Ertl K. (1996) Spin-dependent electron attenuation lengths and influence on spectroscopic data. J Magn Magn Mater 159:103–108
12. Frank J. (1973) The envelope of electron microscopic transfer functions for partially coherent illumination. Optik 38:519–536
13. Müller T. (1995) Bildentstehung im LEEM. M.S. Thesis, Technical University Clausthal, Clausthal, Germany
14. Veneklasen L.H. (1991) Design of a spectroscopic low energy electron microscope. Ultramicroscopy 36:76–90
15. Bauer E., Franz T., Koziol C., Lilienkamp G., Schmidt T. (1997) Recent Advances in LEEM/PEEM for structural and chemical Analysis. In: Rosei R. (Ed.) Chemical, Structural and Electronic Analysis of Heterogeneous Surfaces on Nanometer Scale. Kluwer, Dordrecht, 85–91
16. Chmelik J., Veneklasen L., Marx G. (1989) Comparing cathode lens configurations for low energy electron microscopy. Optik 83:155–160
17. Mankos M., Kolarik V., Veneklasen L.H. (1990) Electron-optical properties of multiple magnetic prism systems. Nucl Instrum Meth Phys Res A 298:189–198
18. Kolarik V., Mankos M., Veneklasen L.H. (1991) Close-packed prism arrays for electron microscopy. Optik 87:1–12
19. Telieps W., Bauer E. (1985) An analytical reflection and emission UHV surface electron microscope. Ultramicroscopy 17:57–66
20. Tromp R.M., Mankos M., Reuter M.C., Ellis A.W., Copel M. (1998) A new low energy electron microscope. Surf Rev Lett 5:1189–1197
21. Rose H., Preikszas D. (1992) Outline of a versatile corrected LEEM. Optik 92:31–44
22. Müller H., Preikszas D., Rose H. (1999) A beam separator with small aberrations. J Electron Microscopy 48:191–204
23. Fink R., Weiß M.R., Umbach E., Preikszas D., Rose H., Spehr R., Hartel P., Engel W., Degenhardt R., Kuhlenbeck H., Wichtendahl R., Erlebach W., Ihmann K., Schlögl R., Freund H.-J., Bradshaw A.M., Lilienkamp G., Schmidt Th., Bauer E., Benner G. (1997) SMART – a planned ultrahigh-resolution spectro-microscope for BESSY II. J Electr Spectr Rel Phen 84:231–250
24. Rempfer G.F., Mauck M.S. (1992) Correction of chromatic aberration with an electron mirror. Optik 92:3–8
25. Rose H., Preikszas D. (1995) Time dependent pertubation formalism for calculating the aberrations of systems with large ray gradients. Nucl Instrum Meth Phys Res A 363:301–315
26. Preikszas D., Rose H. (1997) Correction properties of electron mirrors. J Electron Microscopy 1:1–9
27. Rose H., Krahl D. (1995) Electron Optics in imaging Energy Filters. In: Reimer L. (Ed.) Energy-Filtering Transmission Electron Microscopy. Springer, Berlin Heidelberg, 43–149
28. Bauer E. (1991) The possibilities of analytical methods in photoemission and low energy electron microscopy. Ultramicroscopy 36:52–62
29. Duden T., Bauer E. (1998) Spin-polarized low energy electron microscopy of ferromagnetic layers. J Electron Microscopy 47:379–385; (1998) Spin-polarized low energy electron microscopy. Surf Rev Lett 5:1213–1220

30. Bauer E., Duden T., Pinkvos H., Poppa H., Wurm K. (1996) LEEM studies of the microstructure and magnetic domain structure of ultrathin films. J Magn Magn Mater 156:1–6
31. Duden T., Bauer E. (1997) Magnetic domain structure and spin reorientation transition in the system Co/Au/Co. MRS Symp Proc 475: 283–288; (1999) Exchange coupling in Co/Cu/Co sandwiches studied by spin-polarized low energy electron microscopy. J Magn Magn Mater 191:301–312; (1999) Influence of Au and Cu overlayers on the magnetic structure of Co films on W(110). Phys Rev B 59:468–473; (1999) Biquadratic exchange in ferromagnetic/nonferromagnetic sandwiches: A spin-polarized low-energy electron microscopy study. Phys Rev B 59:474–479
32. Bauer E. (1978) Augerelektronen-Spektroskopie und Mikroskopie. Leopoldina-Symp. "Physik und Chemie der Kristalloberfläche", Halle, DDR, unpublished
33. Bauer E., Telieps W. (1988) Emission and Low Energy Electron Reflection Microscopy. In: Howie A., Valdre U. (Eds.) Surface and Interface Characterization by Electron Optical Methods. Plenum, New York, 195–233
34. Stöhr J., Padmore H.A., Anders S., Stammler T., Scheinfein M.R. (1998) Principles of x-ray magnetic dichroism spectromicroscopy. Surf Rev Lett 5:1297–1308
35. Schmidt Th., Heun S., Slezak J., Diaz J., Prince K.C., Lilienkamp G., Bauer E. (1998) SPELEEM: combining LEEM and spectroscopic imaging. Surf Rev Lett 5:1287–1296
36. Schmidt Th., Bauer E. (2000) Interfactant mediated quasi-Frank-van der Merwe Growth of Pb on Si(111). Phys Rev B 62:15815–15825
37. Schmidt Th., Ressel B., Heun S., Prince K.C., Bauer E., to be published
38. Schmidt Th., Schaak A., Günther S., Ressel B., Bauer E., Imbihl R. (2000) In situ imaging of structural changes in a chemical wave with low energy electron microscopy: The system Rh(110)/NO + H_2. Chem Rev Lett 318:549–554
39. Meyer zu Heringdorf F.-J., Kähler D., Horn-von Hoegen M., Schmidt Th., Bauer E., Copel M., Minoda H. (1998) Giant faceting of vicinal Si(001) induced by Au adsorption. Surf Rev Lett 5:1167–1178
40. Meyer zu Heringdorf F.-J., Schmidt Th., Heun S., Hild R., Zahl P., Ressel B., Bauer E., Horn-von Hoegen M., to be published
41. Schmidt Th., Ressel B., Heun S., Prince K.C., Bauer E., to be published
42. Heun S., Watanabe Y., Ressel B., Bottomley D., Schmidt Th., Prince K.C. (2001) Core level photoelectron spectroscopy from individual heteroepitaxial nanocrystals on GaAs(001). Phys Rev B in print
43. Heun S., Schmidt Th., Ressel B., Bauer E., Prince K.C. (1999) Nanospectroscopy at Elettra. Synchrotron Rad News 12:25–29

9 From Microcharacterization to Macroscopic Property: A Pathway Discussed on Metal/Ceramic Composites

J. Rödel

9.1 Introduction

The aforegoing chapters focused on the description of high-resolution imaging and spectrometry. The current status of the most advanced techniques was described and the prospects for the near and far future were presented. While these achievements are spectacular and noteworthy as a fundamental achievement in basic science in its own right, without correlation to any macroscopic property, they may remain a mere academic exercise without apparent consequence.

As we focus on advances in high resolution, we are more and more in danger of loosing ground, that is, the connection to macroscopic properties. As the material volume we utilize for information becomes more and more minute, its relation to a macroscopic property is more and more detached and the degree of statistical relevance requires careful consideration. A benefit in this respect lies in the quest for advanced miniature devices for microtechnology, where there is a chance that the size scale of relevant components becomes smaller as the investigated volume is reduced by our advanced characterisation techniques.

The solution we must seek is to establish a complete link between microcharacterization and macroscopic property. Naturally, this will entail utilization of techniques at various size scales and collaboration with researchers from different disciplines.

In the following, two examples are provided for the relation of advanced microcharacterization to macroscopic property. Certainly, no claim is attempted that these examples are the most suitable to demonstrate a complete connection. Rather, these are examples where the attempt and the inherent difficulty of providing a complete transfer from atomic resolution to macroscopic property can be demonstrated. Also, these are examples where the technical application is evident.

The examples chosen are from the field of structural mechanics. In particular, interfaces between metals and ceramics are considered, as these exhibit a wide variation in type of interfacial structure, composition [1], and interfacial bonding [2]. Also, the observed mechanical properties, which arise owing to the interplay of work of adhesion and metal plasticity, can be affected by orders of magnitudes. Nevertheless, while several theoretical models for crack propagation along interfaces have been advanced [2], a paucity of experimental data limits their validation. The language we choose is a very qualitative one, as this book addresses readers with a primary interest in microcharacterization. Nevertheless, the bases in mechanics and materials are sound and references point the interested reader to more detailed descriptions of the inherent mechanisms involved.

The first example describes microcharacterization and macroscopic properties (interfacial fracture energy) of metal/ceramic laminates with planar interfaces. Here, the transfer between atomistic characterization and component property appears straightforward, since the microstructure of the specimen appears not to come into play. However, even in this simple example, it is not evident a priori, at which size scale of investigation the crucial input for a microscopic property will be fed into the materials system. As it turns out in this case, it is not the atomistic level. Metal/ceramic interfaces are of interest for the adherence of conducting layers on substrates for the semiconductor industry as well as for joining purposes in general.

The second example describes microcharacterization and macroscopic properties of metal/ceramic composites with interpenetrating networks. These are sponge-like microstructures, where a connecting hard phase (the ceramic) is infiltrated with a ductile phase (the metal). The microscopic material property is provided by the adhesion between the metal and the ceramic, the macroscopic properties of the final part of interest are the fracture toughness and the fracture strength. These materials that are used as cutting tools for metals, in brakes for cars, as heat sinks in the microelectronic industry, and are being investigated for a number of further applications, particularly for the automobile industry.

In both examples, the structure and chemistry of the metal/ceramic interface dictate the stress transfer across the interface, thus control the stress distribution in the specimen and fracture resistance of the interfaces. An interfacial resistance between ceramics is difficult to measure and is only affected to a very small degree by effects conducive to microcharacterization. However, the fracture resistance of metal/ceramic interfaces is dictated by the work of adhesion, and is magnified by metal plasticity. Therefore, in the examples chosen, interfacial structure and chemistry can have huge effects on microscopic properties.

Before these examples can be discussed, a brief outline of our current understanding of interfacial decohesion is required.

9.2 Interfacial Decohesion

In the simple case of complete thermodynamic equilibrium in vacuum, interfacial decohesion between two dissimelar materials identifies the interfacial fracture energy, G_{i}, with the work of adhesion, W_{ad}. The work of adhesion can be written as a combination of two surface energies, γ_{m} (surface energy

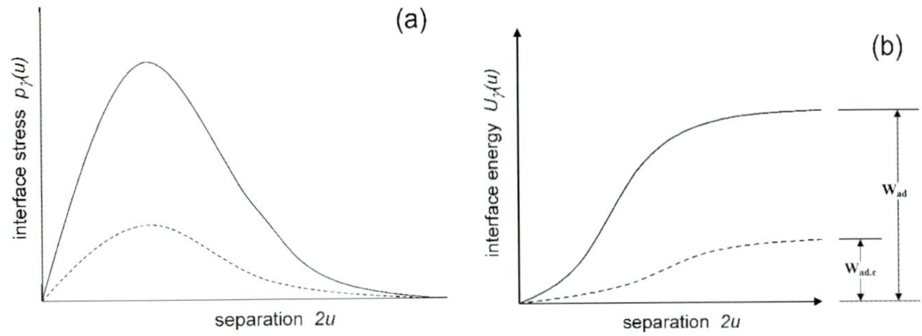

Fig. 9.1. (**a**) Interface stress $p_\gamma(u)$, as a function of separation distance u, for the pure system (*solid line*) and the system containing an additional species (*dashed line*); (**b**) interface energy (energy required to separate both phases up to a certain separation distance) for the pure system (*solid line*) and the system containing additional species (*dashed line*). The modified system may exhibit lower or higher work of adhesion

of the metal) and γ_c (surface energy of the ceramic) and the energy of the interface, γ_{mc} (9.1) [3,4]:

$$G_i = W_{ad} = \gamma_m + \gamma_c - \gamma_{mc} \,. \tag{9.1}$$

Microcharacterization comes into play when the presence of a spurious impurity or voluntary addition of a dopant of a small concentration c is considered. In this case, for thermodynamic equilibrium (that is high temperature and fast diffusion), the interfacial energies are modified according to the Gibbs equation [4,5]:

$$\frac{d\gamma}{d\ln c} = -R_g T \Gamma_c \,. \tag{9.2}$$

Here, Γ_c is the surface excess or the relative adsorption (mol/m^2) of an additional species with concentration c at the particular interface; R_g is the gas constant, and T is the temperature. Adsorption, therefore, always reduces the interfacial energy. The relative changes in the surface energies of the metal and the ceramic, respectively, as well as the change in interfacial energy, determines to what degree the work of adhesion is influenced (lowered or increased) by segregation. Again, it is emphasized, that (9.2) is for thermodynamic equilibrium only, but fracture may occur where the interfacial segregating species is not in thermodynamic equilibrium.

The interfacial fracture energy, G_i, can also be viewed as the energy necessary to pull the materials apart against cohesive forces acting between the two materials surfaces. This will require a certain interface stress, $p_\gamma(u)$, where u is half the crack opening displacement. This interface stress is influenced by the presence of modifying species as well as their concentration [3], Fig. 9.1.

At this juncture, the first complications arise: The thermodynamic quantities formulated in (9.1) and (9.2) are not sufficient to describe interfacial decohesion in real systems. In contrast, non-equilibrium contributions [6] like surface restructuring (particularly in ceramics, to maintain charge neutrality) and acoustic emission enhance the work of adhesion by a factor of two to six in real systems [6]. At the same time, opportunities for high-resolution imaging techniques like scanning probe microscopy (Chap. 7) arise.

Further complication arises from the fact that the interface is rarely planar. Therefore, the actual area of separation is not well defined and the stress state in front of a tilted crack is also different as compared to a crack perpendicular to the loading axes. This effect will also increase the interfacial fracture energy, G_{i}.

The interfacial fracture energy is known to depend on the actual loading geometry (issue III). Specifically, it depends on the ratio of the stress amplitudes associated with either shear stress or tensile stress at the crack tip. This measure is quantified by the ratio of stress intensity factor for shear stresses, K_{II}, and the stress intensity factor for tensile stresses, K_{I}. The phase angle of loading, Ψ, is then described as follows (9.3) [7].

$$\Psi = \tan^{-1}\left(\frac{K_{\mathrm{II}}}{K_{\mathrm{I}}}\right) . \tag{9.3}$$

If decohesion between a metal and a ceramic is considered, the work of adhesion, including irreversible energy terms, roughness, and the phase angle of loading, is augmented by a plastic dissipation mechanism. The stresses at the interface (Fig. 9.1) reach a level, such that plastic yielding in the metal is initiated (Fig. 9.2). The plastic dissipation typically consumes elastically stored energy far in excess of the energy required for bond breakage at the interface. A distinction has to be made with respect to the actual mechanism of decohesion. In the case of ductile failure, the void spacing as determined by microstructural features, influences the fracture energy. For brittle interfacial fracture at sandwich interfaces, the plastic dissipation scales with the thickness of the metal layer. Hence, a determination of the work of adhesion is afforded by extrapolating data for different metal thicknesses down to zero thickness [8]. For the case of Al_2O_3/Au interfaces, the total interfacial frac-

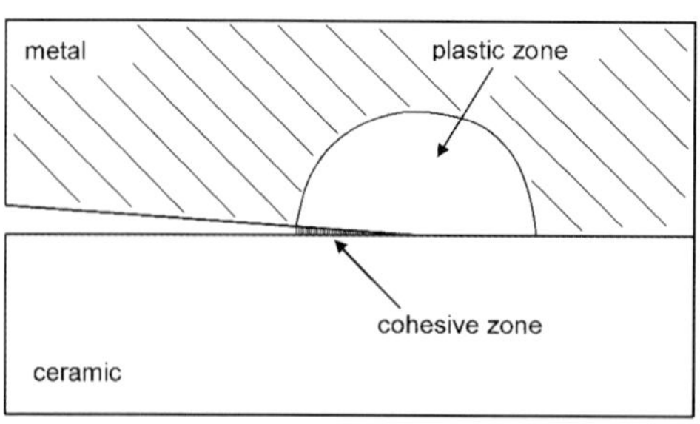

Fig. 9.2. Plastic zone around the cohesive zone in brittle fracture of a metal/ceramic interface magnifies the interfacial fracture energy

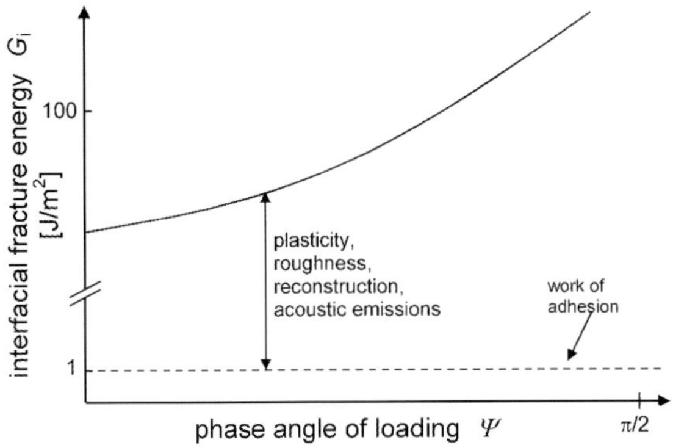

Fig. 9.3. Schematic representation of the dependence of the interfacial fracture energy on the phase angle of loading

ture energy for a laminate with a metal layer of 25 μm thickness was found to be 50 J/m², while the work of adhesion was determined to be 0.5 J/m² [9].

Unfortunately, the correlation between the work of adhesion and the plastic energy, G_p, has not been ascertained in detail. Evans and Dalgleish [8] propose the following equation, which involves the uniaxial yield strength σ_{ys} and the height, h, of the metal layer:

$$G_p = W_{ad}\sqrt{1 + \frac{\sigma_{ys} \cdot h}{W_{ad}}} . \tag{9.4}$$

An exponent n is suggested by Jokl et al. [10] and Elssner et al. [11], with the plastic energy described by [10] (9.5):

$$G_p = kW_{ad}^n , \tag{9.5}$$

with k a prefactor.

If we attempt to include all these further complications, we find that the interfacial fracture energy is higher than the work of adhesion by orders of magnitude and depends on interface geometry, phase angle and non-equilibrium terms (Fig. 9.3).

A final complication arises when one considers the fracture resistance with respect to cracks of finite sizes. The above-mentioned terms are invoked for the case where a short crack propagates for a short distance. In instances where the interface does not rupture along its total area, remaining ligaments may further bridge the crack and lead to R-curve (crack resistance curve) behaviour [3], which describes the dependence of fracture energy on crack length. This behaviour is schematically illustrated in Fig. 9.4. The interfacial fracture energy obtained owing to mechanisms at the crack tip is termed $G_{i,0}$. Conversely, once all the ligaments are broken at a certain distance from the crack tip, this mechanism is exhausted, and the crack length

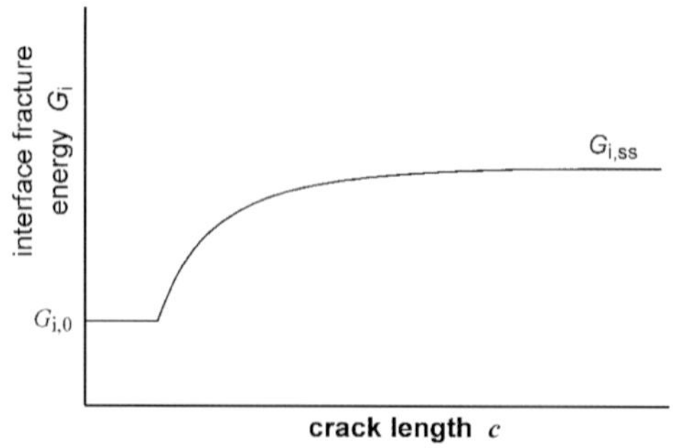

Fig. 9.4. Interfacial fracture energy rises with crack length from tip value, $G_{i,0}$, to steady state value, $G_{i,ss}$

dependent interfacial fracture energy, $G_i(c)$ reaches saturation or steady state (ss) at $G_{i,ss}$.

9.3 Metal/Ceramic Interfaces

9.3.1 Background

If we consider a metal/ceramic interface and contemplate changing the cracking behavior by judiciously adding a dopant, high-resolution imaging and spectrometry can identify the structural and chemical effects of this dopant. Several influences can be envisaged:

1. Influences on the crack resistance of the interface. This may be afforded by either changing the crack resistance in inert atmosphere or by changing subcritical crack growth in corrosive (e.g. humid) environment
   (a) A dopant may alter the work of adhesion at the interface. For the case of complete thermodynamic equilibrium which is rarely attained, this is described according to (9.2) and thereby (9.1).
   (b) A dopant may modify the yield strength of the adjoining metal (for example by precipitate strengthening) and therefore take influence through (9.4) or (9.5).
   (c) A dopant may getter a third element and bind it in an extra phase (e.g. carbon gettering by chromium [13]), thereby negating a detrimental influence of this third element. Several possible modes of taking influence can be envisaged, but mainly the ones ascribed in 1a and 1b are crucial.
   (d) A dopant may enhance dissolution of the ceramic phase, thereby exposing fresh material, which may allow good adhesion.
   (e) A dopant may form a new interphase with constituents of the bilayer and thereby provide a new interface with different work of adhesion.

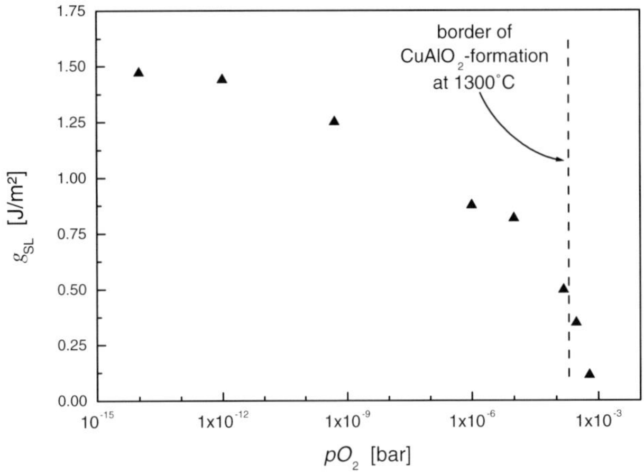

Fig. 9.5. Interfacial energy γ_{sl} at 1300°C is displayed as function of pO_2

2. Influences on the crack driving force. This may be afforded by shielding the crack tip at the interface from the applied stress field.
   (a) A dopant forming a precipitate with high elastic modulus in front of the crack tip will provide load transfer and unloading of the interfacial area in front of the crack tip.
   (b) A dopant forming a weak precipitate can trigger microcracking in front of the crack tip and also crack tip shielding.

In order to highlight some of the concepts described above, the following section presents the results of a model study in an Al_2O_3/Cu laminate. The experimental parameter, which has a dominant influence on the thermodynamics of this system is the oxygen partial pressure, which was varied from $\text{pO}_2 = 10^{-3}$ to $\text{pO}_2 = 10^{-14}$.

9.3.2 Materials and Mechanical Testing

The improved wetting behaviour of the Al_2O_3/Cu system with increasing pO_2 had been ascribed mainly to a change in interfacial energy between the solid and the liquid (Fig. 9.5). This change of interfacial energy was suggested to be due to an adsorption of Cu-O cluster in accordance with (9.2) [14]. For the case that the metal/ceramic interface at room temperature shows remnants of this adsorption, the work of adhesion may be modified and the interfacial fracture energy may be increased. The question of improved adhesion and of segregation of oxygen at the Al_2O_3/Cu interface may be a good test case for high-resolution imaging and spectrometry.

Measurements of the interfacial fracture energy in sandwich type specimens require careful consideration with respect to the loading geometry, the resulting phase angle of loading [7,8], and the projected crack path [15].

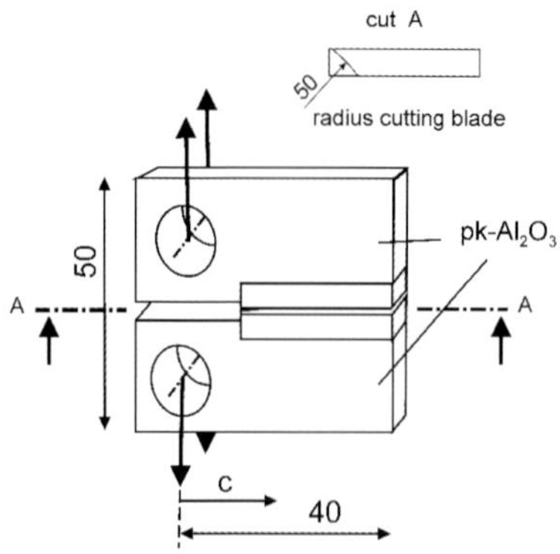

Fig. 9.6. Compact tension specimen with metal/ceramic laminate glued into a polycrystalline ceramic (dimensions in mm)

Since the mismatch of elastic constants leads to a difference between far-field mechanical loading and near-field loading, the phase angle for loading is a function of the materials combination and the type and exact geometry of the specimen used. The phase angle, in turn, influences the interfacial fracture energy. The elastic property difference across the interface finally governs the crack-path selection [15]. As a consequence of the considerations required in the selection of an appropriate fracture mechanical specimen, the manufacturing techniques to produce well-defined interfaces are in variance with the required specimen geometry for mechanical testing. Whereas large compact tension specimens provide a stable crack path [16], interface bonding is preferably accomplished using thin plates. Thin plates, in turn, are easily produced, but lead to test geometries with either unstable crack paths in DCB (double cantilever beam) specimens or problems with cracking perpendicular to the interface in bend bars.

Therefore, a compromise was adopted for the Al_2O_3/Cu system. The interfaces were produced by liquid phase bonding to produce thin bend bars. A solid-state electrolyte cell instrument was employed to set and control pO_2.The bend bars were then glued into a polycristalline alumina holder with a resin to provide a suitable geometry for mechanical testing. Details of the experimental procedure are described elsewhere [17]. Temperature and oxygen partial pressure were chosen such as to provide the specimen with an oxygen content in the metal from 0.032 at.% (corresponding to an equilibrium partial oxygen pressure of 10^{-8} at 1300°C) and 33 at.% (corresponding to the composition Cu_2O).

The external dimensions of the total specimen corresponded to those of a CT specimen according to ASTM E 399 [18]. Utilization of a custom-designed device [19] afforded observation of crack growth either under an optical microscope or in a high-resolution SEM.

9.3.3 Characterisation of Microstructures and Interfaces

The wide variation of oxygen partial pressures during joining affords establishment of a wide variation of oxygen content in the metal, which in turn necessitates microstructural characterization at different length scales, ranging from HRTEM (see Chap. 3) to optical observation.

As we are probing the influence of oxygen on the interfacial fracture energy, chemical analysis with high spatial resolution at the interface is required. Application of parallel electron energy-loss spectrometry (see Chap. 4) provides information on local chemistry as well as local bonding state in direct comparison to the surrounding matrices.

Using conventional TEM, no continuous interphase could be detected at the metal/ceramic interface [20]. Size and location of Cu_2O precipitates depended on the oxygen partial pressure during bonding. In the sample produced at $pO_2 = 10^{-9}$ bar very few precipitates could be found at the interface and none in the bulk metal. The specimen prepared at $pO_2 = 10^{-6}$ bar exhibited no precipitates in the bulk metal, but at the interface. Many precipitates were found in the specimen bonded at $pO_2 = 10^{-3}$ bar (Fig. 9.7). The size of the precipitates in the bulk was in a range from 10 to 30 µm. A similar size of precipitates was also found at the interface, next to small precipitates between 10 and 30 nm. Elemental profiles across the interfaces were acquired for specimens produced at $pO_2 = 10^{-3}$ and 10^{-6} bar in regions were no precipitates were present. These showed that the transition from metal into the ceramic occurred abruptly. In contrast, TEM and PEELS investigations at sapphire/copper interfaces using the rhombohedral sapphire plane and the identical bonding conditions, provide indication for an electron transfer away from the copper. However, it remained unresolved whether this electron transfer is due to oxidation or intermetallic bonding [21].

Light optical microscopy was performed on polished cross sections as well as through the transparent sapphire, where the area fraction of Cu_2O could

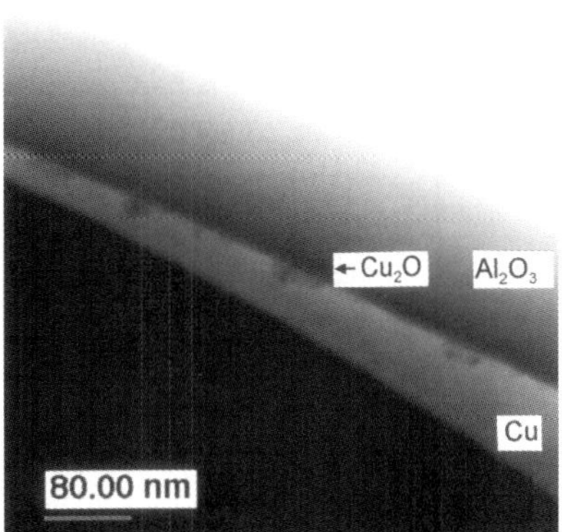

Fig. 9.7. Interface of the sample produced at $pO_2 = 10^{-3}$ bar

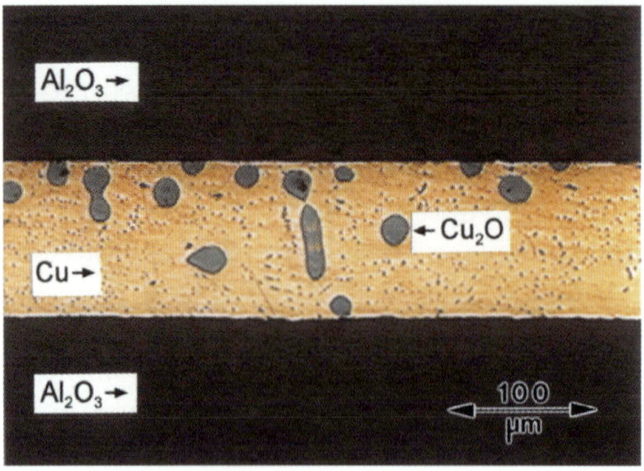

Fig. 9.8. Cross section of an Al_2O_3/Cu interface manufactured at $pO_2 = 10^{-3.5}$ bar at 1400°C and containing 5 at.% oxygen in the copper

be determined [17]. An example of a cross section is provided in Fig. 9.8. Clearly visible are large (10–50 μm) precipitates of Cu_2O, which are located at the interface or in the bulk. Also note, that some precipitates are in the process of pinching off or have just pinched off from another precipitate.

9.3.4 Mechanical Properties

The fracture toughness of the laminates as a function of oxygen content is provided in Fig. 9.9. A value of 1 $MPam^{1/2}$ is obtained for the interfaces containing 0.032 at.% oxygen. With small amounts of oxygen and less than half of the copper layer consisting of Cu_2O the fracture toughness increases strongly up to a peak value of 12 $MPam^{1/2}$. As more and more oxygen is added, more of the brittle Cu_2O is formed during cooling. As a consequence, the fracture toughness decreases again, down to the value of 1.8 $MPam^{1/2}$ obtained for the sapphire/Cu_2O interface.

Although the results from the wetting studies indicated a definite effect on interfacial energies, at least at high temperatures, this trend could not be verified for the mechanical properties of the interface. As no effect of oxygen on the interfacial bonding state could be found, (9.4) and (9.5) are not applicable to explain the increased fracture toughness with increasing oxygen content. Therefore, the answer is sought on a mesoscopic scale with an influence of the crack driving force and not the crack resistance. In order to shed light on the actual crack propagation mechanism, in situ crack growth studies were also performed in HRSEM. Fig. 9.10 demonstrates the influence that Cu_2O precipitates exert on crack direction. Cu_2O has an elastic modulus below that of copper and, therefore, an influence of the precipitate owing to load transfer is not feasible. However, microcracking occurs in precipitates at a distance of 100 to 300 μm in front of the main crack tip and, therefore, alters the stress field and the plastic zone, providing crack tip shielding [17].

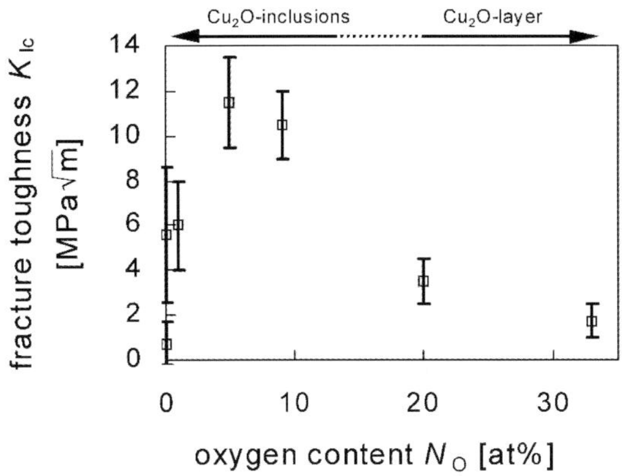

Fig. 9.9. Fracture toughness of sapphire/copper interfaces as function of oxygen content

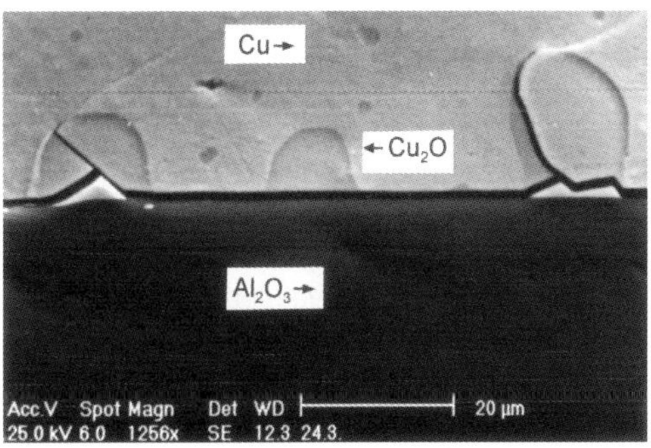

Fig. 9.10. Cracked alumina/copper interface with Cu_2O precipitates demonstrates cracking of the precipitate (crack direction from right to left)

The alumina/copper interface, therefore, provides an example where at high temperature subtle effects of segregation and concomitant changes of interfacial energy are obtained. In contrast, at room temperature, the mesoscopic length scale dominates the mechanical properties of the interface.

9.4 Metal/Ceramic Composites with Interpenetrating Networks

A composite with a complex microstructure as in an interpenetrating microstructure offers even more competing influences if one wishes to track a possible influence of a dopant to an observable macroscopic mechanical effect. The volume content, the average size of the grains as well as the connectivity of each of the phases will have to come into play. In order to separate the influence of interface adhesion from the microstructural influence, processing techniques must be available to alter the interfacial dopant concentration as well as the microstructure over a substantial range. Also, several macroscopic properties need to be obtained. Even then, a deconvolution of interface and microstructural influence without a substantial modeling contribution appears impossible.

In detail, interfacial adhesion as affected by a possible segregant at the interface, which may be detectable by microcharacterization, will affect the following macroscopic properties:

1. The average residual stress in each of the phases since the load transfer across the interface is impeded if the interface is weakened [22].
2. The thermal expansion coefficient of the composite, since a weak interface will exhibit a reduced hysteresis in the thermal expansion behavior [22,23].
3. The toughening effect through the reinforcing phase, since the interfacial fracture energy governs the debond length, which itself affects the closure stress as a function of crack opening displacement [24–26].
4. The strength, since weak interfaces provide sites for crack initiation.

9.4.1 Materials

Three metal/ceramic composites are contrasted in order to highlight the influence of interfacial adhesion on macroscopic material properties. All composites have alumina as the ceramic phase forming a stiff skeleton, with either aluminum [27,28], copper [29] or nickel aluminide [30,31] as the reinforcement phase. The ceramic preforms were prepared using the same route for all materials and therefore afford effective comparison of the influence of all the reinforcing materials. The preparation procedure is described briefly:

Alumina preforms were prepared with porosities varying from 13 to 37% and two different pore diameters of 0.12 and 1 μm. These were then infiltrated with a second phase using a pressure infiltration process which is outlined in detail elsewhere [27]. The resulting material appears to be fully dense with no detectable porosity using displacement techniques or observed during extensive SEM and TEM studies. The three resulting composite materials exhibit various interface properties: The Al_2O_3/Al system is known to exhibit very strong interfaces [27,28]; the Al_2O_3/Cu is described to have an interfacial fracture resistance very much dependent on oxygen activity [17], but a weak interface if no oxygen is present as in our composites prepared by pressure infiltration in a furnace heated by graphite elements. Finally, the Al_2O_3/Ni_3Al exhibits very weak interfaces [31].

In this chapter materials will be described using a nomenclature of a letter followed by a number, where the letter C denotes a coarse microstructure (ligament diameter 1 μm), F a fine microstructure (ligament diameter 0.12 μm), and the number the vol% metal. For example, F24 is a ceramic/metal interpenetrating composite containing 24 vol% of ceramic material.

9.4.2 Characterisation of Microstructures and Interfaces

The microstructures of the interpenetrating network composites are very similar for the three materials considered since the same alumina preforms are infiltrated by a melt. An example of this type of microstructure is provided in Fig. 9.11 as optical micrograph for the case of the Al_2O_3/Ni_3Al. A homogeneous phase distribution can be seen. Furthermore, no remaining porosity could be found.

In the case of the Al_2O_3/Al as well as the Al_2O_3/Ni_3Al [30], the size of the single crystals was determined by taking a series of dark field TEM images. For both these materials the size of the single crystals was found to be substantially larger than the size of the pore channels. Deformation of the reinforcement phase is therefore governed by single crystal behavior. For the Al_2O_3/Al this is considered less relevant, but more pertinent for Al_2O_3/Ni_3Al. Ni_3Al, namely, exhibits pure grain boundary cohesion if tested as undoped polycrystalline material with resulting low fracture strain and fracture toughness [32]. This effect can be negated to a large degree by Boron doping [32]. However, single crystals, albeit showing strong crystallographic anisotropy in fracture strain, exhibit very good plastic deformation, large fracture toughness and high strain to failure [33]. The porous alumina network therefore serves as a constraining framework. The intermittent ligaments of Ni_3Al, which are single crystalline, thus provide most effective deformation and toughening. Figure 9.12, provided as dark-field TEM micrograph, proves that the size of the single crystals is several times the ligament size. This is demonstrated as the (222) diffraction spot for Ni_3Al was chosen for imaging and several grains appear bright in Fig. 9.12. Several representative areas were inspected and gave an average grain size of the intermetallic of 20–30 μm compared to a ligament diameter of 1 to 5 μm.

A lower magnification bright-field image of Al_2O_3/Al shows the microstructure in more detail (Fig. 9.13). No remaining porosity could be detected

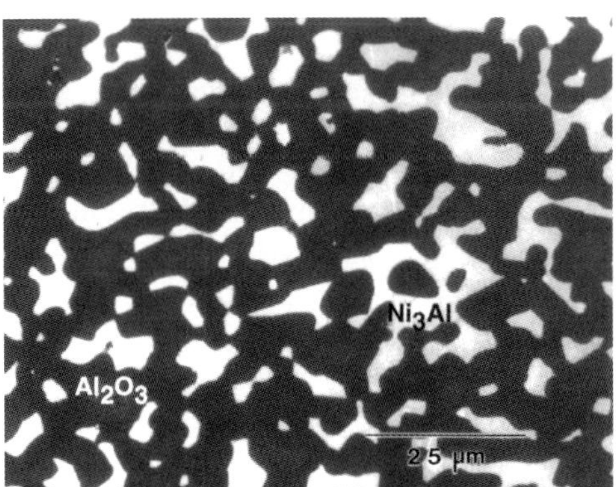

Fig. 9.11. Optical micrograph depicting the microstructure of Al_2O_3/Ni_3Al

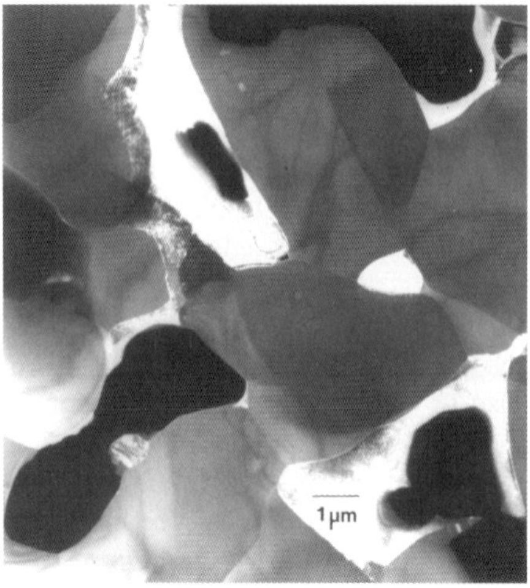

Fig. 9.12. TEM dark field micrograph of the Al_2O_3/Ni_3Al composite. The intermetallic appears bright

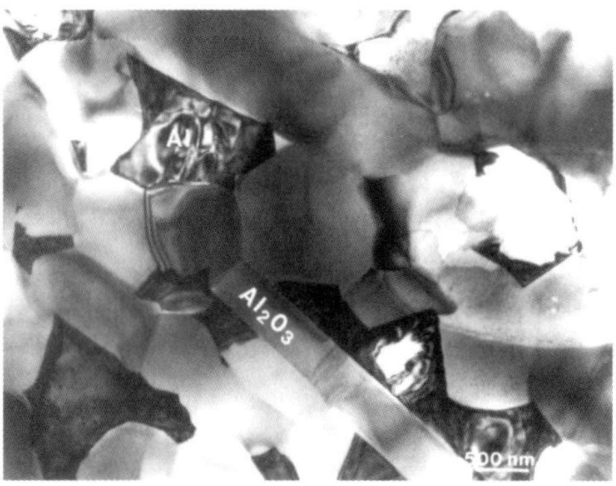

Fig. 9.13. TEM bright field image of Al_2O_3/Al

and in the case of the Al_2O_3/Al no microcracking in any of the phase compositions was found.

The thickness of the metal ligaments has a definite effect on the mechanical macroscopic properties of metal/ceramic composites [27]. On the microscopic level, the length scale can influence microcracking according to (9.6) [34], where D_c is the critical grain size where microcracking occurs, k is a model dependent constant, γ_{gb} is the grain boundary energy, E the elastic modulus, $\Delta\alpha_{max}$ the maximum difference in thermal expansion coefficients and finally, ΔT the temperature over which cooling without stress relaxation took place during processing.

$$D_c = \frac{k\gamma_{gb}}{E(\Delta\alpha_{max}\Delta T)^2} . \qquad (9.6)$$

Fig. 9.14. (**a**) TEM of fine scale and (**b**) coarse scale Al_2O_3/Cu

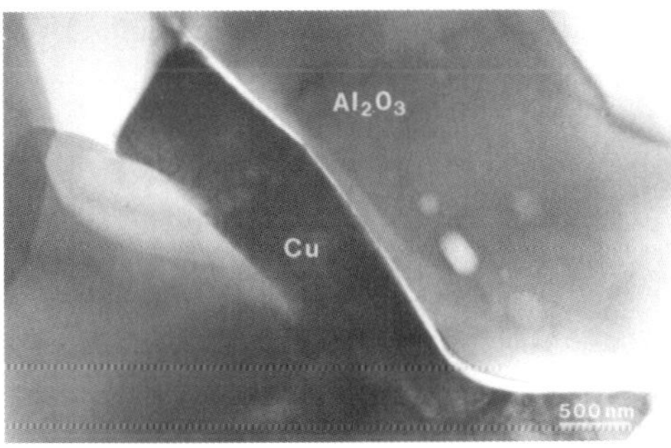

Fig. 9.15. TEM of coarse scale Al_2O_3/Cu, showing microcrack formation

This influence of scale is demonstrated with the Al_2O_3/Cu composite, where no microcracks were found in the fine scale material, but at various sites in the coarse-scale material [29]. This is demonstrated in Fig. 9.14a,b, where the general features of the respective microstructures are displayed.

A large magnification with one microcrack between copper and alumina is provided in the TEM image in Fig. 9.15.

Microcracking, therefore, provides an example, where a small amount of dopant at the interface, which may be accessible to microcharacterization, influences the microcracking criterion through changes in the grain boundary energy as stated in (9.6), but changes in grain size may have a more dramatic influence.

In Fig. 9.14a and b, as well as in Fig. 9.15, no facetting could be found, consistent with earlier work [29]. In contrast, the Al_2O_3/Al composites exhibit facetting at the metal/ceramic interface. The facets are of either prismatic or

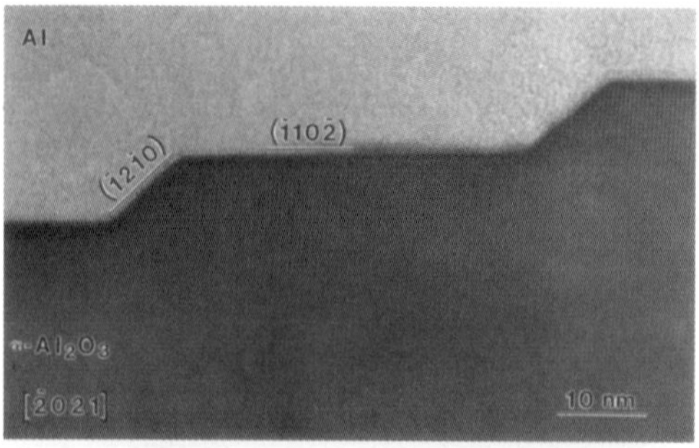

Fig. 9.16. HRTEM of Al_2O_3/Al, showing facetting

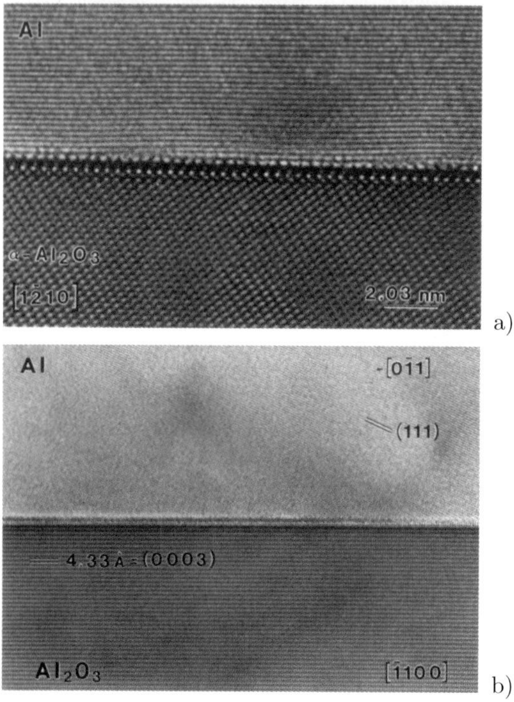

Fig. 9.17a,b. HRTEM showing planar interfaces

rhombohedral character. This effect is demonstrated by Fig. 9.16 and further in Figs. 9.17a and b. It is suggested that facetting occurs where part of the ceramic is dissolved in the metal melt and then reprecipitates in facetted form. This dissolution is an effective means to clean the interface and provide a direct bonding between metal and ceramic. Clean interfaces, in turn, provide stronger adhesion [11]. HRTEM images of Figs. 9.17a and b were both taken at interfaces parallel to the basal plane (0001) of alumina, as this is a plane of low surface energy and found very frequently in this material (e.g. the long planes of the alumina grain marked in Fig. 9.13). Occasionally, lattice planes with weak contrast, could also be imaged in the aluminium. This, however, is not to be mistaken as proof for a fixed orientation relationship between alumina and aluminium. An investigation covering all imaged interfaces showed that no meterrea orientational relationship exists between both phases.

9.4.3 Residual Stresses

A detailed description of the measurement and analysis methods is given elsewhere [22] and will be described only briefly here. Lattice strains in each of the individual phases were determined from shifts in their respective neutron diffraction peaks. Measurements were made in a pulsed neutron reactor and analysed using the so-called "reverse time of flight" method with use of a fast Fourier chopper. Residual stresses were converted into stresses by using Hooke's law. Strain fluctuations were derived from the diffraction peak width and used to provide information on microstresses. The stress free state for the Al_2O_3 and Al phases in each composite sample was obtained by diffraction measurements of the corresponding porous preforms and a sample of pure Al. Samples were machined bars of $5 \times 5 \times 40\,mm^3$.

Residual stresses for different microstructures in Al_2O_3/Al for both phases are provided in Fig. 9.18. Tensile stresses in aluminum are higher for finer ligaments and smaller metal content. The stress level is more than a factor 10 above the yield stress of the corresponding, unconstrained metal as found in a large specimen [22]. This enhancement is due to the high mechanical constraint afforded by the surrounding ceramic, effective as long as a strong interface bonding is maintained. This phenomenon occurs also in thin films [35] and metal matrix composites [36], is qualitatively understood but not accessible to quantitative understanding yet. Although residual stresses were not determined in either the Al_2O_3/Ni_3Al or Al_2O_3/Cu composite, it is surmised that residual stress levels see less of a relative enhancement as

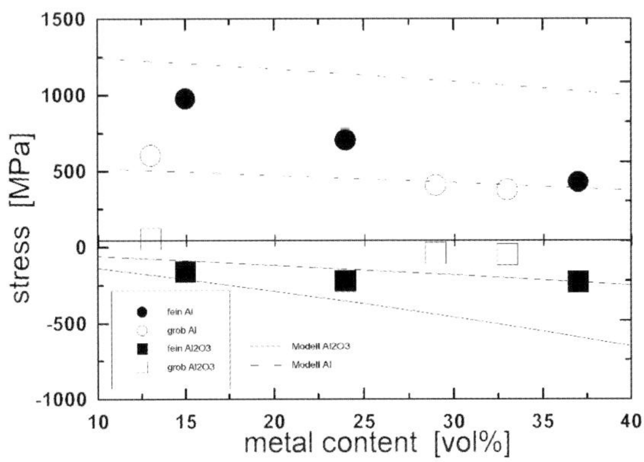

Fig. 9.18. Residual stresses in Al_2O_3/Al for different microstructures

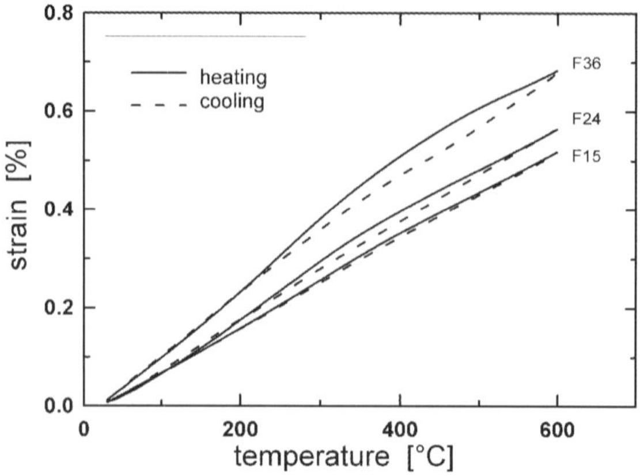

Fig. 9.19. Hysteresis in thermal expansion behavior for Al_2O_3/Al

compared to the yield stress of the unconstrained material in the latter two composites. Weaker metal/ceramic interfaces with attendant microcracking are expected impose to a reduced mechanical constraint on the ductile phase and therefore lead on to a reduced level in residual stress. This issue will further be discussed in Sects. 9.4.4 and 9.4.5.

9.4.4 Thermomechanical Behavior

Thermal expansion behaviour under argon gas is described for the Al_2O_3 /Al [22] and Al_2O_3/Ni_3Al [31] composites. Data for the composites containing aluminium as well as for pure Al_2O_3 and Al were collected over a heating / cooling ramp of 5 K/min between room temperature and 600°C.

A pronounced hysteresis for the Al_2O_3/Al composite is demonstrated in Fig. 9.19. This hysteresis is more pronounced with increasing metal content and increasing ligament diameter. A similar behaviour is also seen in thin films [35] and MMCs [36].

During cooling from the stress-free processing temperature, the aluminium exhibits a considerably higher thermal expansion coefficient than the alumina and is placed under strong hydrostatic tension starting from a so-called viscous temperature (Fig. 9.20). As long as the interface is strong enough to withstand the high stress levels, the stress builds up more and is compensated in the alumina by a compressive stress dictated by mechanical equilibrium [16]. At lower temperature, yielding in the metal sets in, which is simplified as perfectly plastic behaviour in Fig. 9.20. If the material is heated again, linear elastic behaviour sets in, leading to compressive stresses in the metal and yielding in compression. The hysteresis closes at high temperature due to viscous deformation [23].

Using an effective medium analyses (EMA) as described by Kreher and Pompe [37], residual stresses can be computed and compared with residual

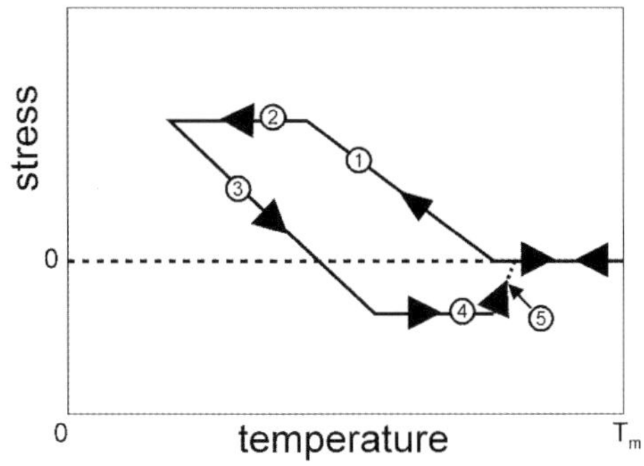

Fig. 9.20. Schematical description of thermal expansion behaviour in metal/ceramic composites with interpenetrating networks

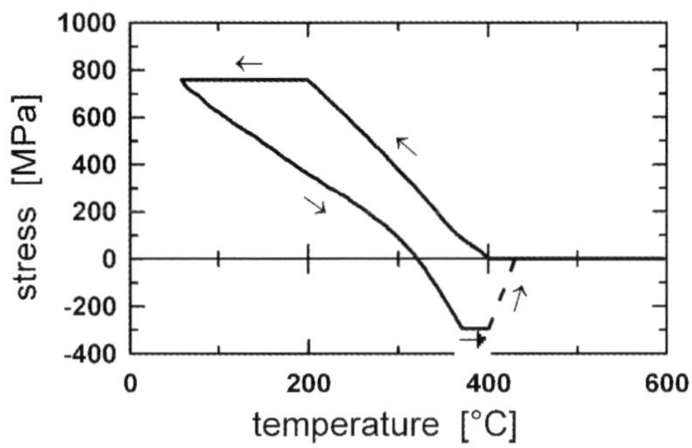

Fig. 9.21. Calculated evolution of residual stresses in Al_2O_3/Al composite F24

stresses as determined via a characterization of lattice strains [23]. As time dependent data can be utilized, temperature dependent residual stresses can be plotted (Fig. 9.21).

A computation based on the effective medium approach demands complete transferral of shear stresses across the metal/ceramic interface. Should this not be guaranteed, this approach cannot be utilized. If any type of hystereses is absent, or very small, this then can also be seen as an indication that the interface does not serve to completely transfer stresses. An example for this behaviour is afforded by the Al_2O_3/Ni_3Al composite as demonstrated in Fig. 9.22. This material is believed to exhibit a weak interface, therefore allows sliding of both phases past each other and thereby providing a stress relaxation mechanism. As residual stresses are not built up, the mechanism as outlined in Fig. 9.20 does not take place and hysteresis is absent in this type of material [31].

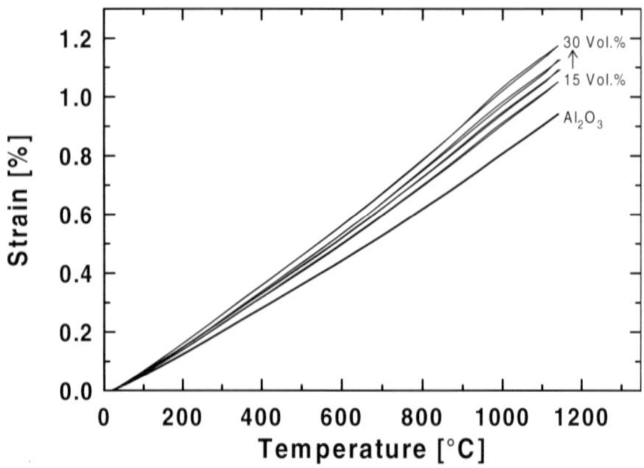

Fig. 9.22. Thermal expansion behavior of Al_2O_3/Ni_3Al

9.4.5 Mechanical Properties

As the atomistic structure and chemistry of an interface will affect interfacial cohesion, it will determine the level of residual stresses and thermomechanical behavior. Macroscopic mechanical behaviour also is affected as long as the creation of microcracking through the existence of residual stresses or by virtue of combination of residual stresses and applied stresses influences macroscopic properties. A simple demonstration of the influence of local interface adhesion on macroscopic mechanical property is provided in contrasting Figs. 9.23 and 9.24. For the system with the strong interface, Young's modulus for a composite can be predicted using standard rules of mixture [38] based on knowledge of Young's modulus for the individual phases. Experimental data for the Al_2O_3/Al composite (Fig. 9.23), therefore, lie in a regime outlined by the theoretical predictions. If the system with the comparatively weak interface, Al_2O_3/Ni_3Al, is considered, the rules of mixture do not apply at all. Therefore, stress transfer across the interface appears not maintained. Rather, microcracking owing to weak interfaces is predicted with composite behaviour as displayed in Fig. 9.24 [31].

Even for cases where residual stresses do not lead to microcracking, interfacial adhesion becomes a crucial parameter once an external load is applied. This becomes an issue for both crack initiation as well as crack propagation. Metal/ceramic composites with interpenetrating network microstructure can be viewed as ceramics exhibiting high toughness with creep properties close to the ceramic [39] and fracture strain similar to ceramics. The primary effect of the metal inclusion is afforded by ductile toughening. This effect is considered in Fig. 9.25.

A crack propagating through the microstructure does not fracture a ductile ligament, but stretches it. The stretched ligament, in turn, provides closure stresses for the crack faces and thereby shields the crack tip from the externally applied load [35]. This shielding effect depends not only upon the

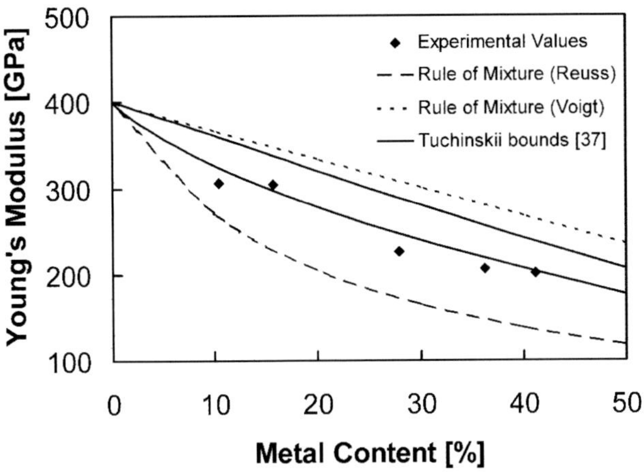

Fig. 9.23. Young's modulus as a function of metal content for the system Al_2O_3/Al

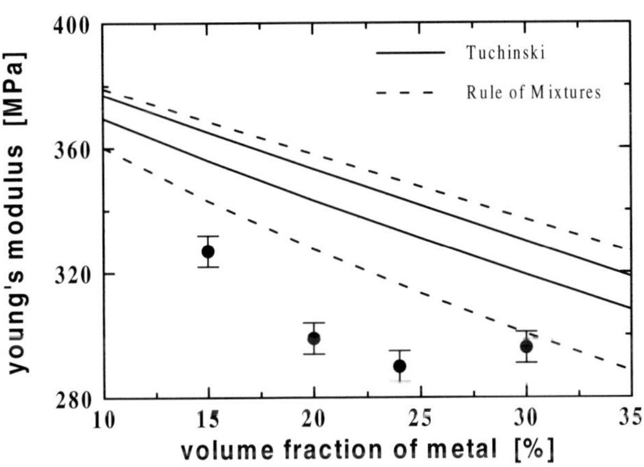

Fig. 9.24. Young's modulus as a function of content of intermetallic for the system Al_2O_3/Ni_3Al

number (volume fraction of metal) and thickness of the ligaments, but also upon the interfacial properties. A high interface strength will lead to small debonding along the interface, high hydrostatic stresses in the ductile reinforcement and, high closure stresses onto the crack faces but early (at small crack opening displacements, thereby small crack lengths) failure of the ductile reinforcement. This behaviour is exemplified by the Al_2O_3/Al with strong interfaces.

In contrast, the Al_2O_3/Ni_3Al composite displays weak interfacial adhesion and strong debonding. Therefore, a larger volume absorbs energy by ductile yielding, but at a smaller closure stress where cracks are still short.

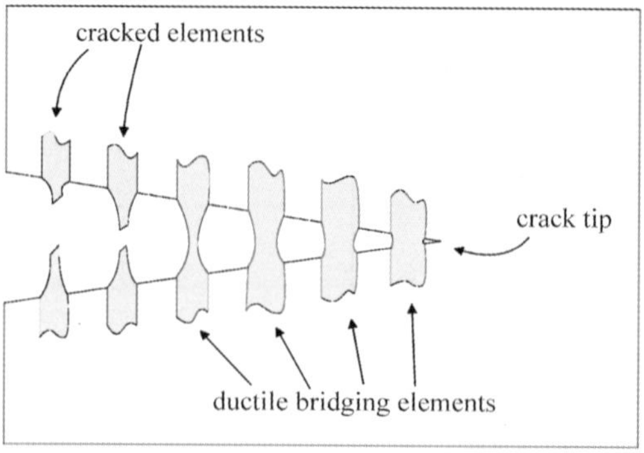

Fig. 9.25. Schematic for crack bridging by ductile ligaments

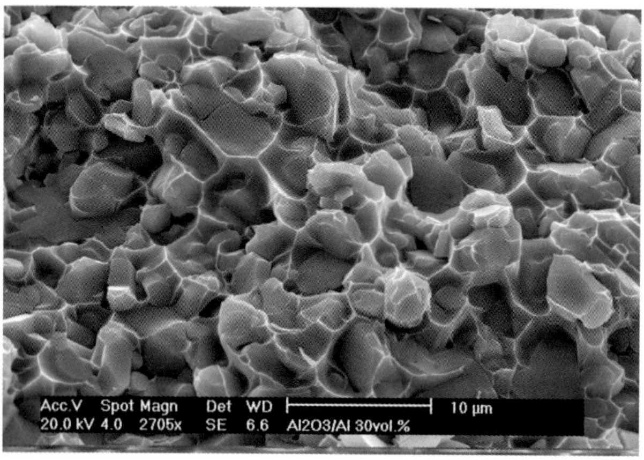

Fig. 9.26. Ductile deformation and limited debonding visualized on the Al_2O_3/Al fracture surface

FEM modeling provides a quantitative understanding for the influence of the debond length on the deformation behaviour and the stress state in ductile reinforcements [26]. Figure 9.28 shows increasing debond lengths L_0 (from top to bottom) and three different load levels at the fiber (from left to right), leading to different levels of fiber extension and volume fraction of voids. $L_0 = 0.3$ represents a strong interface and $L_0 = 0.9$ represents a weak interface. The case of Al_2O_3/Al is thought to be close to the behavior described by $L_0 = 0.3$. A strong interface leads to limited delamination, where there are initially no voids at low stress. As the fiber extends, more voids form in the center of the ligament, thereby reducing the stress in this region. Voiding also occurs close to the interface and final failure occurs along a band from the interface to the area of high void density in the center of the ligament.

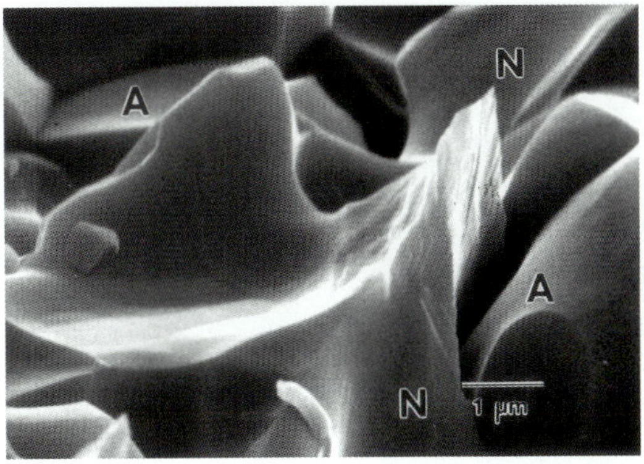

Fig. 9.27. Ductile deformation and debonding in Al_2O_3/Ni_3Al (N describes the Ni_3Al phase, A the Al_2O_3)

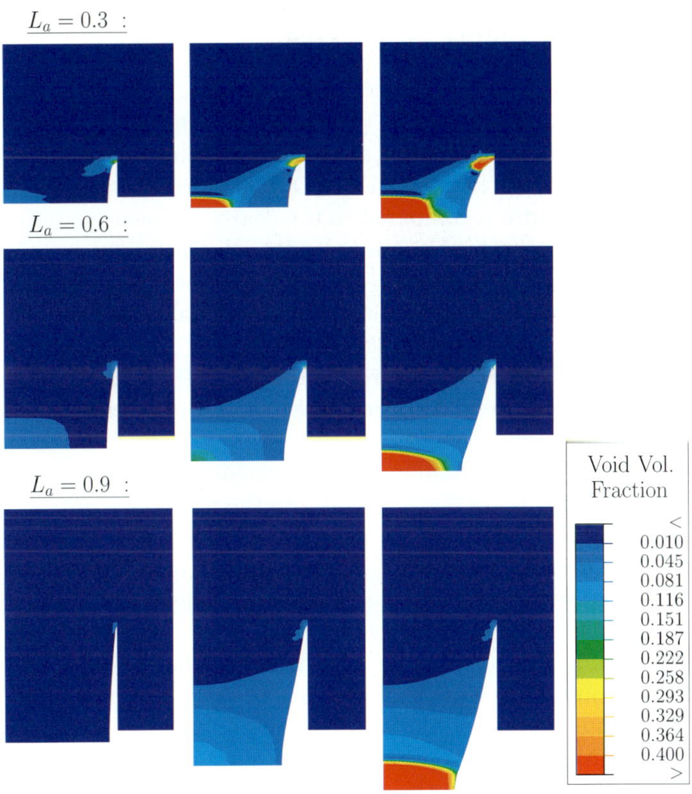

Fig. 9.28. Three levels of increasing delamination (*top* to *bottom*) are shown for three levels of fiber extension. Material parameters were used for the case of alumina and aluminium [26]

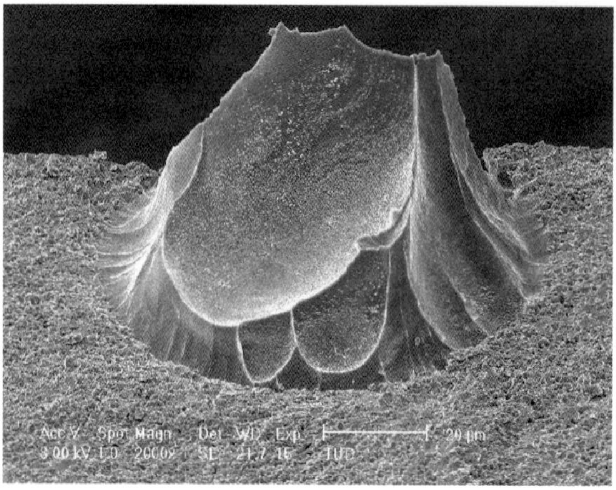

Fig. 9.29. Failure from an artificial cavity filled with aluminium is displayed by showing the fracture surface of an Al_2O_3/Al composite

Processing-related issues do not allow production of the Al_2O_3/Ni_3Al composite with different ligament size. On the other hand, microcracking is influenced strongly by the length scale in the Al_2O_3/Cu system, especially if the fine ligament microstructure is compared to the coarse ligament material (Figs. 9.14a,b). Microcracking in this composite leads to distinct differences in the strength values. Coarse-scale Al_2O_3/Cu was determined to have strength values of 300 MPa, while fine-scale Al_2O_3/Cu had strength values of 550 MPa for composites containing 25% of copper [29]. This drastic decrease cannot be explained by a different degree of reinforcement and different toughening effects but is suggested to be related to microcracks providing efficient crack initiation sites for failure of these composites.

Again, in contrast, aluminium reinforcements provide strong interfaces, even for cases where artificial pores are included into the microstructure and then filled with metal. Failure then proceeds starting from these artificial defects where these can display internal voiding but no microcracking at the interface. Figure 9.29 serves as an example for a metal filled cavity which forms the site for crack initiation and final failure [22].

Al_2O_3/Al, therefore, is a good example for a material that has a strong interface, thereby showing no microcracking owing to residual stresses, effective stress transfer, no observable microcracking under external load, and only little debonding during crack propagation. Failure rather occurs by formation of internal voids (Fig. 9.28). In this case, any adjustment of interfacial adhesion by changing interface chemistry or structure is therefore irrelevant. However, changes in microstructure can have drastic effect on macroscopic mechanical properties. These opportunities afforded by microstructural design were demonstrated by Prielipp et al. [27]. Increasing the metal content, the fracture toughness is increased to a predictable degree, while the strength of Al_2O_3/Al composites varies only very little.

9.5 Outlook: Future Requirements and Developments

We have presented a discussion on the influence that changes in adhesion between two phases can have for the two extreme microstructures, a laminate with no percolation and an interpenetrating microstructure. It became apparent that in macroscopic materials behaviour all length scales, from atomistic adhesion to microstructural scale, have to be considered. While the chosen examples are from the area of structural composites, and functional materials may provide stronger influences of interfacial composition, several conclusions and requirements for the future are suggested:

1. Microcharacterization should be accompanied with determination of a macroscopic property.
2. Modeling should accompany investigations from the atomistic scale to the microstructural scale.
3. The transfer from atomistics to materials properties requires a large degree of curiosity and open collaboration.

Furthermore, on a broader scale, several opportunities for microcharacterization arise, where a better understanding towards the link between microcharacterization and macroscopic properties is contemplated:

4. The non-equilibrium component of fracture is only poorly understood. The difference between work of adhesion and work of separation [6] could be investigated with scanning probe microscopy. This type of investigation may also further our understanding of the anisotropy of fracture energies observed in ceramics.
5. For cases, where interface fracture is determined by the work of adhesion, more details on the bonding state, for example as provided by electron loss near edge spectroscopy (Chap. 5) should lead us closer to a predictability of the influence that small amount of dopants can have on the work of separation.

Acknowledgements

I am indebted to many students and colleagues for valuable discussions: amongst these are Dr. Matthias Diemer (Mannesmann-Sachs), Thomas Emmel, (TU Darmstadt), Dr. Mark Hoffman (UNSW, Sydney), Dr. Achim Neubrand (TU Darmstadt), Dr. Siegfried Skirl (Bosch), Ulrich Stiefel (TU Darmstadt) and Dr. Andre Zimmermann (MPI / PML Stuttgart). I am also indebted to Dr. Frank Ernst, who assisted me to finish this manuscript and to Prof. Manfred Rühle for giving me the opportunity to partake in this volume.

References

1. M. Rühle (1996) Structure and Composition of Metal/Ceramic Interfaces. J. Eur. Ceram. Soc. 16: 353-365.
2. A.G. Evans, J. W. Hutchinson and Y. Wei (1999) Interface Adhesion: Effects of Plasticity and Segregation. Acta metall. Mater. 47: 4093-4113.
3. B.R. Lawn (1995) Fracture of Brittle Solids. University Press, Cambridge.
4. A. W. Adamson (1982) Physical Chemistry of Surfaces. John Wiley Sons, New York.
5. J.R. Rice and J.-S. Wang (1989) Embrittlement of Interfaces by Solute Segregation. Mat. Sci. & Eng. A107: 23-40
6. P. Gumbsch and R.M. Cannon (2000) Atomistic Aspects of Brittle Fracture. MRS Bulletin. 25: 15-20.
7. J.W. Hutchinson and Z. Suo (1991) Mixed Mode Cracking in Layered Materials. Adv. in Appl. Mech. 29: 63-191.
8. A.G. Evans and B.J. Dalgleish (1992) The Fracture Resistance of Metal-Ceramic Interfaces. Acta metall. Mater. 40: S295-S306.
9. I.E. Reimanis, B.J. Dalgleish, M. Brahy, M. Rühle and A.G. Evans (1991) Effects of Plasticity on the Crack Propagation Resistance of a Metal/Ceramic Interface. Acta metall. Mater. 41: 2645-2652.
10. M.L. Jokl, V. Vitek and C.J. McMahon Jr., (1980) A Microscopic Theory of Brittle Fracture in Deformable Solids: A Relation between Ideal Work to Fracture and Plastic Work. Acta Metallurgica. 28: 1479-1488.
11. G. Elssner, D. Korn and M. Rühle (1994) The Influence of Interface Impurities on Fracture Energy of UHV Diffusion Bonded Metal-Ceramic Bicrystals. Scr. Metall. et Mater. 31: 1037-1042.
12. A.G. Evans, M. Rühle, B.J. Dalgleish and P-G. Charalambides (1990) The Fracture Energy of Bimaterial Interfaces. Mat. Sci. And Eng. A126: 53-64.
13. F. Gaudette, S. Suresh, A.G. Evans, G. Dehm and M. Rühle (1997) The Influence of Chromium Addition on the Toughness of γ-Ni/a-Al_2O_3 Interfaces. Acta. mater. 45: 3503-3513.
14. M. Diemer, A. Neubrand and J. Rödel (1999) Influence of Oxygen Partial Pressure and Oxygen Content on the Wettability in the Copper-Alumina System. J.Am.Ceram.Soc. 82: 2825-2832.
15. R.O. Ritchie, R.M. Cannon, B.J. Dalgleish, R.H. Dauskardt and J.M. McNaney (1993) Mechanics and Mechanisms of Crack Growth at or near Ceramic-Metal Interfaces: Interface Engineering Strategies for Promoting Toughness. Mat. Sci. And Eng. A 166: 221-243.
16. A.G. Atkins and Y.-W. Mai (1988) Elastic and Plastic Fracture. Ellis Horwood Limited, Chichester.
17. M. Diemer, T. Emmel, D. Gross and J. Rödel (2000) Fracture Toughness of Copper/Alumina Interfaces. J.Am.Ceram.Soc. submitted.
18. ASTM Standards, Vol. 3.01 (1989) American Society for Testing and Materials.
19. J. Rödel, J.F. Kelly and B.R. Lawn (1990) In situ Measurements of Bridged Crack Interfaces in the SEM. J. Am. Ceram. Soc. 73 :3313-3318
20. M. Diemer, M. Hoffman, A. Neubrand, J. Rödel, H. Müllejans and M. Rühle (1997) The Role of Oxygen Partial Pressure on the Wetting Behaviour and the Interface Chemistry of Copper-Alumina Interfaces Proceedings for 5th International Conference, Joining Ceramics, Glass and Metal, 64-68.

21. U. Alber, H. Müllejans and M. Rühle (1999) Wetting of Copper on α-Al_2O_3 surfaces depending on the orientation and oxygen partial pressure. Micron 30: 101-108.
22. M. Hoffman, S. Skirl, W. Pompe and Jürgen Rödel (1999) Thermal Residual Strains and Stresses in Al_2O_3/Al Composites With Interpenetrating Networks Acta Materialia, Vol 47: 565-577
23. S. Skirl, M. Hoffman, K. Bowman, S. Wiederhorn, J. Rödel (1998) Thermal Expansion Behaviour and Macrostrain of Al_2O_3/Al Composites with Interpenetrating Networks. Acta materialia. 46: 2493-2499
24. M.F. Ashby, F.J. Blunt and M. Bannister (1989) Flow Characteristics of Highly Constrained Metal Wires. Acta materialia, 37: 1847-1857
25. P.A. Mataga. (1989) Deformation of Crack-Bridging Ductile Reinforcements in Tough Brittle Materials. Acta materialia. 37: 3349-3359
26. M. Hoffman, B. Fiedler, T. Emmel, H. Prielipp, N. Claussen, D. Gross and J. Rödel (1997) Fracture Behaviour in Metal Fibre Reinforced Ceramics. Acta materialia, 45: 3609 - 3623
27. H Prielipp, M. Knechtel, N. Claussen, S.K. Streiffer, H. Müllejans, M. Rühle and J. Rödel (1995) Strength and Fracture Toughness of Aluminum/Alumina Composites with Interpenetrating Networks. Mat. Sci. and Eng. A 197: 19-30
28. A. Zimmermann, M. Hoffman, T. Emmel, D. Gross and J. Rödel, (2000) Temperature Dependence of Strength and Fracture Toughness of Aluminum/Alumina Composites with Interpenetrating Networks. J. Am. Ceram. Soc., to be submitted
29. M. Knechtel, H. Prielipp, H. Müllejans, N. Claussen and J. Rödel (1994) Mechanical Properties of Al/Al_2O_3 and Cu/Al_2O_3 Composites with Interpenetrating Networks. Scr. Met. 31:1085 - 1090.
30. J. Rödel, H. Prielipp, N. Claussen, M. Sternitzke, K.B. Alexander, P.F. Becher and J.H. Schneibel (1995) Ni_3Al/Al_2O_3 Composites with Interprenetrating Networks; Scr. Met. 33: 843-848
31. S. Skirl, R. Krause, S.M. Wiederhorn and J. Rödel. (2000). Processing and Mechanical Properties of Ni_3Al/Al_2O_3 Composites with Interprenetrating Network Microstructure. J. Am. Ceram. Soc. Submitted.
32. C.T. Liu, C.L. White and J.A. Horton (1985) Effect of Boron on Grain Boundaries in Ni_3Al. Acta Materialia, 33: 213-229.
33. M.S. Kim, S. Hanada, S. Watanabe and O. Izumi (1988) Strength and Fracture of Single-Crystalline Ni_3 (Al,Ti) and Ni_3 (Al,Ta) Intermetallic Compounds. Acta Metall., 36:2615-26
34 S. W. Paulik, M.H. Zimmerman and K.T. Faber (1996) Residual Stress in Ceramics with Large Thermal Expansion Anisotropy. J. Mat. Res. 11: 2795-2803.
35. D. Nix (1989). Mechanical Properties of Thin Films. Met. Trans. A, 20: 2217-2245
36. T.W. Clyne and P.J. Withers (1993) An introduction to metal matrix composites, Cambridge University Press, New York, NY, USA
37. W. Kreher and W. Pompe. (1989) Internal Stresses in Heterogeneous Solids. Akademie Verlag Berlin.
38. L. I. Tuchinskii (1983) Elastic Constants of Pseudoalloys with a Skeletal Structure. translated from Poroshkovaya Metallurgiya. 7 [247] 85-92

39. Siegfried Skirl (1998) Mechanische Eigenschaften und thermisches Verhalten von Al_2O_3/Al und Al_2O_3/Ni_3Al Verbundwerkstoffen mit Durchdringungsgefüge. Ph.D. Thesis, TU Darmstadt
40. A.G. Evans. (1990) Perspective on the Development of High-Toughness Ceramics. J. Am. Ceram. Soc. 73: 187-206.

10 Microstructural Characterization of Materials: An Assessment

R.W. Cahn, G. Ertl, and J. Heydenreich

10.1 Microcharacterization of Materials

There is no doubt that the microstructural characterization of materials has played an important role for solving important problems in the fields of solid state physics, solid state chemistry, and materials research and technology during the past decades. Based on the close correlation between microstructure and properties of materials the characterization of materials in subnanometer regions (this means down to molecular and atomic dimensions) is one of the key tasks in these fields. In 1985 the Volkswagen-Stiftung responded to this situation with the establishment of a priority program entitled "Microcharacterization of Materials and Devices". At that time electron microscopy (EM) had been the key method for this purpose, and the further development of this method, especially towards a better spatial resolution (for the imaging of both geometrical structures and chemical inhomogeneities), was regarded as the main task. Concerning the special field of the imaging of surfaces, the invention of scanning tunneling microscopy (STM) in the eighties yielded a powerful method for atomic imaging of surfaces, complementary to high-resolution transmission electron microscopy (HRTEM), which is mainly applied to image the interior of materials. Following the STM technique a great variety of scanning probe microscopies, especially atomic force microscopy (AFM), have been established. From the beginning of the nineties onwards not only the powerful methods of electron microscopy (having been applied successfully for many years) but also the rapidly expanding methods of scanning probe microscopies have been widely applied to the investigation of materials, and the priority program shifted its main goals from the development of characterization methods to the application of these methods in topical fields of materials research.

The question arises to what degree the 12 years priority program on the microcharacterization of materials and devices gave positive impulses to the development of topical fields of materials research. The foregoing chapters of this book try to give answers in detail, based on the results obtained from symposia in Göttingen (1994), Stuttgart (1996) and in Weinböhla (1997). This chapter emphasizes the main lines and refers to synergy effects indicated by the collaboration of scientists, partly from different research fields. The assessment comprises three parts: possibilities and limitations of electron microscopy, the potential of scanning probe microscopies, and the impact of both on materials research, concerning model materials as well as complex material systems.

In electron microscopy, first of all *transmission* electron microscopy (TEM), used to investigate of the interior of thin-foil specimens, has increased in importance. However, also surface electron microscopy, especially scanning electron microscopy (SEM) – in competition with scanning *probe* microscopies – and emission electron microscopy, have been further developed, and the priority program responded to all these techniques.

The main directions in TEM are: high-resolution TEM, for imaging of geometric structures down to molecular and atomic dimensions, respectively, analytical TEM (ATEM) for characterizing chemical species in nanoscopic dimensions, and in situ EM for imaging dynamic processes.

In HRTEM there are at least three ways of improving the resolution limit: (i) high-voltage electron microscopes with an accelerating voltage in the range between 1 and 1.5 MeV, which provide a better resolution by shorter electron wavelengths, (ii) correction systems for instruments with an accelerating voltage between 100 and 200 kV, thus improving the resolution by decreasing the electron-optical aberrations of the objective lens – especially the spherical and the chromatic aberration, (iii) holographic systems (biprism electron interferometers) followed by numerical image reconstruction to eliminate the influence of electron-optical aberrations on the available resolution.

Concerning better resolution to be obtained by high-voltage transmission electron microscopes, as early as in 1987 the Volkswagen-Stiftung financed a so-called medium-voltage EM (400 kV) to be installed in Stuttgart [1] as a powerful basis for high-resolution work, especially for the investigation of interfaces in ceramics and metals. Based on this, in 1991, independent of VW support, a 1.25 MeV high-voltage high-resolution electron microscope ("Stuttgart Atomic Resolution Microscope", ARM) was installed, again in Stuttgart [1]. Equipped with its top-entry stage, this instrument reaches a point resolution of 1.05 Å. Like the above-mentioned medium-voltage instrument, the ARM has been used successfully in a universal way for characterizing materials in subnanoscopic dimensions (and also for in situ work, as will be shown below). As to the ARM, the question arose to what extent radiation damage to the materials to be investigated occurred by the impact of the high-energy (1.25 MeV) electrons. Besides this, the high financial efforts for the acquisition of instruments of that type have to be taken into account.

The alternative of using an electron microscope in the 100 to 300 kV range for high-resolution electron microscopy, not by achieving shorter electron wavelengths, but by correcting the electron-optical aberrations of the objective lens, is thus of interest. The project "Subångström Electron Microscopy", treated jointly by Rose in Darmstadt [2], Haider in Heidelberg [3], and Urban and co-workers in Jülich [4] resulted in a corrected 200 kV instrument that achieves a point resolution between 1.2 and 1.3 Å. According to the concept of Rose, the main parts of the correction system for the aberrations of the objective lens are two suitably dimensioned and arranged magnetic sextupoles. As the basis of the whole correction system its electron-optical parameters were calculated by Rose, who – in following the tradition of Scherzer – is a profound specialist in the field of non-rotation-symmetrical correction elements. Based on this concept, Haider experimentally realized the correction system in a microprocessor-controlled way. Here, the precise adjustment of the individual elements of the correction system needed special attention. The application of the corrected electron microscope to topical problems of materials science has successfully been tested by the Urban group, where the above-mentioned parameters could be confirmed with the example of semiconducting materials. The project "Subångström Electron Microscopy", as one of the most successful projects within the priority program "Microcharacterization of Materials and Devices" has now shifted its emphasis on eliminating also the chromatic aberration of the objective lens, especially by employing electron monochromators in the illumination system. So the community of electron microscopists is hopeful to soon approach "Subångström" imaging.

Another important way of improving the possibilities of high-resolution imaging, as mentioned above, is the use of electron holography. The project "High-Resolution Electron Holography Including Numerical Reconstruction", treated by Lichte and co-workers in Tübingen and Dresden, respectively, is related to this [5,6]. Using the interference of an object wave and a reference wave in a biprism interferometer, a hologram is produced, from which the image is obtained by numerical reconstruction enabling one to compensate the influence of the aberrations of the electron microscope. This procedure allows one to obtain a quantitative representation of the amplitude and of the phase of the reconstructed wave in the real space as well as in the Fourier space, which is important for the reliable interpretation of the image contrast of electron micrographs. In practice, the borders of wedge-shaped specimens are investigated, allowing the close neighbourship of an object beam and a reference beam.

Besides the achievement of an optimum lateral resolution, image contrast formation is one of the key problems in HRTEM. To understand contrast formation quantitatively, a reliable description of the interaction processes between electron beam and specimen is necessary. Rose et al. [2] have successfully contributed to this field by a project entitled "Simulation of Energy-filtered Electron Micrographs with Phonon and Electron Excitation". In this project, the different elastic, quasi-elastic and inelastic scattering processes in the specimen have been treated uniquely, thus forming the basis for a description of the partial-coherent image formation in the electron microscope. Considering the importance of a suitable choice of the respective imaging parameters, Phillipp, Hofmeister, and co-workers [1,7] worked out a premise-free method of reconstructing the exit surface wave function in HRTEM in a project entitled “Computer-Aided Optimization of Imaging Parameters”. Closely related to the problems of resolution and image contrast is the quantitative interpretation of high-resolution electron micrographs. As a topical example of the possibilities and limitations of quantitative HRTEM, Schröter and colleagues [8,9] investigated semiconductor interfaces in layer systems. They showed convincingly the correspondence between the growth parameters and the structure of the interface for Si/SiO_2 interfaces, for ternary III/V semiconductor layers, and also for porous silicon.

Under the title “Combination and Application of New Methods of Quantitative High-Resolution Transmission Electron Microscopy (QHRTEM) for the Microcharacterization of Crystal Defects” a close cooperation between 7 scientific groups was initiated in 1996. High synergy effects have been achieved by this cooperation. The joint project emphasized the further development of quantitative high-resolution methods for microstructure analysis and their close relation to practical problems in materials research. The majority of individual projects were related to semiconductors: heterostructures and extended defects (local strain mapping, Schröter and Ahlborn [8]), strained and defect-containing hetero-interfaces (Strunk, Stenkamp [10]), low-dimensional semiconductors, and in particular to quantum structures in strained systems (Phillipp [1]), III/V-heterostructures (including the use of fuzzy logic, Hillebrand [7]). Besides semiconductors, high-temperature superconductors and quasicrystals have also been quantitatively microcharacterized – in a project by Urban[4]. Furthermore, two-dimensional lattice defects in metals and ceramics (Rühle [1]) and the early stages of the phase separation in anorganic glasses (Mader [11]) have been investigated.

Extending high-resolution TEM to high-resolution analytical TEM with possibilities of characterizing variations in chemical composition or interatomic bonding on a nanoscopic or subnanoscopic (atomic) scale in a wide sense constitutes a continuous process, because by quantitative interpretation of electron micrographs, already HRTEM allows the characterization of the chemical species to a certain degree. The basis of this was created some years ago by the pioneering work of Ourmazd and co-workers [12], in which they succeeded in developing the real space analysis methods “Chemical Mapping” and “QUANTITEM”. In a project entitled “Quantitative EM of Inner Boundaries”, between 1997 and 2000, Ourmazd and co-workers [12] have tested their computer simulation programs by characterizing concentration profiles in Si_xGe_{1-x} heterostructures grown by molecular beam epitaxy (MBE) as well as chemical vapor deposition (CVD).

In a narrower sense, analytical TEM means the combined application of CTEM or HRTEM for imaging the specimen morphology, spectroscopic methods like X-ray spectroscopy, including wavelength dispersive (XWDS) and energy-dispersive (XEDS) spectroscopy, and electron energy-loss spectroscopy (EELS) for the local analysis of the chemical composition, and electron diffraction methods – microdiffraction, especially convergent beam diffraction (CBED) for the local analysis of the crystal structures.

Treating the joint project "Chemical Composition with high spatial resolution" Kohl and co-workers [14] and Mayer and co-workers [1] have shown convincingly the possibilities and limitations of the energy-filtering TEM (EFTEM) with respect to detection limits and lateral resolution, related to the theoretical background (Kohl) as well as to the experimental situation (Mayer), also for the topical field of the mapping of the electron loss near-edge fine structures (ELNES), for example for the analysis of interatomic bonding. Application of analytical TEM to the detection of light elements in magnetic and electronic materials and to the characterization of multi-layers for magnetic, optical, and mechanical applications have proved to be successful in two projects of Bauer and Wetzig [15]. Concerning the application of electron diffraction techniques to analytical TEM, in a project entitled "Charge Density Distribution from Energy-Filtered Convergent Beam Electron Diffraction (CBED)" Mayer and Rühle [1] succeeded in measuring charge density shifts of orbital electrons, thus improving the exactness of structure factor measurements. The results were tested on the intermetallic phase NiAl, where bond states could be studied. For the "Determination of the positions of light foreign atoms in crystals by the aid of analytical EM" (project title) Rose and co-workers [2] proposed a theoretical procedure of interpretating the energy-filtered diffraction patterns. The calculation of the cross section is based on the assumption that the secondary electron states are quantized according to the angular momenta, and the wave function of the primary electron in the crystal is described by Bloch waves.

In situ transmission electron microscopy as the third aspect of TEM serves to observe the microstructure and defect structure of materials under the action of defined external influences (for example mechanical or thermal specimen treatments, radiation influence, or reaction with environmental gases). Necessary prerequisites for this are the use of relatively thick specimens (some μm), which are representative of bulk materials, on the one hand, and large specimen chambers in the microscope for mounting attachments for special specimen treatments (for example deformation devices) on the other. Both requirements are fulfilled by high-voltage electron microscopes, which became commercially available in the mid-sixties. These first high-voltage electron microscopes (HVEMs, for example in the 1 MeV range) in spite of short electron wavelengths, have not incidentally been used for high-resolution work. Because of technical difficulties, such as instability of the acceleration voltage in the high-voltage range, the achievable resolution at that time was restricted. Later on, electrically highly stabilized and expensive HVEMs became available for high-resolution work. Independent of the priority project "Microcharacterization of Materials and Devices", the Volkswagen-Stiftung supported the in situ HVEM already in the sixties by subsidizing the supply of a 650 kV high-voltage microscope for scientific work in Stuttgart [1]. Since that time, also in other places in Germany, for example in Halle (Messerschmidt [7]), methods of in situ HVEM (for example the construction of special attachments for specimen treatment) have been successfully developed and applied to observe dynamical changes in the microstructure of materials and to solve important problems in solid state physics and materials science.

Within the framework of the priority program from the wide field of direct surface imaging in electron microscopy the development of low-energy electron microscopes (LEEM), also usable for photo-emission electron microscopy (PEEM), has been pursued very successfully by Bauer and co-workers in Clausthal [16] and by the Rose group in Darmstadt [2]. In a project entitled "Direct Imaging Microscopy Using Low-Energy Electrons", Bauer and his co-workers have developed a LEEM device (since 1987) equipped with spectroscopic facilities providing not only a high lateral resolution but also the capability of chemical analysis with high energy-resolution. The applied spectroscopic methods are synchrotron-radiation emitted photo-emission electron microscopy (XPEEM) and photon-emitted Auger electron emission microscopy (AEEM). After a short test of the microscope at BESSY I in Berlin the device has been used continuously at the synchrotron radiation source ELETTRA in Trieste. In the meantime already five instruments exist in the world that are based on this instrument of Bauer. In the majority of cases these extremely surface-sensitive instruments are used for the imaging of micro- and nano-structures (for example metals) on semiconductor surfaces. Initially closely collaborating in the Bauer project, in a separate project Rose [2] performed thorough theoretical investigations as the basis for the construction of an optimally corrected low-voltage analytical electron microscope (LEEM), yielding a spatial resolution of 1 nm and an energy resolution of 0.2 eV. The presupposition for reaching these parameters is the correction of the chromatic and the spherical aberration of the objective lens. Further developments related to an optimum beam splitter, necessary for the separation of the incident electrons and the electrons reflected at the specimen surface, have not been completed during the priority project. The elaborated concept for a high-resolving analytical LEEM, allowing also photo-emission electron microscopy (PEEM), is the basis for a universal device, which now is under construction within the framework of a program of the Federal Ministry of Education and Research with many collaborating scientific institutions in Germany. Within the next 2–3 years this microscope, being of great interest for surface microscopists, shall be installed at the BESSY II synchrotron radiation source in Berlin.

Besides the development of electron microscopy imaging techniques, as described in the preceding paragraphs, application of these methods to solid materials and their surfaces requires to further elaborate these techniques. In other words, one needs to develop *materials-specific*, practical techniques, including specimen preparation as well as methods of image interpretation, image quantification (including chemical analysis), image processing, and computer simulation of corresponding processes. Within the priority program many projects have been involved in these tasks. These projects particularly addressed the microcharacterization of semiconductors, metals and alloys, and interfaces, precipitates, and particles of different materials. Among these projects, the study of semiconductors (bulk materials as well as layer structures) is the most extensive one.

The results of Wilbrandt's project "Dislocation networks in GaAs" [9] are related to an extended characterization of existing types of dislocations, a work which is important with respect to the fact that it is technologically very difficult to get dislocation-free materials. To carry out the investigations, besides HRTEM Wilbrandt also used the weak-beam dark-field technique, which is especially powerful for characterization of defects with extended strain fields. A joint project of Heime [17] and Kubalek [18] related to the phase separation of quaternary semiconductors. Especially GaInAsP and ZnMgSe have been investigated by the combined application of a large number of characterization methods, including Z-contrast-imaging, CBED, EELS, cathodoluminescence, and corresponding methods of computer simulation. Their nanocharacterization of the phase separation phenomena is based on a reliable calculation of the thermodynamic miscibility gap of the materials (using a program packet developed by themselves). Schröter and Kittler [8,13] succeeded in the structural (HREM) and electrical characterization (electron states) of metal silicides ($NiSi_2$, Cu_3Si) in silicon. The electrical characterization has been based on the use of the EBIC method (electron beam-induced current) of SEM, and the technique of deep level transient spectroscopy (DLTS).

In the field of semiconductor layer structures a current cooperation with high synergy effects was successfully pursued by Stolz [19], Ploog and Däweritz [20]and Tapfer [21] in a project entitled "Ternary Semiconductor Heterostructures". The speciality of this project is the combined use of TEM and X-ray diffraction in all groups. Stolz and co-workers are investigating distortion fields, the elasticity modulus, and the correspondence between lattice strains and the chemical composition of layer structures grown by metal-organic gase phase epitaxy (MOVPE): $(Ga_xIn_{1-x})As$, (GaIn)P/GaAs heterostructures. The work of Ploog and Däweritz is related to MBE-grown layers, especially $Al_xGa_{1-x}As$ grown on (001) surfaces of GaSb and $In_xGa_{1-x}As$ grown on GaAs substrates, with the aim to find optimum growth parameters. The investigations of Tapfer concerning AlGaAs/GaAs heterostructures (MBE), MOVPE-grown GaInAs/GaAs heterostructures, MOVPE-grown (GaIn)As/-GaAs/Ga(PAs)/GaAs superlattices, and GaInP/GaAs superlattices allow the detection of strain parameters and the microcharacterization of the interfaces within the layer systems. For interpreting the results of X-ray diffraction Tapfer developed a general dynamical theory of X-ray diffraction at thin lattice-strained superlattices. A further example of a fruitful collaboration is the work of Wagner [1] and Bill [22] on SiC and GaN layers (amorphous, partly crystalline, crystalline) on Si, 6H-SiC and Al_2O_3 substrates, investigated by HRTEM and ATEM. A special procedure for the formation of SiC layers on a 6H-SiC monocrystal could thus be worked out (patented in the meantime). The investigations within this relatively new project, especially related to GaN, are still under way. Gerthsen and co-workers [23] elucidated growth processes (Stranski-Krastanov growth) in layers of semiconductor compounds, especially InGaAs on GaAs(100) including also quantum dots of this material using cross-section electron microscopy and a suitable algorithm for image interpretation. The aim of the work lies in a general description of the relaxation of the lattice mismatch during the growth process.

Based on the trend to use Si–Ge solid solutions as microelectronic materials in an increasing way, Banhart [1], Peisl, and Metzger [43] and Strunk [10] are still working in projects concerning SiGe layers. Using convergent beam electron diffraction (CBED) Banhart is investigating the lattice strains in LPE grown SiGe layers. As the result of this work it is shown that layers of this type (in a large thickness range) are strained layers without misfit dislocations in the border region to the silicon substrate. A procedure of the rocking profile analysis of the Bragg lines in the CBED diagrams, developed by the authors, can be used also by other research groups. The work of Peisl and Metzger concerns Si_xGe_{1-x}/Si heterostructures, which are investigated by the diffuse scattering of X-rays at grazing incidence. As a result, the strain fields in self-organized Ge/Si quantum dots in multilayers were detected. A newly developed method of iso-strain scattering was helpful for these investigations. Furthermore, the order and shape of incoherent Ge islands on Si(111) has been characterized. The main result of Strunk's research is the explanation of strain-induced island scaling during $Si_{1-x}Ge_x$ heteroepitaxy. As part of the broad field of silicide layers Wetzig and co-workers [15] have carried out the project "Nano-Structure Development of $ReSi_{2+x}$ Layers". The aim of the project is the elucidation of the correspondence between the nanostructure of the layers and their thermo–electrical properties during the transition from the amorphous to the nano-crystalline state. The methods used are analytical transmission electron microscopy (ATEM) on the one hand, and electrical measurements related to thermoelectric power and conductivity, on the other.

Within the priority program the use of electron microscope imaging and electron diffraction techniques for the investigation of metals and alloys involves TEM and CBED. Haessner [25] applied these methods to characterize the texture of deformed polycrystalline metals (for example Al and Ag) with the aim of finding orientation correlation functions. Haessner succeeded in improving the automatization of such analysis procedure. Within a project "Internal Stresses in Materials with Heterogeneous Structure", Mughrabi et al. [10] investigated monocrystalline nickel-based superalloys and polycrystalline copper. The aim of their work was to detect local variations of the lattice parameter in these materials after defined modification of the microstructure by mechanical treatments. Besides the above-mentioned techniques (TEM and CBED) high-resolution X-ray diffractometry was also used. Altogether, a reliable measurement technique of deformation-induced strains has been achieved, leading to the idea of a new project (financed by the Volkswagen-Stiftung) for the investigation of real structural elements, such as turbine blades.

The application of EM methods, especially HRTEM and ATEM, to the microstructural investigation of interfaces becomes increasingly important. Within the project of Ernst et al. [1] $Al/MgAl_2O_4$ and $Ag/MgAl_2O_4$ interfaces have been investigated with respect to both their structure and their bonding states. Since the lattice mismatches between the metal and the oxide is very small in these systems, MBE growth of Al and Ag on $MgAl_2O_4$ substrates yields epitaxial thin films with a *coherent* metal/oxide interfaces to the substrate. Besides TEM investigations the interfaces were modelled by ab-initio density functional calculations. The project of Rühle's [1] and Claussen's [26] groups were concerned with the study of ceramic/metal interfaces. Rühle et al. used especially EELS, including the interpretation of the near-edge fine structure (ELNES) for the investigation of real materials. The main materials of Claussen's investigations were Al/Al_2O_3, Cu/Al_2O_3, and also Ni_3Al/Al_2O_3. Emphasis was laid on the elucidation of the mechanical properties of these materials. The mechanism of the formation of precipitate-free zones in a nickel-base superalloy was successfully investigated in a project of Nembach [27].

Precipitates and particles are especially suitable objects for TEM. In a project "HRTEM of Ni_3Al Precipitates" P.-J. Wilbrandt [9] investigated the early stages of processes of phase separation, based on a detailed knowledge of the size distribution of the nuclei. It is remarkable that within the framework of this project also a concept has been developed for better alignment and a simplified operation of a transmission electron microscope. The project of Mader's [11], entitled "Precipitation in Transition Metal Borides", was concerned with titanium diboride-doped with tungsten boride and chromium boride. It is shown that the precipitates of these dopants are plate-like. HRTEM and EFTEM (energy-filtered TEM) revealed the appearance of the tungsten boride platelets in double monolayers. Interesting in situ TEM investigations of catalysis processes of bimetallic alloy particles (Pd–Ni, Pd–Au, and Pd–Ag on carbon and SiO_2 substrates) are described by Anton [28] These particles are formed by evaporating the components on electron-transparent substrates within a TEM modified for in situ work. Complex catalytic reactions are investigated after a suitable gas atmosphere has been established (oxygen partial pressure: 5×10^{-6} mbar). Also belonging to the field of heterogeneous catalysis, the metal-substrate interaction of Al and Si with Al_2O_3 and SiO_2 substrates and with carbon at elevated temperatures and in reducing gas atmospheres was the subject of a project of Jaeger [29]. The results achieved concern the role and the structure of metal/substrate phase boundaries in composite materials. In closely related projects entitled "Metal dispersions in metal plasma-polymer composites", using TEM and analytical TEM, resp., Baalmann [30] and Jäger [29], succeeded in reproducibly forming metal clusters having diameters between 1.4 and 1.5 nm. For Pd they showed a clear dependence of the lattice constant on the cluster size.

The award of the Nobel Prize in Physics in 1986 to Binnig and Rohrer, the inventors of the scanning tunneling microscope, together with Ruska who have developed the transmission electron microscope, reflected the enormous impact of the scanning probe microscopic techniques on the investigation of various properties of surfaces down to atomic resolution. Different groups in Germany became involved in the construction and application of these techniques soon after their discovery, and the growth and further development of this field is also reflected by the various projects supported within the priority program.

The scanning tunneling microscope subject to further development in different directions, comprising (i) Extension to a large temperature range (from 4 K to 1100 K) and increase of the scanning speed (enabling recording of real-time images for 'video' representation, Neddermeyer [31]), combination with other surface sensitive (for example XPS, AES, and ion scattering, Behm [32]) and bulk techniques(for example positron annihilation, Krause-Rehberg [31]) and (ii) novel experimental techniques, such as probing the light emission introduced by the tunneling current with semiconductor samples (Hartnagel [33]) as well as the unique thermovoltage STM based on variations of the thermovoltage between tip and sample (being at different temperatures), which is sensitive to the local electronic density of states at the Fermi level und thus to the chemical state of the local area probed (Möller [34]).

The 'classical' field of application of the STM in materials science is related to phenomena of the thin film growth as well as the modification of surface properties by chemical reaction (such as oxidation). Accordingly extensive activities were concerned with the mechanism of CVD processes in semiconductor growth (Behm [32], Köhler [35], Henzler [36] and with the epitaxial growth of thin metallic films Wandelt [37] to which topics now a vast literature exists. The application of the STM is not restricted to the solid/gas interface, but may be used in liquids as well and hence represents one of the most powerful methods of modern electrochemistry (Meissner [38], Kolb [39]). Here again, deposition of overlayers or electrolytic dissolution has been studied in detail. Measurements of the local photocurrents in an electrochemical STM with a semiconductor electrode offer interesting prospects for probing the local properties of the space charge layer (Meissner [38]). Most remarkable in this section, however, appears the successful manipulation of single atoms by controlled deposition of tip material onto the sample surface (Kolb [39]).

The atomic force microscope (AFM), although inferior to STM with respect to spatial resolution, is even more versatile in various aspects and was also widely installed in projects of the priority program. Here again several instrumental improvements and new developments can be listed: The resolution may be improved by applying the so-called dynamic mode of operation (Wiesendanger: 28), elastic properties can be probed by implementation of an ultra-sound nearfield microscope (Arnold [51]), local magnetic properties are accessible by the magnetic AFM (eventually also combined with a Kerr microscope) (Schäfer [41]), probing of electrostatic forces enables the study of ferroelectric materials for example with respect to the polarity of grain boundaries or stress-induced polarization (Fink [15]), AFM with acoustic phonons enables the detection of defects in semiconductors by ballistic phonons in the bulk and at the surface (Eisenmenger [41]), and the development of force modulation microscopy can be used for analysis of micromechanical properties even of biomolecules (Gaub [42]). There are also widespread applications of AFM techniques: Characterization of the surface topography as well as fabrication of nanostructures down to 10 nm on semiconductor surfaces by the use of specially developed tips (Kotthaus [24]) and the investigation of phase separation in thin films of polymers as well as their structuring are particularly remarkable achievements in this context (Krausch [44]). Finally, combination of the AFM with other techniques of materials characterization have to be mentioned, such as scanning electron microscopy (SEM, Fuchs: [14]) or scanning electrochemical microscopy (SECM) for characterizing and microstructuring conductive polymers (Gaub [42], Schuhmann [45]).

Apart from the large number of projects dealing with the development and application of scanning probe microscopy techniques also a few studies using other surface sensitive techniques were supported. For example, X-ray diffraction under glancing incidence was applied to the investigation of 'buried' semiconductor interfaces including quantum well nanostructures (Pietsch [46], Forchel [47]), and spot profiles of reflection high energy electron diffraction (RHEED) data from epitaxial semiconductor layers (Däweritz [20]) were analyzed. The remarkable developments of LEEM and PEEM techniques (including spectroscopy) for surface studies have already been mentioned above.

Some projects of the program were focused primarily on improvement or development of characterization techniques, and a number of these have been discussed in the foregoing. Others were focused on specific materials or groups of materials, although even some of these sometimes required novel developments in the characterization methods used. In what follows, we shall examine some selected projects concerned with "realistic" materials, which seemed to us to have achieved valuable insights into practical issues.

The investigation by Mughrabi and Biermann [10] on superalloys has already been mentioned. The unusual feature in this project was that the samples were subjected to realistic stress/temperature combinations and both morphological changes and internal strains were examined. The findings were presented at a major superalloy conference and have received considerable interest.

Al-Li alloys have a superalloy-like microstructure (with precipitates having the same crystal structures as those in conventional Ni superalloys); the microstructure was examined by TEM, together with details of dislocation motion, especially cross-slip between precipitates, in a study by Nembach [27]. The effects of size and volume fraction of precipitates were studied. The methodology here was of a long-established kind, but appropriate to the problems at issue. It is not yet clear whether alloys based on Al-Li will find widespread application in aeronautics, because of various difficulties linked with limited ductility and corrosion behaviour, but probably they will.

An unusual project which was based on a combination of field-ion microscopy (FIM) and atom-probe microanalysis was Frommeyer's [48] examination of the behaviour of ternary alloying elements in titanium aluminides. This family of intermetallics has received enormous attention worldwide in recent years because of their potential combination of lightness, high-temperature strength, and oxidation resistance, which tantalisingly promises use in jet-engine components (though that is slow in arriving). Germany is well to the fore in this swell of research. What Frommeyer and his team have done is to examine the site occupancy of alloying elements, which have been found either to enhance ductility (Cr, V) or to reduce it (Nb). The latter element, unlike the former, was found to be restricted to substituting for Ti atoms, and in-situ fractures occasioned by the high electric field strength in the FIM instrument were found to be linked to Nb-rich sites. These findings were also found to be in accord with electron theoretical computations. There were numerous publications.

A project related to the foregoing is one masterminded by Schütze [49], jointly with Wieser [50]. This research, still in progress, focuses on the oxidation of TiAl and the influence of ion-implanted trace dopants on the process. A particular objective was to examine which dopants can aid the generation of a highly protective alumina film. A remarkable range of analytical techniques was pressed into service to examine the structure and composition of the layers of the oxide and of the oxide/alloy interface, as well as the oxidation kinetics. TEM, STEM+EDX, STM/AFM, EELS, XRD and Rutherford backscattering were all employed. This promises to be a model for the focusing of a wide range of microstructural characterization techniques on a single, well defined problem.

A further metallurgical project pursued by Schreiber [51] was concerned with the optimisation of Ta anti-diffusion barrier layers for interpolation between Si and Cu layers in microcircuits. The suitability of Ta, a very refractory metal, for this purpose has been known for some years but optimisation of the mechanical properties in particular remained a problem. Schreiber applied atomic force microscopy, including a newly developed variant of this technique which he calls "ultrasound force microscopy", to examine the roughness, grain size and homogeneity as well as the yield stress of the layers. Experiments were also done on amorphised Ta(N,O) layers. An imaginative project, this, with a good publication record.

Turning now to ceramics, Telle [52] has been pursuing an unconventional project aimed at the strengthening of diborides (TiB_2, WB_2, CrB_2), which could substitute for W and Ti carbides in so-called "hard metals" (cermets used for metal cutting). By forming solid solutions between the three borides at a very high annealing temperature, strengthening precipitates of $(Te,W,Cr)_2B_5$ can be formed. The resulting microstructures and local compositions are being examined by HREM and EELS. A wholly unexpected early finding is that several thermodynamically unpredicted compounds also appeared, the most remarkable being a metallic glass formed by the known process of "solid-state amorphization". A collaboration with a Californian laboratory has been begun to attempt to make the strengthened borides by means of high-temperature "combustion synthesis" technique. This (as yet uncompleted) project is an example of the range of unexpected crystal (and non-crystal) species which can be generated in hitherto somewhat unfamiliar materials, giving plenty of scope for the exercise of microstructural characterization.

Another ceramic project, jointly run by Birringer [53] and Rosen and Berger [54], focused on different ways of producing nanocrystalline tungsten carbide for incorporation in cermets for cutting tools (for example mechanical milling and co-sputtering). These ways have been compared, with a view to achieving tougher and more wear-resistant tools. The carbides made were investigated microsctructurally, and in co-sputtering experiments, a remarkable byproduct was the discovery of pure tungsten with the A15 structure (the crystal structure of standard metallic superconductors). The nanocrystalline carbide sintered more easily than normal material. The original objective has in the meantime been overtaken by an American industrial group using a spray-reaction technique suitable for large-scale production, but the potential of ceramic projects to deliver the unexpected has been shown once again.

This brings us to polymers. A particular interesting pair of projects, the second still in progress, has been masterminded by Petermann [55], a polymer specialist. These were concerned with metallic films evaporated on to semicrystalline or amorphous polymers, and the objective (using AFM, STM and TEM) was to examine the nature and degree of preferred orientation (texture) in the deposited films. There was in fact substantial texture, generated in part by the crystalline component in the polymer substrates and in part by the surface morphology; the effect of this last variable has been dubbed "graphoepitaxy". To give an idea of this research, one of the numerous papers resulting from these projects was entitled "Lattice Orientations of Evaporated Metals onto Uniaxially Oriented Semicrystalline, Ultrathin Polymer Films". This research was combined with an examination of diffusion of the metal into the polymeric substrate (greater for amorphous than for semicrystalline polymers) and also a separate measurement of the surface diffusion of deposited metals at the metal/polymer interface, which depended on the reactivity of metals. Many papers resulted from both projects. This kind of study must be of practical interest since polymer films in contact with metals are used in various devices, miniature batteries for instance, and also in some microcircuits apparently.

A related project, which was in progress for 5 years, was run by Faupel [56]. This project focused entirely on the diffusion of metals in polymers. Radiotracers were used in conjunction with sputtering to uncover successive layers, and also TEM of cross-sections of polymer/metal bilayers. It was found that some metals (noble metals) diffuse to some extent into polymers even during the deposition process. Reactive metals form sharp interfaces with polymeric substrates while less reactive ones (such as noble metals) diffuse in more readily even below the glass transition temperature of amorphous polymers. Copper conducting layers, as used in microcircuits do not pose a diffusion problem. The many publications include a major invited review on the diffusion of metals in polymers.

Of 49 projects altogether in this program which were judged to be of special material science interest, we judged 27 to be concerned with "realistic" materials as opposed to "model materials"; most of the "realistic" ones were among the more recent projects. This is reassuring: in 1994, when the progress of the entire program up to that year was subjected to a formal review, the assessors concluded that it was important that future projects should be devoted to realistic materials, and this recommendation has clearly been successfully implemented. A number of these involved a combination of characterization techniques rather than just a single one, and this is also an encouraging sign that German investigators have been becoming familiar with a wide range of techniques – and this was one of the original objectives of the whole program.

Contributing Institutions

1. Max-Planck-Institut für Metallforschung Stuttgart
2. Technische Hochschule Darmstadt, Institut für Angewandte Physik, Licht- und Teilchenoptik
3. Europäisches Laboratorium für Molekularbiologie Heidelberg
4. Forschungszentrum Jülich GmbH, Institut für Festkörperforschung
5. Universität Tübingen, Institut für angewandte Physik
6. Technische Universität Dresden, Institut für Angewandte Physik und Didaktik
7. Max-Planck-Institut für Mikrostrukturphysik Halle
8. Universität Göttingen, IV. Physikalisches Institut
9. Universität Göttingen, Institut für Metallphysik (now: Institut für Materialphysik)
10. Universität Erlangen-Nürnberg, Institut für Werkstoffwissenschaften
11. Universität Bonn, Institut für Anorganische Chemie

12. Bell Laboratories Holmdel/USA
13. Institut für Halbleiterphysik Frankfurt/Oder
14. Universität Münster, Physikalisches Institut
15. Institut für Festkörper- und Werkstofforschung Dresden
16. Technische Universität Clausthal, Physikalisches Institut
17. Technische Hochschule Aachen, Institut für Halbleitertechnik
18. Universität GH Duisburg, Fachgebiete Werkstoffe der Elektrotechnik
19. Universität Marburg, Wissenschaftliches Zentrum für Materialwissenschaften
20. Paul-Drude-Institut für Festkörperelektronik Berlin
21. Centro Nazionale Ricerca e Sviluppa Materiali Brindisi (Division of Structural Investigations and Microanalysis)
22. Universität Stuttgart, Institut für Nichtmetallische Anorganische Materialien
23. Universität Karlsruhe, Laboratorium für Elektronenmikroskopie
24. Universität München, Sektion Physik
25. Technische Universität Braunschweig, Institut für Werkstoffe
26. Technische Universität Hamburg-Harburg, Arbeitsbereich Technische Keramik
27. Universität Münster, Institut für Metallforschung
28. Universität Hamburg, Institut für Angewandte Physik
29. Universität Bremen, Institut für Angewandte und Physikalische Chemie
30. Universität Bremen, Institut für Angewandte Materialforschung
31. Universität Halle, Fachbereich Physik
32. Universität Ulm, Abt. Oberflächen und Katalyse
33. Technische Hochschule Darmstadt, Institut für Hochfrequenztechnik
34. Universität Stuttgart, 4. Physikalisches Institut
35. Universität Kiel, Institut für Experimentalphysik
36. Universität Hannover, Institut für Festkörperphysik
37. Universität Bonn, Institut für Physikalische und Theoretische Chemie
38. Forschungszentrum Jülich, Institut für Energieverfahrenstechnik
39. Universität Ulm, Abt. Elektrochemie
40. Fraunhofer-Institut für zerstörungsfreie Prüfverfahren Saarbrücken
41. Universität Stuttgart, 1. Physikalisches Institut
42. Technische Universität München, Lehrstuhl für Biophysik
43. Universität München, Sektion Physik
44. Universität München, Institut für Physikalische Chemie
45. Technische Universität München, Lehrstuhl für Allgemeine Chemie und Biochemie
46. Universität Potsdam, Institut für Festkörperphysik
47. Universität Würzburg, Lehrstuhl für Technische Physik
48. Max-Planck-Institut für Eisenforschung Düsseldorf
49. DECHEMA, Karl-Winnacker-Institut Frankfurt/Main
50. Forschungszentrum Rossendorf, Institut für Ionenstrahlphysik und Materialwissenschaften
51. Fraunhofer-Institut für zerstörungsfreie Prüfverfahren Saarbrücken, Institutsteil Dresden
52. Rheinisch-Westfälische Technische Hochschule Aachen, Institut für Gesteinshüttenkunde
53. Universität Saarbrücken, Lehrstuhl für experimentelle Physik der kondensierten Materie
54. Technion Haifa, Israel Institute of Technology, Department of Materials Engineering
55. Universität Dortmund, Fachbereich Chemietechnik
56. Universität Kiel, Technische Fakultät, Institut für Materialverbunde

Index

Printing: Mercedes-Druck, Berlin
Binding: Stein+Lehmann, Berlin

Springer Series in
MATERIALS SCIENCE

Editors: R. Hull R. M. Osgood, Jr. J. Parisi

1 **Chemical Processing with Lasers**
By D. Bäuerle

2 **Laser-Beam Interactions with Materials**
Physical Principles and Applications
By M. von Allmen and A. Blatter
2nd Edition

3 **Laser Processing of Thin Films and Microstructures**
Oxidation, Deposition and Etching of Insulators
By. I. W. Boyd

4 **Microclusters**
Editors: S. Sugano, Y. Nishina, and S. Ohnishi

5 **Graphite Fibers and Filaments**
By M. S. Dresselhaus, G. Dresselhaus, K. Sugihara, I. L. Spain, and H. A. Goldberg

6 **Elemental and Molecular Clusters**
Editors: G. Benedek, T. P. Martin, and G. Pacchioni

7 **Molecular Beam Epitaxy**
Fundamentals and Current Status
By M. A. Herman and H. Sitter 2nd Edition

8 **Physical Chemistry of, in and on Silicon**
By G. F. Cerofolini and L. Meda

9 **Tritium and Helium-3 in Metals**
By R. Lässer

10 **Computer Simulation of Ion-Solid Interactions**
By W. Eckstein

11 **Mechanisms of High Temperature Superconductivity**
Editors: H. Kamimura and A. Oshiyama

12 **Dislocation Dynamics and Plasticity**
By T. Suzuki, S. Takeuchi, and H. Yoshinaga

13 **Semiconductor Silicon**
Materials Science and Technology
Editors: G. Harbeke and M. J. Schulz

14 **Graphite Intercalation Compounds I**
Structure and Dynamics
Editors: H. Zabel and S. A. Solin

15 **Crystal Chemistry of High-T_c Superconducting Copper Oxides**
By B. Raveau, C. Michel, M. Hervieu, and D. Groult

16 **Hydrogen in Semiconductors**
By S. J. Pearton, M. Stavola, and J. W. Corbett

17 **Ordering at Surfaces and Interfaces**
Editors: A. Yoshimori, T. Shinjo, and H. Watanabe

18 **Graphite Intercalation Compounds II**
Editors: S. A. Solin and H. Zabel

19 **Laser-Assisted Microtechnology**
By S. M. Metev and V. P. Veiko
2nd Edition

20 **Microcluster Physics**
By S. Sugano and H. Koizumi
2nd Edition

21 **The Metal-Hydrogen System**
By Y. Fukai

22 **Ion Implantation in Diamond, Graphite and Related Materials**
By M. S. Dresselhaus and R. Kalish

23 **The Real Structure of High-T_c Superconductors**
Editor: V. Sh. Shekhtman

24 **Metal Impurities in Silicon-Device Fabrication**
By K. Graff 2nd Edition

25 **Optical Properties of Metal Clusters**
By U. Kreibig and M. Vollmer

26 **Gas Source Molecular Beam Epitaxy**
Growth and Properties of Phosphorus Containing III–V Heterostructures
By M. B. Panish and H. Temkin

27 **Physics of New Materials**
Editor: F. E. Fujita 2nd Edition

28 **Laser Ablation**
Principles and Applications
Editor: J. C. Miller

Springer Series in
MATERIALS SCIENCE

Editors: R. Hull R. M. Osgood, Jr. J. Parisi

29 **Elements of Rapid Solidification**
Fundamentals and Applications
Editor: M. A. Otooni

30 **Process Technology for Semiconductor Lasers**
Crystal Growth and Microprocesses
By K. Iga and S. Kinoshita

31 **Nanostructures and Quantum Effects**
By H. Sakaki and H. Noge

32 **Nitride Semiconductors and Devices**
By H. Morkoç

33 **Supercarbon**
Synthesis, Properties and Applications
Editors: S. Yoshimura and R. P. H. Chang

34 **Computational Materials Design**
Editor: T. Saito

35 **Macromolecular Science and Engineering**
New Aspects
Editor: Y. Tanabe

36 **Ceramics**
Mechanical Properties, Failure Behaviour, Materials Selection
By D. Munz and T. Fett

37 **Technology and Applications of Amorphous Silicon**
Editor: R. A. Street

38 **Fullerene Polymers and Fullerene Polymer Composites**
Editors: P. C. Eklund and A. M. Rao

39 **Semiconducting Silicides**
Editor: V. E. Borisenko

40 **Reference Materials in Analytical Chemistry**
A Guide for Selection and Use
Editor: A. Zschunke

41 **Organic Electronic Materials**
Conjugated Polymers and Low Molecular Weight Organic Solids
Editors: R. Farchioni and G. Grosso

42 **Raman Scattering in Materials Science**
Editors: W. H. Weber and R. Merlin

43 **The Atomistic Nature of Crystal Growth**
By B. Mutaftschiev

44 **Thermodynamic Basis of Crystal Growth**
P–T–X Phase Equilibrium and Non-Stoichiometry
By J. Greenberg

45 **Thermoelectrics**
Basic Principles and New Materials Developments
By G. S. Nolas, J. Sharp, and H. J. Goldsmid

46 **Fundamental Aspects of Silicon Oxidation**
Editor: Y. J. Chabal

47 **Disorder and Order in Strongly Nonstoichiometric Compounds**
Transition Metal Carbides, Nitrides and Oxides
By A. I. Gusev, A. A. Rempel, and A. J. Magerl

48 **The Glass Transition**
Relaxation Dynamics in Liquids and Disordered Materials
By E. Donth

49 **Alkali Halides**
A Handbook of Physical Properties
By D. B. Sirdeshmukh, L. Sirdeshmukh, and K. G. Subhadra

50 **High-Resolution Imaging and Spectrometry of Materials**
Editors: F. Ernst and M. Rühle